Principles of Infrared Technology

A Practical Guide to the State of the Art

John Lester Miller

CHAPMAN & HALL

I(T)P An International Thomson Publishing Company

New York • Albany • Bonn • Boston • Cincinnati • Detroit • London • Madrid • Melbourne • Mexico City • Pacific Grove • Paris • San Francisco • Singapore • Tokyo • Toronto • Washington

This edition published by Chapman & Hall, New York, NY

Printed in the United States of America

For more information contact:

Chapman & Hall
115 Fifth Avenue
New York, NY 10003

Chapman & Hall
2-6 Boundary Row
London SE1 8HN
England

Thomas Nelson Australia
102 Dodds Street
South Melbourne, 3205
Victoria, Australia

Chapman & Hall GmbH
Postfach 100 263
D-69442 Weinheim
Germany

Nelson Canada
1120 Birchmount Road
Scarborough, Ontario
Canada M1K 5G4

International Thomson Publishing Asia
221 Henderson Road #05-10
Henderson Building
Singapore 0315

International Thomson Editores
Campos Eliseos 385, Piso 7
Col. Polanco
11560 Mexico D.F.
Mexico

International Thomson Publishing - Japan
Hirakawacho-cho Kyowa Building, 3F
1-2-1 Hirakawacho-cho
Chiyoda-ku, 102 Tokyo
Japan

2 3 4 5 6 7 8 9 XXX 01 00 99 98 97 96

Library of Congress Cataloging-in-Publication Data

Miller, John Lester, 1959--
Principles of infrared technology: A practical guide to the state of the art/by John Lester Miller.
p. cm.
Includes bibliographical references and index.
ISBN 0-412-09861-X
1. Infrared Technology. I. Title.
TA1570.M618 1994 93-32092
621.36'2--dc20 CIP

Visit Chapman & Hall on the Internet http://www.chaphall.com/chaphall.html

To order this or any other Chapman & Hall book, please contact **International Thomson Publishing, 7625 Empire Drive, Florence, KY 41042.** Phone (606) 525-6600 or 1-800-842-3636. Fax: (606) 525-7778. E-mail: order@chaphall.com.

For a complete listing of Chapman & Hall titles, send your request to **Chapman & Hall, Dept. BC, 115 Fifth Avenue, New York, NY 10003.**

Contents

Preface

This book is about general infrared (IR) engineering, technology, practices, and principles as they apply to modern imaging systems. An alternative title to this book with appeal to managers and marketing personnel might be "Everything You Always Wanted to Know about Infrared Sensors, but Couldn't Get Answers on from Engineers." This book is not meant to be a comprehensive compendium of IR (like the *Infrared and Electro Optical Systems Handbook*). Rather, it is intended to complement such texts by providing up to date information and pragmatic knowledge that is difficult to locate outside of periodicals.

The information contained in this book is critical in the day-to-day life of engineering practitioners, proposal writers, and those on the periphery of an IR program. It serves as a guide for engineers wishing to "catch up," engineers new to the field, managers, students, administrators, and technicians. It is also useful for seasoned IR engineers who want to review recent technological developments. Undoubtedly, some may wish to argue specific controversial points or relationships. Others will miss the traditional development of equations and conclusions from first principles. My goal was not to repeat these foundations—they are presented in many other texts. My goal is to define, scrutinize, and appraise contemporary technologies with enough detail for the engineer and scientist and with the clarity and conciseness needed by managers. Therefore, the book functions as a compilation of the state of the art, and provides background information for the reader to separate the practical from the chimerical. It is written from a practical point of view—not the purely theoretical or financial. It contains the perspective of those who design and produce IR hardware on a daily basis. Because engineering is usually the most costly, time-consuming, and critical phase of an IR product, this perspective gives valuable insight to financial officers, marketeers, and others working in the infrared industry.

The book covers passive components and systems for infrared imaging from one to 50 microns. The book is concerned with instruments for astronomy, military applications, meteorology, environmental monitoring, resource management, law enforcement, surveillance, commercial, and industrial uses. It includes material on the manufacturing process, management, cost, schedule, weight, and power.

This is not a book on theory, radiometry, or infrared physics. It does not cover active systems, interferometers, visible cameras, near IR (<1 micron) systems, fiber optics, starlight scopes, or viewers. Detailed range equations and radiometry are not covered. However, select isolated impacts of modern technology on fundamental equations are addressed, as in quantum well detectors or the effects sensor fusion can have on probability of detection. The equations in this book are meant to mathematically illustrate points to nontechnical users. They are also intended to provide references (that are otherwise hard to find in books) for engineers. Therefore, the presented equations are not developed from first principles, but are presented in their final form. There are many excellent books and papers on classic radiometry, range equations, and probability of detection listed in the bibliography. It is assumed that technical readers have some of these texts and are somewhat familiar with the material and nomenclature. For the nontechnical reader, it is sufficient that he realize that modern technology imposes subtle changes on old methods of analysis and figures of merit. The nontechnical reader should remember that if a quantity is a multiplier in the equation, the result will increase and decrease with the multiplier. Obviously, the effect is the opposite for a quantity that is a divider. If the quantity has an exponent (is raised to a power; e.g., it is squared), the value will tend to change faster. By merely looking at the equations, you can understand some engineering and scientific sensitivities.

The heart of effective engineering is knowing what numbers to put in these equations. Another purpose of this text is to help the reader develop an understanding of the numbers to use. In Part II, scaling relationships for weight, power, and cost are given for each major sensor component. In Part III, the particular system attributes of various kinds of IR sensors are discussed, along with examples that can be used as a basis for scaling. This book provides junior engineers with data needed to function at more senior levels by means of scaling relationships, common knowledge, and system examples.

The book is broken into four parts. The first is about managing infrared electro-optical projects. The two chapters in this part point out specific challenges and attributes of the management of an infrared system. Chapter 1 explains the infrared industry as it stands in the 1990s, the challenges facing the industry, and marketing information. Chapter Two concentrates on managing infrared projects and dealing with sub-contractors.

Part II deals with the components or subsystems that form an infrared system. The chapters are arranged in the general order in which a photon would get transformed into data. Five critical subsystems, common to all IR systems, have chapters devoted to them: telescopes, focal planes, cryocoolers, signal/image processors, and pointing mechanisms. Specific systems that are addressed include IR cameras, seekers, FLIRS, specialized airborne sensors, space sensors, and environmental sensors.

Each of these topics is the subject of other texts devoted to them. The reader is referred to the bibliography and chapter footnotes for theoretical and/or in-depth technical information. The chapters are meant to provide the junior engineer, systems engineer, or manager with the basics needed to understand figures of merit, manufacturing processes, scheduling difficulties, costs, weights, and power drivers of the subject. Basic "back of the envelope" relationships are presented. The chapters on hardware (parts two and three) contain a series of data sheets which give pertinent information on current systems. These are useful in searching for a manufacturer or product, as well as for scaling. Taken in their entirety, these allow a reader to develop a feel for the condition and direction of the state of the art.

Part III concentrates on whole systems. These chapters are arranged in order of increasing complexity, from IR cameras to FLIRs, airborne scientific imagers, seekers, and space sensors. For organizational purposes, infrared search and track (IRST) systems are discussed with FLIRs in Chapter 10. Although IRSTs represent a system class to themselves, there is a movement to combine the FLIR and IRST functions. A special chapter included at the end is specifically geared to environmental sensors because of their increasing market share. The subject matter is separated by its basing and complexity. Each chapter covers the technology basics and what is unique about the category, including specialized figures of merit (see Table P.1). Each chapter also contains a representative list of manufacturers and organizations active in producing the IR systems discussed in the chapter. Again, a series of data sheets on specific systems is provided to give the reader a feel for what is currently being produced and what to expect in terms of cost, schedule, and weight scaling.

The chapters on hardware (Parts II and III) contain a series of data sheets that give pertinent information on selected systems and components. These are useful in searching for a manufacturer or product, as well as for scaling. The data sheets do not represent every system or component that is available. Unfortunately, due to the proprietary or classified nature of the data, some chapters have key systems omitted (especially telescopes, image processors, gimbals, and missile seekers). Nevertheless, when taken in their entirety, the data sheets allow a reader to develop a feel for the condition and direction of the state of the art. The prices and delivery should be considered as a general guide that was accurate at the time of this writing. Market pressures are reducing the prices of many components and systems.

The popular figures of merit presented in Part III were difficult to spread among the systems. Some figures of merit, such as NEDT or NEFD, may be appropriately applied to IR system of various genres. Rather than repeat the discussion of a figure of merit, the reader is referred to the chapter that contains the discussion (see Table P.1).

TABLE P.1 Common EO Figures of Merit Used in the IR Industry

Figure of Merit	*Abbrev.*	*Appropriate conditions*	*Location of discussion in this text*
Clutter Equivalent Target	CET	Cluttered backgrounds	Chapter 9
Minimum Detectable Temperature	MDT	Radiance differences	Chapter 12
Minimum Resolvable Temperature	MRT	Human interface with IR sensors viewing high backgrounds	Chapter 10
Minimum Resolvable Temperature Difference	MRTD	Human interface with IR sensors viewing high backgrounds	Chapter 10
Noise Equivalent Delta Radiance	NEDN, NEΔN	Radiance differences	Chapter 12
Noise Equivalent Delta Temperature	NEDT, NEΔT	High uniform backgrounds with temperature differences	Chapter 8
Noise Equivalent Flux Density	NEFD	Low and uniform backgrounds	Chapter 11
Noise Equivalent Irradiance	NEI	Low and uniform backgrounds	Chapter 11
Noise Equivalent Target	NET	Low and uniform backgrounds	Chapter 9

Part IV includes three appendices useful to anyone even remotely involved in the IR industry. Appendix B is a list of accepted IR terminology. This terminology should be used by anyone in the industry. Appendix C is a bibliography of other texts that give more detail about the theoretical subjects presented in this book.

Several topics could have fit into more than one chapter. There is, in fact, a general trend in the industry toward redefining traditional functional allocations among subsystems and distributing requirements across components. Examples include adaptive optics, which fuse optics, mechanics, and processing. They are covered in Chapter 8, under stabilization mechanisms. Optics are also being used for analog optical computing; this is covered in Chapter 7 on image processors.

The same is true of the systems presented in the third section, which is divided into chapters based upon basing and complexity, not application. For example, astronomical instruments can be spaced based, ground based or air based. They are discussed in Chapters 9, 12, and 13, respectively. I apologize to the reader who has difficulty locating the subject of his interest. To mitigate the confusion, Table P.2 is a matrix of popular applications and basing alternatives. Additionally, the detailed index provides a quick path to specific topics.

TABLE P.2 IR Sensor Application and the Corresponding Chapters

Application	*General-Purpose Cameras (Ch. 8)*	*Seekers (Ch. 9)*	*FLIRs and IRSTs (Ch. 10)*	*Space Based (Ch. 11)*	*Environmental and Weather (Ch. 12)*
Archeology	✔			✔	
Astronomy	✔			✔	
Drug interdiction	✔		✔		
Environment	✔			✔	✔
Geology	✔			✔	✔
Intelligence	✔		✔	✔	✔
IRST	✔		✔	✔	
Law enforcement	✔		✔		
Military	✔	✔	✔	✔	✔
Night vision	✔		✔		✔
Support of low-intensity conflicts	✔	✔	✔	✔	✔
Weather	✔			✔	✔

Dedication

To my understanding wife, parents, friends, and mentors.

Acknowledgment

The author wishes to cordially thank all those who diligently reviewed the manuscript: Bruce Allen, Dr. Willard Birtley, Herbert Duvoisin III, Corinne Foster, Dr. Ed Friedman, Dave Hughes, Fred Johnson, Tony Knight, Eric Mauiex, Grant Milbouer, Glenno Miller, Doug Murphy, Dr. Pete Riggle, Tom Roberts, Jim Sommers, Wade Scherer, Larry M. Scites, Don Shelton, Dr. Gerard Steiner, Art Telkamp, Archie B. Treadwell, and Linda Witman. The author is also gratefully thankful to all those who supplied information and releases.

I

Management

1

The Challenge for Infrared Technology

"We try to picture what the products will be and then say, 'What technology should we be working on today to help us get there?'"

John Scully

1.1 THE CHALLENGE

Today's infrared engineers, technicians, marketeers, and managers may be faced with the greatest challenge since Herschel discovered infrared (IR) radiation. Like Stephen Gould's evolution theories, the IR industry is evolving via punctuated equilibrium from the realms of classified vaults to commercialization, medicine, and environmental monitoring. This punctuated equilibrium is a result of external changes to an industry that was already being pushed into producibility, reliability, and cost consciousness by the National Aeronautics and Space Administration (NASA), Strategic Defense Initiative Organization (SDIO), and tactical military users.

The external changes outlined in Table 1.1 reflect the downturn in defense spending, NASA and European Space Agency (ESA) funding stagnation, and growing concern over the Earth's environment, as well as the entry of Soviet infrared technology into the free market. It is an exciting time to be alive, but there is the threat of corporate extinction.

Infrared technology's challenge is to become more suited to commercial applications while giving the taxpayer a bigger "bang for the buck" in government programs. To do this, producibility, affordability, and schedulability must be enhanced across the board. Yields must be improved on every component discussed in Part II of this book, while the sensors' integration and test must become more automated. Gone are the days when 500 percent budget and schedule overruns could be tolerated for the sake of performance.

TABLE 1.1 Effect of External Changes on the IR Industry

Change	*Bad News*	*Good News*
Changing roles of the armed services	Uncertainty Incumbent corporations may be replaced	Potential for upgrades to existing systems in order to meet new requirements Old entrenched corporations may be replaced by newer, more energetic companies.
Declining military spending	Decrease in overall sales	Big ticket programs will tend to be cut while avionics enhancements increase
End of Cold War	Decrease in strategic systems Decreasing military budgets Proliferation of enemies and targets Unclear requirements	Opens up potential markets and vendors/suppliers in previously communist countries Aftermarket upgrading suppliers
Entry of former Soviet and communist countries into the free market	Exceptionally low cost makes it hard to compete with previously communist countries	Larger technology base to draw upon Exceptionally low cost provides competitive edge when teamed.
Everyone is craving more information	Customer emphasis will focus on the end product leaving less room for development	IR can provide more information and should be sold as such, not as a toy technology
Increased commercialization	Must make the transition to production orientation Requires capitalization and human resource investments	Increased sales
Increased emphasis on the environment	Funds from R&D efforts in EO may be diverted to instrument manufacturers	EO technology can play a major role in environmental monitoring and understanding
Product life cycles are becoming shorter	Business as usual does not support rapid IR product design	Inherent changes in the method of doing business will support shorter product life cycles
Trend toward a global economy	Competition will increase in areas that were previously closed to foreign interests	Facilitates international sales and vendor relationships

1.2 THE MARKET ENVIRONMENT

1.2.1 Overall Market

The changes outlined in Table 1.1 augur shifts in the market environment. Market forces will lead the industry—willingly or otherwise—into new paradigms. The environment for infrared components and systems in the nineties can be summed up in one word: volatile. A second word—diversification—might prevail as government and industrial users replace military users as dominant customers. If one doesn't diversify, then one must downsize, as many aerospace companies have chosen to do.

Table 1.2 gives an illustrative outlook for various applications of IR systems. There are many rosy predictions for the detector and IR industry. Recent projections, in light of the defense downturn, still forecast that IR detector sales may be four to six times their early-nineties levels by the turn of the century [Leech and Gutmanis, 1]. Increases in the market may or may not occur, but the emphasis for future markets is not totally away from defense. Rather, it is toward a better balance between defense and other uses. There seem to be plenty of potential users for infrared systems. The problem is that few want to pay the current prices. Every marketing student knows that products are priced high in their early stages, with later decreases reflecting reduced costs, recaptured costs, and consumer desire. Early in a product's history, a few customers will pay any price for the object, so the manufacturer can price it high to cover developmental costs. Once those revenue sources have purchased the product, the price needs to drop to attract other customers. The early stage of high prices and few customers (mostly governments) for infrared sensors is declining.

Moreover, price elasticity for infrared products is increasing as governments worldwide tighten their financial belts. Price elasticity is the classic market force that dictates the relationship between the number of items sold and their price. An highly elastic item will sell far more when the price is low than when it is high. To maximize profits, one sets the price low and accepts a lower profit per sale, but greatly increases volume. Conversely, with an item with little or no price elasticity (e.g., gasoline during shortages, food, and other necessities), there is a less direct relationship between sales and price. Profit, therefore, is maximized by positioning the price very high. In classic government IR contracts, this means continually increasing performance.

IR technology is maturing to a state where it is losing its black magic and novelty from the technological perspective. Today, we are moving from selling major technolical advances to selling comprehensive solutions. To cite an example from another industry, (almost) nobody buys a computer to program or just to play with the technology. Rather, one buys a computer to perform a function. Whether selling to a government, industrial, or commercial user, it will become more and more important to sell information and added capability, and less important to sell the technology itself. One must sell a capability that is better than those of competing companies and technologies; one can not just sell a technology. One must promote an end use. The most obvious end uses for IR are added information and night operation.

When it comes to commercialization, experience can help. Lerner stated, "Don't sell technology, don't sell remote sensing. Sell information that meets the needs of your customer at least as efficiently and more economically than the way they meet those needs today" [Kernstock, 2]. Users other than military and space

TABLE 1.2 Outlook for various IR markets in the nineties

Application	*Prospects for Market Growth in the '90s*	*Enabling Features*	*Representative Current Industry Leaders*	*Current Government Leaders*
Astronomy	Poor to fair—lots of desire but no money	High sensitivity and low cost	Ball CE Hughes IR Labs Matra SSE	ESA NASA Ames NASA Goddard NASA JPL NSF
Commercial auto night vision systems	Market too immature to assess	Low cost/high reliability Uncooled FPAs	Hughes/GM	DOT
Customs and border patrol FLIRs and sensors	Good	Low-cost systems	FSI Hughes Inframetrics Westinghouse	Coast Guard DEA NVEOD Various LEAs
Drug interdiction	Good	Low-cost systems System versatility	FSI Hughes Inframetrics	Coast Guard DEA NVEOD Various LEAs
Earth observing sensors	Good, but funding volatile	Long-life cryocoolers	Ball EDO/Barnes ITT Loral Martin Marietta Matra SBRC SSE Thompson Westinghouse	ESA NASA NOAA
Endangered species monitoring	Uncertain due to high cost of IR systems relative to their effectiveness	Low cost Quiet or no cooling	Inframetrics Mitsubishi	Nature advocacy agencies and groups Waterways Experiment Station (Army Corps of Engineers)
Environmental monitoring	Good	Radiometric accuracy and repeatability Accurate filters	Hughes SBRC Ball Daedalus Hughes LIRIS Matra Perkin Elmer	EPA ESA NASA NCAR
General-purpose cameras	Fair	Low cost High reliability Requirements tuned to customers	Agema Amber/Raytheon CE Inframetrics IR Industries IR Labs Kodak LIRIS Magnavox Mitsubishi Snarnoff	DRA NVEOD

TABLE 1.2 Outlook for various IR markets in the nineties (continued)

Application	*Prospects for Market Growth in the '90s*	*Enabling Features*	*Representative Current Industry Leaders*	*Current Government Leaders*
Gun sights	Fair, but volatile	Miniaturization Lightweight Low or no cooling	Hughes Magnavox TI	NVEOD US Army
Industrial	Fair	Low cost High reliability Uncooled FPAs	Agema FSI Inframetrics IR Industries	
Infrared search and track	Good, but volatile	Large FPAs Clutter rejection Efficient image processing	Martin Marietta Signaal Spar TI Hughes	NVEDO USAF USN
Interplanetary space sensors	Poor	High reliability Dense FPAs Low power consumption	Ball Dornier Hughes Martin Marietta	ESA NASA Ames NASA JPL
Law enforcement	Good	Low-cost systems Lightweight systems	Agema FSI Inframetrics	DEA NVEOD
Medical	Fair	Specialized requirements Low life-cycle costs	Agema Bio-Rad Labs Bio-Rad Meas. BioLase CE GE Medical	NIH
Military FLIRs	Poor in the early '90s, good projections in late '90s, but volatile	Gen II FPAs Image fusion Stabilization	Aerojet FSI GEC Hughes IAI Loral Magnavox Martin Marietta SAT Simrad Spar TI	DRA NVEOD WRDC RADC US Navy Various weapon program offices
Military ground vehicle sensors	Good, but volatile	Miniaturization Light weight Low or no cooling	Hughes Loral Magnavox TI	NVEOD US Army tank and automotive command
Military space sensors	Fair, but volatile	Staring FPAs Long-life cryocoolers	Aerojet Hughes Lockheed Martin Marietta Matra Rockwell SAT SSE Talandic Research TRW Utah State Univ.	ARPA Philips Labs SDC SDIO USAF

TABLE 1.2 Outlook for various IR markets in the nineties (continued)

Application	*Prospects for Market Growth in the '90s*	*Enabling Features*	*Representative Current Industry Leaders*	*Current Government Leaders*
Missile seekers	Poor in the USA, good but volatile in Europe	Low-cost staring arrays Lightweight components	British Aerospace Euromissile Hughes Martin Marietta Matra SEP TI Thompson	French Army NATO SDC SDIO USAF US Army Missile Command USN
Night vision for commercial airlines and shipping	Market too immature to assess	Low cost High reliability Good imaging	Hughes Mitsubishi Sperry TI	ARPA
Perimeter surveillance	Good	Low cost Good imagery High reliability	Agema FSI Kodak Loral Mitsubishi	
Remote pollution monitoring	Good	Spectral agility Low cost High reliability	EDO/Barnes Inframetrics	EPA
Rocket and jet engine diagnostics	Market too immature to assess	Miniaturization Light weight Little or no cooling	Aerojet Perkin Elmer	US Air Force NASA
Tactical Missile warning	Good	Small size Staring FPA High reliability	CE Lockheed Sanders Martin Marietta	Wright Labs USAF USN
UAV/RPV sensors	Good, but volatile	Miniaturization Light weight Low or no cooling	AAI Hughes IAI TI	ARPA NVEOD Program offices
Weather instruments and cameras	Good	Radiometric accuracy and repeatability High reliability	Hughes ITT Matra SAT Westinghouse	NASA Goddard NOAA US Air Force

agencies are usually less sophisticated, less enamored with the technology, and less willing to pay for development and extra capabilities. Medical, industrial, and police users want something that will work, is simple to use, and enhances their capability (not the manufacturer's capability). To compete, IR technology must provide the user with added capability, solve specific problems, or provide a less expensive solution to those problems. In short, the IR industry must make the transition from a "technology push" orientation to a "market pull" orientation.

The Japanese seem to have a sixth sense for selecting technologies to pursue for commercialization. They do not seem to put as much emphasis on end uses as do Americans and Europeans. Rather, they places the emphasis on what has commercialization potential vis-a-vis low cost and high manufacturing rates. The Japanese seem to be more content with letting the market determine the end uses once the product is mature. This philosophy requires a long-term commitment that is rare in American companies. In the eighties, the Japanese emphasis was to develop infrared televisions using Pt:Si FPAs. Judging from recent published papers, current Japanese LWIR FPA development seems to be bypassing doped silicon and, to some extent, even Ir:Si. Instead, they seem to be concentrating on Si:Ge heterojunction and homojunction devices and quantum wells.

1.2.3 Environmental Monitoring

Environmental monitoring is a growth segment for IR products. Infrared systems can be used to monitor gas effluents, smokestack emissions, and thermal pollution of air and water. IR systems have also been effectively employed to monitor the movements of endangered species (as many are nocturnal and shy away from lights) [Campbell, 3]. In general, IR systems can offer environmental enforcers the same covert surveillance aspects that they offer general law enforcers.

There are two segments in the environmental market. First are the environmental researchers, who require highly calibrated, repeatable instruments. Second are the pollution monitoring groups, that require low cost, user-friendly cameras and instruments.

1.2.4 Astronomy

The infrared is a key region of the electromagnetic spectrum for astronomy. Numerous phenomena occur in the IR that are important for pragmatic considerations as well as pure science. Infrared in astronomy opens up several research opportunities to scientists, including:

- *Observation through interstellar dust.* The extinction of light from the center of the galaxy is dimmed by about one trillion times in the visible, but only 16 times in the infrared [McLean, 4].
- *Sensitive observations*. Very old objects from near the big bang are so redshifted that they peak in the IR [McLean, 5]. Observation of these quasars and galaxies provides insight into both the early universe and galaxy formation.
- *Observations in certain absorption bands.* For example, observation in the methane at 2.2 microns reduces the brightness of the giant planets, allowing more detailed observation of rings and moons [McLean, 6].
- *Seeking new planets in our solar systems or others.* IR is uniquely useful for observations of the IRAS-discovered, solid matter disks surrounding stars (e.g., Vega, Fomalhaut and Beta Pictoris). These disks are believed to be the beginnings of solar systems.
- *Dust and cold gas studies*
- *IR spectral study of chemical and astrophysical processes*

- *Studies of brown dwarf stars*
- *Studies of planetary atmospheres*
- *Sky survey for Earth-crossing asteroids and comets with potential to impact the Earth*
- *Astrogeology.* Reflection spectra in the SWIR can indicate the mineralogical composition of heavenly bodies [Cruikshank et al., 7]
- *Mass estimates of comets and asteroids.* This is done through thermal inertia calculation based on their change in temperature [Cruikshank et al., 8].

Chief characteristics astronomers look for in IR Instrumentation are low cost, high calibration, high sensitivity, and high resolution.

1.2.5 Paramilitary

As noted in Table 1.2, law enforcement is one of the greatest uses for IR technology. Law enforcement agencies must prevent crime and apprehend perpetrators. Passive imaging IR gives law enforcement agencies the ability to covertly (without lights) detect humans in dense foliage, detect decomposing bodies, and easily spot campfires [Aikens and Young, 9]. Law enforcement agencies (LEAs) demand real time, clean, detailed images with increased content. Sensor fusion (see Chapter 6) is one answer to this, but it requires stabilization and incredibly high throughput processing. Night vision can allow the officers to conduct covert surveillance at night and approach suspect sites with no lights. Multispectral processing can help identify gaseous effluents from suspect sights. For example, drug processing labs give off a “tremendous amount of gas and vapors” [“IR, Optical Sensors Search Out Drug Labs,” 10]. With a sufficient number of bands, an individual automobile can be “fingerprinted” via the spectral emission from its exhaust, or even by the weathered paint on its roof. A test using 63 bands was only confused by two cars out of 18 via SWIR roof paint reflection [Huguenin and Tahmoush, 11].

Law enforcement uses require eye-safe apparatus. Generally, this means passive IR, erbium lasers, or exceedingly low-power lasers. The passive option is the most politically viable.

LEAs want highly reliable, low cost sensors that solve a particular need, such as gathering admissible evidence or allowing driving without lights. The weight and size of a system is important because the equipment is often based on small planes and helicopters where every kilogram of added drag is noticed [Holton, 12]. Market penetration is difficult because the selling process “usually involves a long sales cycle and willingness to work with less sophisticated but price conscious buyers” [Holton, 13].

1.2.6 Military

Regardless of the diversity of the uses described in Table 1.2, it is evident that much of the future market will still be driven by military applications. As such, it is useful to have a basic understanding of four key US DoD budget categories: 6.1, 6.2, 6.3, and 6.4. Category 6.1 research money is for the acquisition of knowledge. Category 6.2 is for exploratory development. Category 6.3 is for advanced development. Category 6.4 is for engineering development. The distinctions between these categories are often lost when it comes to electro-optics (EO), as EO or IR parts are often sub-systems of a major defense platform such as a ship, missile, or plane. This may result in development of an IR sensor using 6.4 money (instead of 6.2 or 6.3) as a tiny part of a large acquisition. This is because the platform, a big ticket item, may be a 6.4 item.

In the eighties, the IR detector market grew at the astounding rate of around 20 percent per year, which outpaced the 10 to 15 percent growth of the total EO detector market [Kales, 14]. In dollar value, nearly 60 percent of the IR market is military. As a result, a vast stagnation of the growth experienced in the eighties had been expected to occur in the nineties [Kales, 15]. However, this may be offset by the success of IR equipment in Operation Desert Storm. Militaries worldwide are looking to infrared to provide them with night operation, clandestine operations, targeting and tracking capabilities.

As overall budgets decrease, military commanders are forced to do more missions with fewer and less specialized major assets. The money is not available for many new specialized expensive systems, such as new planes, boats, and subs. A less costly alternative is to replace aging avionics packages with ones that enhance practical capabilities and make the big-ticket assets that exist more versatile for additional missions. Electro-optics stands to have a bright future in declining military budgets because it provides increased capability for these big-ticket items at a fraction of the cost of a new airplane or submarine. Upgrading rather than replacing large-ticket items was emphasized by Dr. Charles Hertzfield, director of defense research and engineering, who said, "The Pentagon's technology strategy has been focused into seven broad areas, including increasing modularity so old systems can be easily modified and upgrading key weapons systems to meet new threats" ["Pentagon Identifies 21 Key Technologies," 16]. This tells the IR industry that it must continue down the modularity path. The upgrading of weapon systems is an incredible opportunity for the IR industry as it matures. This can be effective for the military and taxpayers in a myriad of situations including:

- Replacing a single detector missile seeker with an FPA
- Replacing Gen I FLIRs with Gen II FLIRs
- Adding IR sensors to systems that currently do not have them
- Reducing weight and power of EO systems on satellites
- Adding spectral bands and tunable filters to existing sensors

Another military growth area for IR is in radar stealth platforms. It simply does not make sense to equip a stealth plane with a general radar that screams out the plane's position. The answer may be an infrared search and track (IRST) system.

IRSTs can also be used to more effectively detect a stealth plane or ship. For this mission, EO IRSTs are more effectively used as backups to radar because they have a "narrower field of vision and shorter ranges than radar" [Munro, 17] and have decreased performance in poor weather.

Another military use, for multispectral IR, is for identification of friend and foe. This need was made especially apparent after Desert Storm, where friendly fire and accidents claimed 114 of 338 casualties, as well as almost half of the 20 people killed in the invasion of Grenada [Morrison, 18]. IR can aid identification via thermal signatures, shape or tailpipe emissions.

The transition of strategic infrared resources and technologies to tactical military uses is another growing area. As the cold war ends and military emphasis is shifted to low intensity conflicts and quick, minor wars, the focus of strategic technology must transform to supporting the tactical commander. This shift was evident in the Talon Shield program, where the Air Force hopes to provide infrared data to the field commander from several Defense Support Program (DSP) satellites ["While DoD Wants Data Faster," 19].

In traditional US and North American Treaty Organization (NATO) military programs, 20 years can elapse from the conception of a technology to engineering development, during which time the requisite theories, designs and components are worked on [Vatz and Poulsen, 20]. The time required to sell and defend a research program within the US government is approximately linear to the cost of the program, taking about one day for a $10,000 program, and a hundred days for a $1 million project [Hartman, 21]. This indicates that a research program should be taken in incremental steps of less than $1 hundred million, because multi-billion dollar research programs will practically never be approved in an era of declining budgets. There are two exceptions. The first is a perceived national emergency. Missile gaps, bomber gaps, and the Apollo program bear testimony to the fact that large programs can be approved faster if Washington recognizes an urgent need.

The second exception is the opposite: if the research is politically risky, the program's funding may be dragged out indefinitely, regardless of size. Such was the case with much SDI research, which fell in a gray area within the antiballistic missile (ABM) treaty.

As a result of these applications, it is expected that the 6.3 budget category will grow in the nineties as the Pentagon's procurement emphasis shifts from fielded systems to silver-bullet prototypes and the upgrade of existing systems. About 15 percent of the 6.1 budget was allotted to optics and related research in 1980 [Guenther, 22]. It has been estimated that about $400 million of 6.3/6.2 activity is currently earmarked for electro-optics. Additionally, it is predicted that total night vision sales to the Pentagon will increase about 2.5 percent per year, peaking around 2001 with a $1.4 billion market, with tank vision systems consuming almost 10 percent of this ["Future Looks Bright for Night Vision Technology," 23]. The European EO military market is expected to be $300 to $500 million per year in the mid-nineties.

1.3 THE IMPACT OF COMPONENT TECHNOLOGY ADVANCES

Like computer technology, IR technology doubles every five years or so, and proprietary ideas are never proprietary for long. A good idea must be acted upon and turned into a product quickly. Keys to increasing the commercial and industrial uses of thermal imaging equipment lie with developments already underway: reduced costs, enhanced software, improved cooling (reliability, maintenance, and ease of use), improved system reliability, and meeting delivery schedules. More future FLIRs and space-based sensors will contain multiple bands and tunable filters for added flexibility and increased performance.

Reflective optics will use more beryllium and silicon carbide mirrors as their production costs are reduced. These offer the advantages of high stiffness and very low mass. Refractive optics will benefit from new exotic coatings such as diamond-like carbon, making them far more durable.

With project price reductions, mosaic focal planes and Gen II scanners will replace common modules and SPRITEs. This will improve performance and reliability while reducing mass and power consumption. Uncooled focal planes offer the commercial user higher reliability, lower weight, and lower power consumption with a tolerable loss in sensitivity. Staring arrays will push to full format televisions of 640 by 480. Pt:Si is already commercially available, with several companies working on HgCdTe, quantum wells, and InSb full format FPAs.

Image processors will give the user more and more processing power, making the feel of an IR system more user-oriented and friendly. Raw imagery will not be presented: information will. The technology will become transparent to its users, who will get the information they desire, instead of being left to interpret fuzzy videos by themselves. Massive image processing will allow a system to identify an enemy tank or the type of blight infecting a forest. The near future also holds image fusion where images of multiple bands are combined into a single display, offering increased knowledge content in a display. Current efforts in spectral or image fusion include combining MWIR with LWIR. Sensor fusion is also occurring, in which television and IR images are fused in a manner that gives users optimum image content.

Cryocoolers will continue to shrink in size while becoming more reliable. There are a few on the market claiming MTTF of nearly 8,000 hours and a few that can cool an entire FPA for a few watts of input power and less than 250 grams. This will provide users who can tolerate their vibration, noise, and power consumption with more user-friendly systems.

Image intensifiers will lose markets as SWIR, MWIR and, eventually, LWIR sights become smaller and less expensive. Physics dictate that the further into the IR (at least to 12 μm) one goes, the better the night vision will be, although the poorer the resolution for a given size aperture. Lightweight optics, along with uncooled (or minimally cooled) SWIR and MWIR detectors, will dominate in the next century by leveraging the physics with lightweight, user-friendly components.

1.4 HOW TO GET THERE FROM HERE: AN UNCLEAR ROAD MAP

1.4.1 Introduction

The way to meet the challenge of survival is to capitalize on the good news column in Table 1.1 while mitigating the effects of the bad news column. The current paradigm of an endless circle of low levels of production, changing requirements, and high costs outlined in Figure 1.1 must be broken. Traditionally, few components or systems were produced before a customer placed an order. Again, this was tied to the vicious circle of cost in Figure 1.1. A circle of low demand for high performance systems kept production low. This led manufacturers to make units

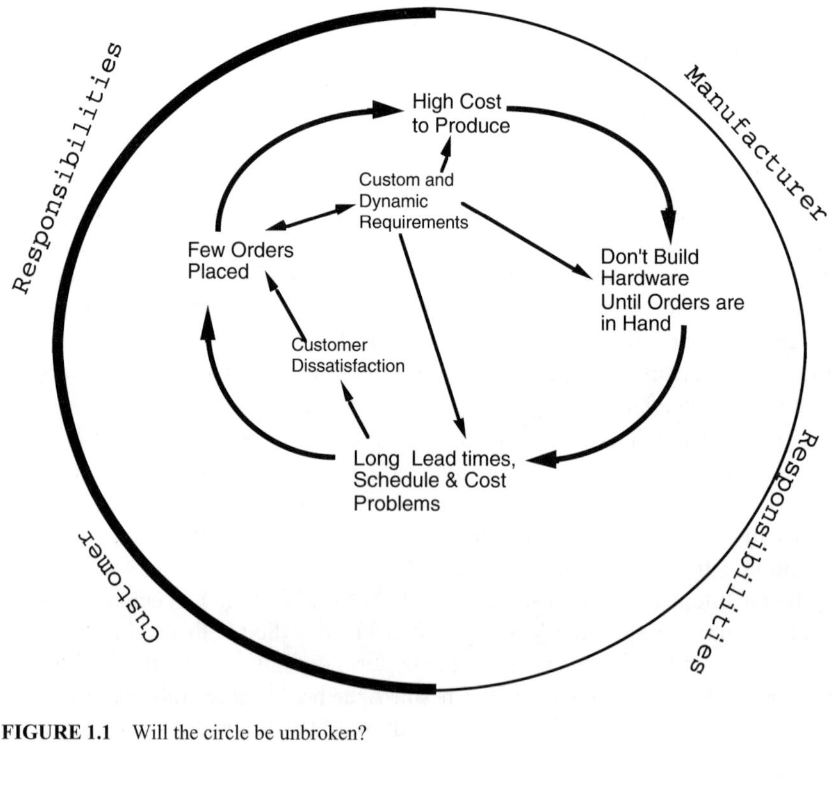

FIGURE 1.1 Will the circle be unbroken?

only when contracts (or orders) were received. This after-the-contract manufacturing leads to schedule delays, cost overruns, and manufacturing rate problems. In turn, customers (mainly governments) placed few orders and even reduced or cut existing orders. This pattern has to be broken if the IR industry is to commercialize successfully.

This is easier said than done. However, if everyone in an organization is aware and empowered to help, change becomes an easier task. Specific actions required to prosper in the nineties include:

- Reduce costs by designing for simple manufacturing and inexpensive components (design to cost)
- Enhance producibility by designing systems with readily available components
- Shorten product development cycles
- Expand the global market for suppliers and customers (i.e., have a global vision for all products)
- Market technologies from government technology transfer programs
- Provide sufficient capital for targeted new development
- Concentrate on long-term investments and strategies, not quarterly profit
- Increase commercialization

1.4.2 Reduction in Cost

Users are usually concerned with life cycle cost (LCC), not just the price tag on the equipment. When marketing a capability, this is the cost that should be stressed to the user. This emphasis will increase as IR systems are sold as information/capability enhancers, not just technologies. Life cycle cost can be lowered by improving reliability, making the depot phase less expensive. The maintenance a system requires can be reduced and made easier.

The cost of the equipment is a strong function of the user's requirements, environments, and desires. Sometimes it is less costly for the user to buy a more capable system than he needs than to design one from scratch. The first attribute that should be addressed is the basic style of the infrared system. Is it a serial scanner, a parallel scanner, or a starer? Common wisdom on the relationship between the cost and style of an IR sensor states that cost increases in large steps between serial scanners, parallel scanners, and starers. However, as the cost of focal planes and image processors are reduced, starers are becoming cheaper than the other architectures for equivalent performance. Because the complexity and cost of optics, stabilization, and gimbals changes in steps as one moves from serial to line to starer, transitions in cost from technology to technology are not smooth. Similarly, performance isn't linear either because the performance of FPAs, gimbals, and optics often decreases as the number of pixels is increased. The manufacturer should thoroughly understand the user's needs and what the user feels is a performance figure of merit. Some unsophisticated users do not care about arcane figures of merit like minimum resolvable temperature (MRT) or noise equivalent irradience (NEI). They simply want a good, clean display on a television screen, which can still make a serial scan the most logical, cost-effective choice.

Additional factors in cost reduction for the equipment are touch labor, engineering labor, and management overhead. The most effective ways to reduce touch labor are to automate and modularize. The easiest way to reduce engineering costs is to provide an environment in which engineers can efficiently engineer (what they are paid to do). They should be relieved from administrative and nuisance work. The easiest way to reduce management overhead is to limit staff who do not produce with clear value added. Reduce capital equipment and design effort by closely basing one new product on another, an older one, or by using a common module approach in the generic sense. Vendors can more economically produce smaller lots by employing materials and resources planning(MRP), just in time(JIT) scheduling, and flexible workers who are trained in several disciplines. Lastly, the use of a shelter or other small company arrangement to do menial and simple work can greatly reduce cost.

1.4.3 Producibility

While producibility is often tied closely to cost, they are fundamentally different concepts. A low-cost item may not be producible (due to shortages or legal constraints, for example), while a high-cost system can be very producible (e.g., labor-intensive activities such as diamond mining). Suitable producibility results when the technology, process, and delivery channels are mature enough to support the desired rates. Enhancements in infrared system and component producibility have been occurring throughout the eighties and nineties. However, large-scale production of many available technologies (such as Z-planes, silicon carbide optics, and large LWIR staring arrays) are still awaiting major customers to amortize development costs and refine manufacturing processes. These technologies need a production champion to fund the nonrecurring costs of mass production.

Meeting rate production of IR systems is always painful. The IR industry has lagged in its process development. The component and technology development have outpaced the production capabilities. This is being addressed by the US government with ManTech and MODIL programs in FPAs and optics. Additionally, ARPA is increasing emphasis on production across the defense industry. ARPA Director Gary Denmen noted, "When we get in trouble on a weapons program, it isn't because someone screwed up the physics… It's because we didn't know how to produce it" [Robinson, 24]. There are classic cases of a prime pressuring a detector house until a perfect detector is made for a prototype unit, only to find delivery problems several years down when the system is in production.

1.4.4 Shortening Product Development Cycle

In today's environment, there is a high risk that a market will not be there when a product is finally ready. Companies must shorten the product development cycle and reduce cost. Decreasing the length of the product development cycle is a change from the government sponsored business as usual.

But this change can occur. For instance, Hughes used stereolithography on a CAD workstation to create a plastic model for the yoke of the An/AAQ-16 turret. The model was used in tooling to make an aluminum casting in half the normal time [Hughes, 25]. Modern workstations can also provide for 3-D animation that can be used to check assembly procedures and verify that the fields of view are not vignetted as a sensor slews around its field of view (FOV) [Hughes, 26].

Teaming with another company or government may at first seem like a contradiction. However, many companies are doing just this. By joining with an industry leader or a potential competitor, development time can be shortened and costs shared.

1.4.5 Global Vision

Before international sales and teaming can occur, companies must become more global in vision. This global vision entails participation in international conferences, understanding international needs, and forcing governments to allow suitable exports. Joint ventures with international firms are excellent ways to capture international markets and suppliers. Such teaming has occurred between many IR companies such as Texas Instruments and Aselsan (of Turkey) to produce night vision devices ["Aselsan of Turkey, TI to Produce Night Vision Equipment," 27].

Providing technology, systems or services normally reserved for domestic uses is another potential business opportunity. The United Arab Emirates asked Litton-Itek to produce a spy satellite with one-meter resolution that would cost between \$100 and \$500 million [Kiernan, 28]. Under the plan, the US would make, launch, and even operate the satellite, with the UAE determining its use (within US national security bounds) and receiving the data [Kiernan, 29]. In essence, the commodity being sold would be more of a service than a product.

1.4.6 Technology Transfer

Technology transfer from the government is another technique used to develop new markets and products. Unfortunately, in America, as in most countries, this can be painful. It can take up to 18 months for a government lab-developed technology to be approved for private industry [Lytle, 30]. It is generally considered that the most successful transfers of technology occur when an individual champions the product or the process, acting like a mini-program manager to get the technology transferred. These people should be from one of the parties of the transfer, be they from within the government, the industry, or a university. Additionally, Norm Augustine pointed out that whenever we speak of technology transfer, we are talking about reducing the emphasis on short-term results for long-term gain and the transfer of people. This is made difficult by stringent conflict-of-interest rules that restrict the movement of people from the government to industry ["Technology Transfer Takes Center Stage," 31]. In the United States, Cooperative Research and Development Agreements (CRADAs) have been implemented to transfer technology. These are available for IR technology from NVEOD, Sandia, and Los Alamos.

Whether a CRADA or internal, research is important and necessary for every government, company, and facility in the IR industry. Most large companies have a research facility located away from their operating units. In the IR industry, this approach seems to have more cons than pros. The "Not Invented Here" syndrome and the accompanying poor transfer of technology is regrettable. While this practice may be the antithesis of concurrent engineering, it persists throughout the world. Scientists and research engineers with experience in production and prototype engineering can more effectively do their jobs. Moreover, co-locating them with the business unit provides a cross-fertilization of ideas with the people making the profit.

Lastly, when considering technology transfer, it is important to realize that the state of the art depends upon the artist.

1.4.7 Sufficient Research For New Products

Gilman shows that the best amount to spend on research is about six percent of total sales for specialty electronics and instruments [Gilman, 32]. One would expect this optimum to be somewhat higher in the IR industry, which is rife with ideas and opportunities looking for lower cost and higher reliability solutions. Gilman also points out several features of the general research process that also apply to infrared prototyping and development [Gilman, 33].

- Research condenses large amounts of activity
- Costs progress logarithmically
- Research activity is a poor measure of research productivity
- Management control and monitoring tend to rely on activity (e.g., man-hours) and not productivity.

1.4.8 Provide Long Term Leadership And Vision

Regardless of the factors cited in the preceding section, the product development cycle for an IR product is very long compared to the rapid advancement of the technology. To effectively have a new product line in the future, years of development and fore-marketing must occur. There must be a commitment from management to short-term funding to ensure the company will be a long-term player. The entire company—scientists, engineers, managers, technicians, and secretaries—should concentrate on long term investments and strategies, instead of the short-sighted quarterly profit.

Although most Asian and European companies take this concept for granted, it seems to be especially difficult for many American companies. This is because American companies often address their management goals on yearly, quarterly, or even monthly sales and profits. This is as ill-advised as taking one's cholesterol measurement a few hours after going on a low-fat diet and expecting to see an improvement (usually several weeks to several months are required). Just as cholesterol drop cannot be measured on an hourly basis, one simply can't base decades of future profit on a monthly assessment.

1.4.9 Increased Commercialization

A market-driven approach is required to successfully commercialize. This is difficult to instill in most defense groups because the people and infrastructure in place were often developed to service defense contracts, not to function within a market-pull environment. It is said that venture capitalists would rather take a good commercial management team with a mediocre technology than an inexperienced team with a superior technology. Unfortunately, any transition from military markets to commercial markets is faced with the latter [Dullen, 34].

Another barrier to commercialization is national security. IR technology is often funded by governments, who tend to keep it classified to give their militaries an edge. Additionally, technology transfer is limited to prevent the spread of high-technology weapons. Although these measures are needed, they have the effect of slowing commercialization and requiring industry to re-invent the wheel time after time. One logical course is for governments to reduce the time needed to declassify a technology or procedure when it feels that there is no longer a security penalty for doing so. This frequently occurs when they discover other governments have developed and are selling the same technology. The Chinese seem to do this well. The drawback may well be the immoral arming of the third world. While there are no easy answers, it is important for managers and engineers to be aware of the considerations.

Penetration into commercial markets is also hampered in many cases when IR technology competes with more mature technologies. An example is in the commercial/industrial security/surveillance field. Here, IR offers great advantages, such as the elimination of area and search lighting (resulting in savings), as well as increased performance. However, IR technology is competing with mature visible cameras and lights. Because the logistic trails for replacement parts, maintenance, and such are not as mature, customers are less likely to adopt the new technology.

Joint ventures with commercial companies is one method used to break into more mature commercial markets. General Motors is working to transfer Hughes' night vision technology to cars by teaming Hughes up with Delco, a company that has long had one foot in aerospace and one in commercial markets [Polsky, 35].

1.5 CONCLUSION

From a corporate perspective, the unstructured artistry and price of infrared technology must be reduced to increase profits. From the government perspective, reliability and life cycle cost must be improved. Industrial and paramilitary users want reliability; user-friendliness and added value. They want to buy end uses that represent improvements at reduced cost, not nifty technologies.

The least painful way to achieve this is through the production of thousands of units. While there are plenty of markets for infrared components such as detectors and coolers, few markets can support such a large production of infrared systems. Therefore, the common module approach to infrared components is as valid today as it was in the early seventies. There are added needs now for higher performance, lower weight, lower power, and inexpensive components. Moreover, the common module philosophy must also be applied to test procedures, fixtures and the assembly of equipment. Although this is all not easy, every step taken toward the modular approach to the production of new generation systems will greatly benefit the entire industry. These steps are occurring at NVEOD, in various manufacturers, and through technology transfer.

Lastly, focusing on a customer and program is important. This focus should be demonstrated before the program is started by gearing capitalization and IRAD to the program area. A Commerce Business Daily (CBD) announcement should not be a surprise. Rarely are contracts awarded to bidders who haven't done months of groundwork before the proposal with the customer.

References

1. D. Leech and I. Gutmanis. 1992. "The US Detector Industry: Prospects for Commercial Diversification," *Proc. SPIE* 1683, 2:12.
2. N. Kernstock. December 19, 1988. "Most Investors Shun High Risk Commercial Space Ventures." *Aviation Week and Space Technology*, 45–56.
3. W. Campbell. November and December 1992. "Panthers Go Underground," *Nature Conservancy,* 7.
4. I. McLean. March 1988. "Infrared Astronomy's New Image." *Sky and Telescope,* 254–8.
5. Ibid.
6. Ibid.
7. D. Cruikshank et al. 1991. "Three Basaltic Earth-Approaching Asteroids and the Source of The Basaltic Meteorites," *Icarus* 89:1–13.
8. Ibid.

9. D. Aikens and W. Young. 1991. "Airborne Infrared and Visible Sensors Used for Law Enforcement and Drug Interdiction," *Proc. SPIE* 1479:435–44.
10. "IR, Optical Sensors Search Out Drug Labs." July 1992. *OE Reports*, 4.
11. R. Huguenin and D. Tahmoush. 1991. "Remote Spectral Fingerprinting for Law Enforcement." *Proc. SPIE* 1479:403–11.
12. W. Holton. December 1992. "Tales from The Dark Side: Shedding New Light on Crime," *Photonics Spectra,* 52–61.
13. Ibid.
14. W. Kales. September 1989. "Will Defense Cutbacks Hurt the Detector Industry?" *Laser Focus World,* 121–2.
15. Ibid.
16. "Pentagon Identifies 21 Key Technologies for Maintaining US Military Lead." May 20, 1991. *Aviation Week and Space Technology,* 57.
17. N. Munro. June 24, 1991. "European Officials: Electro-Optics Will Counter Stealth?" *Defence News,* 44.
18. D. Morrison. April 1991. "Maverick and the Mark 1 Eyeball." *Lasers and Optronics,* 28–31.
19. "While DoD Wants Data Faster." May 11–17, 1992. *Space News,* 2.
20. Vatz and Poulsen. 1980. "Projections of Electro-Optic Trend Correlations." *Proc. SPIE* 260:36–9.
21. Hartman. 1980. "Management in a Government Laboratory: A Case Study." *Proc. SPIE* 260:18–23.
22. L. Guenther. 1980. "Optics Research Contract Decisions." *Proc. SPIE* 260:11–17.
23. "Future Looks Bright For Night Vision Technology." December 1991. *Military and Aerospace Electronics,* 26.
24. J. Robinson. September 30, 1991. "ARPA: Manufacturing Key." *Electrical Engineering Times,* 10.
25. D. Hughes. August 19, 1991. "Growing Use of CAD/CAM Workstations Leading to Paperless Design Process." *Aviation Week and Space Technology,* 44–46.
26. Ibid.
27. "Aselsan of Turkey, TI To Produce Night Vision Equipment." September 23, 1992. *Defense News,* 29.
28. V. Kiernan. November 22-29. "Itek to Limit Spy Satellite Capability." *Space News,* 1, 20.
29. Ibid.
30. D. Lytle. February 1990. "Is Technology Transfer a Flop?" *Photonics Spectra,* 60–62.
31. "Technology Transfer Takes Center Stage." February 1993. *NASA Tech Briefs,* 14–8.
32. Gilman. "Research Management Today." March 1991. *Physics Today,* 42–7.
33. Ibid.
34. R. Dullen. November 1992. "Changes in Defense Industry Will Accelerate Technology Transfer, but Path to Success is Not an Easy One." *OE Reports,* 9.
35. D. Polsky. February 3, 1992. "Diversity, Profits Guide Firms' Guns-to-Butter Conversion." *Defense News,* 12.

2

Management of Electro-Optics

"In scientific research, political savvy, physical endurance, money, and maybe guts can be as important as scientific insight."

Gary Taubes

2.1 INTRODUCTION

This chapter focuses on management principles for the unique environment in which new IR products are created. Much of it is based on too-frequently observed management errors. While some of the chapter's content may seem obvious, common sense is too often forgotten or ignored. The chapter's frankness is intended to underscore the importance of the issues it discusses.

In Rigoletto, Verdi wrote that a woman is "a fickle feather in the wind." One might suspect that he got this phrase from an early IR researcher. Infrared schedules, budgets, and technical performances, likewise, are often fickle feathers in the wind. Managing an IR program can pose several challenges. The components do not always arrive punctually. Experts can be prima donnas. The science is sometimes difficult. IR engineering may be arcane. The state of the art advances daily. Its requirements are subject to frequent and rapid change—especially in government programs. And the cost is seemingly always too high.

There are as many reasons for this fickle nature as there are IR programs. Problems can often be traced to:

- Pushing a technology ahead of its time
- Grossly underestimating the schedule and/or budget, often by failing to realize that advanced sensor design is an iterative process that calls for rework
- Improper test planning

- Too much groupthink, diminishing project effectiveness
- Overworked employees
- Ineffective quality assurance programs
- Immature requirements
- Poor or insufficient front-end system engineering

Let us examine each of these points in detail. There is a rush to incorporate a new IR technology into a sensor or weapons system while it still needs further development. This is analogous to enrolling a sixth grader in college. IR technology is frequently pushed into applications before it has been satisfactorily developed. The program, which must then take on the additional development, suffers cost and schedule overruns. The technology is openly criticized as "unworkable."

Such a situation is a consequence of a poor understanding of the art. And the resulting conclusion is simply untrue. Our sixth grader may excel in college after he finishes high school and becomes more mature. IR managers and engineers must be very careful when using technologies that have not been proven. Risk reduction programs and workarounds must be planned and kept ready for implementation. This problem has surfaced in the use of tunable filters, LWIR staring arrays (as used in the Javelin missile), and Z-planes. IR marketeers would do well to take a hint from Orson Wells and refuse to sell any technology "before its time."

Insufficient program planning is an endemic problem in the industry. An early and detailed program plan identifies potential problems, helps managers and engineers articulate the issues, helps provides a basis for change management, and serves as a reminder of the program's expected progress. Budgetary overruns usually occur when it becomes necessary to solve an unforeseen problem, or to redo a task that either wasn't done correctly the first time or has new inputs late in the cycle. While this may be a caused by a legitimate change in requirements, it is usually the result of poor planning and a lack of margins. A program plan should be created with input from the individuals doing the work. Budgets and schedules should not be arbitrarily cut to get to a predetermined bottom line. This common practice provides the perfect setting for overruns. Unfortunately, it is often felt that the customer has only so much money, and that the entire program should be planned around that cost. Designing to cost is often appropriate and necessary. But the way to reduce cost is to eliminate bells, whistles and secondary functions that can substantially raise expenditures—not to promise to deliver them at a negligible cost.

When a project gets behind, a crisis environment can develop, as did with the Geostationary Operational Environmental Satellite (GOES)-Next satellite instruments project. A schedule-driven philosophy of management and engineering can replace realistic product-driven decisions and schedules. The GOES program was designed to use protoflight philosophy and fly qualification units. The lack of test units, coupled with hardware complexity, resulted in continual modification that raised questions about the system's flight worthiness. One crisis followed another.

Once an organization falls into the pitfalls of crisis management, it is difficult to get out. More disquieting, when a large program falls into chronic crisis management, it can have a domino effect, dragging a whole company into the crisis. The only way out is to produce a reasonable schedule, establish costs to which all parties agree, stick to both, and convenience the customer. One way to prevent (or at least head off) crisis management is to develop a preliminary plan very early in the program, comparing accomplishments to the plan as the project progresses.

Many have traced overruns and problems that surface when a project is well underway to a simple lack of planning. It does no good to work overtime and expend superhuman efforts to get hardware assembled if the sensor must wait for a vacuum chamber for it to be tested. The initial management plan should consist of

- Program personnel organization chart
- Work breakdown structure
- Identification of critical paths
- Functional flow chart
- Budget and temporal spending plan
- Personnel allocation (manloading)
- Detailed schedule and task description
- Security requirements and plan (if appropriate)
- Monitoring plan
- Risk assessment and mitigation plan (including manloading risks)
- Customer communication plan
- Concurrent engineering and/or TQM accommodations
- Definition of test plan with a first-cut at the scheduling of test facilities
- Marketing plan to keep the project sold and develop new customers
- Quality assurance philosophy
- Effective internal review plan (such as frequent "red teams")
- Procedures to disseminate new information
- Detailed list of deliverables, with schedules and assigned responsibilities

Testing and calibration are usually among the last things to be done in the development of an IR system because they can only be done when the entire system is assembled and functioning. Often, an IR system will have experienced several budget and schedule overruns by the time it reaches the testing and calibration phase. As a result, the test and calibration plan is sometimes reduced in scope, lowering the scientific value of the hardware and the confidence in its performance. In infrared systems, special attention must be given to test and calibration throughout the program, beginning with the concept formation stage. Failure to do so will cause schedule and budget problems and, possibly (as with the Hubble), poor technical performance.

Mitigation of this pitfall starts early in the concept phase. One should consider the following:

- Is the system testable as configured?
- What tests are needed?
- How do these tests translate into fielded performance?
- Can NIST support calibration to the desired accuracy?
- Do adequate test facilities exist?

A test plan including facility and manpower budgets and schedules should be completed prior to the Preliminary Design Review (PDR). By the Critical Design Review (CDR), risks should be identified with mitigation and alternative paths defined, instituted and followed. After the CDR, test engineers must work concurrently with design and manufacturing to assure facilities will be ready when the hardware is, and that the tests are meaningful and feasible.

Neither mundane nor gross testing should be overlooked. It was thought nearly impossible to test the Hubble, primarily because it was to be made so accurate that no test set-up could measure its errors. As a result, tests that would have identified the gross error that existed were discounted. The "groupthink" desire to test the system's super accuracy blinded engineers to the need to conduct gross error testing, which may have identified the problem.

The classic subdivision of a prototype program includes up-front concept studies, business development, and funding searches. This is followed by a generic design program leading to a PDR. Following this is a detailed design, leading to a CDR. A fabrication or manufacturing phase follows, in which subassemblies and components are built. Finally, there is the Integration, Assembly, and Test (IA&T) phase. Often a results/data processing phase concludes the prototype development.

The up-front studies and funding development of "Phase Zero" is usually conducted by senior visionary engineers, managers, and marketeers. This is the saddle of creativity that defines the problem and identifies potential solution architectures. The hardware's needs, problems, and missions are defined.

This marks the beginning of a flow-down of nebulous needs, desires, solutions to requirements, and, finally, specifications. The first part of a program should concentrate on detailed system engineering, leading to a system requirements review, where everyone involved agrees that the requirements are understood, properly documented, and traceable to customer needs. A general conceptual design should be generated for a basis for trades and analyses to depart from.

The generic or overall design leading to a PDR is also a creative phase that must be manned by senior creative people who have dirty fingernails and scars from similar projects and related efforts. This leads to a design that is workable, practical, and within the constraints of the total program. This is the place for applied creativity and innovation. Usually some risk reduction tests, demonstrations, and studies are conducted, along with numerous trades and analyses. Depending upon the sensor's complexity and the urgency to get the product to market, this phase can last from three months to a year. During this time, requirements and specifications should be finalized.

The detailed design leading to a CDR is usually twice the PDR phase's manloading, and it usually lasts from one and a half to two times as long as the PDR phase. This phase has applied creativity mixed with detailed engineering. It should be staffed by the same senior level visionaries, with the added manpower supplied by capable junior engineers. At a CDR, about 80 percent of the drawings should be completed. The packaging should be done, interfaces well defined, logistics and supportability defined, reliability fully budgeted, and a detailed bill of material should exist. Critical new designs should all be proven and verified with breadboards. In theory, no parts should be ordered and no fabrication started. However, in infrared sensors, market forces usually dictate taking the risk and starting fabrication/procurement of long lead items during this phase.

The fabrication phase reduces engineering to a key core group, and replaces the junior engineers with fabrication personnel. The technicians are performing the meat of this phase. Its length and duration depend upon the complexity of the subassemblies to be made. Finally, the whole sensor is assembled and tested. The tests must include environmental, functional, and radiometric.

Uncontrolled, self-feeding group thought or "groupthink" can be rife in an IR project. Small projects and small companies seem especially susceptible. Groupthink often occurs in highly cohesive groups when members "tend to suspend their individual critical thinking in favor of judgments on which other members appear to concur. Groupthink results in an illusion of unanimity and in less effective decision making" [Fisher, 1]. Groupthink can occur at any time, from the proposal stage through project completion. One dangerous consequence of groupthink during an IR project is the development of an arrogant attitude that alienates the customer, resulting in a tendency to discount alternative ideas that the competition uses effectively. Concurrent engineering can decrease the likelihood of groupthink by providing varying opinions and viewpoints [Steiner, 2].

Scientists, engineers, and managers working in infrared are a busy bunch. In an informal survey by the author, 68 percent of his calls to these types were answered by a secretary or voice mailbox. Twelve percent were unanswered or busy. A mere 20 percent were directly connected to the correct person. Moreover, it took an average of 4.6 calls before the author and his colleagues made a live connection. This is more attempts than Norm Augustine's informal poll of corporate executives, that showed an average of a mere 3.8 attempts per communication [Augustine, 3].

This means several things to a manager. First, he should not wait until the last minute to try to contact or get any information from an IR engineer or manager. If you can't reach an electro-optics professional, you might try calling the CEO of his company. He might have less to do. Second, it accentuates the need for offices to make it easier to contact personnel. Finally, it suggests that each company should try to reduce the engineers' and managers' loads, allowing them to concentrate as much as possible on the tasks at hand. Toward that end, companies should assure their support staffs are assisting and reducing the engineers' and managers' workloads. Support staffs have been known to bury their fellow workers in needless paperwork, while refusing to respond to legitimate tasks.

One of the most critical support functions is quality assurance. The function of the quality assurance staff is to document the quality of the hardware that the engineers and technicians submit. To do this, quality inspectors must be experts in their fields. There is no value added in having a quality assurance person rubber stamp a procedure if he doesn't know what is going on. For example, a lack of specific electro-optic expertise may have contributed to quality assurance's failure to detect the incorrectly assembled null lens set-up that was used to measure the Hubble's primary mirror figure. Had the quality inspector been an expert optician, he might have questioned the assembly procedure.

The problem facing management is how to properly staff QA departments with EO experts. A large IR house can specifically train and develop optical or cryogenic system quality experts. Small companies can conduct QA as a rotating design engineering position. They may also use consulting houses that focus on this work. An expert on one program could serve as a part-time or temporary QA chief on another program. The only catch for a small company is that the quality engineer must not inspect his own work. There must be an independent check, and QA must report to the same level of management as the engineering department.

Many programs are launched into premature design and fabrication with immature top level requirements. This has several unfortunate effects. First, the requirements are likely to change mid-project, causing redesign, increased budgets, and lengthened schedules. Second, this prevents the detailed early system engineering needed to properly reach a system requirement review or system concept review. Third, immature requirements indicate a questionable need from the customer's perspective, and the life of the program may be in jeopardy. Fourth, immature requirements may indicate a changing situation that may eventually impact program stability. Finally, this may be an indication that the customer is not sophisticated enough to understand the technology. This may result in tension and difficulties with the program.

Proper and sufficient early system engineering is the key to producing a high tech product (such as an infrared system) within schedule and cost constraints. One of the first tasks for system engineering is to flow general goals and desires to top level requirements, and from these top level requirements to detailed requirements and system specifications. Then, system specifications must be divided and partitioned among subsystems, components, and parts. This flow is often time consuming and difficult to accomplish. However, without such an exercise, the final product will not satisfy the customer's needs.

This process must start with a detailed examination of customer needs, and how the project satisfies these needs. Such needs are the basic reason why the project exists. A vision of the project and resulting hardware, software, and data must be conveyed to all personnel. Top level system requirements must be documented and understood. If an environmental research sensor is to gather scientific data, its top level requirements may be to detect a certain chemical species at a certain range with a certain accuracy. It is likely that radiometric errors and performance will directly flow from the science that is to be supported. Allowed errors and uncertainties should be explored, traded and partitioned in a manner that will generate the data that is needed. A first cut at error budgets and performance estimation of a conceptual design should be conducted. Critical technologies, components, and programmatics should be identified for detailed study. All requirements and specifications should be tied to achieving the goals and meeting the error budgets within the system's constraints. The flow from requirements to specifications requires many studies and trades. Throughout this process, the system engineer must draw heavily on experts in many disciplines and fuse their inputs. This must be done with excellence, documented, and communicated before detail design can proceed.

2.2 ORGANIZATIONAL ISSUES

IR personnel are usually more educated and higher paid than average employees. They can also be more arrogant. Sensor design and manufacture incorporates various experts in interdisciplinary fields. These include control theory, semiconductor physics, radiometry, optics, cryogenics, structures, software, analog and digital electronics, and image processing. This is mixed into a witch's brew that yields the magic black art of sensor design. The manager's challenge is to orchestrate a gaggle of these magicians into a finely tuned team.

A lack of qualified personnel or a scheduling conflict can make the organization of an IR program a difficult task. One may lack the staff to logically organize a project by discipline. Or, the staff may not agree to fill positions as a result of personality conflicts or the perception that a given position is somehow "beneath" an assigned person. Remember, you are dealing with sensitive experts. The result is that the most productive organizations are often not logical and not directly out of their management textbooks. Instead, they are based on the available personnel.

Selection of the right person for the right job is, however, of utmost importance. Before the selection can occur, the right people must be identified. Their backgrounds, personalities, and ambitions should be scrutinized against their positions. A common problem to avoid is selection of a person because of excellence demonstrated in past work that was not on an electro-optics project. A great mechanical engineer's experience producing camshafts rarely translates into excellence in opto-mechanical engineering. A new program can be effectively staffed by giving the personnel more responsibility than they had on the previous job, and by clearly defining growth paths for everyone. Final choices should be partially a group decision by those already on the project.

It is usually better to work the program schedule and organization around the right person instead of accepting someone who is a poorer match. If there is an excellent engineer whose specialties happen to include an eclectic mix of signal processing and gimbals, organize both functions under him. Look at the personnel available when making an organizational chart. Let the personnel determine the org chart, not the other way around. Design the program around the available people, project goals, and end product. Availability must be considered, especially with eclectic mixtures. If person A is vacationing in Tibet, it is unwise to assume person B can do both optical train design and cryocooler selection, just because person A can.

When mass producing a widget, simplicity, flow, and cost of production are the most critical aspects of making a profit. In an IR organization, engineering is the key to a successful project. Although production concerns are important, rarely are sensors (and sometimes components) made in an abundance that makes ease of production the primary goal. When this does occur, engineering must still be excellent to profitably mass produce an infrared component or system. Too often, engineers are used to do the work of secretaries, facilities, security or others. Engineers should have the tools and the time to do their jobs. They should be treated with the greatest respect by mangers, support services, and production personnel. Without the above, the engineers will become de-motivated and hindered in the performance of their job. The result can be an inferior design that requires rework or redesign and may not be producible in a timely and cost effective manner. The organization should be geared to satisfying the technicians' and engineers' needs and desires. One way to visualize this is to turn the org chart upside down to see who supports whom and should be doing what for whom.

Most IR product development or prototype sensor systems are managed by a program manager using personnel from a matrix organization. Forming such a project organization is reputed to have several advantages, including allowing a group (even a group of one) to focus entirely on the project, increasing communication flows and allowing an expert in the field to be the manager. Unless sequestered, rarely is a project group really focused entirely on a single project. And sometimes the manager identified is not really an expert in the field. The disadvantages of this approach include increased management complexity and inconsistency in the application of company policy.

A matrix organization is usually formed by a company to support various projects. A matrix organization has personnel reporting up two chains of command. One is typically the functional organization and the other is the project or program organization. An electrical engineer would belong to a department of electrical engineers and have a supervisor in that department who reports to a director of engineering. This same electrical engineer would be assigned to a program, and would concurrently report to a set of managers in the project's hierarchy. Matrix organizations provide more managers with jobs and increase costs. Such organizations are 50 to 100 percent more costly to support than product-focused companies [Dee, 4]. They do, however, provide a focused team to conduct a project, allow a project to use people only when they are needed, allow personnel to pursue two career paths, and allow easier transitions from one ending project to a beginning project.

Middleton conducted a mail survey [Middleton, 5] on the perceived advantages and disadvantages of project organization in aerospace. A similar survey, conducted during the preparation of this text, focused more specifically on the IR industry. Questionnaires that asked managers working on IR systems whether they agreed or disagreed that a given attribute was an advantage or disadvantage were distributed. As Table 2.1 indicates, 92 percent of aerospace managers agreed that "better project control" was an advantage of a project organization in Middleton's survey. Table 2.1 also indicates that 88 percent of infrared managers agreed that "better project control" was an advantage of a project organization in the survey conducted for this book. Tables 2.1 and 2.2 detail the results of both surveys. The survey conducted for this book included some issues not addressed by Middleton (hence the N/As). The survey for this book had 50 respondents.

Aerospace managers and IR managers perceive many of the same advantages of project organization. Both groups viewed "better project control" as an advantage that actually results from project organization. Both groups overwhelmingly did not feel that higher profit occurs through project organizations. There are two instances in which the groups differed by more than 30 percent of the respondents. IR managers responded far more favorably than Middleton's to the effect a project organization has on program security and product development time. Perhaps the disadvantages are the salient factors managers must examine in light of their own

TABLE 2.1 Perceived Advantages of a Project Organization

Major Advantages of Project Management	*Percent of Middleton's Respondents for General Aerospace*	*Percent of Author's Respondents for the IR Industry*
Better project control	92	88
Better customer relations	80	56
Shorter product development time	40	78
Lower program costs	30	44
Improved quality and reliability	26	34
Higher profit margins	24	22
Better control over program security	13	56
Earlier to implement customer defined changes	n/a	56
Better project visibility and focus on results	n/a	56
Higher morale	n/a	44
Easier infusion of SOA technologies	n/a	56
Increased communication	n/a	66

TABLE 2.2 Perceived Disadvantages of a Project Organization

Major Drawbacks of Project Management	*Percent of Middleton's Respondents for General Aerospace*	*Percent of Author's Respondents for the IR Industry*
More complex internal operations	92	56
Inconsistency in application of company policy	80	56
Lower utilization of personnel	40	32
Higher program costs	30	44
More difficult to manage	26	34
Lower profit margins	24	22
Worse technology transfer within company	13	78
Tendency for functional groups to neglect their jobs and let the project organization do everything	n/a	66
Duplication of skills	n/a	66
Decreased communication on non-project issues	n/a	66
Harder to use SOA technologies	n/a	22
Decreased communication	n/a	22

programs. These are areas in which our predecessors have had difficulty. The IR program manger should be on the lookout for these problems. The survey indicates that IR managers cited technology transfer, problems with functional organizations, and decreased off-project communication as the major drawbacks to project organization, whereas more aerospace managers cited complex internal operations and inconsistency in company policy as drawbacks.

Astute manages and engineers will always search for new technology developments within their companies or government agencies to exploit, regardless of their current project. Unfortunately, the problem of technology transfer within the company, so frequently cited by IR managers, is serious. Project organizations tend to build barriers to open transfer within the company based on security, proprietary technology/competition fears, the not-invented-here syndrome, and the general breakdown of communications that is the product of the isolation effect of large projects. In a matrix organization, the functional manager should make technology transfer a top priority.

Project organizations usually cause complex internal operations that may be responsible for many of these drawbacks, and for general support services (such as facilities or publications) operating less efficiently than they could. It also suggests that the project leaders and management will be tangled in a web of conflicting needs, authority, and responsibility in off-project issues. Inconsistency of company policy occurs because some projects are "haves" while others are "have nots," depending on the project's importance, the forcefulness of its managers, and corporate policy. Oddly enough, IR managers deemed complex internal operations less of a result of project organization than their counterparts in aerospace management, in general.

2.3 SYMBOL CONFUSION

It seems to be a fundamental human quality: people tend to confuse symbols for the objects they represent. There is an old adage that one should not eat the menu. This means the word "ravioli" is not a cheese filled pasta. It is only a word that represents the food. The actual thing (in this case, the pasta) will be delivered later. Similarly, voicing love for your spouse, is not loving her/him. It is merely a representation—a symbol of your love. Primitive cultures do this frequently with magic words (e.g., abbra-ca-dabbra) and religious idols that supersede in importance what they are meant to represent. Special importance is placed on special words or the names of gods. Then, social problems arise when these names are confused with the actual things. Confusing the symbol for the article is also common in western civilization. When it occurs in infrared engineering and engineering management, the result is often disaster.

These days, in engineering, TQM and concurrent engineering are frequently cited (ad nauseam, sometimes) but not always practiced. The author was once part of a fifty-person program that excelled at concurrent engineering. Owing in part to this, they won a large follow-on program. The program grew so fast that the company launched a formal concurrent engineering plan. The result was that everyone began saying the words but *not* following concurrent engineering practices. We confused the symbol with the actual practice.

Avoiding confusion between symbols and the ideas they represent should be of prime importance to management. It is important to remember that people make decisions. Cost accounting tools and management techniques only provide information upon which decisions can be made. No Return On Investment (ROI) calculation ever makes a decision. Rather, people who examine ROI calculations do. Gantt charts and program plans are symbols. They are needed, and useful. One cannot merely pronounce the words "radiometric analysis" and have it done! Likewise, one can't merely put a task on a Gantt chart or in the program plan and assume it will be performed. The program manager must be sure the engineer doing the radiometric analysis has the time, tools, and the skills to do it. The engineer doing the work is the needed function, not its appearance on a schedule. This is a simple concept, but one that is often overlooked.

Symbols are often confused with actions in quality assurance programs. A quality assurance procedure is written down and everyone thinks they have a quality program. They don't. They have some paper with directions, ideas, and such.

It is not a program until it is followed by the personnel doing the work. This, likewise, occurs with manuals of engineering procedures and practices.

There is nothing wrong with symbols. They are necessary for our development, evolution, and thought. We would not have language, science, mathematics, engineering, or even this book without symbols. One can't easily order a dinner without a menu! It is a wondrous attribute of being human. We would not be humans without the ability to symbolize. They are, however, detrimental when confused (on an individual or group basis) with the article they represent.

2.4 APPLICATIONS OF FAD MANAGEMENT

Concurrent engineering (CE), integrated product development (IPD), and simultaneous engineering (SE) can be effective, even essential, in IR projects if management and engineering are serious about them and take the time to suffer all the way through them. The approach is to form a team with representation from every discipline. In contrast to conventional approaches, where they are not considered until near the end, operations, test, and quality control engineers are involved from the beginning of a project. Engineers from customer and vendor organizations should also be included. Sometimes organizations believe that concurrent engineering will occur if they co-locate various disciplines together in the same part of a building. While co-location does tend to facilitate concurrent engineering, it is not the only action needed for effective concurrent engineering. Moving people around will not result in concurrent engineering. The engineers must be forced to communicate with each other. Concurrent engineering is sometimes confused with concurrent management, where a group a managers meet and form a team. While there is nothing wrong with concurrent management, it shouldn't be confused with concurrent engineering.

Concurrent management works like concurrent engineering by bringing in representatives from all pertinent management disciplines early in the project. Finance, quality assurance, marketing, and personnel leaders should be aware of the program, and should "buy in" to its progress.

There is one aspect of product development that seems to have escaped the "concurrent" prefix. Research and development is, perhaps, the most important function to be performed concurrently, and the least often exercised. It has become fashionable for large corporations to sequester their scientists away in campus-like settings, hoping they will conduct useful work. The sad fact is that they are also sequestered from the needs of the rest of the company. The engineers trying to figure out how to meet a specification and the marketeers trying to fill a need rarely understand what is going on a few thousand miles away at their company's R&D center. Similarly, the scientist rarely knows the programs these engineers are working on, and, therefor, can't possibly understand the problems that they are having. Small companies don't have the luxury of separate R&D centers. Their scientists and engineers have offices down the hall from the program managers and production engineers. Their research people often participate in proposals and concept formations. As a result, their research is often more focused and salient to the business.

Total Quality Management (TQM) is continuous process improvement yielding a continuing better product. TQM is based on two basic principles: continuous process improvement and customer satisfaction. Both should be considered globally in the IR industry. The process is not just the manufacturing process, but also continuous improvements that should be made in management, engineering, R&D, scientific, clerical, financial, and subcontracting processes.

The global customer should also be considered throughout. The customer may be another company or another division of your company, but that customer also has a customer that should be considered. For example, the customers of a NASA contract aren't just in the purchasing center. Rather, they include NASA headquarters, Congress, and the taxpayer. If every program manager examines his project and is certain he is giving the taxpayers their money's worth, future funding will be easier, morale will be higher, and political support will be greater. In short, everyone will win.

TQM also implies increased training, changing attitudes, and empowering those actually doing the work. To do TQM is to constantly improve every day, week, year, and decade with no end. Process improvement may be slow and the magnitude of its results may diminish as the overall process is improved. An optimum process can really never be achieved in the IR industry, because there will always be changing environment, requirements, and technologies that require adaptation by various organizations (including governments).

Several participants in the IR industry have captured Baldridge quality awards. Texas Instruments Defense Group was a 1992 recipient. Marlow Industries, a major manufacturer of thermo-electric coolers, won in 1991. Their victory was achieved through a self-assessment of their weakness and the implementation of a continuing improvement plan. The cost of quality is offset by the benefits. Behind Marlow's quality drive was "...the conviction that it costs a lot more to find a new customer or recover a former customer than to keep a current customer happy" [Maynard, 6]. This is especially true of IR systems. Finding new customers is expensive and increasingly difficult as traditional business bases decline.

TQM and MRP have been successfully employed throughout the LANTIRN program. Reportedly, LANTIRN was bid at cost. The contractor then "systematically redesigned its original hardware development offerings, importantly reducing the number of processors in the system as well as maintaining close cost controls over what was produced. Its winning strategy was based on long term commitment to engineering and manufacturing excellence that allowed it to take an aggressive bid strategy based on a superior understanding of its future costs" [Dee, 7].

In another IR program TQM success story, the time required to repair test equipment during the test and integration of satellites at TRW was sliced in half, and the reliability of test equipment rose 54 percent as the result of the installation of test points that allowed diagnosis of problems without disassembling the test equipment. TRW and the Air Force are implementing recommendations from a joint government/contractor process action team that will result in a 25 percent reduction in paperwork for monthly management data [TRW, 8].

2.5 MANAGEMENT MARGINS

Good engineering practices include margins on weight, power, and performance. Yet, rarely are margins applied to producibility, cost or scheduling. This is especially true of hardware delivery schedules. Although one cannot plan for a failure, one should reserve between five and fifteen weeks per year and simply label them as margin. These times can be used to catch up with or to get ahead of the promised delivery date. If not used, everyone will look like heroes for delivering early. The program manager should also plan a healthy reserve (10 to 20 percent) in the schedule, manpower, and budget. It may all disappear within a few months, but such reserves provide an increased chance of success and a cushion in case of problems. Additionally, it may allow him to get his resume circulating before things get worse.

Margins should also be applied to rates of production. Rarely can a facility operate smoothly at 100 percent capacity. Margins of 10-30 percent should be applied to assure a cushion exists if a machine breaks or a piece of capital equipment needs overhauling.

Scheduling should contain both margins and buffers. Long lead items must be thoroughly investigated to provide the necessary downtime. Even a common motor for a scan mirror can require a six to eight month lead time. A simple schedule margin proposal that almost any level of management will approve is to order long lead items 20 to 30 percent earlier than the lead quoted by the vendor. Schedule buffers should follow a major task, such as the finishing of a spec, design or test. Buffers must be budgeted, which has the good characteristic of providing flexibility, but the negative characteristic of attracting upper management's attention or the attention of an engineer. To avoid attracting too much attention, buffers should be kept small (never more than a week or two, and never more than a few percent of the total time). The key is to keep them small and plentiful. Attracting attention defeats the buffer, for both management and engineering will surely have ideas on how that time and money could be better used. Usually, the only way to implement such a buffer is to stretch a task by the buffered amount and make it clear to engineering that the work is really to be done early.

Scheduling is severely impacted by the procurement cycle. Ample procurement time should be allowed from the first developmental schedules throughout the program. Procurement cycles usually require specification, RFP packages, quotes, proposals, reviews, and negotiating before a subcontractor can be selected. This can easily take a few months and could be as long as a year.

2.6 MONITORING PROGRESS

The department of defense has developed procedures and methods of cost/performance tracking aimed at "cost plus" programs. Many of these systems are so good that they should be used to internally monitor any project, including firm fixed price, level of effort, commercial, and even IRAD programs. The key for effective use is to prevent the tracking system from consuming too much of the program time while assuring it is giving timely and accurate information.

Among these are Cost Schedule Control Systems (CSCRs), which are monitoring methodologies that provide "real-time" data on the progression of a program during its execution. CSCR provides evolving information on the work achieved and the money spent, along with what was planned. This is an excellent quick indication of a program's health. One cannot know if he is breaking the speed limit in his car unless the speed limit is posted and he has a functioning speedometer providing real time information. Analogously, one cannot know if a program is exceeding budget or underperforming unless he has real time data and a plan to judge it against.

CSCR begins with developing a comprehensive work/spending plan. Work packages or tasks must be laid out from beginning to end. Such packages should be short, usually with no more than two or three reporting periods. The budget must be distributed among packages so that each has an assigned cost determined by the number of engineers, the number of days, the number of trips, and so on. All costs must be scheduled and time phased. Then, all of the budgeted costs assigned to each work package are added. They can then be plotted based on cumulative spending as a function of time, producing a plot like the one shown in Figure 2.1. Thanks to personal computers and spreadsheets, these plans can sometimes be quickly assembled from the proposal.

Some believe that this reporting is impossible for cutting edge technology or R&D programs. This simply is not true. While work packages are more difficult to define for R&D programs, they are not impossible and they are frequently very useful. First, one should realize that work packages do not presuppose a result. They only state that a given level of work must be done. Work packages do not have to predict a given answer or result. They only state that an answer or result will be obtained. Second, small work packages are better than large ones. Breaking an R&D program into tiny packages usually makes defining the work and cost planning easier, not harder. A work package can be written to do a certain quick experiment or even just to prepare a lab. Traditionally, an R&D program may have branches at certain critical points where if one result is obtained the program will go down one path and if another result is obtained, it will go down another path. In these cases, it is best to determine the time and resources needed to reach the branching point and only fully plan to that point, although sufficient planning for gross scheduling and budget on all likely paths is needed. Once a critical point is reached, further planning can occur. The manager should be careful not to spend too much of his resources too early.

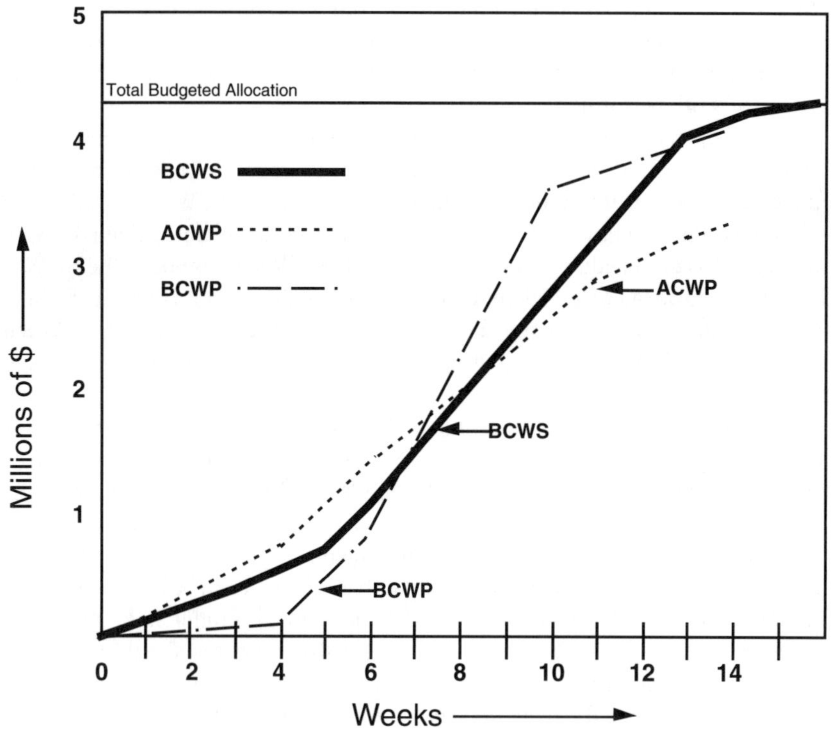

FIGURE 2.1 Cost of a Project as a Function of Time

The complete plan must include all aspects of the statement of work, deliverables, and the customers' expectations. Reserves may or may not be included. Travel, computer, and material costs should be included. The plan must consider engineering and technician availability. Be sure to include subtle costs a program may incur, such as material handling, overtime, phone, postage, etc. It is difficult to determine exactly what trips need to be taken and when. A quick estimate should be used based on past programs and known differences. The plans don't have to be perfect, but should be all-inclusive. Rarely (if ever) are plans completely followed, but that's okay. The important thing at this point is to have a reasonable vision and path to reach the end within cost and schedule. The initial plan gives one a footing to stand on and assess the program. The CSCR allows one to do this while the program is progressing. One can always re-allocate any remaining funds later, or understand and explain other deviations. Without a plan, deviations can't occur and therefore can't be understood or corrected.

The heavy solid line in Figure 2.1 is a fictitious plot of a Budgeted Cost of Work Schedule (BCWS) derived from a program plan. This is the time-phased vision to complete the task within cost and schedule. Now comes the fun part of the

CSCR—determining where the program is each week relative to the plan, as in Figure 2.1. The program manager needs to know at least two things to do this. First, the amount of money spent (or irretrievably committed). And second, the amount of work performed.

Any competent finance department should be able to provide a report on the cumulative amount of money spent on any project/program. This can be plotted on the same sheet as the Actual Cost of Work Performed (ACWP). ACWP is the light dotted line in Figure 2.1. When this line falls on the planned line (BCWS), the program is on schedule and getting the job done. When it is above the BCWS, the program is ahead of schedule. It is behind schedule when it is below the line. When this line falls on the planned line (BCWS), the program is on budget. When it is above the BCWS, the program is over budget. It is underspent when below the line.

The only missing piece is an estimate of the amount of work performed. This can be simply the number of work packages completed, the number of milestones satisfied, or the percentage of some task completed. When a package is completed, credit is taken for the cost originally budgeted for that task. This is sometimes called "earned value." If a change in requirements, philosophy, or direction requires a work package to be partially of wholly re-done, it can be subtracted from the estimation in the future and replotted. Earned value is plotted in Figure 2.1 as Budgeted Cost of Work Performed (BCWP). Estimating this should be as mechanical and quantitative as possible, measuring actual outputs and hard results. Sometimes estimating the work done is somewhat crude and subjective, but even so, it is still better than nothing. When this line falls on the planned line (BCWS), the program is on schedule and getting the job done. When it is above the BCWS, the program is ahead of schedule, and it is behind schedule when it is below the line.

The notional plot of Figure 2.1 assumes a program of about $4.3 million allocated dollars to be completed in 16 weeks. The reserve, profit, and administrative fees should be in addition to (above) the $4.3 million allocation. First, the work is scheduled and cost estimated from the program plan. This gives the BCWS, which is plotted. Each week an estimate of the work performed is conducted and plotted. Also each week, the amount of funds spent is plotted. Figure 2.1 represents the course of events on the 14th week of the 16 week program. This is why the ACWP and BCWP end at the 14th week. These can only be determined after the fact.

This fictitious project starts out in grave trouble. Costs increase beyond what was planned and the project is overspent by the second week. To make matters worse, very little work is accomplished. The project is in the worst possible situation, over budget and under performing. Action must be taken. Clear changes occur by week five. Although the costs are continuing to overrun and the slope has changed for the worse, there are some positive trends. First, things are beginning to get done. The BCWP has increased dramatically. The staff was probably enlarged or a subcontractor's role increased. Second, although cost is increasing, the change in the line's slope was scheduled anyway the following week. These trends continue, and by week seven the work performed matched the cost expected for the work (i.e., the lines plot at the same point). Although over budget for the given week, they are not over budget for the work performed. At this point, the project can be considered on budget and about one week ahead of schedule. By week 10, the program has completely turned-around its previously gloomy fate. At this point it is in the best of all worlds—under spent and ahead of schedule. By week 14, the work is very near completion and the program is under spent by almost $1 million. It is very likely that the project will be completed on or before schedule and for less money than planned.

The BCWP and ACWP form a synergy to let the manager know the status and direction of the program. The difference along the vertical is the cost variance and the schedule variance is the difference along the horizontal. The slope of the line is the direction that the program is heading (and probably will continue to head) unless specific actions are taken. Any deviation should ignite an investigation by the manager, who should take corrective action if needed. The relationship between the costs and work performed and their respective slopes must be constantly monitored. The manager must investigate and understand any deviations.

Costs incurred out of sequence may not be bad. The point here is to realize that they are out of sequence and assess the impact. Maybe a meeting was held early and the travel and labor costs were spent a few weeks earlier than planned. It may not matter. However it may if the meeting was less effective because it was held too early, future costs may result. The important point is that the manager now knows that he is over cost and may need to take corrective action. In this case, it is simply as beginning the meeting later. In more serious situations it may require unpaid overtime, staff adjustments, or other drastic measures. Whatever the measures, they will always be less painful and less Draconian if done immediately, rather than after the overrun has progressed like a cancer.

One criticism of this method is that it is a lagging indicator. It shows where the program was at the close of the last reporting period. This data latency is a major reason for frequent reporting. A data latency of a month can rarely be tolerated, but a week latency is usually not disastrous. Much effort has gone into determining a program's leading indicators, but most are almost meaningless by themselves and have only marginal effectiveness when combined. These include planning and counting the number of people in a program, drawing release status and those about to be released, material orders placed, and engineers confidence level at completing their tasks.

Many IR companies and managers fail to do simple planning and comparing. These tasks do not make program management more difficult. Instead, they make it easier. Their reluctance may be due to a fear that upper management will use the plans to crucify the program manager or make the manager look like a poor planner. Neither of these should be true. It should be realized that no plan is immaculate, but no program can be steered without one.

2.7 CAPITAL EQUIPMENT AND FACILITIES

Capital equipment required to design and build IR systems can be surprisingly expensive. Capital costs required to produce 100 HgCdTe wafers a month are in the neighborhood of $100 million [Luttmer, 9]. The wise manager prepares both short term and long term capital plans with diligence. It is expensive to play the IR game, but the rewards during high military spending can be great. In 1987-88, Aerojet invested more than $40 million, indirectly yielding over $500 million in contracts (mostly for DSP) ["Aerojet ElectroSystems Investment in Sensor Technology Yields $500 Million in Contracts," 10]. Capital outlay is not limited to buying equipment and facilities. It can also be effectively used to purchase a knowledge base. Aviation Week also reports that phenomenology gathering allows a company to bid firm fixed price contacts with confidence. "The GE Electronic Systems Organization has an extensive library of infrared data gathered over a 15 year period," it stated. "This library include 60 hours of clutter data on naturally occurring events collected by a company-owned C-54 with a tail mounted staring sensor in the mid-1980s. Data were gathered on these flights on smoke stacks, illuminated clouds, solar glints, and F-16s operating at both military power and afterburner. Rocket motors were also characterized" [Hughes, 11].

Capital outlay can be used to purchase software that enhances an engineer's capabilities. There is a plethora of canned software on the market that aids in infrared system and/or component design. The engineer and scientist should take a close look at all programs in the general field. The technically cognizant person should do a thorough trade study, considering his total budget, the required hardware and the outputs.

A manger on a limited budget must assess the utility of software in detail. Is it really justified, or is the engineer just making a playpen for himself? One should also explore leasing agreements. If only a few systems are likely to be designed, or only contract monitoring, then a "pay-as-you-go" agreement may be the best answer. In any case, the manager should weigh the cost of the software against the engineering staff's time savings and the customer's accuracy requirements. The manager will almost always find that if he is in the infrared business, he will be required to equip his staff with at least one program in each of the main engineering categories. The manager must also consider the hardware required to run the software. Does he have it in house, or does he need to buy hardware as well?

The manager should never purchase any computer hardware or software without adding the cost of training. Often the vendor will have a training program, that is typically a bargain. The manager must also include in the cost of the time it will take for all employees to be trained. Additionally, throw in several man-days per engineer for them to just play with the program and come up to speed. The latter is important (and will happen anyway), so the astute manager will budget for it from the beginning to avoid overruns. The worst situation is to buy software and hardware and have no one ever use it!

The fundamental task in choosing software is to balance the cost of the software with its capabilities and potential benefits. For proposals, the cheapest and most rudimentary is usually sufficient, while often only the Cadillac of software tools may suffice when it is time to order expensive components based on the program's output.

Software is not the only tool needed to construct an IR project. Assembly of infrared sensors and components usually requires clean rooms and specialized equipment. A clean room's cost depends upon its size and the level of cleanliness to be maintained. Often a good rule of thumb is that a dedicated clean room facility will cost from $500 to $2,000 per square foot.

Testing infrared sensors is not a trivial issue. Special test facilities that cost tens of millions to build may be required. The lack of consistency from one sensor to the next is responsible for part of this expense. The individualization of sensors is a product of their complex nature. For example, each optical element has a few percent difference in scatter, transmittance, or reflectance, and each FPA has individual alignment and uniformity inaccuracies. It is likely that these differences will continue to exist for at least a few decades. The only way they can be made inconsequential is to enforce extremely tight specifications on each stage of the process. The state of the art in manufacturing of IR sensors doesn't support these standards today.

There seem to be three approaches to testing: expensive facilities specialized and dedicated to a particular project; dedicated facilities to generic IR production; and simple, less costly facilities generated to a specific product line. Usually space sensor manufactures take the first or second course while FLIR and camera houses take the latter. Of note are the expensive (from tens to hundreds of millions of dollars), dedicated EO test facilities for sensor systems at AEDC (Tullahoma), Boeing (Seattle), Lockheed (Sunnyvale), McDonnell Douglas (Huntington Beach), and Rockwell (Anaheim). Each of these is a national asset, and should be considered for use before another dedicated facility is built. Recently, portable test facilities have been developed for pseudo-generic systems. Sub-contracting your product's testing often makes sense in light of the capital investment otherwise needed.

Large sensor test facilities can cost from tens to hundreds of millions of dollars. Even minor test facility can be expensive. The basic radiance emitter for testing is the "blackbody." These are devices built to replicate the Plank function over a specified temperature and wavelength range. The best blackbodies mimic the Planck distribution as a gray body with high emissivity and with an error of 1 percent or less. This is done by both assuring a blackish emitting surface and a geometry that mitigates reflection of radiation from other sources and a total surface temperature control to within 0.1 Kelvins.

The calibration of blackbodies is a particularly vexing situation. NIST currently supports limited radiance calibration, but most blackbody vendors still just provide calibration for a temperature sensor located within the blackbody. The problem with a calibration by temperature sensor alone is that it does not consider the inaccuracies imposed by emissivities not quite of unity and reflections of sources not of the temperature sensor temperature. Moreover, the temperature sensor, by definition, is not on the emitting surface. Because it is not on the emitting surface, a slight (but always unknown) temperature gradient or differential may exist. Blackbody systems are rather inexpensive, usually a few thousand dollars for one that operates at ambient and a few tens of thousands for one that operates at cryogenic temperatures. Often the large cost is incurred in accurately using the blackbody to produce a meaningful test. Depending upon the sensor bandpass, the entire blackbody and sensor may need to be in a nitrogen or vacuum environment. Additionally, the sensor's aperture should be filled with the blackbody radiation. This means either a complex and large area blackbody or an optical collimator. Such calibration can be very expensive (e.g., millions for meter-class cryogenic collimators).

From the foregoing, it can be seen that component and system testing can consume a large amount of the total system cost and schedule. It is typical that such testing can comprise 40 percent of FPA labor and 50 percent of space-based sensor labor. If only a few sensors are to be produced, the standard method of custom testing with high amounts of touch-labor is applicable. If, however, a production of tens to hundreds of units is expected, testing should be automated to an appropriate degree. Automated test fixture control and data acquisition are available and suitable for large volume production. Another approach sometimes used by manufacturers of ground-based cameras and FLIRs is to limit individual unit calibration, relying more on manufacturing uniformity than testing to determine individual sensor performance. This may or may not be acceptable depending on the intended use of the sensor and the factors affecting sensor lot uniformity.

2.8 PRICE/COST ESTIMATION

Historically, cost estimates for IR projects have been dismally inaccurate. The most common cause of cost escalation is not poor management, inefficient engineering, or frequent requirement changes. Although these conditions occur, the single greatest cause is that the state of the technology is still immature enough that new unknowns continually escalate the cost. This is a natural consequence of the constant desire to insert the latest technology into designs, coupled with the relatively small production quantities often encountered.

Accepted proposal rules, formats, and government requirements tend to increase the likelihood of cost escalation. Proposals to most governments do not contain the margins that everyone agrees should be there. Sometimes cost escalation is hidden in excess engineering, that usually gets cut back during negations. The process is filled with factors that lead to unintentional underbids. While component quotes are required from vendors to support government proposals, the vendor's expectation of actual obtaining the business often doesn't merit a large investment in finalizing prices at proposal time. Additionally, these vendors and subcontractors do not finalize their price to a realistic number. The "cost-plus" nature of most developmental contracts allows such casual pricing. In fact, cost competition on proposals (but not performance) encourages it. Managers feel their companies will be in a better position if the proposal bid is as low as possible. This leads to optimistic pricing and a bid with no cost margin.

The cost/price estimations in this book really don't deal with the detail required for a proposal submission. The top-level methods presented are based on estimation and scaling for rough, order-of magnitude purposes. These methods are often accurate when scaling a small amount about one variable. They are inaccurate when scaling several variables by factors more than two. Regardless, their relative results are useful for concept formation, engineering trades, and to check a sub's input for sanity.

There are more IR development programs than there are production programs. Non-recurring costs are usually cost drivers, and the largest is usually engineering and management. The component costs in Part II of this book represent subcontractors or vendor prices. These usually only make up 25 percent or less of the total price of the developed system. In programs where production occurs in large quantities, component costs make up a significant part of the recurring costs.

In general, one can scale a production cost from a development or first unit cost by employing "learning curves" derived from percent learning rates. These learning curves can be developed from empirical observation of those who manufacture large lots of commercial products. Manufacturing studies have shown that the "touch labor" on a product seems to decline as the workers gain experience. Each time another unit is made, the touch labor decreases. Regardless of industry, workers appear to experience a predictable decrease in labor each time the number of units doubles. This decrease, expressed as the percentage of the time required to do the next doubling, is called the Percent Learning Rate (PLR), and can be used to determine a learning curve. Often the PLR is called the learning curve. If it takes 10 hours for a technician to assemble a board the first time and his percent learning rate is 90 percent, it will take him 9 hours for the second, 8.1 for the fourth and 7.3 for the eighth. The most commonly used relationship to estimate the production cost of a unit [given in Equation (2.1)] is to take the number of units to the exponent of the log of the PLR divided by log2 and multiply by the first unit cost, or

$$C_n = C_1 N^{\left(\frac{\log PRL}{0.3}\right)} \tag{2.1}$$

where

C_n = cost in dollars of the "nth" unit

C_1 = cost in dollars of the first unit

N = total number of units to be made

PLR = percentage learning rate (in decimal notation)

Equation (2.1) assumes a constant learning curve that is historically justified. However, in IR systems the learning curve usually is steeper in the beginning. It then yields less reduction in cost as the number of units increases. Typical learning curves for IR components and systems are usually in the 85 to 95 percent range. However, some space instruments (like DSP sensors) have demonstrated learning curves of over 100 percent (the more you make, the more they cost). This usually is because of improvements and inflation.

Like touch labor, management, residual engineering, and indirect labor also tend to decrease as more units are made. The process becomes refined and capital equipment is better utilized, reducing the cost as more experience is gained by making additional units. These learning curves can be applied to the total per unit cost, but with less fidelity. For total unit cost there is less evidence to suggest a fixed decease per doubling of production. Additionally, mixing management (which is likely to have a higher learning curve) and engineering (which is likely to have a lower learning curve) muddies the calculation. The material and components are likely to pose different or no (100 percent) percent learning rates. Finally, at some point the capital equipment maintenance, or its replacement, may cause costs to rise. Again, such composite learning curves are usually a higher percentage (indicating less decrease in cost) and less accurate than ones for just the labor.

Customers buying IR products are less and less interested in technology advancement and more and more interested in a return on their investment. They will request total program cost with increasing frequency in the nineties, even when funding a concept development. The total cost of the production phase can be represented by the area under a learning curve. Mathematically,

$$C_t = C_1 \int_{n_b}^{n_e} N^{\left(\frac{\log PLR}{0.3}\right)} dN \tag{2.2}$$

where,

C_t = total cost of the production run

C_1 = cost of the first unit

n_e = total number of units made when the production run ends

n_b = total number of units made when the production run begins

N = total number of units

PLR = percent learning rate

Managers rusty on calculus shouldn't be alarmed by the above equation, as it can be easily evaluated on a spreadsheet using your portable computer.

Learning curves can be applied to the low production relationships in this book, vendor proposals, and quotes. The only caution is that they should not be applied twice to the same effort. Do not apply them to quotes from a vendor who has already applied them! One should also be careful to distinguish between price and cost. Price is 20 to 30 percent higher, reflecting profit, cost of money, overhead, and so forth.

There has been an increasing desire to be able to model the cost of an infrared system for a given program versus engineering parameters. While this seems straightforward, given several of the cost relationships in this book, extreme caution should be exercised. In general, the true cost of an infrared system cannot be expressed in a closed form nor can it be easily modeled. It may have step function sensitivity to requirements and environments. Generic relationships are rarely reliable when attempting to forecast an absolute cost. Their best use is to give the managers, users, and politicians an idea of how cost is sensitive to requirements over an excruciatingly narrow range. One can rarely achieve a direct cost from generic relationships, and the relationships rarely predict the step function nature of several variables. For instance, the cost actually may fall as one moves from a complex SOA scanner to a starer, though common wisdom indicates that starers cost more. Nonetheless, a simple engineering/cost model can be invaluable in the concept formation of a program.

2.9 SELECTING AND MANAGING SUBCONTRACTORS

Subcontractors can be critical to an IR project, as they supply many crucial components. To understand the scope of subcontracting, consider the Infrared Space Observatory (ISO) project, which "has 20 users of hi-rel electronic components who, together, need more than 220,000 parts of more than 300 types which originated from 70 suppliers" [von Hoegen, 12]. While successful subcontract management is a subject suitable for volumes, the following is a brief synopsis of potential pitfalls and solutions.

2.9.1 Selecting a Subcontractor

Like buying a new car or house, choosing a subcontractor is both a demanding and a rewarding task. It is rewarding because you can see and learn a lot about the industry and the development of the technology. It is demanding because of the importance and difficulty of choosing a subcontractor that satisfies, and hopefully optimizes, the program. Proper selection of a subcontractor (or contractor, from the customer's perspective) will make management easier and facilitate program success.

In a perfect world, no Request For Proposal (RFP) would be issued without a finalized set of perfect specifications, and the bidder would never be placed under contract without a completely negotiated specification and statement of work. The world is far from perfect. Rarely does a complete set of well-defined requirements and specifications exist when a contract is awarded, (let alone when the RFP is issued) or when a subcontractor/bidder is asked to commit to a schedule and price.

There is no complete solution to this problem. However, there are several mitigating steps that the prime can perform. When specifications are not firm or detailed, one important subcontractor task should be to aid in finalizing these specifications. They should be well within the state of art for non-development subcontracts, and on the periphery for developmental contracts. When specifications are not firm or TBD, the bidder should be asked to supply the specifications used for the quote. The RFP package should contain a detailed statement defining the bidder's role, scope, and tasks.

One developing trend in the industry is to issue a draft RFP and solicit comments. This is an excellent method for the prime to determine if what he wants is obtainable within his constraints, and which bidders are serious and understand the problems associated with the project. This is also a low-cost way for the primes to assure themselves that they have a good selection of potential bidders. However, this requires substantial nonreimbursable expenditures on the part of the bidders.

Another approach for larger programs is to award several initial study contracts for the subcontractor/bidder. The purpose is for each potential bidder to spend a month or two writing his own set of specifications that meet the higher-level system needs and objectives. This has the drawback of significantly delaying procurement and increasing initial costs. This is especially useful if the prime is somewhat naive on the technology or service being rendered. On the other hand, this extension of concurrent engineering can often save both time and expense over the total program [Steiner, 13].

If the purchase is not of a catalog item, then it is frequently to everyone's advantage to let the bidder know the expected price range so he can properly scope his effort. Some primes (or end use customers) refuse to do this on the assumption that the bidder will propose a price at the high end of the range, without consideration to lower priced options. This is rarely the case. The prime can still designate the price as a important criteria for selection. The bidder knows he has competition and won't want to lose by failing to consider lower priced alternatives.

The response from the bidder should identify major challenges and describe the technical approaches they will take to meet them. This should be specifically requested in the RFP. The challenges might include solving a performance shortfall, making delivery on schedule, or satisfying cost requirements. The defined scope must be in line with that of the prime, or a convincing case must made as to why they differ. This must be carefully examined in light of the other criteria. Following the submittal of the proposal, an exchange of questions and answers can increase understanding of the proposed approach. If it is not completely sound, the bidder should not be considered further and properly notified.

The schedule should be reviewed in detail with engineers, managers, and technicians and/or assembly personnel if appropriate. It should meet the prime's desires, have sufficient milestones and details to allow tracking, and be realistic. If the prime calls for a schedule that is a challenge, the bidder should provide a risk mitigation plan describing how it will meet the proposed schedule and a separate schedule that would be comfortable and low risk. The prime can then examine the latter and try to accommodate it, as much as possible, to reduce mutual risk. Occasionally, surprisingly easy accommodations by the customer or prime can assure subcontractor success.

Like schedule, cost estimates should be realistic and within the scope determined by the prime. When these are mutually exclusive, it should be pointed out. Costs should be compared to the other bidders, and should vary by less than 50 percent or so among the bidders. If it varies by more than this, it is likely that some of the bidders do not understand the job or that the prime failed to properly define it. On the other hand, it could also be that the low bidder has a definable cost advantage due to previous work, synergistic contracts, a willingness to use his own money, or has a facility advantage.

Frequently, a bidder's proposal is evaluated merely on the cost, schedule, and technical approach. There is no argument that these are the most critical characteristics, and deserve the highest weight. However, they lose value if not considered in light of other related criteria. Some suggest that optics subcontractors must be known thoroughly, and also should be evaluated on [Marchi, 14]:

- Technical and manufacturing capability
- Test facilities
- Management (management stability, financial stability, and prior performance)
- Communication

Each of the above is important, but the list might go further to include other critical criteria, including:

- Technical experience
- Understanding program requirements/goals
- Creativity
- Enthusiasm
- TQM /CE emphasis
- Location

2.9.2 Managing a Subcontractor

Technical managers often view managing infrared contracts (or subcontracts) as extremely challenging and frustrating tasks. The history of slipped schedules, cost overruns, and underperformance documents the delicate nature of the technology and difficulty it poses in effectively managing vendors and subcontractors. These difficulties are equally likely to arise from the sensor integrator or a key component supplier. Both must meet schedules and specifications for total program success. Even if the parent organization is completely on schedule, a slip by a subcontractor for a key component can be a disaster.

It is never too early in a project to consider vendors and sub-contractors. On the ISO, a parts procurement plan was set up early in the definition phase. A procurement board was able to reduce the number of electronic parts ordered from 500 to 350. The ISO's centralized parts procurement board was able to reduce non-recurring costs to less than 45 percent of recurring costs (most space programs experience somewhere between 50 and 200 percent) and the total parts cost would have been approximately 80 percent higher if every user had purchased them separately [von Hoegen, 15]. This reduces the cost of the parts, overhead, paperwork, reduces the inevitable confusion, and makes obtaining spares easier. Parts often must be ordered before the design is frozen to make the schedule. This is tricky, dangerous, and a fact of life in most IR projects.

Several situations can cause problems with a subcontractor. The most frequent are:

- Specifications not defined when the prime forces a commitment for schedule and budget
- The subcontractor poorly staffs the program
- The subcontractor has difficulties with his vendors or suppliers
- Marketed technology does not support application and/or schedule
- The prime over-emphasized cost or schedule, causing the subcontractor to take an overly optimistic position when it proposed the effort.
- The subcontractor had little or no competition and perceived other projects as more important
- The prime had unrealistic expectations
- Poor communication through the prime of customer needs and desires
- Ineffective monitoring by the prime
- Poor or nonexistent configuration management by either party
- Subcontractors were not given adequate time to provide a reasonable bid and/or program plan

The largest function in contract management is monitoring; the largest fear is surprise. To mitigate the surprise factor, performance, schedule, cost, and quality must be monitored during all phases of the subcontract. The overseeing organization must implement an effective, efficient monitoring policy geared to adding value to the entire program (or subtracting as little value as possible). This policy should incorporate regularly scheduled teleconferences, letter status reports, and facility visits. The frequency depends upon the size of the effort, the criticality, and performance to date on the contract. Even the smallest effort should have informal telecons and status letters at least once per month, with a couple of visits per year. It is important for the overseeing agency not to be mislead by ambiguity or whitewashed in the status letters. There are several techniques to see through potential subcontractor camouflage. One is to include engineers and technicians (on both ends) in telecons and status letters. Ask them specific questions about their area. Another is to avoid a policy where all communication goes through the management channel. Just call up the cognizant engineer or scientist and have an informal discussion. Such calls should be confined to gathering information so that they don't cause confusion or misunderstood direction. A communication balance must be achieved. To aid this, it is important to agree upon a sufficiently detailed schedule, with frequent discussions planned throughout the life of the contract.

Assuring that the entity conducting subcontract monitoring adds value instead of subtracting it can also be a difficult challenge. The most straightforward way of doing this is to be sure communications travel in both directions during status reviews. Engineers, scientists, and managers should closely monitor the process. The monitoring effort should explain any changes in key technical drivers, funding changes, or flexibility in the requirements. Attempting to provide genuine help to the subcontractor via experts, facilities, or contacts is another way of adding value.

Finally, transforming a vendor to a partner is important. As primes downsize, they won't have the staff of experts to perform vendor selection and monitoring. The vendor relationship may evolve to a partnership transcending projects, programs, and customers. Prime A may automatically consider vendor B because of his previous track record.

2.10 CONCLUSION

This chapter suffers from the "easier said than done" syndrome. Many of the most valid suggestions cost time and money, which may not be available. The manager must review his own program and associated risks and decide where he can afford to implement the suggestions and ideas.

Successful EO projects don't just happen. They are caused by excellence in management, science, engineering, and production. The management emphasis must change as the program matures. Table 2.3 lists some development stages in an infrared program and the key activities that must take place. Management must enable these key activities.

TABLE 2.3 Management Emphasis for Various Stages of an IR Program

Program Type	*Key Activity/Management Emphasis*	*Program Example*
Study	• Creativity • Selling • Flow of desires and needs to requirements	• Shipboard IRST • F-16 IRST
Technology development	• Goals • Creativity • Objectives • Flow of requirements to specifications	• SDI programs • ManTech programs • EOS • FEWS
Hardware prototype and proof of principle	• Interfaces • Schedule • Cost • Creativity • Procurements • Specifications	• Comanche FLIR • Javelin • EOS • LOSAT
Production	• Cost • Rate • Component delivery • Efficient operations • Modifications for new sales	• LANTIRN • Thermographic instruments • Industrial cameras • Nighthawk • DSP

When dealing with other organizations, it must be remembered that each link in the infrared food chain is vital to every other link. Every link requires the respect and cooperation of the others. Customers and primes must act in total good faith and honesty when requesting any proposal submission. Also, the subcontractor must act in complete good faith and honesty in any submitted proposal. Personal and corporate reputations are critical to survival.

The management of electro-optics is a combination of science and art that must include intuition and insights based on a thorough knowledge of the technology and persistent monitoring of its progression. The key to successful management of IR projects can be summed up as effectively using top engineers, performing all transactions with the highest integrity, and employing the state of the art technology within cost constraints, coupled with unrelenting monitoring and control of all aspects of the process.

References

1. A. Fisher. 1980. *Small Group Decision Making.* New York: McGraw Hill, 67.
2. Private Communications with Dr. Gerry Steiner, 1993.
3. N. Augustine. 1983. *Augustine's Laws.* New York: Viking, 74.
4. W. Dee. 1989. "Defense Contractors Must Change to Survive Competition in 1990s." *Aviation Week and Space Technology,* July 17, 99–102.
5. C.J. Middleton. 1989. "How to Set Up a Project Organization." in *Managing Projects and Programs* (complied by the *Harvard Business Review*). Boston: Harvard Business Review Books, 65–84.

6. M. Maynard. 1991. "No Price Tag on Being Best, Winners Show." *USA Today,* October 10, B1–B2.
7. W. Dee. 1989. "Defense Contractors Must Change to Survive Competition in 1990s." *Aviation Week and Space Technology*, July 17, 99–102.
8. Information courtesy of TRW, 1992.
9. J. Luttmer. 1992. "MMST Development for HgCdTe FPA Manufacture." *Proc SPIE,* 1683:24-32.
10. "Aerojet ElectroSystems Investment In Sensor technology Yields $500 Million in Contracts," 1988. *Aviation Week and Space Technology*, January 18, 66–69.
11. D. Hughes. 1991. "GE Missile Warning Systems Rely on Staring Arrays." *Aviation Week and Space Technology*, October 21, 42.
12. M. von Hoegen, J. Minnee, and F. Zeigler. 1991. "Coordinated Parts Procurement For ISO-A Contribution to Cost-Effectiveness." *ESA Bulletin,* 67 (August):29-38.
13. Private communications with Dr. Gerry Steiner, 1993.
14. F. Marchi and E. Wallerstein. 1980. "Management of Shiva Optics." *Proc. SPIE,* 260 (1980):44–49.
15. M. von Hoegen, J. Minnee, and F. Zeigler. 1991. "Coordinated Parts Procurement For ISO -A Contribution to Cost-Effectiveness." *ESA Bulletin* 67 (August):29–38.

II

Component Technologies

3

Infrared Telescopes

"A telescope with good optics and a lousy eyepiece is a lousy telescope. A telescope with good optics, a good eyepiece, and a lousy mount is even worse."

Unknown (overhead projection at Astrofest 1988)

3.1 TECHNOLOGY BASICS

The first component of an infrared system encountered by an entering photon is the telescope. This is the engineer's first opportunity to enhance the system's performance (or degrade it, as occurred with the Hubble Space Telescope). The optical train's (or telescope's) function is to collect photons and place them on the focal plane in a proper configuration. This function encompasses light gathering, focusing, bandpass definition, magnification, stray light rejection, and field of view control. Spiro and Schlessinger point out that, "The chief differences between infrared and visible optical designs are in the properties of materials and the effects of longer wavelengths" [Spiro and Schlessinger, 1]. Two more chief differences can be added: generally much wider spectral bandpasses and consideration of the thermal emission from the optics themselves.

3.1.1 Telescope Types

The earliest telescopes were made from refractive elements. Newton was the first to develop a reflective design. Refractive and reflective optics were the only choices that existed until the last decade, when diffractive optics appeared. Today, refractive optics usually yield a lighter and lower cost telescope if all the elements are under a few centimeters in diameter and can be purchased from catalogs. Refractive telescopes are also less sensitive to misalignment. Beyond a few tens of centimeters in aperture diameter, the lighter and less expensive telescopes have reflective optics. An all-diffractive telescope has yet to appear, but diffractive elements are being used to clean up the beam, color correct, and reduce the number of elements. A key advantage of diffractive elements is their ability to make one-element achromats.

There are various types of reflective telescopes. Cassegrains and prime focus architectures tend to weigh the least, with off-axis and Schmidt designs costing and weighing the most. The field of view (FOV) for an on-axis reflective telescope is small, typically 1.5° or less. If image quality can suffer at the edges, this field of view can be pushed to about 3° or 4° with an increase in cost and delivery time. Off-axis telescopes typically weigh and cost more and have longer delivery times than on-axis optical trains. This is due to the fact that they are more difficult to manufacture and align, and that a more robust support structure is often required. However, in some applications that require more than two mirrors for on-axis designs, or high dynamic environments, off-axis may weigh less when light gathering power is compared instead of primary size. This is because off-axis systems do not have a central obscuration. Off-axis designs have naturally better stray light rejection, larger fields of view, and no central obscuration. In general, weight, cost, and volume of a telescope scale from the largest optical element, and the largest mirror in an off-axis system is usually the tertiary, not the primary.

If a Schmidt telescope is very large, it may cost less than other comparable designs due to the simplicity of grinding its spherical primary. However, any cost savings in the primary must be weighed against the extra cost of the refractive corrector plate. The aperture of a Schmidt is always smaller than the primary. The corrector plate at the aperture determines the telescope's diffraction limit and light gathering power. Schmidts can have much larger FOVs than other on-axis reflective systems with 5 degrees to 15 degrees possible.

Two stage designs, also known as reimaging telescopes, contain more than two mirrors. They have a telescope within a telescope, as depicted in Figure 3.1. The first mirror the light encounters is called the primary. The second is the secondary. The third is the tertiary. The fourth is the quaternary. A four-mirror reimaging telescope forms two images. In this example, one is at the focal plane and the other is near the quaternary mirror. Notice the curve to the focal plane representing the Petzval curvature of the image. These configurations narrow FOVs with high off-axis rejection. Two-stage telescopes are obviously heavier and more difficult to align. Their benefits include high off-axis rejection, four surfaces to allow correction (especially useful if the primary is spherical), and convenient location of a Lyot stop. Two-stage telescopes are popular with IR designers because they may provide warm operation of the primary and secondary mirrors at long wavelengths with cooling required only at the tertiary and quaternary. This is because the entry elements are before the field stop.

Multi-mirror telescopes (MMTs) are telescopes that gather light from more than one primary mirror, as shown in Figure 3.2. The "effective primary" is com-

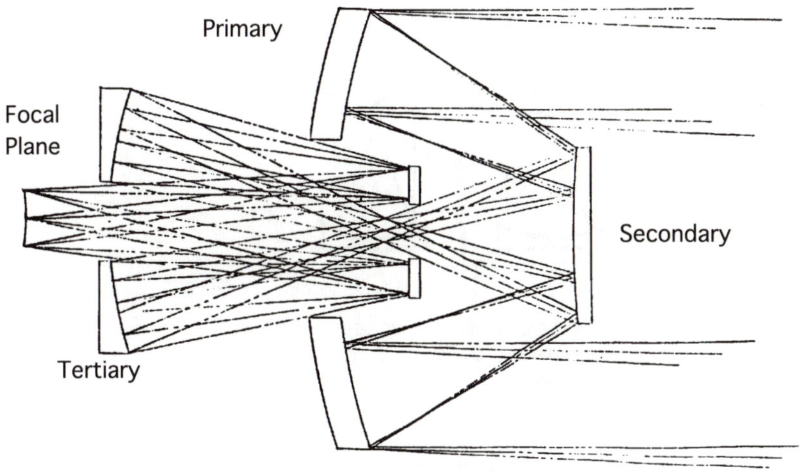

FIGURE 3.1 A Four-Mirror Reimaging Telescope (courtesy of Marion N. Todd, patent in Marion N. Todd's name held by Hughes Aircraft)

posed of several separate mirrors and the telescope combines the light from each into a single beam before it is incident upon the focal plane. It is simply viewing through more than one telescope at once to form a single image. The most famous example is the multi-mirror telescope in Arizona, which combines the light from six mirrors to a single focal plane. Alignment is critical and difficult with this type of design, and the MMTs have to employ real-time alignment schemes using lasers. When it was completed, the Arizona MMT was the largest telescope in the world in light gathering power. Now it is being retrofitted with a single large primary. The diffraction limit of multi-mirror telescopes is limited to the size of the individual mirrors since the light is not combined in phase. The Air Force Weapons Lab in Albuquerque has sponsored much work coherently combining multi-mirror telescopes in phase, as is done with radio telescopes, to achieve long baselines. Wavefront sensing with real time active feedback control has enabled independent telescopes to be combined in phase, yielding the diffraction limit of a single-mirror as wide as the separation between the telescopes, although only in that dimension.

Segmented mirrors are composed of individual pieces (or segments), each with a part of the mirror curvature on it (not the entire curvature), assembled to make a single mirror (e.g., the Keck). Unlike the multi-mirror concept, there is only one overall mirror surface, but many separate hardware pieces. Segmented mirrors lend themselves to active wavefront control, sometimes called phase correction or phase control. This is when each segment is micropositioned in real time to compensate for distortions from the atmosphere, thermal gradients, or misalignment. In theory this works wonderfully. In practice this technology is just breaking even.

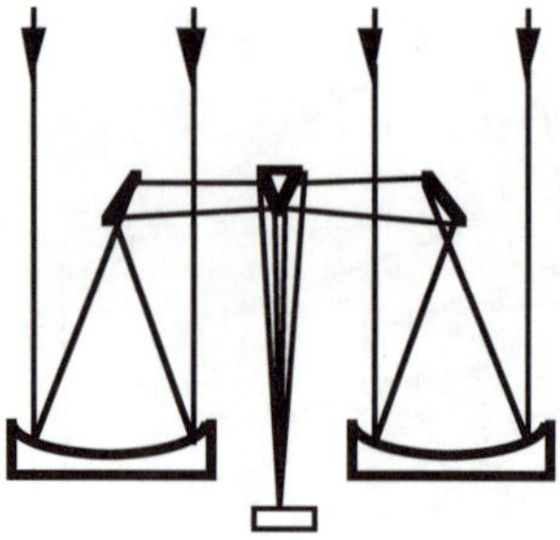

A multi-mirror telescope has separate apertures and combines those multiple apertures onto a single focal plane.

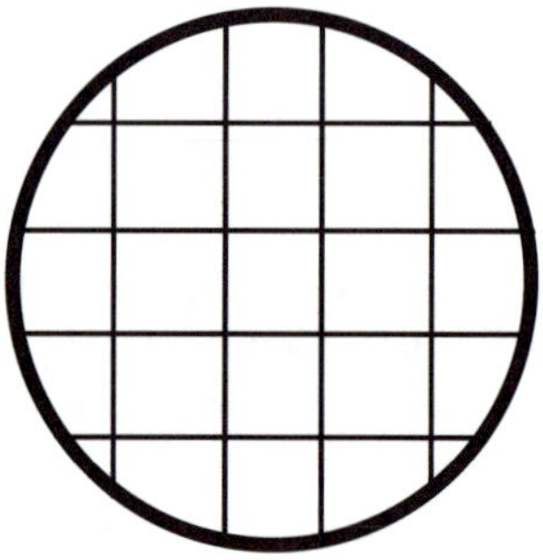

A segmented mirror is a collection of individual mirror parts into a single mirrored surface representing a single aperture.

FIGURE 3.2 Segmented Mirrors vs. Multi-mirror Telescopes

Recently, the National Optical Astronomy Observatories (NOAO) reported a factor of two improvement in RMS surface distortion when trying to correct for thermal gradients in a 1.8 meter mirror [Pearson, Stepp, and Fox, 2]. Segmented mirrors are not necessary to accomplish wavefront correction; metal mirrors with a flexible (few microns) front surface also provide the needed tilting and displacement for phase correction, control, or even conjugation. Care must be taken in optomechanical design, as active wavefront control tends to distort the image more than clean it up. Segmented (or deformable) optical elements must be operated at ambient temperatures. It is difficult with current technologies to cryogenically cool the mirrors and their actuators. This can impose severe constraints on high sensitivity, low background sensors.

3.1.2 Basic Concepts

Without dwelling on optical design, several germinal (and often trivial) concepts must be understood. These include resolution, modulation transfer function, and scatter. Optical designers use three types of focal length: back focal length, front focal length, and effective focal length. Effective focal length is the only one we will discuss. Back and front focal length should be left to optical designers and packagers, as they depend on individual elements and usually not systems.

To understand resolution, one must understand the diffraction limit. When focused, light does not form a perfect point at the focal plane. In its cleanest form, it diffracts to form a disk surrounded by rings. To make a long and complicated story short, the central spot of the diffraction pattern (called the Airy disk) on the focal plane of an unobscured aperture contains most (perhaps 80 percent) of the energy. It can be defined as

$$\text{Spot diameter in microradians} = \frac{244\,(\text{bandpass cut-off in } \mu\text{m})}{\text{aperture diameter in cm}} \tag{3.1}$$

Consider a notional 90 cm, off-axis circular aperture telescope supporting a bandpass of 8 to 11 μm. This example would have a diffraction spot diameter on the focal plane of 244(11)/90 = 29.8 microradians (μrad). Equation (3.1) is valid for diffraction limited circular apertures with no central obscuration. On-axis reflective telescopes have a secondary in the middle of the light path. The effect of this central obscuration is to reduce the size of the Airy disk and the energy in the Airy disk to less than the 84 percent of unobscured diffraction limited systems. Transferring energy out of the central disk to other rings blurs the image. This allows a slightly smaller central disk, at the cost of sensitivity and image quality. For a circular aperture with a central obscuration the Airy disk can be defined as

$$D_b = \frac{244\,(\lambda co)\ (1 - \varepsilon^2)}{D_o} \tag{3.2}$$

where

D_b = angular diameter of the blur in μrad

ε = linear fraction of diameter that is obscured by central obscuration

λco = cut-off wavelength in microns

D_o = diameter of aperture in cm

This is the equation to be applied to all on-axis reflective telescopes, because they all have a central obscuration. Equation (3.1) applies to refractive and off-axis telescopes. To illustrate Equation (3.2), let's assume the notional 90 cm telescope is an on-axis system with a central obscuration 30 cm in diameter. In this case, the angular blur diameter is 244(11)(1 – 0.09)/ 90 = 27.1 μm. This is a slight (about 9 percent) decrease in blur extent.

NASA, among others, has deliberate designs for "thinned aperture" telescopes that have a large central obscuration. This allows for a very narrow central disk with reduced energy in that disk. If light gathering power and low spatial frequency resolution are less important than high spatial frequency resolution, these should be considered. However, the reader should not get the impression that central obscurations are always desirable. The effect of moving energy from the central disk to other parts of the focal plane is a big drawback. It reduces contrast, blurs images of objects other than point sources, and generally lowers the modulation transfer function (MTF).

These equations define the smallest angular size of a spot a system can make on the focal plane. They do not necessarily define the spot's angular size--just what nature will allow. Poorly made optics will result in a much larger spot. A diffraction-limited telescope will approach this, never exceed this and in real life (with alignment errors and such), probably won't meet it. The diameter of the first Airy disk can be though of as the resolution of the optical system (not necessarily the sensor) at a zeroth order of understanding. Many will define resolution as twice the diameter of an Airy disk. The Airy disk is the ultimate limit on optical performance, but not on information resolution. Under some conditions it is possible to separate two Airy disks that are closer than their radii (called the Rayleigh criteria). It is also possible to spread this disk over multiple pixels or scan across it and determine a position accuracy that is less than the Airy spot diameter. Algorithms and techniques for sub-pixel accuracy, sub-diffraction limited position determination are rapidly advancing. The amount of sub-pixel or sub-diffraction limit information that can be extracted is empirically related to the signal-to-noise ratio (SNR). This is caused by the algorithm's call for neighboring pixels to sense the edges of the Airy disk or other rings. For a scanner, the pixel must be read when just a part of the disk is on the pixel.

A general rule of thumb for square pixels in a staring sensor is that the best position accuracy available is on the order of

$$\text{Positional accuracy} = \frac{\text{angular diffraction limit}}{(\text{SNR})^{n}} \tag{3.3}$$

where

SNR = signal-to-noise ratio

n = constant ($0.5 < n < 1$)

If our off-axis notional 90 cm telescope is combined with a staring array and the overall SNR is 10, then positional accuracy can be determined to

$$\frac{29.5}{10} \approx 3\ \mu\text{rad}$$

For a pixel being scanning over a bright target, the amount of sub-diffraction limit information possible is on the order of

$$\text{Positional accuracy} = \frac{\text{angular diffraction limit}}{(12 + (\text{SNR})^2)^{0.5}} \tag{3.4}$$

If our off-axis telescope is combined with a scanning array and the overall SNR is still 10, positional accuracy can be determined to

$$\frac{29.5}{\sqrt{12 + 100}} \approx 2.8\ \mu\text{rad}$$

Under identical SNR conditions, scanning arrays allow interpolation to slightly more accurate positions. However, scanning arrays integrate (collect photons from) the target for a much shorter time, so for a given target, background, range, and FPA noise, staring arrays will have higher SNRs. Detailed simulation is required to trade staring arrays for scanning arrays in any application.

It is also important to be able to calculate the disk's size in length measurements, as opposed to Equation (3.1)'s angular measurements. Length measurements allow one to choose the pixel (unit cell) size on the focal plane. This can be calculated as

$$d = (2.44)\ (\lambda co)\ (f/\#) \tag{3.5}$$

where

d = linear diameter of central airy disk in same units as λco

λco = long wavelength cut-off

$f/\#$ = effective F/number or speed of the optical system

If our 8–11 μm telescope is an f/1.8, the diameter of the blur in μm is 2.44 × 11 × 1.8 = 48 μm. This equation is important because it will usually be used to balance the pixel size on the focal plane with the F number of the optical system. For our example, an FPA detector size of 40–50 μm (common for arrays) will suffice. If a 50 μm pixel is chosen, the Airy disk can be completely contained in the 48 μrad extent. The pixel (or unit cell) size is usually specified to be about the same as the central Airy disk. The higher the performance of the sensor system, usually the smaller the pixels and the smaller the f/#. This is an unfortunate deal by nature, because the pixel size is an important driver of focal plane cost, producibility, manufacturability, and delivery schedule. Likewise, the f/# of an optical system is a driver on its cost, producibility, manufacturability, and delivery schedule. This elementary equation must be analyzed, traded, and properly balanced with respect to the system's requirements.

The blur spot is the Airy disk along with any aberrations that effectively make it larger. It should, whenever possible, be about the same size as the detector (or pixel). This may mean that a field lens right in front of the focal plane, or a lenslet on the focal plane must be used. Matching the two is important because a pixel may exhibit sensitivity variations across its surface. Additionally, most focal planes do not have a 100 percent fill factor. Some, such as monolithic Pt:Si, can be exceedingly low (30 percent), and if the spot size is small, the entire signal may fall on a non-photoactive area of the focal plane.

Three dimensional positioning of the optical elements and focal plane is becoming increasingly important as pixel sizes shrink. It is easy to design a sensor requiring spacing tolerances well beyond that which is possible. Modern engineers must be very cognizant of this. The tolerance of FPA axial positioning can be calculated by

$$\pm \text{ distance from perfect focus} = \frac{D_b f^2}{D_o} \tag{3.6}$$

where

D_b = The allowed angular extent of the blur (typically the larger of the airy disk or pixel size in radians)

f = Effective focal length

D_o = Aperture diameter

If we chose a 50 μm pixel size for the focal plane of our notional telescope and wanted to maintain the Airy disk within a pixel, the allowed angular extent of the blur would be equal to the angular extent of a pixel, or 50 microradians. Since we said it was an f/1.8 system, the effective focal length is 1.8 × 90 = 162 cm. Therefore, the allowable amount of translation error of the focal plane is 0.0146 cm, or about one-seventh of a millimeter, which is difficult but reasonable for typical manufacturing techniques. This equation should always be used to check tolerance, to make sure the intended system can be built. One seventh of a millimeter seems reasonable and achievable within engineering tolerances. However, the temperature ranges and size of the telescope must also be considered. In this case, it is likely that the telescope is a meter or more long and may have to function though temperature swings of ±50° C, which could make such a tolerance difficult to achieve.

Any defocus affects the MTF, which describes the change in modulation caused by the system or the change in contrast. It is usually expressed as a plot of contrast change as a function of spatial frequency. This is an extremely useful figure of merit for any imaging system because it can be measured and reflects real-world errors. Ideally, the area under the curve would be as large as possible, indicating that the optical train caused little degradation to the incoming image. MTF is a function of element tilt, placing, and figure.

As pixels get smaller, resolution, image requirements, and defocus become more important, and can be the limiting factors. It is increasingly important for proper placement of the elements and the focal plane, and the allowable tolerance may drive the entire sensor cost if it is required to a placement of less than 0.02 mm. It should be noted that the tolerance could be a lot looser than this and still have a high cost if the system must operate over a large temperature range. Additionally, a curved focal plane may be required for precise resolution and imaging requirements. Although curved focal planes have been built, they represent a cost and complexity that should normally be avoided.

Another important optical concept to infrared design is scatter, which can introduce considerable error and seriously degrade system performance. Scatter is the process of sending light away at unintended angles. It is related to the cleanliness of the optics and their RMS surface roughness. A rule of thumb for wavelengths of less than 4 or 5 μm is

$$\text{Total integrated scatter} \propto \frac{4\pi s}{\lambda} \tag{3.7}$$

where

s = RMS roughness

λ = wavelength

It should be noted that shorter wavelengths will scatter more than longer wavelengths. This is an important consideration in IR telescopes, which may support observations in wavelengths that vary by a factor of 10 from the shortest wavelength to the longest.

Good figure and low scatter are opposing goals for a polisher. If a technician polishes for good scatter, the result is often a poor figure and vice versa. Scatter is mainly caused by the optical element's roughness. However, dust, pitting, and contamination are prime contributors. They are primarily the products of age. Scattering from dust and pitting can be significant if the dust (or pit) size is comparable to the wavelength. A resonance in scattering occurs when dust is on the order of the wavelength. Many have noted that placing a good coating on a rough surface merely increases the scatter. While often the case for thin films, it is not true for many claddings (or platings), since the thickness is so large that the RMS roughness of the surface depends on the scatter of the clad surface.

Comparing scatter from one surface to another is difficult. First, NIST does not yet support a suitable standard, although one is in the works. Second, scatter values depend on the spatial bandwidth of the measuring instrument. A 10 angstrom (Å) surface is better than a 20 Å surface when measured at the same bandwidth. But the 20 Å might actually be better, depending on how the measurements are taken. It is, therefore, important to know and understand the measurement specifics.

An optical element's figure is the shape of its front surface. Its quality is measured in fractions of a wavelength departure from the desired shape. There is no consensus as to what wavelength, but it is usually quoted at the HeNe laser line of about 0.6328 μm, or the center operating wavelength of the system. Sometimes for an infrared telescope it is quoted at 10.6 microns. Also, it can be stated as a peak-to-valley or RMS measurement. Roughly, a figure of $\lambda/20$ is required for interferometric, phase sensitive, and critical imaging applications; $\lambda/10$ is satisfactory for imaging optics requiring low beam distortion (especially when involving multiple elements); and $\lambda/4$ is adequate for general purposes [Miles Griot Inc., *May 11, 1992*. One must be careful of ads and vendor supplied specifications. To help mediocre optics sell better, some companies manipulate tolerances in their ads. Consider a parabolic mirror advertised as "accurate to 1/16 wave." The rating may mean that no "peak" is higher (+) and no "valley" is lower (–) than 1/16 of a wave relative to the ideal surface in between. But the total peak to valley error—from the top of the highest high to the bottom of the lowest low—would still be as much as 1/8 wave. Because of the nature of reflection, a mirror's defects are doubled on the wavefront (the light converging to the focus), so the resultant image is good to only a 1/4 wave [Ceravolo, 4].

In the U.S.A., "Scratch-Dig" is another method of measuring the surface quality of coatings and substrates. Defined by MIL-M-13408C, it is generally it is more lenient for commercial grade optics and most stringent for high power laser applications. Commercial grade IR optics usually have about an 80/50 or 80/60 measurement; 60/40 for better quality or visible optics; 40/20 for low power laser applications; and 10/5 for high power laser applications [Roberts, 5].

In the infrared, thermal emission from the optics must be considered. Optic's emissivities are typically low, but a 300K lens may dominate the FPA's signal at 15 μm. As a result, it is desirable to have IR optics of low emissivity and high transparency. For long wavelength operation, it is often necessary to cool them.

Generally, clean reflective optics have a low emissivity of 5 percent or less, while refractive elements can be much higher.

Back reflection of cold surfaces can also cause an unwanted image effect in the IR. This is called the Narcissus Effect, because it usually occurs when the detector is seeing either itself or its cold filter. It is typically reduced by using anti-reflection coatings on the inside, defocusing the return cold signal, nulling it out, canting flat surfaces, or notching or darkening the very center of the optical system.

In some military or space applications, the potential susceptibility of IR optics to nuclear effects can be a driver. The "optics can darken, crack, form bubbles and even melt" [Kiernan, 6]. Refractive optics are usually less hard than reflective.

Chopping is frequently used in IR sensors to provide AC modulation of the signal. This allows the large noise floor from background and instruments to be easily subtracted out. This chopping can be accomplished by several methods, the most popular being a spinning wheel, tuning fork, or dithering secondary. Sensors that view highly non uniform backgrounds (e.g., FLIRs) usually employ an internal spinning (or chopping) mirror reflecting a blackbody source. On the other hand, astronomers have the advantage of peering into deep, cold space so they usually use an empty portion of the sky as a reference and chop by jittering the secondary mirror. The secondary is tilted off to the side by a degree or so, which points the field of view away from the object under study, presumably to an empty part of the sky. The astronomers must analyze the chopping area before data is taken to assure that it is, in fact, empty. Actually, some IR sources have been discovered by accident by chopping to a part of the sky that was not void of emissions.

For a hardware example, consider the Keck. When observing in the IR, a specialized smaller secondary is placed in the Keck. The IR secondary is gold coated for higher IR reflectivity and figured for a longer focal length. It is made of solid beryllium [Smith, 7]. The secondary is of a smaller diameter (50 cm, as opposed to the 1.4 meter visible), which prevents the edges of the primary from contributing to the light gathering power. Thus, the Keck is a smaller telescope when used in the IR mode. Removing the edges of the primary from the FOV by a smaller secondary is done to keep stray IR radiation from the warm structure, floor, mirror support, etc. from contaminating the data. Because of its smaller size, high stiffness, and low weight, the secondary is able to chop by dithering back and forth at about a 50 Hz rate.

3.2 COATINGS AND FILTERS

Most optics are coated to provide reflection for a mirror, to limit reflection on the surface of refractive optics, to provide protection, and to provide spectral filtering.

Coatings are frequently the weak link in optics' durability. This is especially true with refractive IR materials, because the coatings must be deposited at lower than optimal temperatures to prevent substrate damage.

3.2.1 Coating Deposition

Whether a coating is to provide reflection, protect the optical element, reduce back reflection, or provide spectral filtering, it may be applied through several processes.

Astronomers frequently use the thermal evaporative coating method to apply a metal to a substrate because it produces the highest reflection. In this process, the metal is heated to a sublimation point in a vacuum chamber. The molecules migrate until they are deposited on the substrate. A quartz crystal monitor is usually used to observe the thickness of the deposition. This process is applicable to metal films of up to about 500 nm in thickness [Hagen and Bickel, 8]. It is typically low cost, and is used to coat the largest astronomical primaries.

Electron beam (E-beam) evaporation is another deposition method. An E-beam is controlled by magnets to sweep across a target material source. The material is evenly heated, vaporized, and deposited on the substrate. Once a layer has grown to its desired thickness, the beam is directed to another material. Major advantages of this process are the control over the deposition it affords, its ability to deposit thicker films, and its ability to coat with multiple materials (by moving them in and out of the E-beam) [Hagen and Bickel, 9]. Growth rates of 1 to 100 Å per second are commonly achieved [Morelli, 10]. The disadvantages include added expense and the possibility of secondary radiation damage from the e-beam.

Ion sputtering is the third popular coating technique. Here, ions from an inert gas are directed to target samples by an electric field. As the ions knock pieces of the target off via momentum transfer, some of them become deposited on the substrate. Metals are usually deposited via DC sputtering and dielectrics via RF sputtering. Advantages of ion sputtering include better adhesion, faster cycle times, and harder films. Its disadvantages include poor rate control, uncertain uniformity, and the potential for higher localized heat on the substrate [Hagen and Bickel, 11].

3.2.2 Specifications and Requirements

Filter manufacturers who understand applications and system problems often write the best specifications. The manufacturers know what they need to know and where they have questions. OCLI recommends the following checklist for writing specifications for square bandpass filters [Morelli, 12]:

1. Substrate
2. Substrate shape, dimensions, tolerances, and critical aperture

3. Environmental requirements
4. Angle of incidence for central ray and half-cone angle of incident energy
5. Operating temperature
6. Center wavelength and tolerance
7. Minimum acceptable peak transmittance
8. Nominal bandwidth in microns or percent of center wavelength
9. Level and spectral range of required attenuation

When this checklist is used, a sample specification may resemble the following [Morelli, 13]:

"Fabricate and coat square bandpass infrared interference filters according to the following specifications. Substrate is to be supplied by the filter manufacturer."

1. Size = 25.00 ± 0.12 mm diameter
2. Thickness = 0.5 ± 0.08 mm; wedge angle ≤ 0.05°
3. Flatness before coating ≤ 1.0 waves @ 0.6328 μm
4. Surface quality = E–E per Mil-F-48616
5. Center wavelength = λ_o = 4.250 μm ±1 percent
6. Minimum bandwidth = 1.7 percent
7. Minimum peak transmittance at $\lambda_o \geq 55$ percent
8. Out-of-band transmittance:
   Transmittance at $\lambda_o \pm 3$ percent of λ_o to be ≤1 percent absolute
   Transmittance at $\lambda_o \pm 4$ percent of λ_o to be ≤0.1 percent absolute
   Transmittance at $\lambda_o \pm 8$ percent of λ_o, and at all other wavelengths
   Outside these limits and less than 7.0μm, to be ≤ 0.01 percent absolute
9. All spectral characteristics to apply at 77 K and an incidence angle of 15°. Spectral scans at these conditions are to be supplied with the filter.
10. Completed filters are to be scanned for out-of-band leaks using a spectrophotometer resolution < 0.1 μm.
11. Completed filters are to withstand the following tests:
    Humidity per Mil-F-48616
    Mild abrasion per Mil-F-48616
    Adhesion per Mil-F-48616

There are three basic abrasion tests for coatings: rubbing the coating with cheesecloth (US Federal Spec CCCC-440 and US MIL-C-48497), rubbing with an eraser (MIL-E-12397 and US MIL-C-48497), and the RSRE (now DRA) Grittington Wiper Test [Hichwa, Klinger, and Swab, 14]. To pass the latter, the coating must have no visible damage after 5,000 passes of a wiper blade with a 40 gram load in a sand and water slurry. Diamond-like carbon is the only typical IR coating capable of passing the Grittington, while ZnS can pass about 50 cheesecloth wipes [Hichwa, Klinger, and Swab, 15]

3.2.3 Reflective Coatings

The effects of temperature, humidity, and pressure can change a coating's properties. This is especially critical in IR, when optics are sometimes cooled. This applies even to bare metal coatings. Some, especially gold, copper, and silver, tend to reflect less UV as their temperature is decreased, which is opposite to their behavior in the IR ["Mirror Coatings: Cold Reflections," 16]. Generally, the coating must be applied to all mirror substrates. The metal coating is usually deposited by evaporation or by one of the more modern techniques discussed above. Gold is generally the best coating for the IR, although aluminum, silver, and beryllium reflect well in defined wavelength regions (IR telescope makers have long been the brunt of jokes for "gold plating" their optics). For long wavelengths, bare metal substrates (especially beryllium, aluminum, and copper) may have such sufficient reflectivity and reduced scatter properties to be used bare.

A well known empirical relationship for the reflectivity of metals in the LWIR can be approximated by

$$R \approx 100 - \frac{3.65\,(\rho)^{0.5}}{(\lambda)^{0.5}} \tag{3.8}$$

where

R = percent of reflectivity

ρ = resistivity of the metal in $\mu\Omega$/m

λ = wavelength of the light in μm

3.2.4 Diamond Coatings

Recently, nature has given IR designers the gift of synthetic diamond coatings. Diamond has a Moh hardness of 10 and an index of refraction from 1.8 to 2.4. This is close to the square root of germanium's index of refraction. Thus, diamond makes an excellent quarter-wave anti-reflection coating. However, the wonderful consequence of using diamond is gaining its surface hardness for the optical element. A diamond-like carbon (DLC) coating on a front surface window provides increased hardness. Pitting and debris from insect impacts is a continual problem with helicopter FLIRs. Dust and small rocks quickly ruin a front Ge optic. The life of both can be extended greatly by employing a diamond coating. Diamond can also be used to "seal" a hygroscopic lens or window to mitigate the effects of water. As with any seal-in coating, pinholes are a common problem. However, the DRA exposed germanium coated with diamond to sea water for more than four months without it showing degradation [Farabaugh, Feldman, and Robins, 17].

The high thermal conductivity of diamond also provides for increased thermal hardness of the reflective, refractive, or diffractive substrate. If connected to a heat sink, the coating can help direct excessive aerothermal, laser, or nuclear-induced heat away from the optic. Unfortunately, because the diamond layer is thin, its heat removal efficiency is usually low for bulk substrates. Diamond can also be placed between the layers of an interference filter or directly onto a focal plane. Analysis has indicated an improvement of at least a factor of six in thermal hardness using this technique, although experimental results also indicate that the damage threshold is less than what the calculations indicate. A silicon substrate was damaged at 1.5 joules per square centimeter instead of the predicted 3.5 j/cm^2, while a measure of free standing diamond films showed a laser hardness to 7 J/cm^2 [Albin, 18]. However, such discrepancy is common of any coating and may be attributed to the substrate quality or coating contaminants.

True diamond is usually monocrystalline and composed only of carbon. Most synthetic diamond coatings or slabs are polycrystalline and frequently have inclusion of hydrogen. These polycrystalline amorphous diamonds are frequently called diamond-like carbon, or DLC. The thermal and electrical properties of DLC are also slightly less than pure diamond. The thermal hardness of a diamond layer is related to the amount of hydrogen in the layer; the purer the diamond, the harder it is. Humans have isotope tricks up their sleeves that nature lacks. Isotope selection has provided additional benefits. Diamonds "almost totally of the isotope C^{12}" have exhibited improved heat conduction and are "at least a 1,000 percent more resistant than natural mined diamond, with a mixture of isotopes, to damage from high power lasers" ["Man-Made Diamond Outperforms Nature's Best," 19].

The basic procedure for diamond coating uses chemical vapor deposition (CVD). Hydrogen and methane are mixed in a chamber with a pressure of from 20 to 100 torr. The gases are then excited via microwaves, hot plasmas, or filaments. The gases must then impinge on a hot surface (700° to 800°C) to properly deposit [Pinneo, 20]. The best growth rate for diamond films is about 10μm per hour [Dullien, 21]. This rate is suitable for coatings, but not claddings. By these means, diamonds can result from swamp gas. The current state-of-the-art with synthetic diamond layers is that they can be deposited over many square centimeters, at thicknesses of up to a couple hundred microns, controlled to about 1 μm.

The atomic oxygen that results in the diamond-forming process is not compatible with zinc sulfide or zinc selenide. However, diamonds can be effectively deposited on zinc selenide and zinc sulfide substrates by means of an intermediate layer. This intermediate layer provides chemical compatibility and the reduces the tendency for unevenness ["Diamond Coating Protects IR Optics," 22]. One technique uses a "microwave plasma-enhanced chemical vapor deposition to pre-deposit diamond onto a chalcogenide glass" which is bonded onto zinc sulfide or zinc selenide ["Can Diamond Coatings Make IR Optics last Forever?" 23]. When DLC is inappropriate, the engineer should consider boron nitride which provides good binding and some similar properties.

Thin DLC coatings were effectively employed on the helicopter-based TADS/PNVS FLIR, saving the Army an expected $28 million over three years by mitigating dust and bug pitting. It is hard enough to allow extensive cleaning of the germanium window. Tests showed no deterioration after 5,000 wipes. The coating is expected to increase the life of the window from three to 20 years. DLC coatings provide for easier removal of contaminants regardless of the environment. The cost of the TADS window was raised a few hundred dollars, about the same as if any coating were added. As with implementing any new technique, there were some production start up problems (in this case, pinholes in the coating), but they were worked out by paying closer attention to the Ge material purity and tolerances. However, DLC coatings can provide great reduction in a system's life cycle cost.

Natural diamond windows are commercially marketed in sizes of a few millimeters for a few hundred dollars, and slightly larger sizes are available at a significantly higher cost. Synthetic diamond is not limited to coatings. Currently, synthetic diamond slabs are grayish and not of gem quality. Nevertheless, a diamond slab can be cut and polished into a window or lens. Synthetic or natural diamond is an excellent choice for an infrared refractive material, owing to its great transmissivity. Synthetic diamond slabs are likely to increase the available size while decreasing cost. One study showed that about 175 kg of synthetic diamond slabs will be made per year in the USA by the late nineties.

3.2.5 Spectral Filters

A spectral filter is used to define a system's bandpass. It is usually located near the focal plane and passes IR wavelengths that are a subset of the optics' transmission and the FPA response. The ideal filter eliminates (mostly by back reflection) all colors of light except the desired bandpass, and is invisible and transparent to this exact bandpass. Bandpass filters are usually optically flat, having no optical power.

IR dielectric filters are usually an alternating "stack" of materials of high and low refractive index. Stacks of 50 or 100 alternating layers are not uncommon. This stack forms an interference which can be designed to pass only a specific wavelength. The stack is deposited as a coating of one material, then a coating of the other. These materials are dielectric in nature and are deposited on a transparent substrate using the coating deposition techniques described earlier. Common low index IR materials include cryolite, magnesium fluoride, cerium fluoride, and silicon monoxide. Common high index materials include silicon, germanium, lead telluride, zinc sulfide, cerium dioxide, and titanium dioxide. The substrate can be any material that allows coating and is transparent to the desired wavelength. To mitigate quality and schedule problems, the substrate should be furnished by the filter manufacturer.

The IR engineer usually wants a filter with as steep a cut-on and cut-off as possible. Such stacked dielectric filters can have sharp cut-ons and cut-offs, depending upon the number of layers. Present OCLI capabilities readily provide cut-off slopes of six percent and state-of-the-art designs with slopes of three percent [Morelli, 24]. Slope is defined many ways--the engineer should be certain of the definition of any specification. Common ways to define the slope are from the 1/e to the 1/e points or from the 50 percent to the 50 percent transmission points.

Although high transparency is desired, the more layers the lower the transmission. Broadband IR filters can be made with reasonably steep cut-ons and cut-offs that transmit 90 percent of the inband energy, while very narrow laser-line bandpasses may only transmit ten percent. There has been much recent improvement in very narrow bandpass, high transmission filters, so higher transmission numbers are likely to be common in the near future. Both of these filters are designed to have the light encounter them perpendicular to the stacks. If the filter is tilted or if the light bundle is a steeply converging cone, bandpass shifts will result. These filters may also be temperature-sensitive and must be specified at the operating temperature. The filter should be specified over the entire possible range of detector response. Otherwise, the filter may pass light several microns from the bandpass. This is not always as trivial as may sound at first. As a filter's angle of incidence is increased, it will shift to a shorter wavelength. An increase in temperature will shift the bandpass to longer wavelengths. This necessitates conveying these requirements to the filter manufacturer via well-defined, lucid and clear specifications.

The most exciting advances in filters involve tunable filters and rugates. Tunable filters can be remotely (away from the instrument) altered in real-time to define bandpass shape or change wavelengths of transmission. There are five types: Fabry-Perot, temperature tunable, acousto-tunable, tilting, and circular variable. The latter two are not true tunable filters but are worth mentioning for the spectral agility they provide.

Several marketed tunable filters are essentially miniaturized Fabry-Perot interferometers. The key to their tunability is the use of piezoelectric materials to change the spacing of the end mirrors. They can weigh many kilograms with their control electronics, and have specific bandpasses.

The acousto-optic tunable filter (AOTF, sometimes TAOF) uses the principle of a Bragg acousto-optic interaction in a bi-refringent crystal, not coating properties. Piezoelectric materials play a key role, as they are used as a mechanical transducer to generate acoustic waves in the filter material. These acoustic waves form bands of compressions and rarefactions which behave like a diffraction grating. Changing the RF excitation frequency changes the spacing of the compressions and rarefactions, as if one were changing the line spacing on a grating. A variety of filter materials have been identified for various bandpasses, including: SiO_2 for 0.13 to 2.5 μm, $CaMoO_4$ for 0.4 to 4 μm, TeO_2 for 0.35 to 5.5 μm, and Tl_3AsSe_3 for 1.23 to 17 μm. They have been made to operate from the UV to the long wave

IR. They can be made small enough to weigh only a few hundred grams including electronics, but often their full implementation requires a few kilograms with all the associated controls and electronic drivers.

Changing the temperature of a filter medium can also alter bandpass. Temperature tunable optical filters (TTOFs) are more difficult to implement with an infrared sensor than AOTFs, and are less efficient.

Low technology "tuning" by tilting an interference filter or using a circular variable filter (CVF) should not be discounted. Although limited in their abilities, they also are of a low weight and boast excellent producibility. Tilting an interference filter makes the light travel through its thin film layers a longer distance, shifting the center line and bandpass. CVFs are interference filters whose center line is a function of position. Their bandpass depends on the width of a slit that admits light. They are easily producible, with manufacturers offering them as standard parts costing about a thousand dollars each. These methods are crude for imaging applications because a spectral gradient will exist across the focal plane.

A rugate is a complicated interference filter. Rugates are "multilayer reflectors with a continuously (temporally or spatially) varying refractive index profile. Reflection of light occurs at a designed wavelength without the associated harmonics" [Carniglia, 25]. Typically the term refers to a notch-filter in the spectrum to stop transmission of a particular laser line, or a filter that passes two separate regions of the spectrum. The index usually varies as a sine function. Their cost is high, lead time long, and producibility questionable. However, much work continues to mature this technology.

3.3 REFRACTIVE IR OPTICS

Almost every system made has some transmissive optical elements, including sensors with reflective afocal telescopes. Even specialized all-reflective telescopes often have a transmissive filter. Any cooled IR FPA operating in the atmosphere must have an uncooled window to prevent condensation.

Transmissive optics for IR are made from materials not commonly encountered by an optical designer accustomed to working in the visible spectrum. As listed in Table 3.1, there are numerous materials available for IR refractive optics, but there never seems to be a perfect one for any application.

One feature of refractive elements is that their power (the way the affect the light) depends upon the wavelength. This occurs because the index of refraction is a function of wavelength. Usually, materials exhibit higher indexes at shorter wavelengths, so the shorter wavelengths are refracted (bent) more than the longer wavelengths. The relationship between index of refraction and wavelength is called dispersion. A high dispersion material has a large difference, is often difficult to use, and requires additional optical elements to correct for this color aberration. Conversely, a low dispersion has only a slight change in index of refraction as a function of wavelength. Infrared bandpasses are usually much wider than vis-

TABLE 3.1 Common IR Refractive and Window Materials

Material	*Useful Transmission Range (μm)*	*Typical Maximum Size (cm dia.)*	*≈ SOA Maximum Size (cm dia.)*	*Notes*
AgBr	0.5 to 30	2.5	20	Cold flows at room temp.
AgCl	0.4 to 22	2.5	20	Cold flows at room temp.
Amorphous glass	0.3 to 20 (depending on mixture)	1.0	40	
As_2S_3	0.7 to 12	1.0	5.0	
BaF_2	0.2 to 10	5.0	20	Water soluble
CaF_2	0.3 to 8	5.0	20	Water soluble
CdS	3 to 11	2.5	5.0	
CsBr	0.3 to 30	5.0	100	Water soluble
Csl	0.5 to 50	2.5	50	Water soluble
Diamond	0.3 to 100	0.3	Larger substrates in development	Perfect
GaAs	2 to 13	4.0	10	
Ge	2 to 19	5	50	Large index of refraction Soft Absorption changes with temp.
KBr	0.25 to 26	4.0	20	Water soluble
KCl	0.3 to 20	2.5	100	Water soluble
NaCl	0.3 to 25	2.5	100	Water soluble Shock sensitive
Sapphire (Al_20_3)	0.4 to 6	5.0	30	Expensive
Silicon	1.5 to 9	10	60	
ZnS	0.4 to 14	5	80	Visible transmission facilitates alignment Abrasion resistant
ZnSe	0.6 to 16	5	80	Commonly used Soft

ible, so dispersion and color correction are serious concerns. The IR user is cautioned that all optical properties of a material are essentially dependent upon temperature. If the element is to be cooled, the properties may change, and change greatly.

Back reflection is also an issue with IR systems. Unpowered flat windows are frequently tilted to reduce back reflection, but this causes difficulties with chromatic aberrations and the optics' manufacturability.

Several exotic materials for lenses and windows are not listed in Table 3.1. Spinel ($MgAl_2O_4$) has faded in and out of favor with IR sensor designers. It should be considered when durability and hardness are a concern, and the bandpass is less than 5 μm. It has excellent transparency to 5 μm and is so hard that it is used as transparent armor. The material is especially applicable in environments of high contamination, such as a space sensor near a thruster or a FLIR window. Attention has also been given to $CaLa_2S_4$ due to its hardness and wide transmission (from 2.5 to 16 μm). It is a suitable replacement for germanium when a lower index of refraction is desired. Oak Ridge National Lab is experimenting with a new series of glasses that are applicable to mass production of lenses, prisms transmitting to 2.8 μm, and mirrors. The lead-phosphate glasses have a low melting point allowing direct casting to near net shapes. There is also working on alkaline lead iron glasses, in which the coefficient of thermal expansion will match that of the various metals of metering structures. Calclognide glasses are also commercially developed for some IR applications.

Another development on the horizon is Gradient Index of Refraction materials (GRINs) for IR. Although fabrication issues abound, they can be usefully employed to reduce the number of lenses and the Petzval curvature [Mann, 26]. Regardless of material, debris from furnaces can form inclusions in the substrate, so the furnace must be clean along with the entire polishing process.

3.4 REFLECTIVE OPTICS

Three categories of reflective mirror substrates are used in IR: special glass, metals, and composites. Although still common, glass is being used less and less, owing to its brittleness, mass, and cost. Typical low expansion glass substrates include fused silica, cer-vit (now difficult to find because it is no longer manufactured), Corning's ULE, and Schott's Zerodur. The most common lightweight metals for IR mirror substrates are aluminum and beryllium. Copper, molybdenum, and titanium are used in some special applications. The most promising composites are silicon carbide and metal matrix composites, such as SXA. Generally, coatings must be applied to all mirror substrates.

Depending upon the application, IR systems can experience temperature swings of one hundred degrees, requiring that the telescopes be athermal in design or active focusing. Compensation is achieved by matching the coefficient of expansion of the mirrors/lenses with the metering structure. This is most often accomplished in reflecting telescopes by making everything out of one material. However, there is still a concern during changes of temperature. The mirror and support structure must not have areas of vastly different temperatures across their surface or it may be detected by the FPA and passed as clutter or increased noise. Also, thermally induced figure distortions can result. The mirror should not change so fast as to mechanically damage the coating, metering structure or supports. A figure of merit frequently used for a mirror material suitability under thermal flux is

$$\frac{(\text{Young's Modulus})\ (\text{Thermal diffusity})}{(\text{Density})\ (\text{Coefficient of thermal expansion})} = \text{FOM} \tag{3.9}$$

On a separate note, arrays of mini-mirrors are being developed. Under a ARPA contract, Texas Instruments is employing micromachining to hinge and actuate a mirrored surface of less than 20 microns per edge [Kaye, 27]. This silicon substrate array of two million moving mirrors can modulate the brightness of an image. It may be useful for adaptive optics or photonic computers [Stix, 28].

3.4.1 Glass Substrates

In the late eighties and early nineties, successful manufacturers of large ground-based astronomical telescopes have proven great advances in large glass technology. Table 3.2 contains a smattering of facts about several new large advanced astronomical telescopes. Contrast these to the Mt. Palomar primary, which is 5.08 m in diameter, 60 cm thick, and weighs 17 tons.

Although astronomical ground telescopes are made to be spectrally versatile (from near UV through IR), their advances are germane to large IR systems. The ingenious mirror blank architecture and polishing techniques apply directly to IR telescope manufacture. This is especially true of the pragmatic considerations of weight and cost. Mirrors larger than Mt. Palomar's five-meter or CIS's six-meter must be lightweight to accommodate a practical metering structure and servo systems and to prevent disfiguring under their own weight. Also, lightweighting is needed for the practical considerations of manufacture, transportation, and servicing. Traditional glass reflective telescopes have diameter-to-thickness ratios of 6:1 or 8:1 [Martin, Hill, and Angel, 29]. However, advanced technologies for lightweighting can reduce the effective weight to less than would be indicated by using an 8:1 ratio for weight calculation.

The Keck is a segmented mirror in which the primary is made up of smaller pieces that fit together. This affords an aspect ratio comparable to a1.8 meter segment diameter, not the entire ten meter diameter. Therefore, the mirror is very thin for its overall size. The Keck is polished by applying force to the mirror blanks in

TABLE 3.2 New Technology Glass Telescopes

Project	*Organization*	*Aperture Area (m^2)*	*Primary Design*	*Weight*	*Approx. Cost*
Very large telescope	European Southern Observatory	210	Four telescopes with 8.2 m glass ceramic meniscus	Each segment is 175mm thick and weighs 23 tons[1]	$200 million
Columbus	University of Arizona, Italy	110	Two primaries of 8.4 diameter borosilicate honeycomb sandwich	320,000 kg total[2]	Not available at time of writing
Kecks (Two on Mauna Kea)	California Institute of Technology, University of California, Keck Foundation	76	36 segments of 1.8 m hexagonal Zerodur meniscus segments (per telescope)	Each segment weighs 400 kg	$87 million for the first and $93 million for the second[4]
Magellan	Carnegie Institution, Johns Hopkins University, University of Arizona	50	8 m dia. borosilicate honeycomb sandwich	14 tons[3]	$150 million
NOAOs (two locations)	National Optical Astronomical Observatories	50	8 m borosilicate honeycomb sandwich	260 metric tons total	Not available at time of writing
Japanese large telescope	National Astronomical Observatory of Japan	44	7.5 m dia. zero-expansion meniscus	60,000 lb for primary	Not available at time of writing
MMT conversion	Smithsonian Institution, University of Arizona	33	6.5 meter borosilicate honeycomb sandwich	Not available at time of writing	Not available at time of writing

Sources:

1. Martin, B., J. Hill, and R. Angel. March 1991. "The New Ground Based Optical Telescopes." *Physics Today,* 22–30.
2. Magrath, G. "Optical Astronomy Looks to the Future." November 1990. *Astronomy,* 35–43.
3. Martin, B., J. Hill, and R. Angel. August 1991. "Go for Keck II." *Sky and Telescope,* 119–120.
4. How, J., E. Anderson, D. Miller, and S. Hall. 1991. "High Bandwidth Control for Low Area Density Deformable Mirrors." *Proc. SPIE* 1498:148–161.

a way that deforms the blank. A spherical surface is polished into it and when the force is removed; the resulting shape is the desired conic. For the Keck, this was done when the mirrors had a circular parameter. They were cut to their hexagonal form after polishing. Final polishing of the Keck Two's mirrors includes ion milling. Ion milling holds great promise for low cost, highly accurate optics because surface figures approaching 1/100λ (@ HeNe wavelengths) can be achieved in an highly automated process, and in one run though the ion milling chamber [Roberts, 30].

The Columbus has a creative casting idea pioneered by Roger Angle at OSC. The casting for the Columbus mirror is done by spinning the molten glass as it cools. This produces a parabolic shape to an accuracy of about 1 mm [Martin, Hill, and Angel, 31]. The mirror requires additional figuring to bring it into tolerance, but this rough figure eliminates the need to grind out "some 20 tons of glass at $30,000 per ton" [Martin, Hill, and Angel, 32].

The Airborne Surveillance Testbed's fused silica mirrors provide another example of lightweighting glass. About 70 percent of the weight is removed, with the removal as close as 1/4 inch from the mirror face on the other side ["Eye in the Sky," 33].

3.4.2 Metal Optics

Metals are frequently used for mirrors in IR optical systems. Metallic crystalline structure, ductility, and softness limit the quality of the surface finish of metal optics. Due to the longer wavelength, the optical quality of mirrors can be more lax in the IR than in the visible or UV parts of the spectrum. Hence, while often being too coarse for imaging in the shorter wavelengths, metal optics are appropriate in IR systems. Nickel is an exception. It is still common for a thick (.03 to .15 mm) electroless plating of nickel to be deposited on a metal mirror surface to be polished.

Metal optics lend themselves to figuring by diamond turning and numerical control grinding techniques. It is possible to grind off-axis segments, aspherics, and dramatically low f/# mirrors. Numerical control needs less human intervention and has the potential for large scale manufacturing of inexpensive, complex optical elements. Typically, diamond turning leaves the surface somewhat rough. It is sometimes is used to only get a rough finish, then the element is post-polished by more conventional methods for final accuracy. Reports on the diamond turning of a nickel cladding with less than 10 Å RMS roughness have been published [Taylor, 34], but 20 to 40 Å is more indicative of the state of the art. Common materials that can be diamond turned are those that are not ferrous and do not form carbides from the hot diamond tool tip, although special procedures are being developed for carbide forming materials. Applicable materials include aluminum, zinc sulfide, zinc sulfur, silicon, SiC, germanium, and beryllium but not glass.

Aluminum mirrors are the most common in IR. Their drawback is a lack of specific stiffness. Aluminum must be thick and heavy to have ample stiffness for most dynamic environments. Also, aluminum has a high CTE (≈24 ppm per Kelvin above 200 K), and it exhibits a high slope with temperature. Mirror aluminum is usually a high grade alloy such as 1100, 356, 606T, 6061-T6, or 7075. Aspect ratios of 1:3 to 1:6 are usual for aluminum mirrors, making them lighter than glasses, but heavier than beryllium or silicon carbide. Aluminum mirrors are particularly appropriate for applications that combine low dynamics and low cost applications.

Thermal cycling and machining for surface stability is the key to good aluminum mirror performance. Such procedures tend to be held proprietary. Yoder summarizes a generic manufacturing procedure for an aluminum cryogenic mirror, as described by Fuller, et al. [Yoder, 35], [Fuller, Forney, and Klug, 36], [Reiss, 37]:

1. Receive and inspect raw stock.
2. Rough machine, approximate the surface by a centered spherical cut.
3. Solution heat treat at 950°F. Quench in glycol/water solution. Cycle between liquid nitrogen and boiling water.
4. Perform final machining, machine mounting surfaces.
5. Prepare optical surface.
6. Polish to proper figure surface (in this case a hyperbolic surface).
7. Pre-plate heat treatment.
8. Apply electroless nickel plate; coat all surfaces with equal thickness.
9. Post-plate heat treatment.
10. Fine polish the nickel surface.
11. Test to verify figure.
12. Thermal cycle from ambient to operating temperature as many times as necessary until there is no surface figure change between cycles.
13. Test and repolish if necessary.
14. Repeat steps 9, 10, 11, and 12 until specification at temperate is met.
15. Apply coating (e.g., electrolytic gold).
16. Perform final optical testing assuring the coating did not change the figure enough to be out of specification.

The iteration between temperatures must be done to achieve the figure at the operating temperatures. For cryogenic mirrors like the one described, it was taken to 77 K, which was deemed to be close enough to its 20 K operating temperature. This iteration is time consuming, labor intensive, costly, and necessary. Depending upon application, it must be done on other metals as well. For example, the IRAS beryllium mirror went through a similar cycle of test at temperature, warm up, polish, and retest at temperature. Stress relief during manufacture is important for any metal mirror, and must be carefully applied.

One of the best metals for optics from a mechanical and nuclear hardness perspective is beryllium (see Table 3.3). Beryllium has a high stiffness-to-mass ratio making it an excellent choice for rapid scanning mirrors. Moreover, it is light, with a typical solid density of 1.8 g/cc, making it suitable for honeycomb (or egg-crate) and other lightweight designs. Its reflectivity in the LWIR is high. Its drawbacks are that it is costly to manufacture, there are few suppliers, it has high optical scatter at short wavelengths, it has microyield problems, and there have been reports of thermal and temporal instabilities. There is increasing system-level desire to share apertures among sensors, and cryogenic beryllium usually makes for a poor quality visible mirror. Even with its high stiffness, engineers have managed to fracture and/or bend it. In one recently publicized case, beryllium mirror damaged was caused by the epoxy hygroscopic thermal control epoxy that had been used. It absorbed water, expanded and bent the mirror.

Beryllium's expense is due to many factors. The first is supply. The only nations that produce beryllium are the USA, Canada, Confederation of Independent States, Brazil, the People's Republic of China, and South Africa. Fortunately, a domestic supply exists in a large mine near Delta, Utah. Bertrandite is mined in the Spor Mountains, from which beryllium hydroxide is extracted and converted to metallic beryllium. The mine in Utah can supply a few cubic meters per month if required. Today, rates of production are much lower than this.

Second (and more important) is the toxicity of Be dust. Be dust causes an allergic reaction in three to ten percent of humans. This allergic reaction, called beryIliosis, is fatal. OSHA standards require all Be machining to be done in closed atmosphere, dust-controlled environments. OSHA standards limit exposure to under two micrograms per cubic meter of air in an eight hour day. It can take several years for a company to receive all the government licenses necessary to machine beryllium.

TABLE 3.3 Properties of Beryllium

Property	*Value*
Z	4
Atomic weight	9.02
Latent heat of fusion	560 BTU/lb
Specific gravity	1.85 g/cc
Melting point	1,285°C
Electrical conductivity	40.7% of IACS
Magnetism	Dimagnetic -1×10^6 CgS
Visible reflectivity	0.5
Reflectivity at 10 μm	0.98
Sonic velocity	41,300 ft/s
X-ray transparency	17 times aluminum
Specific heat @ 0°F	0.4 BTU/lb°F
Young's Modulus	4.4 psi $\times 10^7$
Specific stiffness	6.5×10^7 psi
CTE @ 20°C	11.5×10^{-6}/°C

Third, virtually all beryllium in the western world comes through Brush-Wellman, although there is increasing competition from companies in the Confederation of Independent States.

Scatter is high at short and near-infrared wavelengths. Bare beryllium is not suited for any applications in which the wavelength is less than 2 μm. However, the desired stiffness and lightweighting attributes can still be realized by simply coating the beryllium. It is common to coat beryllium blanks with nickel to a thickness of 25 to 200 μm. This can be polished to visible wavelength optical quality, and has no effect on any mechanical property of beryllium except its nuclear hardness. The current state-of-the-art in polishing bare beryllium is to about 10 Å RMS.

Manufacturers responded to the scatter concerns of the optical community by producing special optical grades of beryllium powder specifically designed for mirror blanks. The size and shape of the powder is controlled during manufacturing, and contaminants are reduced. The size and shape are controlled by creating the powder using gas jet impact, coupled with close process monitoring. The result is a random orientation of CTE and polycrystalline strength on the mirror surface. This reduces the microyield and microfracture problems during polishing that are the root of the scatter. If the optics are to be plated for low scatter, the mirror blank need not necessarily be made from the expensive grades. One should consider using the less expensive grades for the blank, with plating of the more expensive grades.

Contemporary advances have been made in the areas of lightweighting and producibility. The egg-crate structure one often sees on the back of glass mirrors can also be implemented on beryllium mirrors. In fact, from a manufacturing perspective, it is easier to produce these voids in beryllium than in glass.

Most Be mirrors made in the seventies and eighties were machined from a vacuum hot pressed (VHP) block of beryllium. This was expensive, due to the fact that the large amounts of machining required created a large amount of dust. This technique also produced high microfracture, causing high scatter. Additionally, anisotropy and CTE inhomogeneity resulted, which caused warped mirrors.

The current preferred technique is hot isostatic pressing (HIP). "HIPping," which has been around since the late sixties, is currently practical for optics due to the development of optical grade, impact-generated Be powders (e.g., I-70A). The HIPping process consists of making steel (usually AKDQ) cans or containers that approximate the mirror's shape and size. These cans must withstand high pressures and temperatures. To reduce mirror weight, copper blocks are machined to the size and shape of the desired back voids. The can is filled with high grade Be powder and the copper blocks in the desired egg-crate arrangement. The can is then welded shut, leak tested, allowed to outgas, and pressurized at high pressures (103 Mpa or 15,000 psi) and temperatures (700-1,300°C). When employing temperatures above 845°C, care must be taken not to form an intermetallic phase between the copper and the beryllium. The can is opened and the mirror blank with the imbedded copper blocks removed. It is then submersed in nitric acid to leach away the copper. The result is a beryllium mirror blank with voids where the copper was. The blank must be machined, stress relieved, and ground to the optical specification. Although this process is done routinely, it is far more difficult and involved than most manufacturing process for silicon carbide.

The strongest lightweight design has voids in the center of the blank but has a backplate. Contrary to popular belief, this can also be accomplished with HIPping. The manufacturer need only place the copper in the center of the blank and leave a small hole to the back to allow the acid to leach it out. Alternately, the beryllium can be bound to the back.

The HIPping technique has produced mirrors with a crude figure on them. This near-net shaping is more applicable to mechanical structures, but does provide benefits for optics as well. By reducing the required machining of the optical surface, expensive material and labor are saved.

Using HIPping, Hughes Danbury Optical Systems has reported excellent mirror CTE homogeneity and lightweighting to 13 percent of solid weight [Paquin, 38]. The size of the mirror is limited by the size of the autoclaving furnace. The largest facilities are somewhat less than 2 meters in diameter. The mirror shrinks during HIPping, so the largest blank that can currently be made is slightly larger than a meter. Facilities can be specially made to accommodate larger optics, if needed. A rule of thumb for shrinkage is that the substrate will shrink by about 20 percent.

A less expensive process for forming blanks from high grade Be powder is the CIP/Sinter/Form. Cold isostatic pressing (CIP) of the powder at 414 Mpa (60,000 psi) precedes baking in a vacuum sintering furnace. Forming of the near net shape is finished by a 1.36 Gg (1,500 ton) hydraulic press. Then, the blank is polished. This process appears to be the least expensive, ranging from 20 to 50 percent less costly than other processes.

3.4.3 Silicon Carbide and Composite Optics

There are significant efforts, developments, and achievements in the manufacture of mirror blanks from advanced composites. Ceramic-like composites (most notably silicon carbide) promise to replace metal and glass in the majority of applications. The reader is cautioned that the type of silicon carbide (SiC) or composite used, its impurity levels, and the manufacturing process can greatly impact the performance.

Silicon carbide provides high stiffness and low weight, and holds the potential for low cost. Unlike beryllium, silicon carbide requires a reflective coating throughout the IR spectral region. SiC technology promises to be a suitable, low cost and healthy replacement for beryllium and a stronger, stiffer replacement for aluminum. Ergo, a community-wide debate is raging between the applicability of silicon carbide verses that of beryllium.

The properties of SiC lie somewhere in optical desirability between aluminum (or glass) and beryllium. However, SiC better lends itself to mechanical design and construction than other materials. Properties vary from one type of SiC to another, so Table 3.4 should be taken as a general guide. Several of its key properties are highly dependent on the quality of the SiC and the bulk manufacturing process.

Recent tests have concluded that SiC is an excellent mirror substrate for harsh environments. Electron beam surface damage thresholds of 0.5 cal/cm^2 have been measured, along with no appreciable surface degradation from a six hour exposure to atomic oxygen at a flux of 2×10^{18} atoms per square cm per second [Goela, 39]. Unlike metal mirrors, silicon carbide can retain its figure when heated to high temperatures, making exotic coating easier. It is also more suitable for cooling in IR applications, where metals and aluminum have cold creep problems that distort their figures. Goela has measured only a 0.005 μm RMS change when cooling a SiC mirror from room temperature to –190°C [Goela, 40].

The K/A figure of merit for thermal stability (thermal conductivity divided by the coefficient of thermal expansion) is 0.26 for solid silicon carbide. Although lower, this is close to the 0.65 figure of beryllium and the 0.4 figure of aluminum. It also compares favorably to about 0.14 for glass. SiC has a low CTE, which reduces with temperature and, unlike beryllium, there appears to be no problem with instability with thermal cycles.

The material availability of silicon carbide is not an issue. Silicon and carbon exist in abundance, and unlike beryllium there are numerous bulk SiC suppliers. There are also several processes to make silicon carbide. Chemical vapor deposition (CVD), hot pressing, and sintering are the most common. Again, different manufacturing process will result in different performance.

The first step in the CVD process is to make a graphite mandrel and deposit SiC on it. Alternately, one may make an SiC mandrel and coat it with a film to prevent sticking. SiC is then deposited in the mandrel. Silicon carbide (being a hard material) is naturally difficult to grind. Microfractures introduced during grinding tend to cause scatter. Silicon carbide is not reflective in IR wavelengths, so it must be coated or clad.

Optical scattering from silicon carbide is generally lower than that of beryllium. However, if cladding is done on the substrate, the amount of scatter depends

TABLE 3.4 Common Silicon Carbide Properties

Property	*Value*
Young's Modulus	311 GPA
Density	2.9 g/cc
Mirror areal density	5 to 25 kb/m^2 (depending on stiffness, process, and expense)
Thermal conductivity	1.56 W/cm-hr K
Elastic modulus	325 Gpa
Coefficient of thermal expansion	2.3 ppm/K at 300 K

upon the properties of the clad. It is often far better than bare beryllium. Properly clad silicon carbide is entirely applicable to visible wavelengths. Samples have been made with a 5 Å roughness. Scatter is not effected by a surface microyield phenomena, as in beryllium. Instead, it is caused primarily by micro-defects resulting from O_2, H_2O contamination of the hydrogen supply (a reactant in the process). Small amounts of O_2 can cause impurity formation. Additionally, debris from the furnace can make inclusions in the material. The moral is that for good quality and low defects, use clean gases and a clean room. The engineer should also be aware of the possibility of pinholes in coatings and claddings. Although most manufacturers have processes which eliminate these, they have been a repeated problem with silicon carbide.

SiC is porous, allowing cryogenic fluid to flow into it. This may provide for uniform cold optics. It can act like a sponge, absorbing LN_2 when immersed in it. NIST has used this property to experiment with blackbodies.

Silicon carbide is not the only ceramic material suitable for mirrors. Optical grade mixtures of metals with ceramics have been recently made into mirror substrates. Metal matrix composites are usually made by combining composite fibers or particulates and small (< 10 micron in diameter) metal powders. This can provide tailoring of thermal performance and improved stiffness, improved plastic yielding, and reduced weight compared to glass and metal substrates. Additionally, easy to machine matrix compositions can be made.

One such matrix, SXA, is an aluminum/silicon carbide. SXA has a typical density of 2.96 g/cm^3, a CTE of about 1.1×10^{-5}/K, and a thermal conductivity of 1.2 w/cm-hr K. SXA is less expensive and has a stronger microyield than beryllium, is easier to machine than silicon carbide, and has a thermal coefficient that can match that of a nickel cladding. Figure 3.3 (photograph courtesy OCA Applied Optics) shows a metal matrix scan mirror for the Leopard tank FLIR made by the Optical Corporation of America. This also illustrates the egg-crate back design, which is a common way to provide high stiffness with low weight. SXA foam can support a facesheet which is polished and coated to form the reflecting surface. The foam is sealed by depositing a substance on it. A SiC face is bonded to the foam. An optical grade material is deposited on the face for polishing. The University of Arizona has made an entire f/6.5 Cassegrain telescope based on a foam 19.3 cm diameter SXA primary weighing only 1.6 kg [Geiger, 41].

Graphite epoxy has been made into mirror substrates. It is not suitable for polishing, so a metal, gel or SiC surface must be applied and polished. Graphite epoxy offers lightweight features with high stiffness. A 6:1 aspect, 6.4 kg, 65 cm diameter nuclear hardened mirror that could survive a 2,000 rad/s^2 acceleration with a l/20 surface figure at 10.6 microns was constructed [Sultanat and Forman, 42]. Water absorption and desorption (outgassing), always a concern when using graphite epoxy, can be disastrous for an IR telescope with cold dewars and/or optics.

United Technologies has developed a graphite reinforced glass matrix named Thermally Stable Composite (TSCTM). It has a near-constant coefficient of ther-

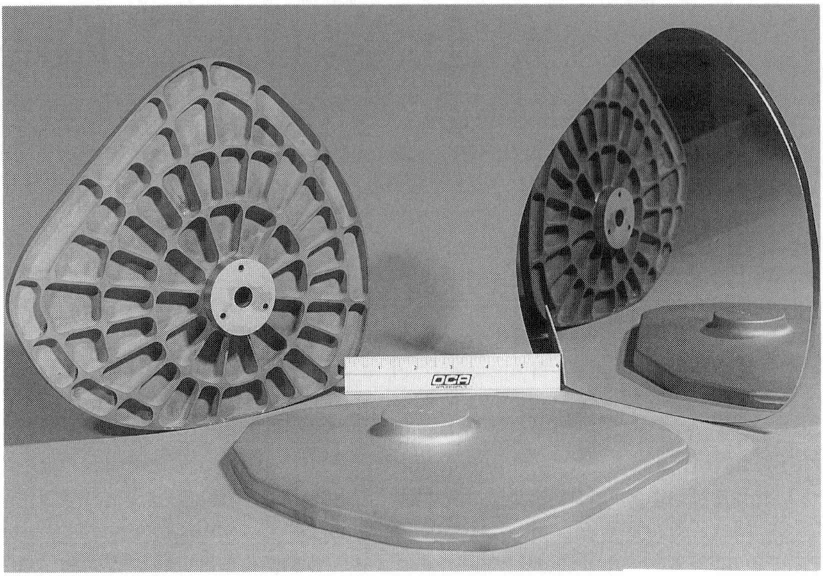

FIGURE 3.3 A Lightweight SXA Mirror (photo courtesy of OCA Applied Optics Inc.)

mal expansion with temperatures from –150 to + 200° F. It boasts low weight and high stiffness. The surface accuracy, when pressed, is about 2 μm. The blank can be clad with a thin layer of copper and polished to ≈ 20 Å RMS. It exhibits an areal density of 10 to 15 kilograms per square meter, depending upon required stiffness [UTOS, 43].

3.5 DIFFRACTIVE IR OPTICS

In addition to reflection and refraction, diffraction can be used to control light. Holographic lenses (elements) are growing in use in the visible, but have almost no current uses in the infrared because the materials suitable for IR applications don't easily lend themselves to holographic processing. However, the recent development of binary optics is germane to IR. Like holographs, binary optics essentially use the diffraction theory to condition light rays. Patterns are placed on the optical element (usually using photolithography) that diffract the light in a desired manner. These photolithographic layers have a differential thickness that is some number of bits times a basic distance, which is a fraction of a light wavelength.

Binary optics are so-named because a "staircase approximation to the ideal continuous phase profile is implemented in the substrate using a binary (power of two) mask-and-etch sequence, in which the number of phase levels generated is a power of two" [Cox, Fritz, and Werner, 44]. "Binary optics utilizes high resolution lithography and ion beam etching techniques to transfer binary or multistep relief patterns onto dielectric or metallic substrates. The relief structure of the lenslet functions as a highly efficient Fresnel phase zone pattern" [Goltsos and Holz, 45]. This manufacturing technique, based on the photolithogaphy and etching of the semiconductor industry, lends itself to low cost, high volume production. This manufacturing technique does result in one significant drawback: "It produces not the curved and angular surfaces of conventional optics but rather a series of steps that approximates the desired shape" [Veldkamp and McHugh, 46]. This can be seen in Figure 3.4 (figure courtesy Honeywell). The binary optic approximates a "Kinoform." The closer the binary optic approaches a smooth transition from one step to another, the more efficient it is. The diffraction efficiency (h) of a binary lens is fundamentally characterized by [Goltsos and Holz, 47]:

$$\eta = \left[\frac{\sin(\pi/m)}{\pi/m}\right]^2 \tag{3.10}$$

where

η = diffraction efficiency

m = number of phase levels of the lens

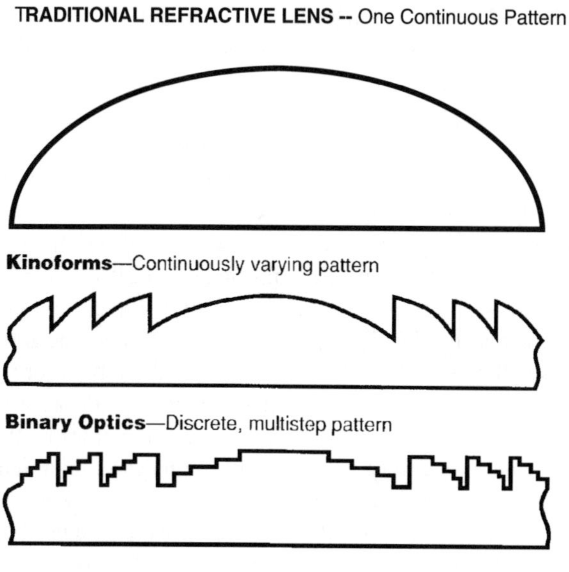

FIGURE 3.4 Binary Diffraction Optics (courtesy of Honeywell)

The state of the art is a four- to eight-mask layer sequence during manufacture, providing from 32 to 64 phase levels. The resulting decrease in diffractive optic efficiency is three to eight percent lower than the ideal [Cox, Fritz, and Werner, 48]. By using diffraction, binary lenses are sensitive to the wavelength of light and are only efficient over a narrow wavelength bandpass. Binary optics also tend to have very high scatter for an optical element. Currently, the smallest features that can be produced on a binary optic are about one micron, making them most appropriate for long wavelength and lower f/# IR systems [Nobel, 49]. Common uses of binary elements are to tilt a ray, provide chromatic correction, and to clean up the final blur spot.

A simple, inexpensive, or poorly-figured lightweight lens (or mirror) can be used along with a binary element to produce the same results as a heavy, costly, well-figured mirror. A binary element can be specifically designed to clean up the error produced by the inexpensive optical element. A spherical large mirror can be ground, providing a low cost for the size but high spherical aberration. This spherical aberration can be cleaned up by employing a binary optic. In the laboratory, a spot size has been reduced from 600 μm in diameter to 3 μm. However, this is at the cost of having to produce a binary optic to match each sloppy mirror or lens. Also, when used in a reflective system, the binary optic ruins the good chromatic image offered by the mirror.

Hybrid elements are refractive elements with a diffractive pattern on the back that reduces chromatic or aspheric corrections. The diffractive effect of binary optics produces a color dispersion 180 degrees out of phase from that produced by a lens. This allows a single, high dispersion lens to be color corrected by applying a binary diffractive pattern to the back of the element.

Corrector plates can be made for both geometric and chromatic aberrations, and binary corrector plates have been made for Schmidt telescopes [McHugh and Zweig, 50]. Binary optics can also be designed to accomplish beam splitting, beam-combining, and laser beam clean up. They can also be used to provide beam steering by making a binary element that tilts the beam. The beam is moved (scanned) by physically moving the element.

Binary optics are being deposited directly on focal planes. It is possible to place a lens on every pixel, in the same array fashion as an FPA. The microlens can act as an optical diffuser to evenly spread the beam or direct the light to the active area, thereby increasing the fill factor. Such a microlens focuses the incoming photons to a smaller area. Since these are applied directly to (or very close to) a focal plane, they are usually 20 to 100 μm in diameter. Because they collect all incoming light and direct it to a single active area, they must be the same size as the center-to-center unit cell (pixel) spacing. In this case, microlenses are analogous to the detector immersion lenses of the sixties, only they are arrays of lenses, one for every pixel. In this application, microlenses can be used to increase low fill factor focal planes to close to 95 percent or so. This is especially useful for monolithic Pt:Si FPAs, which have low sensitivity and low fill factors. A microlens can be made from classic refractive elements (such as ZnS and ZnSe) or diffractive elements.

The horizon for microlens applications continues to expand. Because microlenses concentrate light on small area, they can also be used to free real estate on an FPA for signal processing. It has been estimated that a 256 by 256 element FPA using microlenses to allow on-focal plane signal processing can perform a billion operations per second, roughly equivalent to 100 Sun SPARC stations [Brown, 51]. This represents the fusion of optics, focal planes, and signal processors.

There are three ways to model diffractive lens. The first and best is to buy a computer program that accommodates these optics. However, this technology is so new that many programs (especially the inexpensive versions) do not support them. Other approaches are to specify and optimize a phase polynomial that describes the wavefront to be produced by the lens, or model the diffractive element as a thin refractive lens with an ultrahigh index of refraction [Faklis and Morris, 52].

3.6 ADVANCED MATERIAL METERING STRUCTURES

The metering structure (sometimes called the spider assembly or telescope support structure) is the metrological support for the optical elements. This structure determines the tilt and spacing of the optical elements. If it sags or bends, image quality will suffer.

The structure should be designed to have minimum obscuration and minimum induced scatter. Any struts should be curved, cut, and/or oriented to reduce any contribution to scattered light. Moreover, they should support the secondary in such a manner that temperature gradients will not tilt, translate, or move the mirror's placement along the optic axis. Sometimes the design is eased by allowing the secondary to twist or rotate about the optical axis. Rotation (as opposed to tilting or decentering) has no effect on image quality. The struts may need to be curved in multiple dimensions disperse scatted light. The internal surfaces of the structure should be coated with an IR absorptive material. To maintain optical alignment, the metering structure should exhibit a high modulus of elasticity, high damping capacity, and a coefficient of thermal expansion matched to the optics.

The metering structure must be clean and maintain its cleanliness. Outgassing of water is a concern for IR systems with any epoxy structure, but not with metal or composite structures. The concern is a more serious problem for IR systems due to the cold surfaces and potentially cold optics. Eutectic coatings have been developed to prevent outgassing from graphite epoxy. The problem is that these coatings must be meticulously applied, for a small pinhole will defeat their propose. Vacuum baking the epoxy reduces further outgassing if future absorption can be avoided. Epoxy structures have been used on several space-based telescopes including the Hubble. The Hubble has a metering structure based on rings connected by 48 tubular elements arranged in triangles, all made from fiber enforced epoxy. "It was designed to hold alignment of the mirrors to 3μm despace, 10μm decenter, and 2 arcsec tilt at a first mode frequency of 24 Hz" [Yoder, 53]. Thermal drifts may cause a LOS variation that may need to be accommodated. This is true for a low orbit spacecraft that may cross the terminator every hour or so, as seen with the Hubble.

Recapping, the metering structure should be strong, stiff, exact, inexpensive, easily produced, and of low weight. In addition, it should not outgas or induce scatter, and should match the optics' thermal properties. Recent promise exists that many of these attributes can be simultaneously satisfied with advanced composites. Table 3.5 gives a comparative list of materials that currently may be used.

3.7 WEIGHT ESTIMATION

To determine the weight of either a reflective or refractive telescope, one should determine the weight of all optical elements, the weight of the metering or support structure, the weight of the baffles, and the weight of the aperture cover.

One can scale the weight of the optical elements by the following relationship:

$$\text{Weight} = K_m D_o^n \tag{3.11}$$

where

K_m = constant depending upon stiffness and density

TABLE 3.5 Properties of Advanced Materials Suitable for Metering Structures

Material	*Density (g/cc)*	*Elastic Modulus (GPa)*	*E/p*	*CTE (ppm/K)*	*Notes*
Aluminum	2.7	69	25	≈24	Inexpensive
Beryllium	1.8	280	155	11.5	Expensive
SiC	3	393	127	4.3	
Graphite/Mg	1.94	234	120	1.74	
Graphite/PEEK	1.75	280	160	−1.15	
P/100 Al	2.5	343	137	0.478	
Graphite epoxy	1.2–1.7	255	136	−1 to 4	Absorbs and desorbs water
SXA (SiC, Al matrix)	2.96	145	49	11	
Titanium	4.4	110	7	9	

D_o = diameter of primary mirror or objective

$n \approx$ constant of ≈ 2.7 (empirical data suggest $2 < n < 3$)

Again, this scaling relationship is for telescopes of similar types and identical materials. Aspherics cost more than spherics, and off-axis elements usually weigh 20 to 30 percent more than on-axis ones. This same relationship holds approximately true for entire telescope assemblies. This generalization can be made because the support structure and additional elements usually vary in the same manner as the primary. As the primary gets larger, they all scale. The K_m depends upon the available development, required stiffness, material, and effort expended to provide lightweighting, and should be calculated by referring to a similar design familiar to the engineer. Barring that, Table 3.6 is useful for ball-park estimation of absolute numbers. These are roughly valid for reflective optics from 5 to 100 cm in diameter and refractive optics from 0.5 to 20 cm.

If our notional 90 cm aperture telescope had it's primary mirror made from silicon carbide, one could estimate the weight of the mirror (based on lightweight technologies) as $10 \times 0.9^{2.7} \approx 7.5$ kg. This relationship seems to be less accurate for refractive elements. The weight of a refractive element can be estimated by determining the volume and multiplying by the density. One can assume a 10:1 to 20:1 diameter to thickness ratio. Volume estimating is not always easy because of the odd shapes and thickness required by optical prescription. On the other hand, if one is using a refractive telescope, the aperture is probably small and the actual elements are not likely to be a weight driver. However, a window thickness is driven by pressure differentials or dynamic loads. As a result, Ge or ZnS widows can be a weight driver, and their weight should be estimated carefully based on volume and required support. It should be noted that IR refractive materials are semiconductor metals and have densities that are usually larger than visible refractive materials, which are doped glasses.

The K_m constants in Table 3.6 are crude approximations and should be used carefully. They are heavily dependent upon the mirror stiffness and density, which

TABLE 3.6 Expected K_m for Various Materials

Material	*Typical ≈ Km* *(D_o in meters and M in kilograms)*	*Km For advanced effort* *(D_o in meters and M in kilograms)*
Aluminum	300	30
Beryllium	100	12
Fused quartz	400	150
Germanium	300	250
Silicon carbide	100	10
Zinc selenide	250	200

are functions of material, manufacturing process, and design. Table 3.7 illustrates the importance of stiffness and low density. The table's basic format and first five entries were adopted from Roger Paquin [Paquin, 54]. One can see that silicon carbide and beryllium provide the lowest weight for a given stiffness and lightweighting requisite.

The concept of areal density is often used to determine an optics' weight and/or to scale it. Typical advanced lightweight mirrors are on the order of 5 to 15 kg/m^2. To estimate the weight of a mirror, calculate the projected light gathering area (πr^2) in square meters and multiply by the areal density. Areal density depends largely on the bulk of the material behind the faceplate. This, in turn, depends upon the required figure accuracy, stiffness, mechanical design, and manufacturing process, and is useful for a small range of mirror sizes. Consequently, specifying a single areal density for a material can be misleading.

In general, the percent of bulk material behind a mirror that can be removed for lightweighting increases with the optics' size. Therefore, one should assume less lightweighting (higher areal density) for smaller optics and more lightweighting (lower areal density) for larger optics. Larger optics can be as much as 80 or 90 percent lightweighted, which means their weight is 10 to 20 percent of what the weight would be if it were a solid material. Due to these caveats, scaling from areal density can only be effectively used when scaling less than a factor of two or so in diameter, with the same material and same manufacturing process.

Metal honeycombs, solid gel and composite foam can be used to support faceplates of mirrors. This allows great stiffness and a small fraction of the weight, since foams typically have about 10 percent of the weight of solid bulk material. In the CIS, metal mirrors have been formed with faceplates and backplates supported by metal foams, reducing the weight by 2.4 to 4.2 times. The foam density ranges from 0.3 to 1.7 g/cc depending on the material [Reiss, 55].

TABLE 3.7 Weight Comparisons for Mirrors of Various Materials

Material	*Density*	*E (Gpa)*	*Weight of 1 m D_o Mirror (kg)*
Beryllium	1.85	287	15.8
Zerodur	2.53	91	64.8
Fused quartz	2.19	72	77.4
ULE	2.21	67	85.3
Aluminum	2.7	68	86.9
Silicon carbide	2.9	311	7.8*
SXA	2.96	145	14.5†

*Based on 10 kg/m^2
†Based on 18.5 kg/m^2
Source (entries 1–5): Pacquin, R. September 1986. "Hot Isotopic Pressed Beryllium for Large Optics." *Optical Engineering* 25:1003–1008.

The weight of a mirror made from silicon carbide varies greatly depending upon cost, design, stiffness, and lead time. Mirrors with honeycomb backs typically range from 7 to 20 kg/m^2. Mirrors can also be made from a silicon carbide foam backing bonded to a smooth surface of silicon carbide. This provides an extreme lightweight mirror with good stiffness at only a few kilograms per square meter. Solid SiC can be thinned, with aspect ratios of over 50:1 achieved.

Metal and composite optics have been traditionally considered the lightest weight. However recent advances in glass technology challenge this traditional wisdom. Eastman Kodak built a 43 cm ULE glass mirror with an areal density of 15 kg/m^2 using frit bonding and glass fusion welding techniques [Carts, 56].

The metering structure's weight depends upon materials, required stiffness, and the weight and size of the optical elements. It can be a large weight driver (the support for the Keck weighs 200 tons). If engineered for low weight, the structure can weigh about as much as the optical elements it is supporting. For example, the Mars Observer Camera has a 35.0 cm aperture and a weight of 9.1 Kg, which includes the optics and graphite epoxy structure (but not the electronics). Using the lightest structural materials, beryllium and advanced composites, the metering structure may weigh as little as one fourth (or less depending upon application and design) of the optical elements' weight in low dynamic environments where stiffness is not a major concern. Usually, off-axis metering structures weigh 30 to 100 percent more than two mirror, on-axis reflective designs. However, the elimination of the central obscuration can allow a smaller aperture (for the same light gathering power), which can offset this weight increase. Supports for refractive optical elements usually weigh slightly less due to a more compact packaging that naturally lends itself to high stiffness.

Unless also employed as a structural support, baffles and sunshades do not have to be as stiff since their entire function is to support an infrared black coating and reduce stray light. The weight is then determined by the volume based on baffle design and the density of the material. If the baffle designs are not known, assume the entire sunshade length is 1.5 times the separation from the primary to secondary in an on-axis reflective system, or 1.5 times the aperture of a refractive or off-axis system. Assume the diameter is 20 percent larger than the aperture and that there is a series of annular rings every 1 cm for small telescopes and every 10 cm for large telescopes. These annular rings extend close to the clear aperture and prevent unwanted stray light from reaching the focal plane. Calculate the volume of these rings by assuming a thickness of 0.5 mm (if weight is critical and shock is minimal, try 0.25 mm). For aggressive, low weight calculation, assume they are made from advanced composites that weigh about 2 grams per cc.

To design for low weight, the engineer should attempt to reduce the mirror stiffness and keep all elements the minimum size that the requirements permit. From the conceptual stage of the program, engineers should investigate materials, processes, and availability to select the one that is within his manager's budget and yields the lowest telescope weight.

3.8 COST ESTIMATION

To roughly estimate the cost of either a reflective or refractive telescope, one should determine the costs of all optical elements, the metering or support structure, the baffles, the aperture cover, and the integration and alignment costs. Learning curves seem to be a function of the size of the optics—the larger the optics, the less gain in learning experienced. Learning curves for beryllium have been over 95 percent, but may decrease with new manufacturing techniques. Silicon carbide is expected to have learning curves in the 80-90 percent range, which is similar to most small, high-grade refractive optics. The learning curves are also a function of the telescope complexity and alignment difficulty.

Cost is very sensitive to specification and tolerances. One should be careful not to over specify any tolerance, especially ones that really don't matter or can be handled by simple mechanical adjustments of the mount (like bevels, or diameter tolerances). Generally, cost increases by a factor of two to four as tolerances decrease linearly. A quarter wave optic costs four times what a half wave optic does [Janosky, 57].

Ancient industry lore states that an optical element's cost is proportional to the diameter raised to a power divided by the surface roughness raised to another power, or

$$\text{Cost is proportional to } \frac{KD^n}{\sigma^m} \tag{3.12}$$

where

K = constant

D = diameter

$n \approx 2.5$ $(2 < n < 3)$

s = RMS surface roughness

$m \approx 1.5$ $(0.5 < m < 3)$

This association seems to hold well for passive infrared optic and telescope assembly costs. As a rule, the cost of the entire telescope assembly tracks the cost of the optical elements approximately linearly. In Equation (3.12) one could approximately substitute bandpass cut-on wavelength (l) for the RMS roughness. The exponent (m) of the RMS roughness tends to be less than that for small catalog mirrors and tends to be more than that for large developmental mirrors. It also is a function of substrates. Some are a lot easier to figure than others. It can take ten times the labor to figure a cryogenic beryllium mirror than a silicon surface. Cost is also a slight function of f/# for f/#> ≈ 2 and a stronger function for f/# < ≈2. Therefore, the above equation can be refined for IR systems, rewritten as

$$\text{Cost is proportional to } \frac{D^n}{\lambda^m f/\#^x} \tag{3.13}$$

where

D = diameter of largest optical element

n = constant ≈ 2.7 (2<n<3)

λ = bandpass cut-on

m = another adjustment constant ≈ 1.5 (1 < m < 3)

f/# = f number

x ≈ 0.5 (0.2 < x < 1) for f/# > ≈ 2.5

x ≈ 2 (2 < x < 5) for f/# < ≈ 2.5

This scaling relationship holds value for identical types of optical elements and telescope assemblies with like types of tolerances and similar types and numbers of elements, similar fields of view, and in similar quantities with the same delivery schedule. Excessively small or large optical elements do not follow this scaling relationship. The equation is most useful for estimating the cost of one telescope based on another (scaling the cost based on small changes in parameters). It doesn't hold for a reflector versus a refractor, an imager versus a discrete detector, a two-mirror design compared to a four-mirror design, or a design with more aspherics or an on-axis verses an off-axis. An off-axis usually costs 1.3 to 3 times as much as an on-axis, but again, the better stray light rejection and lack of central obscuration can offset some of the cost increases. Moreover, as automatic testing and grinding progresses, the cost difference of off-axis and aspherics will began to shrink. The above equation is really meant to be used for scaling and not to calculate absolute costs. However, a rough absolute number can be gained by using Table 3.8 for entire telescope assemblies. These are approximately valid for reflective sizes of 5 to 100 cm and refractive from 0.5 to 20 cm, for little or no qualifications in small numbers. The level of specification can add greatly to cost: a space-qualified telescope can cost ten times what a commercial grade version would. Areal cost (analogous to areal density) for fully space-qualified optics can range from $10 to $25 million per square meter.

It is more difficult to provide back-of-the-envelope scaling relationships for refractive telescopes, since they typically have several materials and may have a few elements that drive the entire cost. The best way to estimate the cost is to develop a rough design with a ray-tracing program to determine the number of elements and their material, then look in a catalog. The above relationships can be used to scale the ROM cost of the lens needed from the most similar one that can be found in vendor catalogs. One should be careful not to exceed the maximum diameter or minimum roughness. Maximum diameter is usually determined by the boule that can be grown and roughness is often determined by material physics. As previously mentioned, one should be careful not to overspecify any tolerance, as cost is a strong function of specifications.

For a custom-made lens, the manufacturer has to make a custom null lens, custom tooling and, perhaps, custom test fixtures. These can actually cost more than the actual lens. Circular custom-made refractive optics in IR tend to cost about \$4,000 to \$7,000 apiece for nominal (<100 mm) diameters, regardless of material. Delivery is usually 12 to 18 weeks depending upon the material, tolerance, size, shape, and coating. Odd-shaped mirrors and lenses cost a lot more and take longer to deliver because it is always difficult to cut optics. Large windows (more than 20 cm) cost about \$30 per square centimeter in low volume. Smaller catalog elements are typically \$100 or so each.

The above scaling relationships do not consider the contour (gross shape) of the surface. The type of conic (or irregular faces) greatly affects price. Spherical surfaces are the cheapest. Ellipsoids, paraboloids, and hyperboloids are increasingly more expensive. The most expensive are compound surfaces of more than one conic (fourth order aspheres). This is due to the difficulties they present in testing, grinding, and polishing. Since a sphere is the least expensive optic, a mirror's departure from a sphere is an indication of increasing cost. This asphericity can be given by [SORL, 58]

$$A = \frac{kD^4}{4096f^3} \tag{3.14}$$

where

A= asphericity

k = constant

D = diameter of the element

f = para-axial focal length

This equation "indicates the amount of material to be removed from the 'best-fit-sphere' when generating the surface." The more material that has to be removed, the higher the cost. It has been estimated that with automatic polishing of large optics, it can cost $30,000 per kilogram of glass to be removed. The removal of bulk glass to figure large glass optics cost about $30,000 per ton [Martin, Hill, and Angel, 59]. Polishing costs ten times as much as bulk figuring.

The expense of low production sets of beryllium optics can be from $10 or $20 thousand to $100 thousand for a small telescope with a 10 to 15 cm aperture. Fast steering mirrors cost substantially less, but are still more expensive than other materials. Unmachined block beryllium (not optical grade) costs about $1,000/kg, and machining costs are high.

The cost of silicon carbide mirrors is a current point of argument, mainly because mass production has not yet occurred. Estimates place the cost from 1/10 to 2/3 of that of beryllium in quantities of thousands. However, at the time of this writing, this cost differential does not exist. Generally, a set of silicon carbide optics will cost about the same as a set of beryllium. One can expect to pay about $50,000 for a five-inch primary and secondary coated with aluminum. Recently, Oak Ridge purchased a series of optics of different materials in which silicon carbide was substantially more costly than beryllium. With fairness to the silicon carbide promoters, it must be noted that these optics were made to a design suitable to beryllium and not silicon carbide. The manufacturers contend that they could have met the optical and mechanical performances for a lower cost with silicon carbide if they had the freedom to design the mirrors to suit the production and material [60]. Novel design elements, such as a closed back, yield lower weight and cost for SiC compared to the same stiffness in a Be design. The cost is expected to fall to about a tenth of conventional ULE glass. The general folklore is that SiC should run between 20 percent and 60 percent of the cost of a Be mirror in small quantities. This lack of substantial price differences between Be and SiC for units of less than 50 is due to immature tooling and mandrel set-ups. When the industry matures, the cost for small quantities is expected to drop further.

There are many qualitative reasons to suppose that silicon carbide will cost far less than beryllium in production. It is not poisonous, supply is abundant, the manufacturing process is simpler, and there are several suppliers. Although many silicon carbide mirrors have been made, it is still awaiting the day it will be used in a large production system. Likewise, it has been estimated that stiff, lightweighted mirrors of SXA can be produced for about a third of what similar beryllium mirrors currently cost [Pellegrin, 61].

Glass is relatively inexpensive, especially when comparing large sizes (more than a meter or so). Evidence of this is in Table 3.2. Another data point is that the DoD is paying $60 million for a 3.67 meter telescope facility to be built on Maui, while the telescope assembly is expected to cost only $19 million ["Phillips Laboratory Hires Contraves for Telescope," 62].

New ground-based astronomical telescopes have employed new techniques to control costs. Building a large furnace and spinning it allows a natural rough formation of a parabola. Roger Angle estimates that the cost of an 8 meter blank will range from \$2 to \$3 million 1988 dollars, which is less than half of the traditional estimates ["Angle Builds Them Bigger," 63].

Coating costs range from trivial to extremely expensive, in some cases costing more than the optical element itself. The costs depend on the quantity, difficulty of application, materials used, and stack size.

Binary optics are expensive in specially made, low-production lots, but they are expected to less than double the cost of a refractive element in production. However, their cost should be balanced at a system level, because they can greatly reduce the number of elements or make the prescription of other lens and mirrors far less expensive. Binary optics can be made by ion etching techniques borrowed from the microelectronics industry. This allows low cost production of binary correctors to clean up aberrations from a low cost lens. It has been estimated that the use of binary hybrid elements can reduce the number of elements in an IR system by 33 percent for an f/2.4 system, but the performance of a hybrid design was greatly limited by chromatism in an F/1 design [Fritz, Allen, and Cox, 64].

The design of the metering structure impacts cost and producibility. Curved struts and excessive lightweighting can greatly increase production difficulty. A ballpark estimate of the cost for most designs and small quantities of a non-qualified, uncoated, on-axis reflective metering structures is approximated by

$$\text{Cost} \approx K_{sc} D_o^2 N_e^{1/2} \tag{3.15}$$

where cost is in 1994 dollars and

D_o = aperture diameter in cm

K_{sc} = cost coefficient of
≈≈ 100 for PEEK, Gr:mag, aluminum
≈≈ 150 for graphite epoxy, SiC
≈≈ 400 for beryllium

N_e = number of optical elements

Alternatively, in very large volume, first-cut estimates of the cost of unqualified structures made from SiC, PEEK, Gr:Mag, or graphite epoxy would be on the order of

$$\$\ \text{cost} \approx 30\,(D_o \text{ in cm})^{1.5} N_e^{1/2} \tag{3.16}$$

This large cost reduction and steep learning curve is due to the developed tooling. Moreover, the actual material is inexpensive, so it is not a strong function of the size once the tooling and procedures are developed. However, tooling and mechanical engineering is a strong function of the size.

The baffle costs are driven by the design for a small number of systems and material and coating costs for a large number of systems. As such, baffles should experience low learning curves (e.g., 80 percent). Typically, baffle designs run from $10,000 to $100,000, and about the same to test them. In mass production the materials and costs can be estimated at 25 cents to $1 per gram.

The United States holds a slight lead in research-grade large optics, thanks to the University of Arizona, Litton-ITEK, Ball, Kodak, Hughes Danbury, and UTOS. This lead is being rapidly reduced by Japan, France, West Germany, and England. The United States holds a following position in the manufacture of small standard components. Imported optical components comprise 65 percent of total US consumption by value. Oak Ridge National Lab has started a project to help boost the sagging optical industry. Their goals include

- Cutting lead time on large mirrors from the current 9 to 36 months to 3 to 6 months
- Reduce mirror costs by 50 percent of current R&D
- Reduce mirror costs to 10 percent of today's process

They believe these goals are plausible with currently expected advances. Lead time can be reduced by stockpiling materials, shortening polishing time, automatic in-situ testing, and reducing the production steps. Mirrors can be made more affordable by making more precise near net shapes (3 percent error reduced to a choate 0.1 percent), reducing production steps, automated precision machining and polishing, using statistical process control, and accepting poorer reflectivity and scatter where possible.

To design for low cost telescopes, the optical engineer should examine the materials and processes available and attempt to use all spherics (or ellipsoids and paraboloids) with high f/#s. The system engineer should attempt to balance the spectral bandpass and attempt to minimize susceptibility to scatter and the size of the optical elements. If cost is more important than weight, a trade should be conducted in the early concept stage. It is possible that adding more optical elements can actually reduce total cost. This apparent contradiction occurs because more surfaces allow more chances to clean up and concentrate the beam. The result could be that each surface is easier to make within tolerances, or that the materials are less expensive or easier to work with. This applies to refractive, reflective, catadioptric, and diffractive systems.

References

1. I. Spiro and M. Schlessinger. 1989. *Infrared Technology Fundamentals.* New York: Marcel Dekker, 111.
2. E. Pearson, L. Stepp, and J. Fox. February 1988. "Active Optics Correction of Thermal Distortion of a 1.8 Meter Mirror." *Optical Engineering* 27, 2:115-22.
3. Milles Griot, Inc. 1991. "Mirrors: Materials and Specifications." *Photonics Handbook.* Laurin Publishers, H-341–H-343.

4. P. Ceravolo. December 1989. "Commercial Telescope Optics: Buyers Beware!" *Sky and Telescope,* 564.
5. Private communication with Tom Roberts, 1993.
6. V. Kiernan. September 1991. "Nuclear Tests Advance SDIO Optics Technology." *Defense News,* 40.
7. B. Smith. September 1992. "Inside the World's Largest Telescope." *Photonics Spectra,* 72–84.
8. J. Hagen and E. Bickel. 1991. "Selecting Thin Film Deposition Sources." *The Photonics Design and Applications Handbook.* Laurin Publishing, H-85–H-87.
9. Ibid.
10. D. Morelli. 1991. *Interference Filter Handbook.* Santa Rosa: OCLI, 33.
11. J. Hagen and E. Bickel. 1991. "Selecting Thin Film Deposition Sources." *The Photonics Design and Applications Handbook.* Laurin Publishing, H-85–H-89.
12. D. Morelli,. 1991. *Interference Filter Handbook.* Santa Rosa: OCLI, 46.
13. Ibid., 70.
14. B. Hichwa, R. Klinger, and P. Swab. 1991. "Thin Film Durability: A Critical Parameter." *The Photonics Design and Applications Handbook.* Laurin Publishing, H-92–H-93.
15. Ibid.
16. "Mirror Coatings: Cold Reflections." 1992. *Science News* 139:139.
17. E. Farabaugh, A. Feldman, and L. Robins. 1988. "Growth of Diamond Films by Hot Filament Chemical Vapor Deposition." *Proc. SPIE* 969:24–31.
18. S. Albin, et al. March 1989. "Laser Damage Threshold of Diamond Films." *Optical Engineering* 28:3.
19. "Man-Made Diamond Outperforms Nature's Best." August 1990. *Laser Focus World,* 11.
20. M. Pinneo. "CVD Diamond: Improving on Nature." August 1990. *Lasers & Optronics,* 49–53.
21. R. Dullien. December 1991. "Diamonds and Diamond Films: Applications Are Just Beginning." *OE Reports,* 1, 10.
22. "Diamond Coating Protects IR Optics." March 1991. *Photonics Spectra,* 46–49.
23. "Can Diamond Coatings Make IR Optics Last Forever?" April 1991. *Military & Aerospace Electronics,* 13.
24. D. Morelli. 1991. *Interference Filter Handbook.* Santa Rosa: OCLI, 79.
25. C. Carniglia. 1987. "A Comparison of Various Rugate Filter Designs." *Laser Induced Damage in optical Materials, NIST Special Publication 756,* 272–277.
26. A. Mann. May 1992. "Infrared Zoom Lens In The 1980s and Beyond." *Optical Engineering* 31, 5:1064–1071.
27. Kaye. July 1991. "Infinitesimal Wonders To Come." *Photonics Spectra,* 64–65.
28. G. Stix. November 1992. "Micron Machinations." *Scientific American,* 106–117.
29. B. Martin, J. Hill, and R. Angel. March 1991. "The New Ground Based Telescopes." *Physics Today,* 22–30.
30. Private communication with Tom Roberts, 1993
31. B. Martin, J. Hill, and R. Angel. March 1991. "The New Ground Based Telescopes." *Physics Today,* 22–30.
32. Ibid.
33. "Eye In The Sky." November 3, 1987. *Vectors* 24:2–9

34. J. Taylor et al. September 1986. "Surface Finish Measurements of Diamond Turned Electroless Nickel Plated Mirrors." *Optical Engineering* 25, 9:10013–10020.
35. P. Yoder. 1986. *Opto-Mechanical Systems Design.* New York: Marcel Decker.
36. J. Fuller, P. Forney, and C. Klug. 1981. "Design and Fabrication of Aluminum Mirrors For a Large Aperture Precision Collimator Operating at Cryogenic Temperatures." *Proc. SPIE* 288:104–110.
37. R. Reiss. December 1991. "Instrument Design." *OE Reports,* 7.
38. R. Paquin. September 1986. "Hot Isostatic Pressed Beryllium for Large Optics." *Optical Engineering* 25, 9:1003–1008.
39. J. Goela et al. 1990. "Chemically Vapor Deposited Silicon and Silicon Carbide Optical Substrates For Severe Environments." *Proc. SPIE* 1330:25–37.
40. Ibid.
41. A. Geiger. 1990. "Metal Matrix Composite Foam: A New Material For Sandwich-Construction Mirrors." *Proc. SPIE* 1303:546–553.
42. J. Sultanat and S. Forman. 1990. "A Graphite/Epoxy Composite Mirror For Beam Steering Applications." *Proc. SPIE* 1303:477–483.
43. Information courtesy of UTOS, 1993
44. J.A. Cox, B. Fritz, and T. Werner. 1991. "Process Error Limitations on Binary Optics Performance." *SPIE Proceedings,* 1555.
45. W. Goltsos and M. Holz. November 1990. "Agile Beam Steering Using Binary Optics Microlens Arrays." *Optical Engineering* 29, 1:1392–1397.
46. W. Veldkamp and T. McHugh. May 1992. "Binary Optics" *Scientific American,* 92–97.
47. W. Goltsos and M. Holz, November 1990. "Agile Beam Steering Using Binary Optics Microlens Arrays." *Optical Engineering* 29, 1:1392–1397.
48. J.A. Cox, B. Fritz, and T. Werner. 1991. "Process Error Limitations on Binary Optics Performance." *Proc. SPIE,* 1555.
49. M. Nobel. July 1991. "Diffractive Optics: Growth Technology for The 1990s." *Lasers and Optronics,* 47–52.
50. T. McHugh and D. Zweig. 1989. "Recent Advances in Binary Optics." *Proc. SPIE* 1052:85–90.
51. C. Brown. November 12, 1990. "Microlens, Optoelectronics Coalesce." *Electrical Engineering Times,* 35.
52. D. Faklis and G. Morris. December 1991. "Diffractive Lens in Broadband Optical System Design." *Photonics Spectra,* 131–134.
53. P. Yoder. 1986. *Opto-Mechanical Systems Design.* New York: Marcel Decker, 430, 443.
54. R. Paquin. September 1986. "Hot Isostatic Pressed Beryllium for Large Optics." *Optical Engineering* 25, 9:1003–1008.
55. R. Reiss. October 1991. "Instrument Design." *OE Reports,* 9.
56. Y. Carts. August 1992. "How to Buy Custom Optics That Meet Your Specifications." *Laser Focus World,* 91–100.
57. M. Janosky. November 1990. "Development of a Lightweight Active Optic System for a Spaceborne Relay Mirror Application." *Optical Engineering* 29, 111: 1328–1332.
58. Equation and descriptions adapted from SORL product brochure, "Telescopes, Beam Expanders and Collimators."

59. B. Martin, J. Hill, and R. Angel. March 1991. "The New Ground Based Telescopes." *Physics Today,* 22–30.
60. Open forum discussion at the Boulder Damage Symposium, 1989.
61. R. Pellegrin et al. 1989. "Design, Manufacture and Testing Of A Two-Axis Servo Controlled Pointing Device using a Metal Matrix Composite Mirror" *Proc. SPIE* 1167:318–328.
62. "Phillips Laboratory Hires Contraves for Telescope." January 13–26, 1992. *Space News,* 17.
63. "Angle Builds Them Bigger." August 1988. *Sky and Telescope,* 128–129.
64. T. Fritz and J. Allen Cox. 1989. "Diffractive Optics for Broadband Infrared Imagers: Design Examples." *Proc. SPIE* 1052:25–23.

Appendix 3A

Data Sheets for Representative Systems, Components, and Materials

---

Brush Wellman Optical Grade O-50

Manufacturer: Brush Wellman Inc., Elmore, Ohio

Intended Application: Various

Qualification: Up to S-level

Description:

Brush Wellman developed a proprietary process to extract beryllium from the mineral Bertrandite. The process is used in the world's largest beryllium mine in Delta, Utah. At current consumption, the supply from this one mine should last well into the next century. A new grade of optical Be has been developed (O-50) which has drastically reduced the microfracture of the polished surface. This is done by controlling the size and shape of the powder during the production of the powder. Also, the content of BeO is reduced to less than 0.5 percent. When used with hot isostatic pressing, this new grade may provide for the use of bare Be for wavelengths as short as visible with scatter properties comparable to silica.

Characteristics:

Price: ≈ \$1,100 per kilogram

Information courtesy of Brush Wellman Inc.

CERAFORM®

Manufacturer: United Technologies Optical Systems, Inc. (UTOS)

Intended Application: Various

Qualification: Various

Availability: Mirrors made to order

Description:

UTOS uses CERAFORM silicon carbide to produce stiff, lightweighted mirrors with integral backplates. They are made to a near net shape then polished and coated. One typical application resulted in a 89 percent section stiffness increase and a 37 percent weight decrease for the same section envelope as beryllium.

Characteristics:

Mirror Surface Figure: 0.03 μm to 0.6 μm RMS

Mirror Surface Roughness: 10 Angstroms (RMS)

Largest Mirror Size Available: 60 cm

Information courtesy of United Technologies Optical Systems

Leopard 1 Tank Sight SXA Mirror

Manufacturer: OCA Applied Optics, Garden Grove, California

Intended Application: Belgian Army, Ground Vehicle, Tank

Qualification: Military

Description:

This large scan mirror retrofit is made from SXA by OCA. It scans a pulsed laser and infrared imager. Figure 3.3 is a photograph of the mirror.

Characteristics:

Material: SXA and optical grade aluminum/silicon carbide metal matrix composite made by Advanced Composite Materials Inc.

Diameter: 275 mm

Areal Density: 18.5 kg/m^2

Inertia: 0.17 kg/m^2 in elevation and 0.018 kg/m^2 in azimuth

Resonant Frequencies: 130 to 300 Hz

Temperature Range: –40 to +60°C

Flatness: Eight fringes after endurance tests

Coating: Hughes Danbury laser resistant coating providing 70 percent reflectance at 0.78 μm

Information from Pellegrin et al. 1989. "Design, Manufacture and Testing Of A Two-Axis Servo Controlled Pointing Device using a Metal Matrix Composite Mirror" Proc. SPIE *1167:318–238*

SORL Large Aperture Telescope TLB 8.3-36

Manufacturer: SORL, Chelmsford, Massachusetts

Intended Application: Various, meant for ground lab use

Availability: ≈ 3 months ARO

Description:

This is a large generic telescope designed for LIDAR applications. It is a light bucket providing broadband performance from the UV to LWIR. The large blur makes it inappropriate for imaging.

Characteristics:

System f/#: 8.3

Aperture: 91 cm

Blur Spot Diameter: 2.5 cm

System Focal Length: 762 cm

Back Working Distance: 28 cm

Mirror Coating: Protected Aluminum

Mirror Substrate: Soda lime / Pyrex

Outer Dimensions: 99 cm in diameter, 152 cm long

System Weight: 104 kg

Approximate Price: $45,000

Information courtesy of SORL

Spirit III IR Telescope

Manufacturer: SSG

Intended Application: SDIO

Qualification: Space

Availability: From concept to hardware in ≈ 24 months

Description:

This off-axis telescope was made for low background, long wavelength use. It has an internal scanner. It has an athermalized design for a low scatter of <10-5 PSSR per detector subtense at five degrees.

Characteristics:

Aperture: 38 cm diameter

Wavelength: 2 to 28 μm

FOV: 3° Az, 1°El

f/#: 3.3

Optics Temperature: 30 K

Baffle Temperature: 70 K

Optical material: Aluminum substrate with a nickel cladding

Weight: 145 kg

Information courtesy of Sensor Systems Group

STABLCEL®

Manufacturer: Advanced Composite Materials Corporation

Intended Application: Various

Qualification: Various

Availability: Made to order

Description:

STABLCEL® foam is a composite made of aluminum alloy reinforced with 15 to 25 percent silicon carbide particulate. STABLCEL® foam has a density of about 10 percent of that of solid Al. It is non-toxic and machinable to complex shapes. Thin face sheets of SXA, aluminum matrix composite are bonded to the surfaces of the foam to make very stiff, stable, sandwich-construction mirror substrates. The front SXA, composite face sheet is plated and polished to form the mirror surface.

Characteristics:

Density: 9 to 18 percent relative density of aluminum

CTE: 8.6 to 10 ppm/°F

Elastic Modulus: 250 to 300 ksi

Achieved Areal Density: 4.8 kg per square meter

Information courtesy of Advanced Composite Materials Corporation

---

SXA, Optical Grade Composite

Manufacturer: Advanced Composite Materials Corporation

Intended Application: Various

Basing: Various

Qualification: Various

Availability: Billet, forgings, plate, and sheet

Description:

SXA, optical grade composite is composed of a 2009 aluminum alloy reinforced with 30 volume percent silicon carbide particles. This material exhibits excellent dimensional stability due to its low density, high elastic modulus, high microyield strength and microcreep resistance, high thermal conductivity, and relatively low thermal expansion. The coefficient of thermal expansion is matched to electroless nickel, which is commonly plated onto the SXA, composite's substrate to form the polished reflective surface, eliminating bimetallic bending stresses. SXA, Optical

Grade Composite is a higher performance alternative to aluminum and a lower cost alternative to beryllium for metallic optical components.

Characteristics:

Density: 2.9g/cm3

CTE: 7.8 ppm/°F

Elastic Modulus: 17 Msi

Information courtesy of Advanced Composite Materials Corporation

4

Focal Plane Arrays

"Any sufficiently advanced technology is indistinguishable from magic."

Arthur C. Clarke

4.1 TECHNOLOGY BASICS

4.1.1 Definitions and Architectures

The basic function of a sensor is to collect photons and convert them into useful information or data that can be converted into useful information. Once photons are collected and focused by the telescope, they must be counted. Accomplishing this requires that photons be converted to something that can be electronically sensed. Infrared detectors are the transducers that convert infrared radiation to a measurable electronic signal—essentially the retinas of the IR sensor. It is here that the photons and optics end and the electronics begin. In other words, the detector is the vital interface between the optical domain and the electrical domain. This is where the photon's journey through the sensor ends by knocking loose electrons or raising the temperature of some device that changes its electrical properties (resistance). This is the job of the detector.

A detector changes electrical properties as a function of incident light. Like the human retina, detectors are fragile, difficult to obtain, costly, complex, and application specific. Unlike the retina, they aren't attached to such an adaptive signal processor as the eye-brain system and cannot be manufactured by unskilled laborers in nine months. The near infrared (0.6 to 1.1 microns) can be effectively detected with visible silicon focal planes, usually red enhanced, or with a photomultiplier tube. These, and vidicons, will not be discussed because they are old outgrowths of visible detection technology related to television cameras.

As with much of the infrared industry, the terminology is both misleading and confusing. The focal plane is the location in the optic axis where the light is focused for an imaging system. Additionally, accepted terminology calls a series of detectors placed there the "focal plane," and an array of such detectors is a focal plane array (FPA) or, in technological slang, the "chip." A collection of focal plane arrays or the FPA and some associated hardware may also be referred to as a focal plane assembly (FPA). Occasionally, a focal plane is curved about the Petzval surface using more than one chip in a curved structure. Sometimes, if more than one group of chips exists, the group is called a module. Each detector in the array is called either a pixel, unit cell, detector, or element (take your pick). When referring to an FPA, "unit cell" or "element" are more proper than "pixel" (resulting from confusion with display pixels) or detector (sometimes the entire FPA is called a "detector"). Often, "unit cell" refers only to a pixel on the readout structure. Nevertheless, almost everyone uses pixel for the smallest element of an FPA.

The conversion of light to an electronically measurable change may happen via several physical phenomena. The most common mechanisms are a heat-generated change in resistance (a bolometer), a change in voltage (photovoltaic), a current being generated, or a conductance generated by a solid state process (photoconductivity). Regardless of the mechanism of this transition from "photon" to "electricity," the electronic signal must be gathered, read out, and amplified. The actual method of conversion will be discussed under each material type. But, briefly, these include:

1. Classic semiconductor processes
   - Photoconductive (PC): generation of electrons which change the material's resistance
   - Photovoltaic (PV): two-electrode generation of a voltage difference by the photon creating pairs of electrons and holes
2. Novel semiconductor processes
   - Internal photoemission: a semiconductor photo-emissive effect that occurs in Schottky barriers
   - Quantum wells: which form a PC quantum trap
3. Non-semiconductor effects
   - Bolometric: the change in resistance with an increase in temperature
   - Pyroelectric: a temperature induced change in electrical properties caused by a change in polarization and/or surface charge caused by an increase in temperature

Large arrays are usually made from PV semiconductors. PV lends itself more to large arrays because of its lower thermal power dissipation. PV operation also provides a square root of two improvement in noise over PC, simpler biasing, and more accurately predictable responsivity [Vincent, 1]. Conversely, PC offers high gain, high impedance, and easier manufacture of single elements. Large arrays from bolometers and pyroelectrics are just emerging from the laboratories. Bolometers and pyroelectrics have the advantages of having their sensitivity independent of the wavelength of the incident power and of useful room temperature operation, in some circumstances.

A semiconductor has a valence band (normal electron state for highest energy electron) separated by an energy bandgap from its conduction band. Electrons can jump the bandgap by thermal excitation or by absorbing energy from a photon. A photon can only generate electrons and holes—thereby being detected—if it contains greater energy than this energy bandgap (if the photon shorter in wavelength than the cutoff). Thus, longer wavelength detectors have smaller differences between the valance and conduction band; that is, smaller bandgaps. This also means that they usually require more cooling because it is easier to generate noise via thermal excitation of the electrons across a small (or narrow) bandgap than a larger one.

For all semiconductor detectors (PC, PV, Schottky barriers, quantum wells), the cutoff wavelength can be determined from the energy bandgap in the material by

$$\lambda_{co} = \frac{1.2}{E_g} \tag{4.1}$$

where

λ_{co} = cutoff wavelength in microns, where the response is 50 percent of peak

E_g = energy gap in electron volts

There are several semiconductor detector materials each with its own E_g and cutoffs (see Table 4.1). At wavelengths above the cutoffs, the materials usually become transparent and can be used as refractive optical materials.

As evident from Table 4.1, many materials are manufactured into detectors and a few into arrays. InGaAs is a semiconductor material that detects in the SWIR. The lead salts are photoconductive semiconductors that detect in the SWIR to MWIR. InSb (usually pronounced *ins-bee*) is another semiconductor that can be PC or PV and has useful detection out to about 6 microns. There are also some InSb "strained lattices" operating out to about 10 microns. HgCdTe is a semiconductor alloy that can be either PC or PV. Doped silicon is a photoconductive material providing LWIR detection. For detection above ≈ 25 μm or so, there are two low-sensitivity choices: bolometers, which have a flat response to very long wavelengths (50 microns), and pyroelectrics, which do not require cooling. Recently, quantum wells have been made into large-scale arrays for detection of IR radiation.

TABLE 4.1 Semiconductor Detector Materials and Their Cutoffs

Material	*Bandgap Energy (eV)*	*Wavelength Cutoff (μm)*	*Characteristic Operating Temp*
Si	1.08	1.15	300
In_x GaAs/InP with X = 0.53	0.73	1.7	280
GE	0.66	1.88	280
In_x GaAs/InP with X = 0.8	0.49	2.5	270
HgCdTe	tunable	3 to ≈18	50 to 250
InAs	0.41	3	77
PbS	0.41	3.02	270
PbTe	0.32	3.88	270
PbTe	0.27	4.6	200
PbSe	0.29	4.28	250
PbSSe	0.17 to 0.42	3 to 7	100 to 200
PbSnTe	0.3 to 0.06	4 to 20	40 to 180
InSb	0.18	6.89	80
SI: Zn	0.31	4.0	100
Si:In	0.16	7.8	25
Si:Bi	0.069	18	20
Si:Ga	0.065	19	15
Si:As	0.049	25.3	10
Si:B	0.045	27.6	15
Si:Sb	0.039	31.8	15

A plethora of materials is available to match the large number of applications, requirements, and desires. When choosing an FPA, many system, cost, and political issues must be considered, as pointed out in Figure 4.1. Moreover, different materials respond to different wavelengths and call for different operating temperatures. In addition to different sensitivities, spectral response, and cooling, military tactical requirements have needs that conflict with strategic military, science, and commercial uses.

Most of these materials are formed into detectors that can be grown as arrays on a single substrate. A brief historical summary can highlight the evolution and difference between the architectures of common focal plane arrangements. The top of Figure 4.2 shows an approximate timeline for each of the architectural arrangements. The second level is a notional schematic of the arrays. The lowest part of the figure shows a representative scan pattern and gives a representative integration time to cover a 640 × 480 television pattern. Finally, the sensitivity of the sensor is estimated for this canonical case. For the sake of this illustration, the

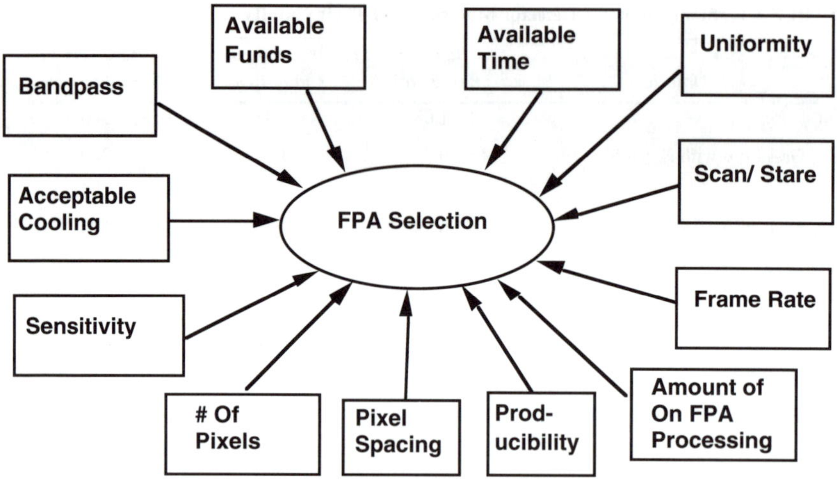

FIGURE 4.1 Important Considerations in FPA Selection

sensitivity of 1 is normalized to the serial scan. The sensitivities are not quite the square root of the increase in integration time. This is due to uniformity, scan efficiencies, and lower average unit cell sensitivity. However, one can easily see why the trend has been to the full framing area arrays.

In the sixties, focal planes often consisted of a single detector, or at most a few discrete detectors, placed on a common cold finger. These serial scans are the oldest and employ a single detector element to form an image by being scanned in two dimensions. Serial scans have high uniformity across a scene because the same detector at the same field position of the telescope generates the output for the entire FOV. Serial scans suffer from a short integration time (limiting sensor sensitivity) and complex scan mechanisms.

The next genre is a line scan, which is a row of detectors scanned perpendicular to its length. This allows longer integration times and easier scan mechanisms. The disadvantages included high cooling loads and the need for uniformity correction. Most linear arrays arrange the unit cells in a staggered fashion to eliminate gaps between rows, unlike the notional drawing of Figure 4.2.

The next technological step was time delay and integration (TDI) on arrays that usually subtend the whole field, in one direction, with an integrated readout. This has the advantages of high sensitivity and resolution but the disadvantages of an exacting scan mechanism coupled with concerns about producibility and cost. TDI increases the signal (adding linearly) by the number of TDI elements (n). The noise (which combines via root sum squared) is increased by the square root of the number of TDI elements, leaving an SNR improvement of the square root of "n." Uniformity correction is still an issue for most materials—nonuniformity reduces the SNR increase to less than the square root of n. Additionally, the telescope

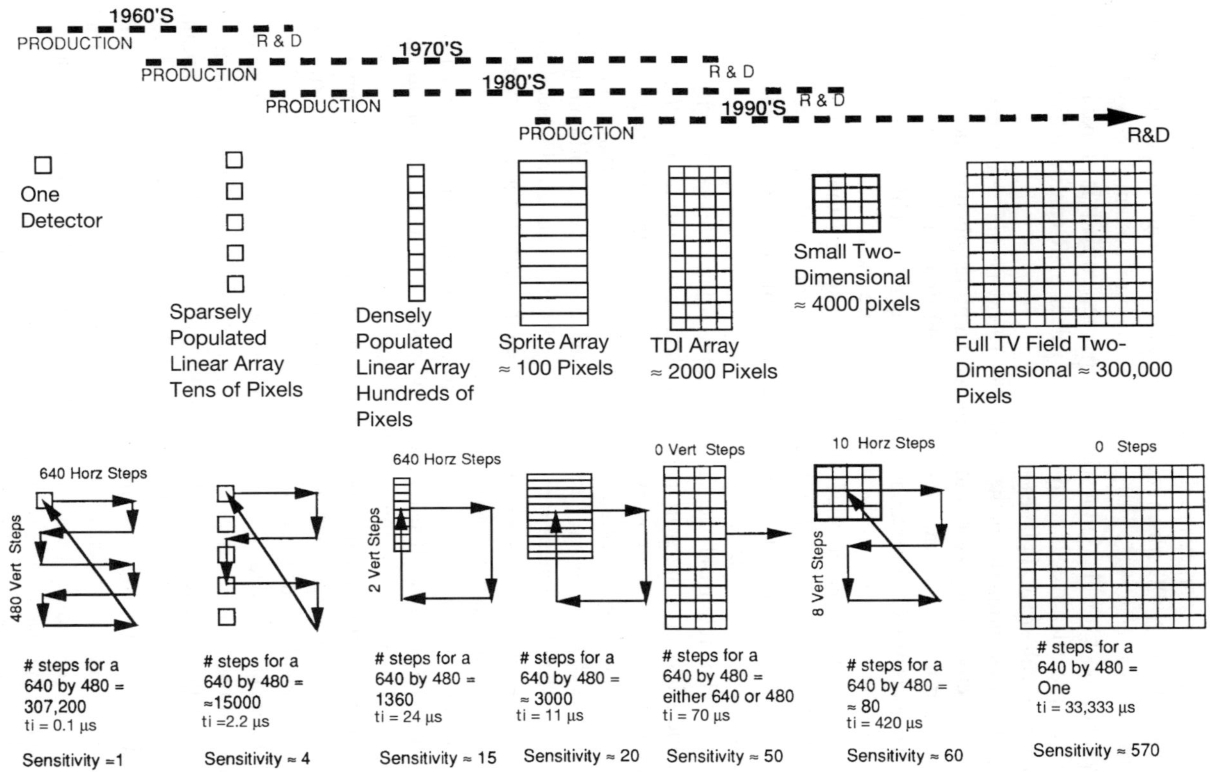

FIGURE 4.2 Generalized Evolution of Focal Plane Arrays

needs to provide a flat field over several degrees and a few centimeters at the focal plane. This is more difficult than telescope designs for single elements. Additionally, the scan across the TDI must be linear to reap the full benefits, and linear scans of the needed accuracy are difficult to achieve. One should be attentive in specifying TDI. The SNR, which increases only as the square root of the number of TDI, must be traded against a large increase in control system complexity, scan linearity, opto-mechanical accuracy and timing complexity. It is probably not the cost- and weight-effective solution if the equivalent SNR can be achieved with larger optics or colder focal planes.

A signal processing in the element (SPRITE) array is similar to a TDI in practice. The "effective" integration time is increased by lengthening the element in the direction of the scan so the target spends more time on the detector. The IR generated signals move to the readout under the influence of a bias at the same rate as the optical scan rate across the detector. The rate the carriers move is determined by the electrical bias applied, and for imaging systems the bias can be quite high. A SPRITE array is usually a small number of long, thin HgCdTe elements (called filaments of bars) arranged in parallel. Each filament can be equivalent to 7 to 14 discrete high D* elements [Leftwich, 2].

This "self-TDI" requires fewer electronics, fewer leads, less sensitivity to uniformity, and a less exacting scan mechanism than does a normal TDI. However, it suffers from loss of resolution, which can be important in clutter rejection and uncertainty in line of sight location important for fire control and search and track. The SPRITE technique also suffers from sensitive, nonlinear, and complicated gain controls when implemented in an imaging system. This usually has the disadvantage of severe crosstalk and bleeding when the gain is high, or a "washed-out" image when the gain is low or moderate.

The seventies saw advances in charge coupled device (CCD) and charge injection device (CID) readouts that supplied the opportunity to place and multiplex thousands of detectors on a single chip assembly. Early arrays were 16×16 or 32×32 arrays of infrared detectors which were bump bonded (see section 4.2.3) onto a CCD with center-to-center spacing of a few hundred microns. Consider the small two-dimensional array of Figure 4.2 that is stepped over the entire field to produce the image. This has the advantage of increased sensitivities and the disadvantages of requiring overlap, a jerky scan mechanism, and a large flat optical field. The designs produced several advantages, including higher sensitivity, the ability to stop and concentrate on a given area, and new clutter rejection, detection, and automatic target recognition methods. Some focal plane array chips can be placed immediately next to each other with only one or two pixel widths of dead space between. The ability to do this is called its "buttability." Spacing is driven by chip layout and lead routing. Small arrays can usually be designed to be butted on two sides, allowing a contiguous two-dimensional array configuration.

Eventually, a dream was born in the infrared engineers' minds of making a camera that would not be scanned or stepped. This required extending the array to television-like sizes in both dimensions. It would stare at a scene, and the focal plane would have ample unit cells (pixels) to form a detailed image on a television display (e.g., 640 × 480 or 512 × 512 pixels). Producibility, uniformity, and affordability problems postponed this step for decades. Even now, most cameras employ some scanning. The final example is a pure full-field starer, providing exact time correlation across the entire scene, no scanning parts, and the highest sensitivities. However, its drawbacks include extensive signal processing, large calibration files for uniformity correction, yield concerns (for some materials), and a large flat field for the telescope to produce.

"First-generation" (or common modules) systems have few detector elements (less than 200), several electrical leads (one for each detector), and room temperature preamplifiers. They also require several cubic centimeters of support electronics per detector. Second-generation devices have 16,000 to 64,000 elements per focal plane, and they operate with 25 leads and several cubic inches of support electronics for the entire array [Botts, 3].

Focal planes can be made multi-color in the infrared as can color TV CCDs in the visible. There are two techniques for doing this. One is to make a filter in a regular pattern that places a different bandpass on each detector. This has the disadvantage of splitting the telescope's airy disk (the circular focal plane diffraction pattern from a point source) over multiple pixels and can cause an angular shift. The other technique is to sandwich focal planes of different materials (or doping together) one on top of the other. Each material then responds to different wavelengths.

One of the keys to FPA uniformity and performance is to have a small sensitivity dependance on operating temperature (not the temperature dependance on incoming flux). Ideally, the sensitivity should depend on nothing that can have gradients or change from operation to operation. A small sensitivity to FPA temperature allows higher tolerances, permits easier and more producible thermal designs, and increases FPA uniformity. Unfortunately, most materials change sensitivity rapidly as a function of small temperature changes (near the operating temperature), and some even change bandgap as a function of temperature.

One must be careful about using an FPA made for low background application for a high background one, and vice versa. Saturation (see Section 4.1.2) is likely to occur when using a device designed for low background applications in a high background one. Lower than expected sensitivity is likely to occur in the reverse situation.

The detector time constant can be a concern for systems that scan the detector rapidly. It is usually defined as the time it takes for the detector's output to fall to the 1/e (or 37 percent) point of the peak response after illumination has been removed. The time constant should be much less than the dwell time.

The center-to-center spacing is the distance from the center of one unit cell to the center of an adjacent cell. Often, this is different from the active area that is the actual detector size. For radiometric and imaging purposes, it is desirable to have as much active space and as little dead space as possible. Sometimes the "dead area" is not completely dead. If not masked off in the manufacturing process, it may respond to some radiation, causing blur and crosstalk.

Crosstalk may be another concern when there is a large difference in signal from one unit cell to another. There are two types of crosstalk. One is the optical blur that is produced by the telescope spreading the photons across more than one detector. The other is an electrical crosstalk caused by the photogenerated carriers bleeding over into neighboring detectors. The detector array and readout both can contribute. Crosstalk is sensitive to the physical layout of the detector array, readout, signal leads, and amplifier configuration [Wimmers, 4]. The crosstalk is highest to the four neighbors on each side, and less on the diagonals. Crosstalk having 1 percent or less of the photocarriers can be achieved for non-saturated operation [Wimmers, 5]. Wimmers has reported spectral crosstalk dependance in InSb with crosstalk increasing as the wavelength increases [Wimmers, 6]. Channel stops can be incorporated into the material between pixels to reduce crosstalk and increase dynamic range.

Radiation effects are a concern in space applications. These include increased noise, passivant charging, and reduction in readout efficiency. These can be mitigated by shielding, thinning the materials, isolating active regions from the bulk material, software circumvention, and coincidence techniques.

The weight of the focal plane is usually a trivial contributor (probably only a few grams) to the sensor system weight. However, the performance of the focal plane is usually a system weight driver because its sensitivity may determine the size of other sensor system components, as listed in Table 4.2. The complete focal plane assembly may not be trivial when one considers all the devices and support structures needed to operate and align it. A typical common module dewar weighs a few hundred grams, while the next generation dewars are likely to weigh only slightly more. Today, these weights can be trimmed largely with the proper investment in schedule and cost.

4.1.2 Readout Multiplexers

Once a photon's energy becomes an electrically measured change, the change must be collected and registered. This function occurs in the multiplexing readout circuitry. A multiplexer is like a manifold; it takes many individual outputs and combines them into fewer outputs that are more easily accommodated. However this analogy isn't perfect, since manifold randomly combines, while a multiplexer only combines in a manner that allows separation. For example, a 640 × 480 FPA would have 307,200 pixels to read out. A multiplexer would arrange these in time and place the output on perhaps 4 or 8 output lines. This multiplexing (usually shortened to "mux") has several advantages over reading out every detector independently. These include reduced power consumption and fewer amplifiers and

TABLE 4.2 FPA Performance Impacts on System Level Attributes

FPA Performance Parameter	*Effect to Sensor System Weight*	*How it May Affect Sensor Level Power*	*Notes*
Array size	The larger the FOV, the less weight needed for the gimbal.	Large arrays require large processors that increase power.	Achieve a balance.
Radiometric sensitivity	Increases sensor system weight as sensitivity decreases due to larger aperture size.	Increases sensor system weight as FPA sensitivity is decreased because of the need to point and stabilize a larger aperture.	Only significant for non-BLIP and non-clutter limited cases.
FPA sensitivity to temperature	Increases total system weight if the FPA is more temperature sensitive due to increases in thermal ballast and cryocooler control.	Increases total system weight if the FPA is more temperature sensitive due to increased cryocooler control.	Achieve a balance based on cost, producibility, and availability.
Uniformity	Poor uniformity increases system weight due to additional optics cooling, processing, and blackbodies.	High and stable uniformity requires less correction, reducing processing and blackbodies.	More important for high backgrounds and high SNR.

lines from the cold finger to the outside, warm world. The mux may be a separate integrated circuit (IC) connected to the array as a hybrid, as illustrated in Figure 4.3 (courtesy of SOFRADIR), or integral to the same IC as the detector (monolithic). Key to infrared readouts are large charge capacity (to handle the noise) and large amounts of multiplexing (to keep leads—and therefore heat conduction—to a minimum).

Hybridization provides for separate manufacturing process optimization (performance and production techniques) of the detector array and the multiplexer. The standard method is to grow indium bumps on each unit cell of the detector array and each address of the mux. Figure 4.4 is a photograph of actual indium bumps that SOFRADIR deposited on a silicon mux with 50 μm spacings. The array is aligned to the mux through an infrared microscope viewing through the mux. After being aligned, the array and mux are physically pressed together, sometimes while adding heat. The result is a cold weld contact between each detector and each address on the readout. Problems can occur with the bumping, including shorting unit cells, delaminating ("de-bumping") with thermal cycles, and incomplete electrical contact. The latter results in dead unit cells. However, current bumping technology yields and techniques are so well advanced that hybrid-

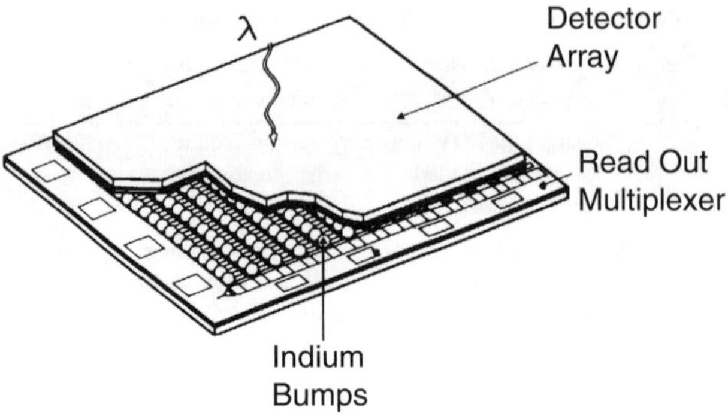

FIGURE 4.3 Hybrid Focal Plane Array (courtesy of SOFRADIR)

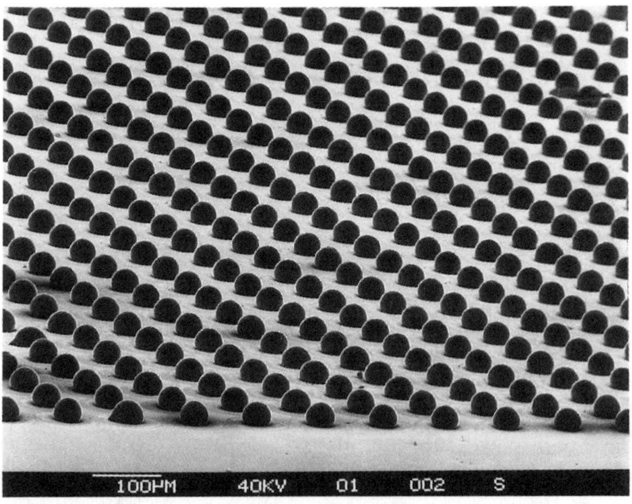

FIGURE 4.4 Indium Bumps on Silicon Multiplexer (courtesy of SOFRADIR)

ization is rarely a performance, cost, or producibility issue for arrays of 256 × 256 or less.

Monolithic approaches use planer readouts of the same material as the detector array. Monolithics typically suffer from low fill factor and a more complex manufacturing process. The latter is because the readout circuitry has to be made out of a material that doesn't lend itself to IC manufacture, and the standard process must be altered. Additionally, optimization of the readout must occur in the same manufacturing process as the detector array and usually results in a compromise.

Direct readouts (DROs) can be made that address individual unit cells and eliminate the purely sequential readout process that is sometimes referred to as the "bucket brigade." These are usually implemented by mating the detector array to an array of field effect transistor (FET) switches.

Most readout devices have a readout noise associated with them. This is an additional and unique noise source to those of the detector. This can be expressed as "KCT" (see below) or reset noise and can be greatly reduced by correlated double sampling. This noise is usually between 50 and 1,000 electrons and is a function of multiplexer speed, temperature, well size, and device fabrication. One can estimate the readout noise from the output switch/amplifier noise and the capacitance reset noise using the following equation [Monin et al., 7]:

$$\sigma_r = \frac{(KCT)^{1/2}}{e} \tag{4.2}$$

where

σ_r = readout noise in electrons

k = Boltzmann's constant

T = temperature in Kelvins

C = capacitance in farads

e = electron charge of 1.6×10^{-19} coulombs

There seems to be an approximately linear relationship between readout noise and well capacity for like detectors/readout combinations [Monin et al., 8]. Readout noise may or may not be included in the figures of merit from suppliers.

Saturation of the well, or trap, is the filling up of the charge storage device with electrons (usually charging a capacitor). This factor must be considered for high background or long integration time applications. The readout circuit can handle only a finite number of electrons at one time. If the detector generates more than this finite number, then saturation will occur. Often, focal planes optimized for low backgrounds will saturate when attempting to use them in a high background implementation. The total number of electrons that can be stored is directly proportional to the gate area of the MOS capacitor. Typically, a well can hold 10,000 to 25,000 electrons per square micron of space, so a 50 micron by 50 micron well can be designed to hold 25 million electrons before saturating.

As the saturation level is approached, blooming crosstalk and nonlinearity effects become stronger. Once the well is saturated, no additional information can be obtained from that unit cell, and the information in the surrounding cells is likely to be corrupted. The level of the saturation can be found by

$$R_{Sat} = \frac{(e_b + e_t + e_{ro} + e_d)}{T_{sat}} \tag{4.3}$$

where

R_{Sat} = ratio of the number of electrons in a well and the saturation number of electrons

e_b = number of electrons generated by the background, including thermal emission from optics and cavity

e_t = number of electrons generated by the target

e_{ro} = number of electrons generated by the readout noise

e_d = number of electrons generated by all noise mechanisms internal to the detector

T_{sat} = total number of electrons the multiplexer can hold for a single unit cell

Since this is not a noise quantity, root sum square (RSS) summation* does not apply to Equation (4.3). This is not a "noise" issue. The "well" has a finite capacity, which provides a fundamental limit on how long it is possible to integrate under a given set of conditions. It is possible, under some circumstances, to read out the detectors many times during a dwell. The charge that is multiplexed out can be added (analogically or digitally) to simulate a longer integration time. This is the splitting up of the integration time to an on-FPA and off-FPA component. This also has the advantage of temporally oversampling to allow advanced signal processing functions such as electron motion compensation for jitter. The drawbacks to this procedure are twofold. First, every time the detector is read out, the readout noise is present. With high sensitivity detectors, this can be a strong noise source. Second, it requires a large amount of electronic and signal processing hardware. Regardless of type, the readout can contribute significantly to nonlinearities. This is especially true of CMOS readouts, as commonly used with InSb.

The multiplexers may be of one of several types including, CCD, CID, or a switched FET. Each has its own drawbacks and advantages.

*The square root of the sum of the squares.

Often, engineers working in the visible part of the spectrum will call their focal plane a "CCD," meaning charge coupled device. This is because silicon (nature's gift to the engineer) is both the detector and the multiplexer in the visible. In the infrared, silicon can only be used for the multiplexer. A CCD is a series of gates which form a potential well in (usually on the surface of) a semiconductor. When a voltage is applied, the well has positive potential and attracts the electrons generated by the detector. The electrons are moved about by lowering the potential in the well next to it. They flow into the next well like dumping water from one bucket to another, as illustrated in Figure 4.5 (recall the "bucket brigade" concept). The next clock cycle moves the charges into the next well by making the electric potential field in the next well more positive than the one in the well currently storing the charge. Electrons, being negatively charged, will move to the more positive field. The CCD is clocked systematically to generate deeper wells in the direction of electron travel and eliminate those behind it. The electrons are separated from those of any other pixel, usually by multiple wells between pixels. As such, a "bucket brigade" is formed, and the electrons are transferred out of one "bucket" at a time. Often, CCDs will incorporate a source follower amplifier. This converts charge to a voltage by changing the voltage on an output diode.

Charge transfer efficiency is the efficiency with which the electrons are moved about, and it is typically as high as 99.99 percent. However, with a CCD, there are many transfers to the final readout, so the cumulative efficiency is appreciably less. The unit cell furthest from the readout may have its electrons transferred a thousand times, for an overall efficiency of only 90 percent. Big arrays are often divided into subarrays, each having a CCD operating in parallel and employing more than one readout line.

A CCD can be implemented in a buried channel (the electrons flow in a defined path not on the surface of the chip) or a surface channel (the electrons flow in a path along the surface) architecture. Buried channel devices have higher transfer efficiency and clocking rates while maintaining lower noise. CCDs at low temperatures have sufficiently low noise that preamplification is usually not required on every unit cell.

CCDs have the undesirable characteristic of relatively high crosstalk and blooming. Blooming is a form of crosstalk in which a well saturates and the electrons spill over into neighboring pixels. This effect is commonly seen in visible CCDs when viewing a bright source against a dim background such as a street light at night. Antiblooming circuitry can be incorporated that essentially channels the excess electrons away from the wells. The CCD manufacturing process can require as many as 10 masking steps.

The charge injection device (CID) is another popular multiplexer. The charge is collected and stored on two coupled MOS capacitors per unit cell. They are read out by x-y coincident voltage addressing techniques in either a destructive or nondestructive manner. That is, they can read out a line (any line) with a process that

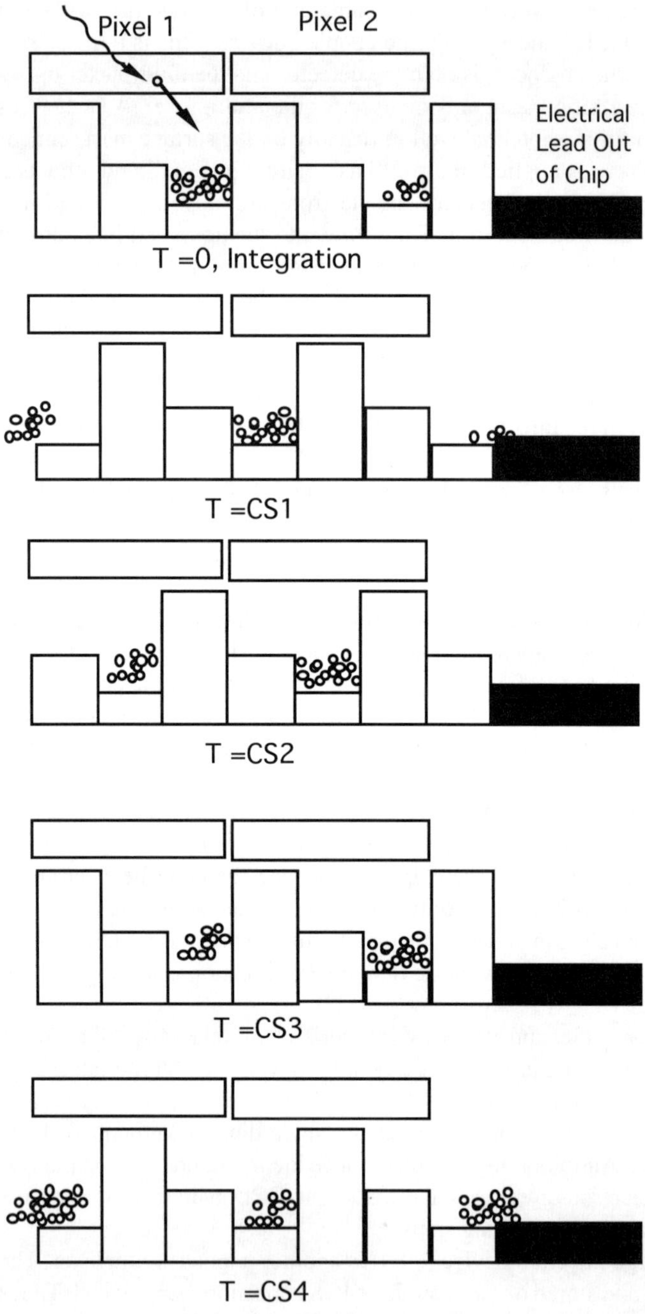

FIGURE 4.5 CCD Function, Moving Electrons with Each Clock Cycle

does not destroy the charge memory. CIDs have the advantage of being able to read out a single unit cell or one area of interest, and this can be accomplished in a nondestructive manner. In contrast, a CCD must read out the entire array each time and destroys all well memory when it does so.

Readouts are easiest to make in silicon, although they can also be made from GaAs, HgCdTe, and InSb. GaAs provides the fastest clocking and increased nuclear hardness, while InSb and HgCdTe provide a monolithic FPA approach. Silicon readouts are commonly available with deep wells, meaning larger integration capacity. A problem is that every array usually requires a custom readout. These are not especially difficult to manufacture but usually take several months (even a year) and several hundred thousands dollars.

4.1.3 Dewars and Packaging

It is important that the temperature of the focal plane be accurately maintained spatially across the focal plane and temporally during operation. This is because several detectors have strong sensitivity and spectral dependance on temperature.

The packaging of an FPA is a critical performance concern and tends to be a cost driver in production The increase in the size of FPAs leads to an increase in the cold shield and FOV which, in turn, leads to larger thermal loads and more stressing packaging considerations [Chatard, 9].

The focal plane's sensitivity is affected through its packaging. The ideal package would be rugged and provide protection, uniform and stable cooling, a perfect vacuum, complete cold shielding, complete thermal isolation, perfect electrical connectivity, modular replacement, easy testing, and lack of induced noise. Focal plane packages are never rugged and must be handled with care. Their vacuum nature implies special clean room precautions to prevent contamination whenever opened.

Even a 0.1 Kelvin differential across HgCdTe can cause an unacceptably large increase in nonuniformity. Heat load depends on packaging, because most of the heat load is from parasitics and radiation. For example, a 256-unit cell PbSe array at 193 K was made which dissipated only 0.5 mW, but the wires dissipated 75 mW, a CMOS mux of 25 mW, and 35 mW for radiation effects [Kreider, 10].

Cold shielding is the implementation of a cold baffle near the FPA that operates as a field stop (i.e., restricts its FOV). It also restricts the amount of stray thermal radiation from the telescope and surroundings. A good cold shield is about 90 percent efficient at keeping out unwanted broadband radiation, while most applications fall in the 50 to 70 percent efficiencies.

Often, some electronics are mounted in or on the same cold package to reduce noise and locate the analog parts close to the detector. The packaging can induce noise in the electronics output via microphonics, ground loops, and improper shielding.

The entire package should never be heated above its “bake out” temperature and should be kept within 10 or 20 degrees of the minimal eutectic temperature of the detector material. These temperature limits must be maintained during transport, thermal testing, and storage.

Dewars are expensive. In production, the dewars often cost more than the focal plane and are more labor intensive. Development of a custom dewar for a tactical application and not advancing the SOA can easily cost half a million dollars. A typical dewar package is shown in Figure 4.6 (photograph courtesy of Amber Engineering). The cold finger is the piece on the lower left. The focal plane mounts to the top and is electrically connected to the leads and readout connections. The vacuum is maintained by pacing the center cylindrical piece around the cold finger and evacuating it through the horizontal port. Also shown in the figure is a split Stirling cryocooler and casing for the assembly.

One standard package is the “lab” dewar. This design provides liquid nitrogen cooling of a focal plane in a compact arrangement suitable for lab or development use, but it is not convenient for field or operational use. This is because of its reliance on liquid cryogen availability and the need to be filled and refilled. These units are usually cylindrical, 15 to 20 cm tall, and 6 to 10 cm in diameter. They quickly cool the FPA using liquid cryogen and usually last several hours.

Most American tactical military applications now use the common module dewars and focal planes. This is a dewar assembly with tens of photoconductive (PC) HgCdTe [Balcerak et al., 11], each dissipating a few milliwatts per detector. In the

FIGURE 4.6 Components of a Dewar Package for a “Second Generation” Linear Array (photo courtesy of Amber Engineering)

U.S.A., common modules are 8 to 12 micron PC HgCdTe arrays with optional lengths of 60, 120, and 180 discrete elements with up to 300 pins [Haystead, 12]. The United States national common module production peaked at about 1,000 per month and over 2,000 were delivered in 1991 [Rotolante, 13]. The common module FPAs are still largely made by artisans, which is appropriate for small quantities but not for large-scale production. Common modules (often called "common mods") cost somewhat less than approximately $15,000 each. Development is underway on a new dewar configuration for larger photovoltaic (PV) FPAs that employs readout multiplexers. These are about the same size as a common module but have fewer electrical lines. A photograph of a SADA dewer appears in the next chapter.

Transistor outline (TO) packages are sometimes used for single detector elements or very small arrays. They are small cylindrical cans with the leads coming out the back, and they look like overweight transistors. Thermoelectric coolers are usually the only cooling integrated into these "cans." They are convenient when the detector is to be placed on a circuit card or where space is extremely limited.

There is no standard package for space applications. Space provides the perfect vacuum so a contained dewar isn't needed. However, the FPA must still be protected, cold shielded and attached to a cooling source.

4.1.4 Overall Producibility and Cost Issues

Producibility and cost are highly material dependent and are discussed under each type of FPA. However, some generalizations and basic principles apply to most, if not all, FPAs. FPAs typically require about 3,000 manufacturing steps. With the exception of common modules, advantages associated with economies of scale rarely have occurred due to low rates of production at any given facility. Producibility of any array is strongly affected by sensitivity, dead pixels, uniformity specifications, material availability (at proper purity), array size, readout availability, and testing.

Many groups are active in FPA research, production, and use. However, a few companies tend to dominate the industry. In addition to instrumentation and dewar assemblies, the Santa Barbara Research Center (SBRC) is one of the leaders in InSb and HgCdTe staring focal planes. Their facility can produce ≈ 500 arrays per month in InSb and ≈ 300 arrays per month in LWIR HgCdTe [Hughes SBRC, 14]. Texas Instruments is another leader, producing over "35,000 of 80,000 detector dewar assemblies in 60-, 120-, and 180-channel configurations delivered to various US and foreign government agencies" [Whicker, 15]. SOFRADIR's capacity for hybridization is about 250 arrays per day, and it has automated test equipment to test 2,500 detectors per hour [Chatard, 16]. Moreover, the German company AEG delivered its 10,000th common module in 1991 [Rotolante, 17].

The cost of focal planes is a complex issue, especially if one is purchasing them in low quantity. It is driven by market forces, requirements, and the specifications of previously made arrays. In the mid eighties, the market forces seemed to push a trend toward lower priced arrays by the Schottky barrier producers. Therefore, prices are tending to drop as the processes become more defined and manufacturers, due to the downturn in the defense industry, have a smaller backlog. The reader is cautioned that the prices given in the data sheets at the end of this chapter are meant only as an approximate and relative guide. If a given FPA is of interest, contact the vendor.

The Schottky barrier price reductions were followed by a reduction in prices of InSb. Manufacturers were trying to recapture the market share that they had lost, and now the makers of HgCdTe seem to be attempting to match the dewar-to-dewar price. The cost of a developmental array is usually dominated by FPA yield, readout yield, and hybridization yield. Once 10 to 100 have been made, the readout and hybridization yields tend to drop in cost importance (except for very large arrays). After hundreds or thousands have been made, as with the common modules, the dewar costs tend to dominate as the detector material becomes less and less significant. The cost breakdown of first generation systems usually finds 75 to 85 percent of the cost being in the dewar ["Detector Mini-Magazine," 18].

The often used "cost per unit cell" (or pixel) is not necessarily a good figure of merit. Cost per unit area can dominate for very large or very small numbers of elements. For large arrays, the total area of an array often determines the boule and wafer yield as well as cost because the physical size limits how many fit onto a wafer. This manifests itself in unit cell sizes that are inversely proportional to the array size. Arrays of 32 by 32 usually have ≥ 100 μm unit cell spacing, whereas 128^2 and 256^2 have about 50 μm spacing. Larger arrays usually have ≤ 25 μm unit cell size.

Generally, the major yield loss in mature microelectronics fabrication occurs in the photolithographic masking steps and is dependent on the chip area. The following classic equation from the microelectronics industry can be used for estimating yield in focal plane arrays:

$$Y_m = \frac{1}{(1 + DA)^m} \tag{4.4}$$

where

Y_m = chip yield

D = defect density per masking step (a low value indicates a few defects per square cm)

A = chip area

m = number of masking steps

Focal planes do exhibit learning curves. The cost of the 120-element common module has dropped from $30,000 to $16,000 in about five years. Additionally, the Maverick IR dewar and detector has dropped from $10,000 to $2,000 in about four years [Rotolante, 19]. For two-dimensional arrays, learning curves are likely to be faster for quantum wells, Schottky barriers, and InSb, than HgCdTe because of the difficulties involved in working with the associated materials. Generally speaking, for production runs of over 1,000, one could conservatively estimate FPA learning curves at 90 percent or, optimistically, at 80 percent.

One last caution is in order when considering FPA cost. The price is a strong function of what has been done already by each manufacturing organization. Because of nonrecurring engineering, changing even a benign specification may cause the price to jump due to the need for redesigned masks, read-outs and electronics. If one is in the market for a low-cost array, the best thing to do is contact several manufacturers to see what designs they currently produce and then try to design a system around what is available. The data sheets following this chapter give an indication of what is currently available.

4.1.5 Generic FPA Sensitivity Figures of Merit

Sensitivity is a function of how efficiently an element transduces electrons compared to how much noise it generates in doing so. FPA performance frequently drives sensor design. Describing the performance of a focal plane for subsequent system-level performance estimates or comparing one to another is a tricky task. There are several figures of merit (see Table 4.3) for focal plane sensitivity, including: NEP, D*, R_0A, quantum efficiency (QE) and noise, responsivity (R), NEΔT, and C_1 and cutoff. NEΔT is really a system-level figure of merit, and it is also discussed in Chapter 9. C_1 and cutoff defines quantum efficiency for a Schottky barrier FPA and, along with the number of noise electrons, is the most appropriate figure of merit for these arrays.

The issue of sensitivity in infrared focal planes is constantly revisited in sensor design and production. There is a great need to standardize the characterization methods and figures of merit among manufacturers, materials, and users. Different materials and applications have different appropriate measures, but increased standardization would facilitate commercialization and user friendliness. It also would remove some of the black art associated with radiometric calculations. There is no free lunch (a universally applicable principle) with regard to detector figure of merit. One must be cautious when comparing figures of merit from one material to another, from one vendor to another, and from one set of operating conditions to another. Almost any figure of merit is useless, and no valid comparisons can be made, without knowing how the measurement was made. The reader should be clear that these figure of merit are for focal planes only—not total sensor sensitivities. These FPA figures of merit in and of themselves are independent of the optics but not necessarily the bandpass or sampling rate. Again, the exception

TABLE 4.3 FPA Sensitivity Figures of Merit

Figure of Merit	*Symbol*	*Units*	*Typical Use*	*Cautions*
C_1 and cutoff	C_1	C_1: dimensionless emission factor Cutoff in μm	Schottky barriers	Fill factor Cavity tuning Noise electrons
Noise equivalent delta temperature	NEDT or NEΔT	Kelvins or °C	FLIR and thermographic system FPAs	Optics Flux Spatial frequency
Noise equivalent irradiance	NEI	W/cm^2	All	Spatial frequency Spectral
Noise equivalent power	NEP	W	Silicon, doped silicon, HgCdTe Pyroelectrics, bolometers	Temperature Flux Detector size
QE and noise	QE or η	Percent of QE No. of electrons, for noise	Quantum wells, InSb	Spectral Fill factor
Resistance area product	R_0A	Ω-cm	HgCdTe, InSb	Temperature Bias Illumination
Responsivity	$\mathcal{R}$	A/W	Pyroelectrics, bolometers, doped silicon	Fill factor
Specific detectivity	D*	Jones $cm\text{-}Hz^{1/2}/W$	InGaAs, InSb, HgCdTe, doped silicon, pyroelectrics, bolometers	Spectral Temperature Flux Blackbody temp. Field of view λ peak

to the above is the NEΔT, which is a system figure of merit sometimes applied to an FPA. Unfortunately, D* can also be a function of the F#.

Responsivity is usually given as amperes of output for one watt of input (or volts per watt). It is simply the output of the detector in amperes, normalized to an input of 1 W.

Noise equivalent power (NEP) is the incident power on the focal plane that produces a signal exactly equal to the average noise signal (SNR = 1). It is measured in watts. Therefore, if a detector has an NEP of 10^{-14} W, a total power (not flux) of 10^{-14} W must be concentrated on the detector by the optics to produce a signal-to-noise of one, and 10^{-13} W to produce a signal-to-noise of 10. The lower the NEP, the higher the sensitivity.

Some people like to see a number get larger as sensitivity increases, thus detectivity (D) was born, and it is merely the reciprocal of NEP. Responsivity, NEP, and detectivity are sensitive to the size of the detector and the rate at which it is sampled. Accordingly, another desire existed in the IR community to normalize detectivity to the detector area and sampling rate. In response, the concept of specific detectivity was born, using the symbol D* (pronounced dee-star) and units of Jones (cm-root hertz per watt, or cm-Hz$^{1/2}$/W). The fantastic dimensions of the Jones are derived from the definition.

The D* figure of merit has several subtleties associated with it. When a D* is quoted, one does not know if fill factor effects, background noise effects, 1/f noise effects, and readout noise were included. The D* is usually quoted at the peak spectral response, which is usually at a wavelength 1/2 to 1 micron shorter than the quoted cutoff wavelength. From there it falls almost linearly with wavelength. Although D* is normalized for sampling rate, it does not provide an analytic representation of 1/f noise. Regardless of whether the quoted D* includes 1/f noise, the effects will not be properly accounted for by merely changing the integration time in the equations. Sometimes D* is quoted at a given blackbody temperature, such as 500 K. At other times if is quoted using a specific bandpass or by just considering a narrow bandpass of laser light. All these conditions should be understood.

In addition to the above, two different D* s are usually quoted—one for detector noise limited cases, and the other for background limited noise cases. The detector limited D* derived from Equation (4.5) should be used for low backgrounds and noisy detectors, while the background limited in performance (BLIP) D* of Equation (4.6) should be used when the shot noise from the background dominates. When in doubt, calculate both and use the lower. The detector noise limited D* is defined as

$$D^* = \frac{(A_d \times \Delta F)^{1/2}}{NEP} \tag{4.5}$$

where

D^* = specific detectivity in cm-Hz$^{1/2}$/W

A_d = area of detector in cm^2

ΔF = noise bandwidth. If unknown or in doubt, assume it is equal to $1/2t_i$, where t_i is the integration time

NEP = noise equivalent power in W

Let's say we want to include a focal plane behind the 90 cm notional telescope we used for examples in Chapter 3. If you recall, we selected a pixel size of 50 μm.

If it were a system with a 1/30 second integration time, and a vendor quoted a weighted average NEP of 2×10^{-13} across the bandpass, then the D* would be

$$\frac{\left[(50\times10^{-4})^2 (\frac{1}{2\times0.33}) \right]^{1/2}}{2\times10^{-14}}$$

which equals 9.7×10^{10} Jones.

D* also can be defined based on the background for a given application. When the sensor system and focal plane are limited by the random noise from the background, it is called background limited in performance (BLIP). A resultant D^*_{BLIP} can be calculated by

$$D^*_{BLIP} = \left(\frac{\lambda}{hc}\right)\left(\eta\frac{A_d}{2\Phi}\right)^{1/2} \tag{4.6}$$

where

D^*_{BLIP} = D* limited by the background

λ = wavelength

h = Plank's constant

c = speed of light

A_d = area of the detector

η = quantum efficiency (from zero to one)

Φ = total combined flux from all background sources incident on each detector in photons/cm²/second. This depends upon the FOV, bandpass, and viewing conditions.

If the background is expected to vary over the bandpass, or if the focal plane has a strong spectral sensitivity, then Equation (4.7) must be integrated as a function of wavelength. This is most easily done on a computer spreadsheet using numerical integration methods. If the D^*_{BLIP} dominates, then increased sensitivity can be achieved only by changing the quantum efficiency, bandpass, pixel footprint, or mission dependent viewing geometries. Equation (4.6) can be rearranged and the constants grouped for a more easily used equation. As pointed out by Spiro and Schlessinger, a background flux can be easily estimated, above which the detector is BLIP, and no increase in R_0A or D* will add to system sensitivity [Spiro and Schlessinger, 20].

$$\phi = 1.27\times10^{37}\left(\frac{\lambda}{D^*}\right)^2\eta \tag{4.7}$$

where

ϕ = background flux that yields BLIP operation for the given D* and λ

λ = wavelength in microns

D* = specific detectivity in Jones

η = quantum efficiency in ratio of zero to one

Following our example heritage, if our focal plane has a D* of 9.7×10^{10} and a quantum efficiency of 0.7, then the level of background that dominates at 10 μm is

$$1.27\times10^{37}\left(\frac{10}{9.7\times10^{10}}\right)2\,(0.7) = 9.4\times10^{16}\ \text{photons}/\text{cm}^2/\text{s}$$

Ideally, the detector noise D* (based on R_0A or NEP) should be a factor of two or so lower than the D^*_{BLIP}. If D* as defined by Equation (4.5) is much higher (by a factor of 10) than D^*_{BLIP} as defined by Equation (4.6), the FPA is too sensitive. In this case, perhaps some cost and producibility savings can be realized by relaxing the specifications. If D* is lower than D^*_{BLIP}, then increased system sensitivity can be gained by reducing focal plane noise. Remember that, regardless of the generation mechanisms, the readout must handle all noises.

Several bastardized figures of merit are sometimes derived from D*. Unfortunately, they have as many meanings as there are engineers. The "System D*" is one, and it is a sensor system figure of merit that includes optic contributions and other non-FPA factors. Others use System D* to mean the needed D* on the detectors to meet sensor system-level sensitivity requirements. Still others use System D* to include all deleterious noise sources so that it may be inserted into generalized NEI equations.

Usually, the D* is quoted at the peak. The spectral sensitivity of D* depends on the detector material. With doped silicon, InSb, or HgCdTe, D* usually peaks just below the cutoff wavelength and falls off linearly with decreasing wavelength (increasing ν). This is because the photon's energy is linearly increasing, and more power is incident with fewer photons. Were D* defined in terms of photons per second instead of power, the redefined responsivity and D* ideally would be constant over the wavelength of the detector. Since classic D* falls off linearly with decreasing wavelength, the easiest way to consider this is to apply a weighted average D* through your bandpass. The weighting depends on the characteristics of the object to be detected.

R_0A is the resistance area product in units of ohm-centimeters (Ω-cm). To be meaningful, R_o and A are usually quoted together because they are interdependent. The larger the R_0A, the better, because it is easier to mate the detector material to readout, and the detector noise will be lower. The subscript "0" indicates that the measurement is taken at zero bias. R_0A is similar to D*, and a D* is often determined by the manufacturer from measured R_0A curves. Essentially, the R_0A determines if an application of a focal plane is BLIP or not. If it is not BLIP, then R_0A determines the noise level. It is (relatively) easily measured and provides a good in-process test when making focal planes. Its value at cryogenic temperatures can be extrapolated from testing at room temperatures by determining the reciprocal slope of the current-voltage (I-V) response.

Ignoring background considerations, (non-BLIP) D* can be calculated from R_0A by

$$D^* = \frac{\eta\lambda}{hc}\left(\frac{R_0A}{4KT}\right)^{1/2} \tag{4.8}$$

where

D^* = specific detectivity in cm-Hz$^{1/2}$/W (Jones)

η = quantum efficiency in a fraction between zero and one

λ = wavelength

R_0A = representative resistance area product for the given temperature

K = Boltzmann's constant

T = temperature in Kelvins

h =Planck's constant

c = speed of light

Or, by combining constant terms in Equation (4.9) and adjusting for empirical results based on the references and data sheets of this chapter, D* can be estimated from R_0A by

$$D^* \approx 3\times10^{10}\eta\lambda\left(\frac{R_0A}{T}\right)^{1/2} \tag{4.9}$$

where

D^* = specific detectivity in cm-Hz$^{1/2}$/W (Jones)

η = quantum efficiency in a fraction between zero and one

λ = wavelength in microns

R_0A = resistance area product in $\Omega\text{-cm}^2$

T = temperature in Kelvins

If the fictitious vendor of our notional example tells us that the FPA's R_0A is 12 Ω/cm^2 at 70 K, then we can verify that the D* should be about 9.7×10^{10} at 11 μm by

$$(3\times10^{10})\ (0.7)\ (11)\ \left(\frac{12}{70}\right)^{1/2} \approx 9.6\times10^{10}$$

R_0A depends upon dopings and mixtures, so the longest response wavelength should be used in this calculation.

One will notice that the empirically based Equation (4.9) gives a lower D* than does Equation (4.8). These are very useful equations, tying together R_0A, D*, quantum efficiency, and temperature. If focal plane delivery problems develop, tradeoffs based on parametric curves of each of these can prove useful. However, the reader is cautioned not to assume sensitivity is merely a function of $T^{1/2}$. Quantum efficiency and cutoff may be a function of temperature, and R_0A is always a strong function of temperature.

Photovoltaic InSb and LWIR HgCdTe exhibit a curious photoconductive effect that tends to decrease the R_0A with illumination. Under normal operating conditions, however, this PC effect in PV devices is not a significant source of noise [Schoolar, 21]. Because of their strong sensitivity dependence on wavelength and their detection mechanisms, R_0A and D* are not proper to describe quantum wells and Schottky barriers.

Another figure of merit for sensitivity is "quantum efficiency," often abbreviated η or QE, and the associated noise measured in electrons. Quantum efficiency is simply the fraction expressed as a decimal from 0 to 1, or as a percentage of the number of electrons, which are generated by a number of photons. The noise is the number of electrons that are generated by means other than photon conversion (e.g., dark current, Johnson noise, 1/f, readout, etc.).

As is usual with IR engineering, there are several subtleties associated with this measurement method. The user must know the quantum efficiency as a function of wavelength. Again, does the number include fill factor effects? Is an antireflective coating assumed? The noise side of the problem includes uncertainties about the noise as a function of flux, FPA temperature, integration time, readout, and so on. As with D*, this method can be used effectively to evaluate a detector, but only if all these factors are clearly understood.

Quantum efficiency depends on antireflective coatings back-reflecting the photons, absorption coefficients at the operating temperature, how many photons get through to the active detector area, and carrier diffusion length. Quantum efficiencies and associated noise are easy to use. One merely calculates how many photons are incident on a detector in the integration time (this step takes all the work). Then one multiplies this by the QE and compares it with the noise electrons. The quantum efficiency is independent of the photon flux and integration time and usually depends only on the wavelength. The other factors enter only when one wants to calculate the associated noise with a flux, integration time, or bandpass different from that quoted.

NEΔT (or NEDT) is defined as the change in scene temperature that produces a change in signal that is equal to the noise from the focal plane. Staring array FPAs can be designed with a NEDT of a few milliKelvins [Balcerak, 22]. This is an important and proper figure of merit for FLIRS or thermographic cameras that are designed to image just such a temperature differential. The reader is referred to Chapters 9 and 10 for additional discussions. NEΔT depends on attributes not associated with detectors, such as the optics' F#, and is more appropriately applied to systems.

The noise equivalent irradiance (NEI) at the focal plane is the FPA output noise normalized to its responsivity. An NEI (in volts) for focal planes can be defined as [Nelson et al., 23]

$$\mathrm{NEI} = \frac{C_s V_n}{\eta A_d e \eta_i T_i G_{elc}} \tag{4.10}$$

where

NEI = noise equivalent irradiance

C_s = effective sense node capacitance

V_n = FPA total noise in volts

η = quantum efficiency

A_d = active area of the unit cell (not necessarily the spacing)

e = electronic charge (1.6×10^{19} C)

η_i = circuit injection efficiency

T_i = integration time

G_{elc} = electronic gain beyond the sensing node

Schottky barriers sensitivity should be defined by parameters that characterize the internal photo-emission and residual noise. These are C1, cutoff wavelength and total noise. Knowing the C1 and cutoff permits the calculation of the quantum efficiency at any wavelength by

$$\eta(\lambda) = \frac{1.24C_1}{\lambda}\left(1 - \frac{\lambda}{\lambda_c}\right)^2 \tag{4.11}$$

where

$\eta(\lambda)$ = quantum efficiency at a given wavelength

C_1 = Fowler emission constant (usually between 0.2 and 0.4 for Pt:Si)

λ = wavelength in microns

λ_c = device cutoff wavelength in microns of the device, which is determined by the Schottky barrier potential

If a particular application for our notional example sensor requires high-resolution, low-sensitivity imaging at 3.9 μm, a Pt:Si FPA would be a natural choice. Another vendor is contacted who specifies a C_1 of 0.32. The quantum efficiency of the FPA at the desired wavelength is

$$\left(\frac{1.24 \times 0.32}{3.9}\right)\left[1 - \left(\frac{3.9}{5.7}\right)\right]^2 = 0.0101 \text{ or approx. } 1\%$$

Nonuniformity and 1/f are two focal plane noise sources that no figure of merit properly addresses (although NEDT, if properly applied, comes close). Unfortunately, they often are the driving noise sources. Many detector materials (especially HgCdTe, but not usually Pt:Si) suffer from 1/f noise. This noise has yet to be fully defined. Published research has linked it to IR flux levels, bias, surface trapping effects, inadequate passivation, and contact technology [Piotrowski et al., 24]. Its causes are largely unknown, and a general analytic expression for it does not exist. However, closed analytic expressions often can be "curve fitted" to real data for a given array or type of array. Additionally, the NEP or D* can be generated as a function of frequency to show 1/f noise effects. This should be done if 1/f noise is expected to be a concern, which it often is for integration times of less than 1/50 of a second.

System designers using the figures of merit of Table 4.3 usually assume all pixels in an FPA have the identical figure of merit. This is categorically false; pixel-to-pixel variations (usually referred to as nonuniformity) do exist. This lack of pixel-to-pixel consistency results from several sources, including detector current variations caused by compositional nonuniformity in detector material and readout electronics, variation in unit cell geometry, unit cell to readout injection efficiency variations, and passivation irregularities [Gopal and Dhar, 25].

The result of this lack of uniformity will be increased fixed pattern noise from all background sources and variable response as a target crosses from one unit cell to another. For high background implementations with high-nonuniformity FPAs where high SNRs are desired, this can be the driving noise source. It has been suggested that the maximum SNR is a function only of the nonuniformity and becomes spatial noise limited [Mooney et al., 26] as follows:

$$\mathrm{SNR}_{max} = \frac{1}{\mathrm{U}} \tag{4.12}$$

where

SNR_{max} = maximum useful value of the SNR

U = nonuniformity

Nonuniformity can be reduced by performing one-, two-, or three-point correction. This is done by having the FPA view known source(s) occasionally. Then the correction is applied by the signal processing electronics. The blackbody should be representative of the viewed spectrum and flux. Usually, a temperature of +12 K and –12 K over the scene mean effective temperature works well [Scribner et al., 27].

Unfortunately, uniformity correction is a nonlinear process. The more "points" used to correct for nonuniformity, the less residual nonuniformity will leak through the correction. However, the more "points" used, the more blackbodies and processing required. Although nonlinear, many cameras and systems only employ only one- or two-point correction and yet achieve excellent results. This is because the calculated linear correction, although not perfect, is sufficient to greatly reduce the effects of the nonuniformity and, in some cases, even reduce them to a negligible effect. One- and two-point correction usually suffices with general imaging systems where the scene dynamics are limited, absolute calibration is not required, and the systems use well behaved FPA materials such as Pt:Si or InSb.

The type and frequency of the correction depends on the operating conditions, allowable nonuniformity drift, and the FPA material. Usually, even an uncompensated 0.1 percent drift across the FPA is objectionable [Nelson et al., 28]. Some functions, such as point target detection, are less sensitive to nonuniformity than other algorithms (e.g., edge enhancement or blur compensation algorithms that enhance the nonuniformities and can become confused) [Nelson et al., 29].

HgCdTe seems to require the most frequent uniformity updates, and Pt:Si the fewest. It has been shown with HgCdTe that the nonuniformity increases in proportion to the logarithm of the elapsed time [Scribner et al, 30]. As a result, scanning systems often view the blackbody once or twice every scan (at the scan edges), while staring systems are "chopped" at regular intervals.

4.2 MERCURY CADMIUM TELLURIDE (HgCdTe)

4.2.1 Detection Mechanism

HgCdTe (commonly pronounced "mer-cad-telluride") is the quintessential IR detector material, so it will be discussed first and used as a basic introduction to general engineering principles concerning manufacturing, producibility, and performance. Its advantages include high sensitivity, cutoff tunability at the time of manufacture, and high operating temperatures. Because of these, HgCdTe often seems to be the material of choice. The material can provide detection from 2 to 26 μm, but large arrays have been built only to a cutoff of 10 or 11 μm. Operating HgCdTe at warmer temperatures naturally increases the thermally generated noise. The increased noise also manifests itself as a decrease in dynamic range. The biggest drawback with HgCdTe is that the material is difficult to manufacturer which, in turn, leads to high costs, delayed schedules, nonuniformity, and unfulfilled performance promises. These problems increase as cutoff wavelength increases.

HgCdTe can be configured as either a photoconductive or photovoltaic device. The PC process operates by a photon transferring an electron from the valance band to the conduction band to cause a change in measured conductance. The PV process employs electronic circuitry that senses both the electron and the resulting "hole." Thus, a PV device offers a 1.4 (square root of two) increase in sensitivity for detector noise limited conditions. However, this is often offset by more experienced manufacturing processes of single-element PC HgCdTe allowing higher R_0A at a given temperature and cutoff wavelength. The PC mechanism dissipates substantially more heat than the PV detection mechanism when implemented in HgCdTe—several milliwatts compared to microwatts or less in the latter. Therefore, PC HgCdTe is usually employed in designs utilizing few discrete elements.

HgCdTe is an alloy whose bandgap may be tuned during manufacture. This allows detector optimization for the desired optical bandpass. This adjustment of the bandpass cutoff results from the relative concentration of mercury to cadmium and the operating temperature. The concentration is defined by "x," which is the cadmium concentration, and "1 – x," which represents the mercury concentration.

It has been reported [Hansen et al., 31] that the energy gap (cutoff wavelength) can be estimated from the temperature and x concentration by

$$E_g = -0.302 + 1.93x + (5.35\times10^{-4})T(1-2x) - 0.81x^2 + 0.832x^3 \quad (4.13)$$

where

E_g = energy bandgap energy in electron volts

x = relative concentration between Hg and Cd in the alloy, where x is the decimal concentration of Cd

T = temperature in Kelvins

while Chu, Mi, and Tang [Chu et al, 32] estimate it as

$$E_g = -0.295 + 1.87x - 0.28^2 + (6 - 14x + 3x^2)(10-4)T + 0.35x^4 \quad (4.14)$$

where

E_g = energy bandgap in electron volts

x = relative concentration between Hg and Cd in the alloy, where x is the decimal concentration of Cd

T = temperature in Kelvins

Both Equations (4.13) and (4.14) are plotted in Figure 4.7 for two different temperatures. The Hansen equation for a 75 K temperature is the top line, with the Chu equation at 75 K the next lower. Conversely, the bottom line is the Chu equation at 200 K, and the one above it is the Hansen at 200 K. Both equations track each other rather well for x concentrations of 0.275 or more. However, the equations diverge at lower x concentrations. (The reader is advised that the program used to generate the plot employed a linear interpolation between the points, so the plot isn't extremely accurate between points.) Curves at temperatures of 75 and 200 Kelvins are plotted. One notes that the cutoff wavelength is lower for the higher temperature. This curious feature does, in fact, mean that the wavelength response of a HgCdTe detector can be extended by reducing temperature. However, on the negative side, this property contributes to the uniformity problem.

A more common semiconductor property is that the dark current (noise) tends to increase as the bandgaps decrease, due to the increased ease an electron has in

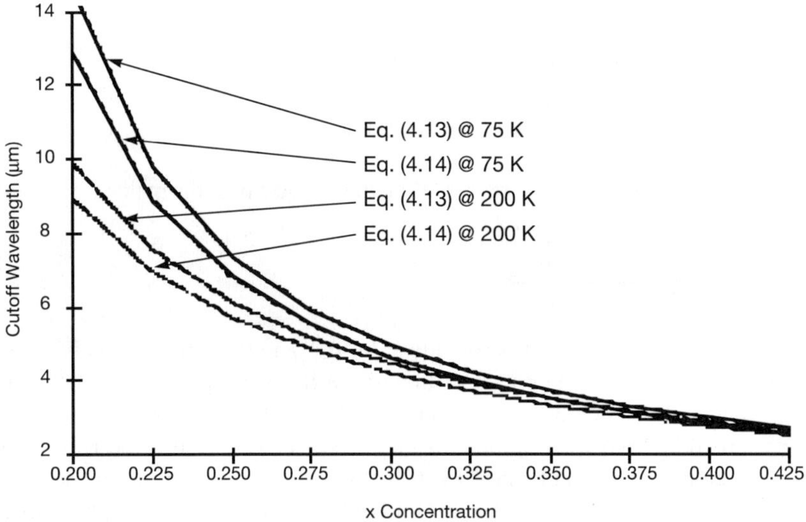

FIGURE 4.7 Cutoff as a Function of x Concentration for HgCdTe

jumping the narrower energy bandgap. It is called "dark current" because it can be measured by observing the FPA output with no radiation input to the sensor (dark). Because of dark current, D* tends to decrease with increasing wavelength and temperature. This is especially true of HgCdTe. Most efforts toward making more sensitive HgCdTe are involved with increasing the detector's resistance area product (R_0A). This is because HgCdTe sometimes is not chosen for background limited applications. When it is chosen, increasing the R_0A results in operation at a higher temperature to save power and weight. R_0A can be increased by thinning the detector to a thickness smaller than the minority carrier diffusion length [Rogalski, 33]. In practice, R_0A depends on x concentration, thermal equilibrium carrier concentration, and temperature. At lower temperatures, the R_0A is a function of "generation-recombination" noise, while at higher temperatures the predominant contribution becomes diffusion noise, which degrades R_0A more rapidly [Vural, 34]. It is usually in this diffusion noise dominated domain that additional focal plane cooling has the greatest sensitivity payoff. Additional cooling of a focal plane in the generation-recombination dominated region usually isn't worth the system complexity, except for ground-based astronomy applications. The systems engineer often wishes to know how the R_0A changes with temperature, since temperature is the only variable that can be controlled once the array is made. Empirical data from recent papers yields a reasonable curve fit for this as

$$R_0A = A \times 10^{BT} \tag{4.15}$$

where

R_0A = expected resistance area product

A = a constant in the temperate range

B = another constant representing the sensitivity of the material to changes in temperature

T = temperature in Kelvins

For state-of-the-art MWIR, A is roughly 5.7×10^8, and B is approximately –0.037 between 100 and 150 K. For state-of-the-art LWIR, A is typically 2×10^7, and B is approximately –0.066 between 50 and 100 K (note the minus sign). These constants should be used with caution, as they were derived from a "best fit" curve based on empirical data published by different manufacturers (from the papers in the reference and the data sheets at the end of this chapter) and for slightly different cutoffs and different array sizes. They are likely to change over time as manufacturing process, controls, and techniques mature.

Unfortunately, HgCdTe has both an R_0A and a cutoff dependance on temperature. The sensitivity can change by a factor of approximately two for every five Kelvin change or so. This necessitates holding the FPA to within a few millikelvins across the surface and not letting it drift by more than a 0.1 or 0.2 K before viewing a blackbody.

R_0A is independent of junction area, provided the junction's dimensions approach the diffusion length of the minority carrier. For HgCdTe, this is about 15 μm, so for unit cell dimensions of less than 20 μm, using R_0A to calculate sensitivities must be done cautiously.

HgCdTe can be made as either homojunction or heterojunction devices. In a heterojunction, the bandgap of the n material does not match that of the p material. Heterojunctions have the advantage of reducing surface leakage currents and isolating the field junction space charge from the surface and reducing the GR noise [Balcerak et al., 35].

4.2.2 Manufacturing Process

There are several detector architectures. The most popular include an indium bump-bonded hybrid approach with or without mesas, a monolithic approach, and a via hole hybrid approach. The best applicable manufacturing process is highly dependent on the architecture. This discussion represents a specific manufacturing flow, and the reader is cautioned that it is presented only to illustrate engineering principles. Any given foundry may reverse some steps or do things quite differently. Figure 4.8 shows such a typical process flow for the manufacture of hybrid

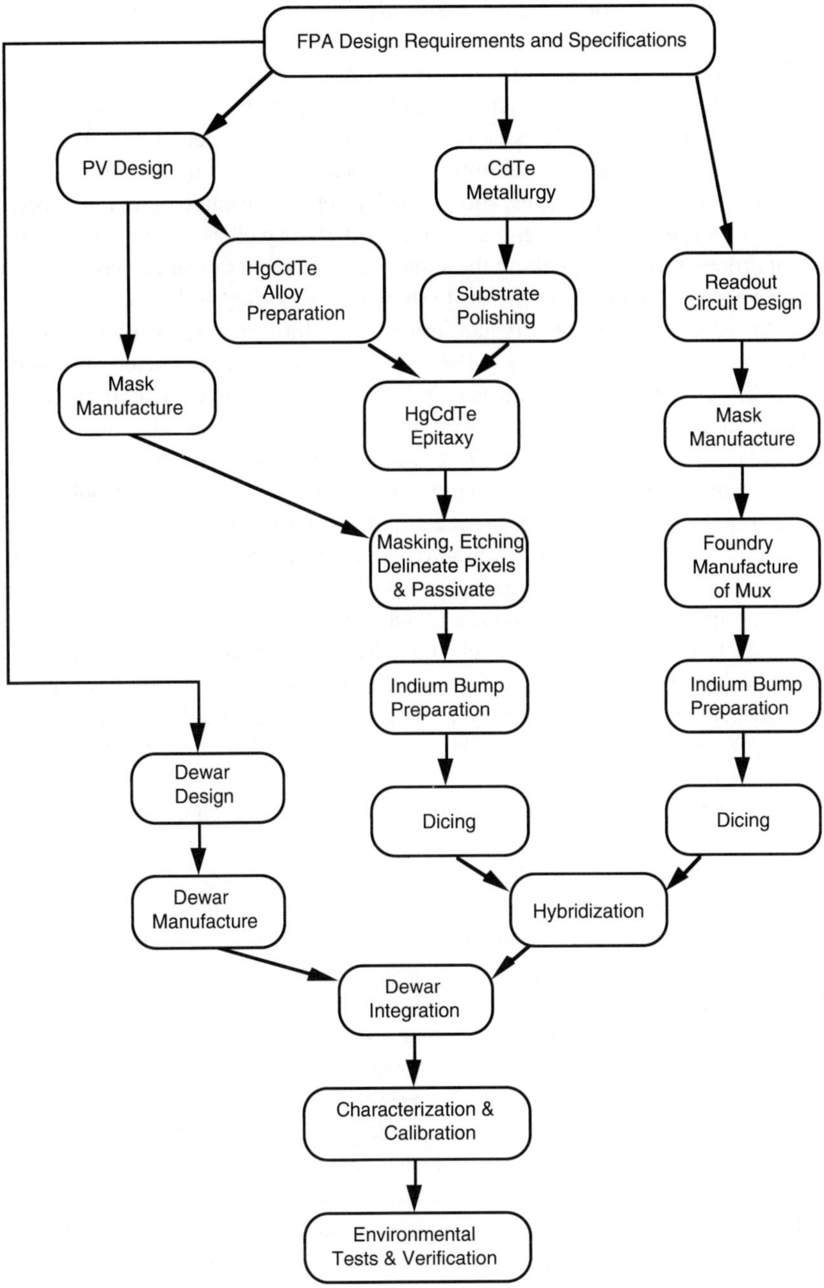

FIGURE 4.8 Representative Process Flow for HgCdTe FPA/Dewar (adapted from drawing courtesy of SOFRADIR)

HgCdTe arrays. The following discussion also applies to other materials described in this chapter, as the general processes are often similar.

Manufacturing HgCdTe is an elusive, proprietary art practiced in expensive facilities. Detector array manufacturing calls for exceedingly clean facilities with clean room usually from "class levels" of 100 to 100,000 depending on process step, laminar flow benches, photolithography stations, handling equipment, accurate saws, infrared microscopes, and expensive deposition equipment. For success, add to this a cadre of skilled and educated workers as well as practical managers.

All processes start by making the substrate. Different companies have their favorite substrates, including silicon, InSb, GaAs, CdZnTe, and CdTe. The CdTe must be grown in a single-crystal format, in a furnace. Liquid phase epitaxy (LPE), molecular beam epitaxy (MBE), molecular organic chemical vapor deposition (MOCVD), and vapor phase epitaxy (VPE) are commonly used processes. CdTe is about as hard as talc, with a hardness of one, so special care must be taken to avoid damaging the surface. CdTe boules are smaller than silicon (usually only 2 to 5 cm in diameter), so special handling equipment must be used. Boule purity and ampoule alignment in the furnace are important for quality substrates. Once cooled, the substrate boule is sawed into wafers and prepared. Preparation usually includes polishing and cleaning, regardless of the material.

For producibility reasons, the use of substrates other than CdTe has been promoted for large arrays. Sapphire substrates have been used for SWIR and MWIR arrays, as have GaAs and silicon for LWIR arrays [Bailey et al., 36]. For sapphire substrates, the HgCdTe detector layer is deposited on the sapphire with a thickness between 8 and 12 μm. This procedure also includes a ZnS or CdTe passivation [Bailey et al., 37].

GaAs substrates, which are transparent to longer wavelengths, can also be used. GaAs has the same thermal mismatch concerns as CdTe, but it is easier to obtain suitable substrates (and thus is more producible) than CdTe. To solve the thermal mismatch concern between the silicon readout and the detector substrate, MOCVD can be used to grow the proper mixture of HgCdTe [Bailey et al., 38].

Once the substrate is processed, the actual detector layer of HgCdTe must be developed. This is typically done with LPE, VPE, or MOCVD. Superlattices of HgCdTe have been proposed to eliminate some of the material processing problems. Mixing of the HgCdTe is a critical step because the x concentration must be kept exact. The bandgap and sensitivity are related to the x concentration. It is important to note how the x concentration's derivative changes with respect to the cutoff. Examining Figure 4.7, one finds that the lower the x concentration (longer wavelength), the greater the change in cutoff wavelength that will occur for a given incremental change in x. For example, if x varies by 0.025 at a concentration of 0.4 (going from 0.4 to 0.425), the cutoff will shift by a mere 0.3 μm, while it will shift almost 5 μm for a 0.025 change (going from 0.2 to 0.225) in a longer wavelength x concentration of 0.2. Mathematically, the derivative $\partial\lambda/\partial x$ varies, being greater at longer wavelengths.

Therefore, a five percent variance in x from pixel to pixel has little effect on cutoff at x levels for the SWIR. However, for LWIR, the same variance at x levels has a several micron cutoff difference. The detectors will have a peak sensitivity just short of their cutoff. If each detector has a a cutoff that varies even half a micron, their responsivity will vary greatly to a signal through a fixed bandpass. The curve in the LWIR region of Figure 4.7, along with the curve of D* values, requires precise material control for uniform sensitivity. This need to control the material so accurately is the fundamental "heart of darkness" of making long-wavelength HgCdTe. The x concentration must be controlled from cell to cell and from array to array in a precise manner that is driven by the state of the art in manufacturing process and technology.

Several companies use LPE to mix the concentration and set the x. It is in this step that slightly different concentrations and/or weights of the constituents can cause an ever so slight variation in x. In LPE, the x concentration is determined by carefully weighing the constituents then transferring the material to the LPE "boat" (a holder for the goop). The boat is then heated in the furnace to several hundred degrees Celsius until a liquid phase is reached (hence, "liquid phase epitaxy"), following which it is slowly cooled. The drawbacks to LPE are poor surface morphology and difficulty in controlling layer thickness and interfaces. This is far more important in PV HgCdTe than in PC [Balcerak et al., 39].

An alternative to LPE is VPE, which allows lower processing temperatures (e.g., 200° C) that can improve crystal quality and give better control over x concentration and interfaces [Balcerak et al., 40]. VPE employs a source wafer in proximity to the substrate. Heat is applied to the wafer, and growth occurs via transport and condensation on the CdTe. The alloy is created by diffusion and tends to have a graded x composition [Balcerak et al., 41].

Other techniques, such as MBE and MOCVD, have the advantages of allowing multilayer growth and thereby establishing the p-n junction. MBE is a refined, high-tech version of the astronomer's optical evaporation deposition. The substrate is held in a vacuum chamber in direct line of sight of shuttered ovens. The ovens heat the Knudsen sources, which are ultra-pure versions of the elements to be deposited, and the atoms migrate to the substrate via their thermal energy. Ion gages monitor the flux, which can be controlled by the shutters and temperatures. MBE has the advantages of well defined layer interfaces and thin deposition layers. MBE machines are costly, require skilled operators (usually the scientist or engineer himself), and growth rates are slow—usually between 0.01 to 1 nm/s [Joyce and Foxon, 42].

MOCVD mixes compounds in their vapor phase. The compounds are pyrolized in a flowing hydrogen atmosphere. The substrate is usually heated by an RF source for deposition of the compounds.

Once the HgCdTe detection layer is formed, the detectors must be delineated. This part of the procedure employes photolithography and etching. This separates or insulates each cell from the surrounding material, ion implantation, and metallization to form electrical contacts on each detector. The detectors can be separated using three-dimensional geometry and etching mesas (like the geographic formations of western Colorado) or by insulating planer structures from each other.

The surfaces must be electrically and chemically passivated. Passivation of the surface is important for several reasons that include control of tunneling and junction leakage and prevention of decomposition (Hg evaporation) of the alloy over time. The mercury is unstable in a HgCdTe alloy and tends to "evaporate" out, changing the detector's properties. This step is also critical for good detector performance because it reduces surface recombination velocity and increases R_0A. Surface preparation is the key to good passivation [Balcerak et al., 43]. Passivation has also been linked to 1/f noise performance. Passivation is often accomplished with the deposition of a layer of SiO_2 or ZnS. Passivation techniques and surface preparation prior to passivation are proprietary to each manufacturer.

At this point, several arrays have been built on a wafer. The prepared arrays must be cut out of the wafer, and the process is called "dicing." The cut array is referred to as a "die." Frequently, the next step is to thin the material to 10 to 15 μm, which can be done by sawing, laser erosion, or chemical etching.

Finally, for hybrids, the readout multiplexer must be attached as indicated in Figure 4.3. This can be done by making the HgCdTe into an electric circuit or, more commonly, by "bumping the array" onto a silicon or GaAs multiplexer. This bumping forms a hybrid and is called "hybridization." The bumping is typically accomplished by growing indium on the detector's electrode contacts and on the multiplexer contacts. Figure 4.4 shows indium bumps on a silicon readout, ready to be attached to HgCdTe. The array is aligned with the multiplexer using an infrared microscope while pressure, and sometimes slight heat, is applied. The indium bumps fuse, forming a cold weld. There is a thermal mismatch between the HgCdTe and silicon. The indium bumps can flex and accommodate some stress caused by this thermal mismatch. This usually limits the physical size of an array to less than 1.5 cm on a side. Edges are where bumping problems usually arise. Indium bump bonding has enjoyed much improvement in the eighties and is typically a high yield step. The process gives few problems for arrays of 256 × 256 or less and unit cells of 50 μm or greater. The problems that do arise with this step include indium running onto a neighboring unit cell, thereby shorting the two together, and bond delamination through thermal cycling.

With monolithic devices, the above process is altered and made more complex by making the readout multiplexer concurrently using the same die and wafer as the detector. Monolithic HgCdTe is a problem for LWIR, and the yield cannot be separated from the detector yield. However, since the hybrid process is not used, delamination through thermal cycling is not a concern.

CdTe boules and the resulting wafers are not large in diameter compared to silicon and GaAs boules. CdTe boule diameters are usually 4 cm or smaller. This limits the number of arrays obtainable from a given wafer. This also limits the size of an array to approximately 2.5 cm or less on a side. Smaller boules and wafers have another producibility disadvantage in that they are incompatible with generic silicon handling and processing equipment that is designed and built for the integrated circuit industry. This has led some foundries to search for substrates other than CdTe.

Techniques that do not bump bond the HgCdTe to the silicon readout have been developed. Instead, they grow the HgCdTe directly on silicon readouts. These silicon readouts are on standard 4 inch (10.2 cm) silicon wafers, and holes are drilled or etched through the HgCdTe (one per unit cell) to the readout circuitry. The sides of these "via holes" are metalized to provide the electrical connection. The hole areas are tiny compared to a pixel, so there is little fill factor loss. Growing the detector array directly on the silicon wafers allows the use of standard silicon handling and processing equipment for greater producibility and affordability.

In-process testing is important to prevent the additional expenditure of time and capital in processing defective arrays. On the other hand, in-process testing is also expensive and time consuming. A balance must be achieved between the expense of in-process testing and the savings it generates by reducing defective array processing. Most manufacturers employ tens of tests on each array during manufacture. Plotting the R_0A versus temperature is a good diagnostic to indicate material quality and surface effects [Vural, 44].

4.2.3 SOA (Size, Sensitivity, Uniformity, Bandpass, Etc.)

Array size, sensitivity, and uniformity in HgCdTe are highly dependent on the x concentration and, therefore, the cutoff wavelength. As the "x" concentration increases (i.e., cutoff wavelength decreases), it becomes easier to make larger, more sensitive, and more uniform arrays that can operate at increasingly high temperatures.

A problem with large-scale arrays is the physical size of the array. Typically, uniformity and stoichiometry can only be maintained over a 1 to 2 cm linear length or so. This, along with practical unit cell size, limits the total number of cells or pixels in the array. This is not a fundamental limitation—merely one based on the processing technology. As technology progresses, array sizes continue to grow. Also, strips of HgCdTe can be aligned together on the same multiplexer or in the same dewar to produce larger effective arrays.

Nevertheless, several developmental large-format arrays have been reported, including the following:

- SBRC has made a 960 × 4 LWIR hybrid connected to four silicon readouts on a single motherboard [Norton, 45]. The company also has fabricated a 640 × 480 LWIR, 128 × 128 pixel 6 μm cutoff array on CdZnTe, and GaAs and Si substrates with 35 μm square unit cells ["HgCdTe Infrared Arrays Grow on Si-Based Substrates," 46; Nordwall, 47].

- Rockwell has made 64 × 64 LWIR HgCdTe on GaAs. Another Rockwell hybrid, 256 × 256 with a 4.9 μm cutoff, had rms nonuniformities of less than 4 percent, and the company has produced a 640 × 480 SWIR [Adams, 48].
- Texas Instruments has made a 128 × 128, 8 to 10 μm HgCdTe using VIMIS technology [Norton, 49].
- Other reports of large arrays include scanning array sizes of 1,000 × 16 [Balcerak, 50], 480 × 4, and 960 × 4 PV devices, and "on-chip" TDI multiplexing reducing the number of leads out of the dewar [Haystead, 51].

Several manufacturing and producibility programs have yielded promising results. A U.S. government ManTech program produced several 128 × 128 arrays, and another U.S. producibility program is producing 960 × 4 and 64 × 64 LWIR arrays [Rotolante, 52]. SBRC is marketing arrays of 480 × 4 HgCdTe [Adams, 53], and Rockwell has commercial lines of 128 × 128 and 256 × 256 SWIR and MWIR arrays.

A good rule of thumb is to specify FPA average sensitivity a factor of 5 to 10 below the theoretical maximum for HgCdTe for the given x concentration (cutoff) and temperature. This will improve yield and reduce cost to a level where the array has a chance of being made in a timely and cost-effective manner.

LWIR HgCdTe suffers from high dark current (or leakage current). This is due to surface leakage and tunneling that can be mitigated by improved passivation, material purity, and heterojunction fabrication. All HgCdTe suffers from 1/f noise. With current technology, it is seldom a driver with frame rate times of less than 1/60 of a second or so. However, for astronomical applications, it is often the driving noise for a readout of less than 10 Hz.

Generally, one can expect D* values for SWIR PV on the order of 1×10^{12} @ 120 K, 1×10^{11} @ 100 K for MWIR, and almost 10^{11} @ 80 K for LWIR. However, some PC HgCdTe detectors have demonstrated D* values of 2×10^{12} in the 8 to 12 μm bandpass [Tebo, 54]. Quantum efficiencies for HgCdTe are typically 50-60 percent without an antireflective (AR) coating and 70 to 80 percent with an AR coating.

Typical LWIR uncorrected uniformities are about 10 percent to 20 percent across an array. These can be reduced about an order of magnitude, with the best HgCdTe corrections reducing nonuniformity to the 1 to 2 percent region.

Cutoffs depend on temperature and x concentration. Most tactical genre LWIR HgCdTe arrays, whether PC or PV, do not really respond in a fashion suitable for practical use out to 12 microns as is often stated. Most "LWIR" PC HgCdTe arrays have a maximum response in the range of 11 to 11.8 μm @ 77 K. The cutoff of the PV types depends upon the individual array, manufacturer, and temperature. Most tactical PV arrays have a shorter cutoff between 10 and 11 μm. Specially made detectors and have been reported as long as 48 μm [Reine, 55], and Belov offers commercial single-element, liquid nitrogen cooled HgCdTe detection to 26 μm with a D* of 1×10^{10} [Tebo, 56]. For a few thousand dollars, several manufacturers offer commercial single-element PC HgCdTe with responses out to 18 μm.

The TICM II British common module uses an eight-element linear array of SPRITE HgCdTe, 62.5 μm wide and 700 μm long, with a tapered end to offset decreasing carrier mobility [Davis, 57]. Sprite arrays usually have high uniformity and large D* values. A small Sprite array with a D* of 2.4×10^{11} Jones has been reported [Braim, 58].

4.2.4 Producibility, Cost, and Power

Planet-wide production of small PC arrays is easily several thousand per year, and probably less than a hundred PV arrays. However, several companies are preparing for large-scale PV array production for American (e.g., LOSAT, TSIP,TWS) and European military (e.g., Triget) programs, and the current worldwide production capacity is in the thousands per month. Major array manufacturers include AEG, Amber Engineering, Hughes SBRC, Loral, Philips, Rockwell International, SOFRADIR, and Texas Instruments. All of the array manufacturers are capable of making discrete elements, as are many additional manufacturers including EDO/Barnes, EG&G Judson, Graseby Infrared, North China Institute of Electro Optics, Servo, and Vigo.

If one is merely developing a new FPA using existing technologies, the FPA can be developed for a few million dollars. The purchase of a copy of an existing array (no new masks need be made) can usually be done for a few tens to hundreds of thousands of dollars. Learning curves show substantial cost reduction occurs for production runs of several hundred to several thousand detectors. Many studies have shown that, if mass produced, arrays could be made for as low as a few thousand dollars each and even meet the elusive $1 per pixel goal. For instance, Loral has produced a 64 × 64 LWIR FPA for $9 per unit cell and expects the cost to drop to $1 per unit cell in large production [Hughes, 59]. Unfortunately, since no large-scale production of large HgCdTe arrays has ever existed, these estimates are based on forecasts and studies, and they are open to criticism. Such criticism was recently directed toward the Javelin HgCdTe FPA, which was to cost an average of $12,500. But the Government Accounting Office (GAO) reported that this would require a decrease of "62 percent between development and initial production from an estimated cost of $62,000 to $24,000, which appeared optimistic for HgCdTe" ["GAO: Cost of Javelin Focal Plane Array Has Been Underestimated," 60].

Yields are hard to define but are low for semiconductor manufacturing—typically less than 1 percent from the wafer to working hybrid for LWIR HgCdTe. The focus for low-cost, large-scale production is on larger boules, automated testing, and statistical process control. Automated testing and handling can greatly help producibility. For example, SOFRADIR's capacity for hybridization is about 250 arrays per day, and the company has automated test equipment that can test 2,500 unit cells per hour [Chatard, 61].

For any material, yield (and hence, producibility and cost) are very sensitive to specifications. The specifications that most affect producibility and cost are the percentage of out-of-spec unit cells, uniformity, and sensitivity. For instance, increasing the acceptable defect lines from 2 to 4 percent can increase yield by 4 to 40 percent [Chatard, 62].

Additionally, the FPA cost (for any material) depends on three key user-generated issues, which are as follows:

1. How many do you want?
2. How much are you pushing the state of the art?
3. How much development effort is there?

The complete development of a new format, new x concentration, and pixel size with low uniformity can run $10 million or more. Critical cost drivers for HgCdTe arrays include substrate defects (dead pixels), uniformity, and R_0A. If the specifications are reduced slightly, there frequently is a large reduction in cost due to higher yields and fewer tests required [Chatard, 63]. The ingenious IR engineer can drastically reduce costs by incorporating system design performance margins to allow array specification in a cost-effective manner.

The good news is that the user can trade costs and the number of FPAs that can be made per month against other system parameters. For example, the FPA's required sensitivity (R_0A) can be reduced by a factor of two for a 40 percent increase in aperture area. The FPA will cost less and be more producible. Depending on system constraints, this could be a cost-effective solution. For low background cases, optic cooling can be traded for uniformity. Another system design tradeoff to improve HgCdTe producibility is cooling weight, cost, and power versus FPA sensitivity.

4.2.5 Appropriate Applications for HgCdTe

Because of HgCdTe's high sensitivity at higher temperatures and its spectral tunability, it has a myriad of applications and is the most versatile of all detector materials. It is the most common material in producing military FLIR systems, and it finds its way into commercial cameras and space instruments. It can be made for either high background or low background implementations.

HgCdTe is useful for SWIR and MWIR detection at temperatures high enough to use thermoelectric or radiative cooling. If high sensitivity is needed with a low background, the HgCdTe in the MWIR and SWIR is an excellent choice. In the MWIR, 80 Kelvin HgCdTe can outperform 20 Kelvin doped silicon.

HgCdTe is also likely to be the material of choice for high background applications of 14 microns or less when cooling to (or below) 80 Kelvin can be tolerated. LWIR HgCdTe is not the material for large LWIR arrays if cost or schedule is a driving factor.

Shipping and environmental constraints could limit HgCdTe use. To reduce the risk of outgassing contaminants, the dewar assembly should never encounter temperatures in excess of the bake-out temperature. Limiting the temperature to which the dewar can be exposed is a major concern for applications where the FPA may be packaged inside in a hot instrument or FLIR located in the tropics.

Another potential application limiter for HgCdTe is the required exact cooling control and frequent blackbody reference frames. HgCdTe is an excellent thermometer, as both its noise and cutoff are a function of temperature. This affects the entire sensitivity curve and requires precise control of the FPA temperature. A 1 Kelvin, control with 0.01 variance across the array is not uncommon. This usually translates into added system-level weight and power to control temperature and provide needed radiometric uniformity corrections.

4.2.6 Miscellaneous Tertiary and Quaternary Material Advancements

Several other tertiary and quaternary detector alloys have been made that mimic HgCdTe's tunability, sensitivity, and manufacturing process.

HgMnTe is a mixed alloy akin to HgCdTe, and it detects IR radiation in both the PC and PV mode. Renewed interest in this tertiary alloy has been sparked by NASA for EOS. Brimrose commercially markets PV single-element versions for a few thousand dollars each.

Another HgCdTe-like tertiary alloy material that has been made into IR detector arrays is lead chalcogenide $Pb_{1-x}Sn_xSe$. Lead chalcogenide epitaxially grown on silicon offers potential for large arrays. It has several advantages over HgCdTe, including stronger optical absorption (yielding higher quantum efficiencies), increased process fault tolerances, and less severe uniformity problems [Hoshino et al., 64]. This material has fewer homogeneity problems for long-wave cutoffs. For instance, at 12.4 μm it's sensitivity is five times less dependent on composition inhomogeneities than comparably mixed HgCdTe [Hoshino et al, 65]. However, PbSnSe suffers from limited frequency response resulting from high diode capacitance and high thermal coefficients of expansion [Norton, 66], which is likely to limit its array size. Hoshino reported a linear array with a 10.2 μm cutoff, 59 percent quantum efficiency, a D* of 5×10^{10} Jones @ 85 K, and a nonuniformity of 3 percent [Hoshino et al., 67]. The Soviets have reported PbSnTe/PbTeSe heterojunctions operating at 77 K and 8.5 μm with an R_0A of 1.2 Ω-cm^2 [Tetyorkin et al., 68].

Another detector material can be made by replacing the Sn with Eu. The Institute Fuer Physikaliche Messtechnik has made $Pb_{1-x}Eu_xSe$ with a 10^{10} D* at 6.7 μm and 95 Kelvin [Halford et al., 69].

4.3 INDIUM GALLIUM ARSENIDE (InGaAs) FAMILY

4.3.1 Detection Mechanism

InGaAs is the tertiary (three-element) version of InAs. InGaAs responds to SWIR radiation from about 1 to 2.5 microns. Its family includes $In_xGa_{1-x}As$ and In_x-$Ga_{1-x}As/InAs_xP_{1-x}$ materials that can be lattice matched to produce bandgaps in the SWIR.

Like HgCdTe, InGaAs can function in either the PV or PC mode. The PC semiconductor process transfers electrons from the valence band to the conduction band, whereas the PV mode produces a sensed voltage from the transference of an electron from the valance band to the conduction band and the subsequent creation of a hole in the conduction band. The PV mode employs the hole as well as the electron, and thus shows sensitivity improvement to the square root of two.

The InGaAs bandgap can be controlled by controlling the indium (X) concentration. The bandgap of $In_xGa_{1-x}As/InAs_xP_{1-x}$ can be controlled by controlling the indium and P concentration. The more indium added, the lower the bandgap. InGaAs can be made to detect to 3 microns by using exceptionally high amounts of indium.

Around 5 pA, dark current is typical for this material. However, InGaAs can exhibit what is called "popcorn noise." This consists of spurious noise emissions of 200 to 1,000,000 electrons. The emissions are random and unpredictable. Some pixels have popcorn noise, and some do not. It has been tied to a high number of crystal defects near array edges [Joshi et al., 70].

4.3.2 Manufacturing Process

Manufacturing includes forming junctions, defining and isolating unit cells, placing electrical contacts on each pixel, passivating, and applying antireflective coatings. It is important to accurately control the indium concentration in InGaAs. This can be accomplished by vapor phase epitaxy or molecular beam epitaxy. The long-wavelength (> 1.9 μm) devices are often formed from graded layers on $InAsyP_{1-y}$, with the P concentration being graded. Vapor phase epitaxy (VPE) is suited for depositing these graded layers. Published literature describes a structure having 15 layers with y increasing from 0.0 to 0.68, which changes the lattice parameters and yields an E_g change from 1.31 to 0.65 eV [Ban et al., 71]. A subabsorption layer is deposited between the graded layers and the absorption layer, the latter with an E_g of 0.45. A silicon nitride layer is deposited as a diffusion mask for zinc doping and passivation. The zinc doping produces the p-n junction. Finally, a cap layer identical to the subabsorption layer is deposited over the absorption layer. InGaAs must be made in class 100 clean rooms.

4.3.3 SOA (Size, Sensitivity, Uniformity, Bandpass, Etc.)

Devices made from these materials consist of linear arrays. Only recently have two-dimensional arrays appeared. The largest available linear sizes are 256 to 300 elements in a line and 128 × 128 for mosaic arrays. Uniformity varies from manufacturer to manufacturer and can be as high as ± 20 percent, depending on the indium concentration. As with HgCdTe, the nonuniformity tends to increase with increasing indium concentration.

Again, an empirical and fundamental semiconductor property is that dark currents (noise) tends to increase as the bandgap decreases, and such is the case with the InGaAs family. Dark currents associated with 100 μm unit cells and the In doping of 0.8 are in the range of 100 to 1,000 nA [Olsen, 72].

These devices have typical responsivities of 0.9 A/W and an 86 percent QE [Olsen, 7473 Inc. has made a heterojunction $In_{0.8}Ga_{0.2}As/InAs_{0.6}P_{0.4}$ with a 50 percent QE and $5mA/cm^2$ dark current density out to 2.6 μm [Tebo, 74]. Others have produced devices with D* values greater than 1×10^{12} [Ban et al., 75].

4.3.4 Producibility, Cost, Power

Yields for these materials are high, and 50 percent quantum efficiency can be achieved with $In_{0.8}Ga_{0.2}As/InAs_{0.6}P_{0.4}$ at a 93 percent yield [Olsen et al., 76]. Single-element InAs detectors sensitive to almost three microns can be purchased commercially for under $2,000. Power dissipation is minimal with a large array consuming a only few tens of milliwatts.

4.3.5 Appropriate Applications

Suitable application of these detectors are anywhere low-cost, uncooled detection of less than 2 μm is required. Line scanners often use InGaAs.

4.4 LEAD SALTS (PbS and PbSe)

4.4.1 Detection Mechanism

Ordinarily, PbS and PbSe materials are intrinsic photoconductive thin films. They were developed in the forties and fifties to provide uncooled high sensitivity in the SWIR. Numerous manufacturers of single elements and linear arrays exist, with just a few making two-dimensional arrays for custom orders.

The basic detection mechanism is a photoconductive effect. As mentioned before, this is where a photon with an energy greater than the bandgap will be converted into a hole and electron that will migrate under a bias to a collection area. The electron produces a measurable change in conductance. These detectors are unresponsive (do not produce the PC detection phenomena) until energized by the application of an electrical bias. They can be operated in a mode in which the bias is turned on and off to control integration. A recent development has led to PV operation by growing the detector on a lead electrode to produce a Schottky diode that does not require a bias. As usual, the PV mode allows for larger arrays, lower thermal dissipation, and an improvement in sensitivity.

4.4.2 Manufacturing Process

Lead salts are usually chemically deposited from a solution or evaporation onto a substrate to form a polycrystalline film-like layer one to two microns thick. Etching is used to delineate the desired unit cell size and shape [Vincent, 77], with quartz being the usual substrate. The quartz is then diced and metalized to form the contacts, and indium bump bonds are placed on the detectors. Finally, they are then mated to a readout.

EDO/Barnes uses VPE to grow the lead salt epitaxially on a BaF_2 substrate. The lead salt film is sensitized with a chlorination process, and a series of masks are used to metalize contacts. This process lends itself to multicolored detectors by depositing PbSe and PbS. Tin also can be added to extend the cutoff [EDO/Barnes, 78].

4.4.3 SOA (Size, Sensitivity, Uniformity, Bandpass, Etc.)

These devices often are made as discrete detector elements, but linear and two-dimensional arrays can be made. A 32×32 is a common two-dimensional size, but larger developmental arrays have been made. PbS has a D* of 1×10^9 uncooled and 5×10^9 when cooled. PbSe with a specific detectivity to 2×10^{10} when cooled has been reported [Tebo, 79]. However, with a peak at 2.4 μm, these detectors often exhibit D* values of 1×10^{11} [Resso and Harris, 80]. Uniformity is similar to that of HgCdTe—on the order of 10 to 30 percent uncorrected with a approximately 30 percent quantum efficiency [Norton, 81]. PbS linear arrays of 6,000 50×50 μm cells have been made in Russia, and they are working on 256×256 and 512×512 arrays [Khryapov, 82].

4.4.4 Producibility, Cost, and Power

Lead sulfide and lead selenide arrays are highly producible. Litton estimates that 6,500 arrays per month can be made at their existing facility [Litton, 83]. The weight of a 256×1 package made at Litton was 65 grams with a BeO base, and it requires less than 150 mW at 195 K [Kreider et al., 84]. Such arrays can be obtained for a few thousand dollars each, with prices increasing nonlinearly with array size.

4.4.5 Appropriate Applications

The lead salts are appropriate for high-sensitivity detection in SWIR and MWIR scanning systems, with minimal cooling and cost, where high resolution is not demanded. They have been used in thermal imagers, FLIRS, and space applications.

The lead salts are fragile. Vincent cautions that they should be protected from humidity in excess of 50 percent, acids (even saliva can pit them), and bright visible light [Vincent, 85].

4.5 SCHOTTKY BARRIERS

4.5.1 Detection Mechanism

Schottky barriers are made of heavy metal "dopants" in a thin silicate layer (film layer of a few monolayers of approximately 10 to 20 Å) on an n-type bulk silicon. Although there are several types, the most common are palladium silicide, with a cutoff of 3.2 μm; platinum silicide, with a cutoff of 5.7 μm; and iridium silicide, with a cutoff of 10 μm. Additionally, some advanced work is occurring with silicides beyond 10 μm.

The detection mechanism is internal photoemission of a carrier from the thin layer when excited by a photon. "The barrier is an uphill potential at the interface that, in dark and low temperatures, allows few electric carriers to move from the metal into the semiconductor. The photoemission process is common to photomultiplier tubes and is analogous to 'kicking balls over the ridge of a hill'" [Silverman et al., 86]. Some photo-generated carriers migrate through this barrier (or layer) and are then collected by the bulk silicon. One reason for their overall low quantum efficiency is the transport mechanism of the photo-generated carriers. Only a small fraction are emitted directly into the active collection area of the bulk silicon. The others are scattered by the electrode walls, grain boundaries, the silicide lattice, and by carriers near the Fermi level [Shepard, 87].

Quantum efficiencies decrease approximately linearly with wavelength and are about 4 percent for Pt:Si @ 2.5 μm, decreasing to 0.5 percent @ 4 μm. Ir:Si typically has less than 1 percent QE at 5 μm, and it decreases as wavelength increases. Figure 4.9 is a plot (courtesy Hughes Aircraft) of the typical quantum efficiency of a Pt:Si array. The linear line of the QE can be easily calculated using Equation (4.12), if one knows the C_1 and cutoff wavelength. Note that the curve dips away from the linear result using the Fowler equation, Equation (4.12). This is because of a selective "tuning" of the cavity's optical properties to enhance response at a given wavelength. The result is a dip that reaches a minimum at half the wavelength for which the device is tuned. Because of this strong quantum efficiency

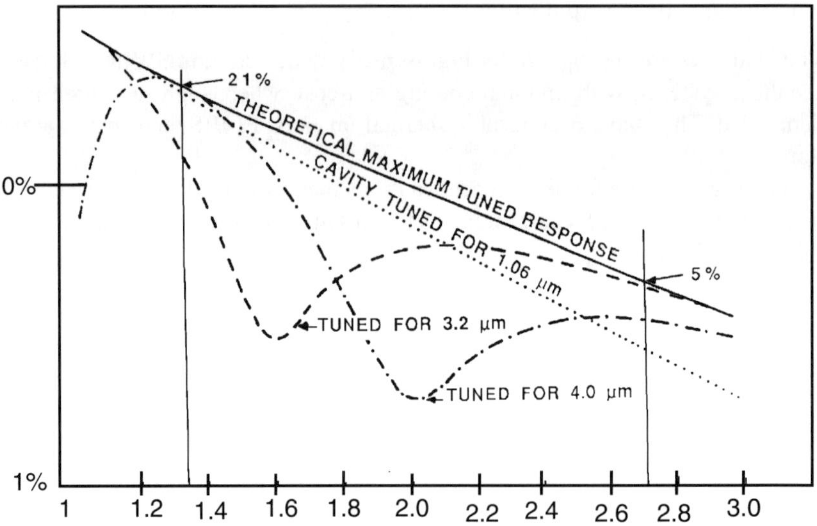

FIGURE 4.9 Quantum Efficiency for a Typical Pt:Si as a Function of Wavelength (courtesy of Hughes Aircraft)

dependence on wavelength, the integral of the curve should be evaluated to determine sensitivity within the user's bandpass.

The target and background spectral distribution must also be considered, given that the FPA will generate more electrons to an input of "X" watts at 3 µm than to an input of "X" watts at 5 microns. If the target is a room-temperature greybody emitter, then the sensitivity dependence of Pt:Si changes in the opposite manner from the blackbody curve, making detection more difficult. One must differentiate the target exitance by wavelength. The user may want to write a spreadsheet program dividing the bandpass into 100 increments ($d\lambda$), calculating the incident photons as a function of λ and the device quantum efficiency, and then multiplying them while performing a numerical integration. The final result will be the number of signal electrons. Division by the number of noise electrons then yields the SNR.

Because of the Schottky's low sensitivity, fill factor is especially important. Some companies have developed detector arrays that employ indium bump bonds to a silicon readout. As with all detector arrays, this allows increased fill factor (80 percent as opposed to 40 percent for monolithic) and separate process optimization of the detector and the mux.

The emission factor, C_1, is proportional to the absorption of light and fill factor, and inversely proportional to the electrode thickness. Increasing electrode thickness from 9 nm to 78 nm resulted in approximately a factor of ten increase in C_1 [Kimata et al., 88]. Spectral variation is caused mostly by conservation of transverse momentum during barrier transit [Shepard, 89]. Figure 4.9 demonstrates the sensitivity as a function of wavelength.

Schottky barrier detectors require colder temperatures than other types of detectors, for the same wavelength. The barrier height determines the dark current, yet the QE at the barrier height is zero. Thus, the useful wavelength cutoff for a Schottky is where a photon has an energy substantially greater than the barrier. This means that a Schottky will always require more cooling than a constant QE device. For example, a 4.5 μm HgCdTe has dark current based on an energy level equal to a 4.5 μm photon, while a 4.5 μm Schottky has a dark current defined by a 5.8 μm energy level.

Since the Schottky barrier can be formed on a CCD, it is possible to provide an architecture whereby it is front illuminated with the incident light striking the silicide layer first. By doing this, the CCD can respond as well as the silicide layer. Such arrays have been fabricated, and they have excellent quantum efficiencies in the visible and UV portions of the spectrum.

4.5.2 Manufacturing Process

Schottky barrier arrays utilize standard silicon production capital equipment and procedures that yield exceptional yields and producibility for these arrays. Platinum is vaporized at 2,000° C and deposited on a silicon integrated circuit a few feet away. The silicon integrated circuit may be a CCD or CID for a monolithic, or a masked-off silicon substrate for the array in the case of a hybrid. The Pt has cooled during its travel of a few feet and forms a metallic electrode containing platinum and silicon. The layer is on the order of 10 Å thick. This electrode is a chemical compound rather than an alloy like HgCdTe [Hughes, 90].

4.5.3 SOA (Size, Sensitivity, Uniformity, Bandpass, Etc.)

Platinum silicide is the largest infrared array available. Tebo reports that several American companies are working on full framing Pt:Si devices. These companies include Sarnoff, Kodak, and Hughes [Tebo, 91]. One can expect that each of these companies is close to reproducing Mitsubishi's 1,040 × 1,040 array and will market similar arrays [Hughes, 92]. Large scanning arrays also have been produced, including one which has 2,048 × 16 elements [Norton, 93]. An example of a typical array has been reported as a 256 × 256 with an 87 percent fill factor, < 10 noise electrons per second (e–/s) dark current, < 60 e– total noise at 60 K, and < 200 out-of-spec unit cells in the center 212 × 212 [Fowler et al., 94]. This unit was developed for astronomical applications.

The sensitivity at a bandpass can be calculated from Equation (4.12). Representative cutoffs are from 5.5 to 5.7 microns, and 0.3 is a typical C_1 value. The total noise is usually 100 to 300 noise electrons per 1/30 s of a frame. NEDT at television frame rates of 0.03 K have been reported [Hughes, 95].

Uniformity is high, approaching <1 percent uncorrected nonuniformity for Pt:Si. Pellegrini reports an uncorrected nonuniformity of 0.5 percent. With correction, the fixed pattern ratio can be dropped to 0.05 percent [Pellegrini, 96]. Single- and double-point correction works well with Pt:Si. Cases have been reported with an average improvement in deviation from the array mean by a factor of 12 for single-point correction alone, yielding total nonuniformities of less than 1 percent [Murguia and Ewing, 97].

The bandpass is a function of the acceptable quantum efficiency as derived from Equation (4.12). Usually, it isn't wise to operate an FPA near the edges of acceptability, or a Schottky barrier within a micron of the cutoff.

Several longer-wavelength Schottky's have been reported that utilized iridium silicide. These require more cooling (usually to around 40 Kelvins) and have a far lower quantum efficiencies. Sarnoff Research Center has produced an Ir:Si with a C_1 of 0.04 eV and a cutoff of 8.6 μm [Shallcross et al., 98]. Also, Tsaur has reported a 128 × 128 Ir:Si array with a cutoff of 9.4 microns, and a 512 × 512 Ir:Si with a 7.3 micron cut and a C_1 of 0.04 [Tsaur et al, 99]. Mixing Ir with Pt to form a Schottky device has met with some success. A device with a cutoff of 9.2 μm was made at Lincoln Labs, with a C_1 about three times larger than the value expected if Ir alone were used [Tsaur et al, 100].

4.5.4 Producibility, Cost, and Power

The Schottkys are highly producible due to their in-process use of silicon IC processing techniques, equipment, and wafers. The easiest large arrays to make are the Schottky barrier diodes made from Pt:Si (SWIR and MWIR), Pd:Si (for SWIR), or Ir:Si (for MWIR/LWIR). Most efforts are now directed toward Pt:Si because it offers the most practical compromise between usable spectral bandpass and sensitivity.

Yields can be as high as 15 percent because silicon wafers use a simpler manufacturing process and less complex handling equipment. Yields of 70 percent or so have been realized for making the Pt:Si on a pre-made CCD. Using silicon wafers allows low-cost production and processing of several arrays simultaneously. It is possible to process 27 arrays in a 488 × 640 configuration on a single four-inch wafer, and hybrid Schottky barriers require as few as five masking steps. The combination of fewer masking steps and several arrays per wafer enhances producibility while reducing cost. The ease of working with silicon is evident from the fact that Mitsubishi makes Pt:Si detectors and silicon memory chips on the same production line [Hughes, 101].

Large arrays are expected to cost about $1,000 each in mass production. They currently run about $40,000 each for small orders. Additionally, a penny a pixel (unit cell) costs have been projected [Hughes, 102].

4.5.5 Appropriate Applications

The key to appropriate utilization of Schottky barriers can be summed up as high background applications with desired high resolution. Due to its low sensitivity and high uniformity, Pt:Si is appropriate only for high background conditions. These devices are also useful for highly cluttered conditions, when sensitivity is not a driver but clutter is. This is because their large format allows high-resolution sensor design. These devices are most acceptable in the SWIR and MWIR due to their acceptable quantum efficiencies.

These focal planes are inappropriate for low background, high sensitivity requirements. They also are inappropriate if cooling to below 80 K is a problem. Concerns with these arrays include exceedingly poor sensitivity, low fill factor, large cooling requirements (50 to 100 K), and lack of a track record in real-world applications.

Another advantage of the Schottky is that, once cooled below a certain level, their sensitivity does not change significantly with operating temperature. This makes uniformity less sensitive to minor temporal or spatial temperature variances. However, obtaining this temperature level is critical, because the output is complete noise until a certain temperature is reached, beyond which the noise drops rapidly. Schottky barrier detectors should not be operated near this temperature, which is often close to that of liquid nitrogen for Pt:Si.

4.5.6 Miscellaneous Related FPA Advancements

Long-wavelength IR detectors have been developed that utilize a Schottky barrier-like photoemission over a heterojunction. These are frequently referred to as pseudobarriers or heterojunction internal photoemissions (HIPs). "The mechanism of detection consists of the generation of holes by absorption of photons in the p+ $Si_{1-x}Ge_x$ photoemitter, followed by internal photoemission of the holes over the potential barrier at the $Si_{1-x}Ge_x$/Si heterojunction. By selecting the Ge content, x, of the photoemitter, one can set the height of the potential barrier" ["$Si_{1-x}Ge_x$/Si-Heterojunction Internal-Photoemission Detectors," 103]. The SiGe layer is similar in behavior to a Schottky barrier, but it is a semiconductor. This results in a relatively narrow spectral response band.

These arrays have manufacturing and detection processes similar to that of Pt:Si and may be implemented as a direct replacement in a system (if thermal noise from the optics allows). MBE and CVD are usually used to deposit the SiGe on multiplexers. The advantages of these devices is their ease of fabrication (similar to that of Pt:Si) and their thermal/mechanical compatibility with silicon, which facilitates connections to multiplexers. Unfortunately, quantum efficiency is typically less than 1 percent (the QE is about 0.8 percent at 10 μm), and dark current is rather high. Their low sensitivity and excessive cooling requirements are current technological challenges.

Large-format arrays of 400 × 400, with a 9.3 μm cutoff operating at 50 K, have been reported [Tsaur et al, 104]. JPL has demonstrated 16 μm cutoffs, and Lincoln Labs has published data on a 322 × 244 SiGeSi HIP FPA with an 18 μm cutoff [Tebo, 105]. Additionally, the Japanese are working on a long-wave monolithic mixture of Pt:$Si_{1-x}Ge_x$ homo- and heterojunction barrier detectors. The barrier height decreases with the x concentration, and arrays have been built with a concentration of 0.2. A strained SiGe lattice layer is formed by MBE [Yamaka, 106].

4.6 INDIUM ANTIMONIDE (InSb)

4.6.1 Detection Mechanism

InSb is an equally sensitive alternative to HgCdTe for MWIR applications with sensitivities from 2 to 5.5 μm. Generally, InSb arrays must be cooled to around 80 K. However, some single-element detectors have been operated effectively at 100 K. Although easier to produce than HgCdTe, InSb is still more difficult to make than Pt:Si. In the seventies, InSb lost a great deal of research effort to HgCdTe development, and in the eighties much commercial interest was lost to Pt:Si. However, the material is finding a niche in the marketplace as a cost-effective alternative to HgCdTe for high-sensitivity MWIR applications that require good, corrected uniformity.

Like HgCdTe, InSb can detect though either the photoconductive semiconductor process of varying the conduction with illumination intensity or the photovoltaic mode where a voltage across the detector is generated from incident light. Since the InSb energy bandgap is 0.22 eV, it can detect only below about 5.6 microns.

Dark current increases rapidly above 77 Kelvin, making the material practically useless for IR detection much above approximately 80 Kelvin for arrays. Since it is easier to control the manufacturing process and produce higher R_0A with a single element, some have been made to operate as high as 120 Kelvin. Work is progressing to increase the useful temperature of InSb to 100 and even 120 Kelvin. This would allow a cryocooler to work less and improve total system reliability. Recently, a technique to suppress thermal noise in semiconductor detectors has been developed that may allow warm operation of InSb. The technique uses electromagnetic carrier depletion (EMCD) to suppress the auger generation-recombination noise [Djuric, 107]. Uncooled InSb has been demonstrated by this technique, and near-BLIP performance is predicted for LWIR (x = 0.18) HgCdTe operating with a TE cooler at 225 to 250 K [Djuric, 108].

On the other hand, InSb can be cooled to liquid helium temperatures to provide the high sensitivity required in astronomy. Lacombe has reported that SAT, a French company, has produced InSb with read noise of 600 electrons rms, total noise of 1,800 electrons rms, crosstalk of <5 percent, and response uniformity of about 10 percent [Lacombe et al., 109].

4.6.2 Manufacturing Process

InSb focal plane architecture can be either monolithic or hybrid. The monolithic InSb has the multiplexer (often a CMOS CID or CCD) and the detector array made from the InSb material. This monolithic version frequently suffers from reduced IR cutoff wavelength and nonlinearities in the CMOS readout. Alternatively, the hybrid method uses indium bump bonding to mate an InSb array with a multiplexer made from another material, which is usually silicon. In this and other hybrid constructions, the thermal expansion rates of the two materials must be considered. These rates must be close enough over the expected operating *and* storage temperatures to avoid exceeding the stress limits of the bond.

The manufacturing process for an InSb array is similar to that of other detector materials. Gaseous diffusion and a surface etch can be used with InSb to define p-type mesas on n substrates for pixel delineation. The p-n junction can also be formed by ion implantation. Photolithography can be employed to define the active areas and provides increased uniformity over other etching processes [Wimmers, 110]. Once the mesa (or planer unit cell) is defined on the InSb wafer, passivation occurs. Metallization of contacts follows, and then indium bumps are grown. The wafer is diced (cut) to yield individual arrays. If needed, it is then hybridized, thinned, and back-surface processed. This entails passivation to produce a low surface recombination velocity and application of antireflective coatings to prevent wanted photons from being reflected before being absorbed. It is important to keep fabrication process temperatures low. This maintains the InSb surface condition to avoid surface leakage. Using anodic oxide and alumina for the surface passivation allows the process temperature to be lower than 100° C after junction formation [Fujisada et al., 111]. This allows a greater selection of materials and reduced thermal stress. Monolithic InSb arrays with a CID built into the InSb substrates have been made with only five photomask steps [Pocock et al, 112].

4.6.3 SOA (Size, Sensitivity, Uniformity, Bandpass, Etc.)

The state of the art in InSb array size is 640 × 480 [Norton, 113] with larger formats (including 1,000 × 1,000) in development. Currently, 256 × 256, 128 × 128, and 64 × 64 configurations are commercially available.

An R_0A of 2×10^6 Ω-cm^2 is commonly achieved at 77 Kelvin and can be increased to 5×10^6 with only 100 mV of reverse bias [Wimmers, 114]. By lowering temperature to 50 Kelvin, R_0As of 10^{11} Ω-cm^2 have been reported [Wimmers and Smith, 115]. Low background D* values exceeding 10^{12} Jones and quantum efficiencies of close to 90 percent are possible, and nonuniformity of only one or two percent is common.

4.6.4 Producibility, Cost, and Power

InSb producibility falls between the difficulties of producing HgCdTe and the simplicity of producing Pt:Si. It is a fragile material that is used for very little other than infrared focal planes. Thus, it needs special production equipment and specially trained personnel. However, it is likely that most current manufacturers could produce hundreds of arrays per month if a contract required them to do so.

The early nineties had arrays commercially available from several vendors for a few tens of thousands of dollars, and InSb is expected to achieve approximately $0.25 per pixel pricing in mass production.

Like HgCdTe, InSb comes in single-element PC or PV arrays. Also like HgCdTe, single PC elements tend to take a few tens of milliwatts per detector, while the pixels in an array require only a few microwatts.

4.6.5 Appropriate Applications

InSb is the detector of choice for high sensitivity MWIR applications where producibility is a concern. InSb is well suited for MWIR thermal imaging, seekers, and narrowband applications such as spectroscopy. It is appropriate for high and low backgrounds, depending on the individual device. It is not suitable if most of the useful energy falls beyond 5 microns.

4.6.6 Strained Lattice InSb

The bandgaps of Equation (4.1) can be made (tricked into being) lower by mechanically straining the lattice. A strained lattice of InSb and $InAs_{1-x}Sb_x$ can be produced by alternating the two materials one atop the other. They have a slightly different lattice constants that are close enough to allow us to grow one material on the other, but different enough to compress the InSb and apply tension to the InAsSb layer. This "strain" can lower the bandgap (E_g) and provide a mechanism for detection out to 12 microns—far beyond the normal 5.8 μm cutoff [Osbourne, 116].

Strained InSb superlattices have shown a detectivity within an order of magnitude of HgCdTe at 7 μm. A PV detection out to 14 μm, with a D* greater than 1×10^{10}, has been observed [Tebo, 117]. Strained lattice InSb FPAs are made as a superlattice (like quantum wells) with MBE or MOCVD laying down the successive layers. Strained lattice InSb should be almost as producible as the normal InSb and cost only slightly more, if produced in large quantities.

4.7 QUANTUM WELLS

4.7.1 Detection Mechanism

Quantum wells (sometimes abbreviated QWIPs, for quantum well infrared photodetectors) are superlattices. Quantum wells work through a quantum mechanical interaction between photons and electrons in a quantum well structure. The detection mechanism can be approximated and system-modeled by the semiconductor photoconductive effect. However, the impedance is so high that the power dissipation is low compared to PC HgCdTe.

A "quantum well" is formed by sandwiching a superlattice with a bandgap between two other superlattices of a different bandgap. The inside superlattice (between the others) has a bandgap (e.g., GaAs) which is less than the bandgap of the other lattices (e.g., AlGaAs). Potential barriers and wells result from these alternating layers of different semiconductor materials, and the movement of carriers is restricted. A slope in potential is produced by an applied bias. There is an optical transition between a quantized bound state in the valance region of the well and an extended upper state that overlaps the conduction band. One "quarky" complication is that the selection rules require the photons to transition the well in an oblique (abnormal) path, so they must be tilted to the normal focal plane. This can be done by placing a diffraction grating at the rear of the detector, physically tilting the FPA, or using a tilted mirror behind each pixel. Quantum efficiency is inherently low, so antireflective coatings and reflectors are required at the base to produce a double path.

The bandpass center and width are tunable at the time of manufacture. The response of the devices is strongly spectrally dependent. Figure 4.10 (courtesy of Martin Marietta) shows two plots of the relative responsivity of specific quantum well devices. Figure 4.10a shows the strongest spectral dependance and is for a single-band quantum well, while 4.10b is a plot representative the relative response for multiple wells on the same FPA. A double quantum well structure allows a wider bandpass, which is more useful for thermal imaging. Advances are occurring whereby multiple wells can achieve a flatter, broader response over 3 to 5 microns in width. However, the strong spectral dependence is clearly evident for either case.

Since the quantum well detector is a fundamentally different detection mechanism from other conventional detectors, it must be modeled differently. A suitable way to model it is to generate a response curve, determine the photons (not watts) as a function of wavelength, and integrate over the response curve. Then compare the results with a measured dark current for the given operating conditions. Frequently, the engineer does not have available the FPA desired to conduct these measurements. In that case, to approximate the response, assume a Gaussian spectral response centered on the center wavelength and a 2 to 4 micron width from half max to half max.

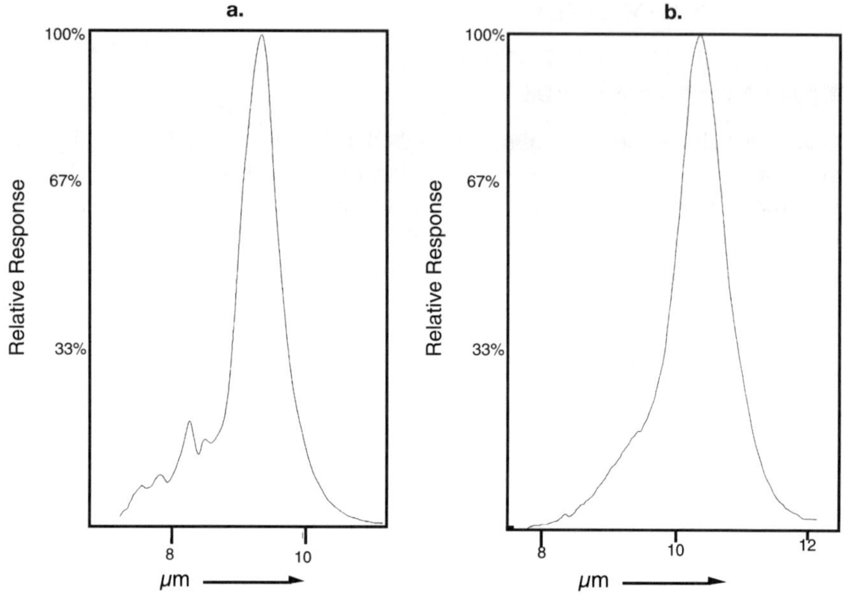

FIGURE 4.10 Spectral Plot of Response for a Representative (a) Quantum Well and (b) Double Quantum Well (courtesy of Martin Marietta)

Some alternative superlattice structures have been proposed by JPL and the California Institute of Technology. One is the hole impeded doping superlattice (HIDS) for detection of radiation in 8 to 17 μm when cooled to near liquid nitrogen temperatures. These are made by MBE using spatially oscillating doping of InAs on Si or GaAs substrates ["Hole-Impeded-Superlattice LWIR Detectors," 118]. Another uses a layer of LaN grown epitaxially on silicon. This could have a bandgap of 0.1eV, corresponding to 12 μm.

4.7.2 Manufacturing Process

The manufacturing process is relatively simple for quantum wells. The capital equipment is identical to that used by the silicon integrated circuit industry, and few masking steps are needed. Mesas can be defined using standard etching techniques, and the arrays are usually hybrid bump-bonded to a silicon multiplexer.

There are at least two challenges in the manufacture of quantum wells. The first is growing a lattice of one material on a lattice of a different material. For quantum well detectors, this is frequently accomplished through MBE (or MOCVD) deposition of individual atoms (of the growing lattice) on a GaAs substrate at an arrival frequency that determines the composition of the alloy. The temperature of the substrate must be correct for the proper adhesion [Rowell, 119].

The second challenge is to maximize the photons that encounter the active area at a diagonal. This can be done by building a grating in the rear of the unit cell. Another method is to form mesas at an angle. The mesas can be defined by standard photolithographic processes to isolate the unit cells. Then they can be grooved by a blade cutting at an angle to form a "V" so that the IR radiation enters the device at a 45 degree cant. The surface roughness of the groove can be made with a 1 μm roughness [Kobayashi et al, 120].

4.7.3 SOA (Size, Sensitivity, Uniformity, Bandpass, Etc.)

Quantum wells have a distinct advantage over all other materials in that they can be made in large formats and long wavelengths, such as 512 × 512 element arrays with center wavelengths beyond 14 microns. In fact, quantum wells developed at AT&T Bell Labs have been reported to approach the sensitivity of HgCdTe in 128 × 128 LWIR arrays. GaInSb/InAs arrays that operate in the LWIR at warmer temperatures than HgCdTe may be suitable for manufacture in 512 × 512 formats ["Hughes Eyes Superlattice Crystals for IR," 121]. However, high sensitivity is hindered by large dark current and quantum efficiencies of less than 30 percent. Dark current can be somewhat reduced by cooling to 40 or 50 Kelvins. Generally, quantum well detectors have sensitivities about one order of magnitude lower than HgCdTe for approximately 12 μm or less cutoffs. D* values of 1×10^{10} Jones and higher at 77 K have been reported by Levine et al. [Kobayashi et al., 122]. Typical LWIR D* values are about 0.5×10^9 cm-Hz$^{1/2}$/W @ 70 K with responsivities of 30,000 V/W. Other reports include a D* of 2.8×10^{10} Jones at 8.3 μm with a 7 V bias and a 128 × 128 QWIP with NEDTs of 10 millikelvins and cutoffs of 9.5 μm [Levine et al., 123].

Quantum well uniformity is exceptionally high compared to most IR detector arrays. As with Schottky barriers, uniformity approaches one percent uncorrected.

Placement of the bandpass center is tunable at the time of manufacture—theoretically, from 2 to 25 microns. The bandpass is not wide—often less than 3 microns. There have been discussions and lab demonstrations of sandwiching two or three quantum well structures on top of each other to double or triple the bandpass width. Straining the lattice can move the band farther into the IR to 12 μm at 77 K [Hughes, 124].

4.7.4 Producibility, Cost, and Power

Quantum wells have not been manufactured in large quantities. Because they employ silicon and GaAs manufacturing procedures, it is expected that they will be much more producible than HgCdTe, especially for long-wavelength, large-format arrays. This is due to GaAs process maturity, lack of passivation requirements, simple pixel delineation, and lower sensitivities to nonuniformities in the process [Levine et al., 125]. Additionally, there is a large personnel base to draw from, as compared to the base for other IR materials. GaAs can be made on wafers as large as 3 or 4 inches, which allows a large number of arrays per wafer, and the wafers are inexpensive at approximately $200 each. The dominant cost for large-scale arrays in production is likely to be the multiplexer. As a result, quantum wells hold the promise of achieving the elusive $1 per unit cell goal for LWIR devices.

Quantum wells operate in a photoconductive mode, so their power dissipation is higher than that of photovoltaic devices. However, the materials are more efficient and the bias is lower than in HgCdTe, with less than a milliwatt of dissipation per channel achievable. However, quantum wells must operate at colder temperatures than HgCdTe.

Commercially available quantum well arrays will be making their debut soon—perhaps before the publication of this book. Their strengths will be in applications demanding high-resolution, low-sensitivity LWIR imaging. Their cost is expected to be far lower than HgCdTe and may be only $5 thousand to $20 thousand higher than similar InSb or Pt:Si arrays.

4.7.5 Appropriate Applications

The advantage of quantum well detectors is in their ease of manufacture for production of long-wavelength, large-format devices. Thus, quantum wells are most useful for long-wavelength applications where dense arrays are required but sensitivity is not at a premium. Another prerequisite for using quantum wells is a lack of concern for cold, powerful cryocoolers. A long-wavelength sensor viewing a high radiance and highly cluttered background is an example of an excellent application. Quantum wells are still developmental and not suited for low-risk, off-the shelf applications.

4.8 DOPED SILICON (Si:X)

4.8.1 Detection Mechanism

Doped silicon can detect though an intrinsic or extrinsic process. The intrinsic detection process occurs when photons excite carrier pairs across the bandgap of a semiconductor, and it is the mechanism that generates the visible response of pure silicon. The extrinsic process occurs when photons excite the dopant impurities to produce free carriers across a bandgap in a semiconductor, and it is responsible for the dominant infrared detection mechanisms of doped silicon.

Doped silicon arrays combine high sensitivity and long-wavelength detection. Their largest drawback is that excessive cooling is required. System designers have been reluctant to pay the cryogenic price of using doped silicon. There are great system advantages of operating arrays at a "balmy" 80 Kelvin or so. These include low-cost and highly available cryogens, and far more efficient and longer life cryocoolers than those applicable to colder temperatures. Therefore, the development of doped silicon has been stymied by required cooling to far below the comfortable 65 to 80 Kelvin range.

Extrinsic detection is a bulk effect rather than a surface effect. The resultant bandgap of a silicon lattice can be altered by placing (doping) impurities into the bulk material to reduce the bandgap energy. The impurities are either an "n" or "p" carrier type. The dopants can detect via a n-type donor mechanism (whereby electrons transition the bandgap) or a p-type acceptor (where holes transition the bandgap). Extrinsic detection is a photoconductive process. Common doping materials and their associated cutoffs are listed in Table 4.4.

Extrinsic doped silicon FPAs provide high sensitivity with relatively easy production for small arrays. Arrays of 64×64 are commonly made, with responses to approximately 24 μm. Their drawbacks include extreme cooling requirements (< 20 K) and hysteresis to high backgrounds.

A curious feature of doped silicon infrared detectors is that an unintentional doping may result from the manufacturing process. A given process to dope silicon with indium may include a step in which some boron is added to the silicon. This may act like a dopant and produce a slight response at wavelengths far longer than the indium cutoff. If not blocked by an optical filter, this can result in radiometric errors and calibration inaccuracies.

As previously mentioned, these devices must operate at low temperatures because their carrier densities are related to FPA temperature and are a function of exp(1/KT). For example, for Si:In, there can be as few as 1,000 to 10,000 carriers per cubic centimeter below approximately 20 K and 10^{13} at 70 K. Hence, these FPAs must operate at the excoriatingly low temperatures of 20 K or less rather

TABLE 4.4 Common Dopants for Silicon

Dopant	*Donor or Acceptor*	*Typical Cutoff (μm)*	*Usable Temperature (K)*
Zinc	acceptor	3	120
Indium	acceptor	7.4	50
Gallium	acceptor	16	25
Aluminum	acceptor	18	20
Bismuth	donor	18	15
Arsenic	donor	23	15
Antimony	donor	25	12
Boron	acceptor	27	10
Phosphor	donor	30	5

than the more desirable 70 K. D* values above 1×10^{13} can be achieved below 25 K for low backgrounds. As usual, there is a BLIP D* and a detector noise limited D* where thermal carriers dominate.

Dielectric relaxation can occur in extrinsic silicon. This causes the curious effect of high responsivities at low frequencies and low responsivities at high frequencies. Another peculiar feature is that a hysteresis can occur in high background conditions.

Silicon detectors can be made in a blocked impurity band (BIB) structure, which is inherently radiation hard due to its thin active region [Leotin and Meny, 126]. The thin active layer is only a 10 to 20 μm epitaxial detection layer grown on a highly conductive substrate. A second, blocking layer is grown over the first to prevent "hopping" currents and to isolate the detection layer [Jackson, 127]. There is one contact on each layer. The sensitivity is also enhanced by reduction of crosstalk and dark current by virtue of blocking the hopping conduction of the impurities.

4.8.2 Manufacturing Process

The manufacturing process is easier than HgCdTe due to the basic silicon material employed. Yields are usually high, on the order of 10 percent or less. The largest complication in the manufacturing process is designing and building electronic circuits for operation at such low temperatures.

These focal planes can be monolithic or hybrids. Monolithics can be made by growing an epitaxial layer of opposite polarity on a silicon CCD. Hybrids allow the separate process optimization of the detector from the multiplexer and are usually hybridized via indium bump bonds.

BIB detectors can be made via chemical vapor deposition (CVD) epitaxial growth. An active layer is grown and doped, followed by a pure blocking layer growth. Finally, a flash annealed n+ layer is applied to provide a transparent contact for incoming infrared [Leotin and Meny, 128].

4.8.3 SOA (Size, Sensitivity, Uniformity, Bandpass, Etc.)

Sizes are typically small for the doped silicon detectors—usually 64 × 64 or 128 × 128. However, to compensate for this, buttability is usually more common than with other detector materials. This includes four-sided buttable devices. A series of 32 × 32 hybrids of various dopings and buttability was reported in the early eighties [Pommerrenig, 129]. Additionally, Aerojet has made a PC 32 × 32 CID Si:Bi operating at 12 K, and Rockwell makes a PC Si:In that operates at 45 K [Dereniak and Crowe, 130]. SBRC makes 32 × 32 and 64 × 64 astronomical Si:Ga arrays with cutoffs from 14 to 17 μm [Monin et al., 131]. Impurity blocked conduction (IBC) 10 × 50 and 192 × 6 arrays have been reported with nonuniformities of less than 1.5 percent and D* of 6.7×10^{12} Jones [Norton, 132]. Extrinsic doped silicon arrays for astronomy have been made in 58×62 formats with Ga, Sb, and As as the dopants [Norton, 133]. Additionally, a bismuth doped silicon with a 29 μm cutoff and a D* of 1×10^{14} was recently reported from Russia [Khryapov et al., 134].

Sensitivity for these arrays is high, and they are almost always BLIP when viewing anything but deep space. In general, the D* values of Si:As and Si:Ga lie between 10^{13} and 10^{14} at 10 and 20 Kelvins, respectively [Botts, 135]. Quantum efficiencies can be expected to lie between 10 and 50 percent [Norton, 136].

4.8.4 Producibility, Cost, and Power

Like most IR arrays, those made from doped silicon suffer from producibility problems. The doping process is an exotic technology, and in-process testing must be at extreme cryogenic temperatures. Although troublesome, several companies have produced tens of usable arrays in a timely manner. The electronics associated with a doped silicon array are not easy to make. There low operating temperatures require special circuits and techniques specifically designed for cryogenic operation. Also, thermal stress is a more serious concern because they need to be cooled more than many other array materials.

Because of those producibility concerns, along with the lack of production rate programs, doped silicon is an expensive array technology. In mass production, it the cost would be reduced, but not as dramatically as with Schottky barriers or quantum wells.

4.8.5 Appropriate Applications

Doped silicon is appropriate for applications where cooling is not a major concern, such as ground-based astronomy or lab work. It is most appropriate when high sensitivity is required and there are low backgrounds. Its advantage over quantum wells is its high sensitivity. There is also a sensitivity advantage over HgCdTe at wavelengths greater than about 7 microns. Doped silicon arrays provide a cost and producibility advantage over HgCdTe at wavelengths greater than roughly 12 microns.

4.9 BOLOMETRIC ARRAYS

4.9.1 Detection Mechanism

The basic bolometric detection mechanism is a change in resistance with heating of the detector element. Bolometric arrays are making a comeback with the introduction of novel techniques using superconductivity and micromachining. A bolometer heats up as it absorbs power (watts from photons), producing a change in a measurable electrical property (usually resistance). Bolometers provide a relatively flat response to submillimeter wavelengths. Historic drawbacks include small arrays, large center-to-center spacing, slow response, low sensitivity, and (often) great cooling demands (possibly < 4 K).

Through recent advancements, micromachining can provide isolation of the bolometric unit cell in an array and provide uncooled LWIR response rivaling conventional cooled detectors. In 1987, the Army's Center for Night Vision funded Honeywell to develop uncooled bolometric arrays [Lytle, 137].

4.9.2 Manufacturing Process

Bolometer arrays can be made by defining unit cells in a bolometric material, providing electrical connections and a readout. As in the previously discussed detectors, this can either be a hybrid or monolithic structure.

A micromachining process is used to provide spatial isolation of the active area of the unit cell from the surrounding bulk material. This prevents conduction heat loses. These bolometric cells must be operated in a vacuum to prevent convective heat losses. Although used in a vacuum, they don't require cooling.

4.9.3 SOA (Size, Sensitivity, Uniformity, Bandpass, Etc.)

Thin-film resistance bolometers have been made with a D* in a vacuum of 4.5×10^8 Jones [Liddiard, 138]. Honeywell has been working on a series of room-temperature bolometer arrays employing micromachining isolation of the active area of the unit cell. The company has produced a 245×328 with 50 μm unit cell sizes, exhibiting a 0.1 K NEDT with f/1 optics and a 30 Hz update [Miller, 139].

4.9.4 Producibility, Cost, and Power

The producibility of these arrays is questionable because they are currently developmental, with no large-scale production occurring. There seem to be no material or process obstacles to high production rates. The basic materials are not rare, cryogenic circuits do not need to be made for the room temperature bolometers and, when utilized, superconductivity occurs at temperatures where IR detector designers are comfortable (approximately 70 K). Silicon micromachining has high yields and appears to present no producibility concerns. It is also heavily funded for commercial noninfrared applications, which will rapidly add to its maturing producibility. The cost may easily be less than a dollar per unit cell in large-scale production.

4.9.5 Appropriate Applications

Room temperature bolometric arrays are appropriate for high background (or high clutter) low sensitivity cameras where cooling can not be tolerated because of weight, power, or reliability concerns. These are especially useful for detection of radiation longer than 12 microns where cooling is limited. Applications include space-based astronomy, personal night vision devices, driving aids, missile warning sensors, surveillance cameras, night skiing [Cota, 140], and thermal sights. These devices can form the basis for a low-cost, lightweight industrial surveillance camera. Their cost advantage over traditional visible systems stems from their independence from light. Billions of dollars per year are spent on industrial lighting for nighttime security. Future costs in terms of capital, pollution, and power plant construction can be saved by merely switching to unilluminated passive IR night vision for night security systems.

4.10 SUPERCONDUCTORS

4.10.1 Detection Mechanism

With the advent of high-temperature superconductors, a vision was born to use their sharp resistance change to bolometrically detect infrared radiation. These devices are a type of bolometer. When a superconducting material transitions into superconductivity, it does so with a steep gradient near its critical temperature (T_c). The change in resistance can be several orders of magnitude within a couple of Kelvins. Superconductive infrared detectors rely on this large gradient in resistivity caused by a small change in temperature. Common materials considered for superconducting Infrared arrays are $DyBa_2Cu_3O_7$ and YBaCuO.

If a superconducting material could be made to absorb infrared radiation and change temperature, the resulting change in resistance would easily be sensed by electronics. A concern in their application is maintaining the temperature, especially across an array, since they experience a measurable output for a change of only a few milliKelvins. The change in resistance of YBaCuO is usually less than an order of magnitude, making resistance change measurement simple and easing thermal control.

4.10.2 Manufacturing Process

As with any detector, the detecting material must be deposited in a pure and controlled form on a substrate. Then, electrical connections are made and a readout scheme implemented. As with any bolometer, the detector should have a tiny thermal mass and must be isolated from surrounding heat loads.

Epitaxial superconductor layered films can be formed by MBE, VPE, or multilayer electron beam. Valco describes a $YBa_2Cu_3O_{7-8}$ detector of 15 layers on a $SrToO_3$ substrate [Valco et al., 141]. It is formed by depositing molecules of Cu, Y, and BaF_2, in that order. They are annealed at 900° C for 15 minutes then baked at 450° C for 6 hours. The films were 0.5 μm thick with a T_c of 85 K. Photolithography was used to metalize electrical connections. Valco also has described a method of laser ablation where a target of YBa_2CuO is ablated by a laser beam of 1.5 j/cm^2/pulse onto a 630° C substrate in a chamber of 170 millitorr of oxygen [Valco, 142].

YBaCuO must be deposited at a high temperature (approximately 700° C). This tends to damage readout circuitry and limits the materials that are available for substrates and multiplexers.

4.10.3 SOA (Size, Sensitivity, Uniformity, Bandpass, Etc.)

As with any bolometric process, these factors have almost constant sensitivity to the input power over a large bandpass.

Work has been ongoing on large arrays of YBaCuO. An effective quantum efficiency of 15 percent, a responsivity of 800 V/W, and a time constant of 1 ms have been reported [Johnson, 143]. Using YBaCuO and TiBaCaCuO, SBRC has built several thin-film superconducting detectors that exhibited "nonequilibrium superconducting" properties similar to photoconductive detection ["World News Breaks," 144]. These films were operated at 10 to 30 Kelvin and enabled a projected D* of 1×10^{14} [Tebo, 145]. Moreover, Kruse has forecast producible large superconducting arrays using the microbolometer approach [Kruse, 146].

4.10.4 Producibility, Cost, and Power

High-temperature superconductors do not suffer from cryogenic electronic designs any more than standard IR detectors, as they operate close to the same (LN_2) temperature. Their materials are common, and process control tolerances are not as critical as with HgCdTe. Therefore, it is expected that high-temperature superconductors will cost less than HgCdTe in production. Such cost projections are speculative, however, due to the current immaturity of development.

4.10.5 Appropriate Applications

These arrays are highly repeatable and are being employed by NIST for standards. High-sensitivity applications at wavelengths greater than 12 μm are appropriate when the cooling can be supported (e.g., astronomical applications). At present, these arrays are highly developmental and should be used only in developmental systems.

4.11 PYROELECTRICS

4.11.1 Detection Mechanism

Pyroelectrics produce a change in magnetic effects when energy (watts) is absorbed and an incremental temperature change occurs. Like bolometers, they have reasonably flat response to submillimeter wavelengths.

In a pyroelectric material, a change in temperature produces a current. This effect can be employed as an infrared detector by allowing a temperature change due to the absorption of electromagnetic radiation. Pyroelectrics are not responsive to a steady light input, since the pyroelectric effect is produced by change. They may be chopped (intermittently illuminated by a physical shutter) to produce an changing (ac) signal from a steady-state scene. To ensure that a temperature change can be generated by the incident radiation, care must be taken with thermal loading the pyroelectric device. Thermally mounting the detector to a heat sink would eliminate any useful output.

The bandpass is dependent on the absorption of power. Therefore, its response as a function of wavelength, is relatively flat, and can extend from the UV to 50 microns or so. Typical materials do not absorb lower than 5 microns. For these applications, an absorbent coating may be applied to allow their use.

4.11.2 Manufacturing Process

Pyroelectric manufacturing depends on the material, manufacturer, and application. Common pyroelectric detector materials are triglycine sulfate (TGS), lead zirconate titanate (PZT), $PbToO_3$, $LiTaO_3$, and $LiNbO_3$. Usually, a boule of the material is grown, and single crystals are formed, cut, and mated to a readout. Mesas have been used to thermally isolate arrays and provide conduction paths to a readout [Carlson et al., 147]. However, TGS is usually produced as a continuous thin film laid down by evaporation or RF sputtering. The unit cell is defined by the electrode that contacts the TGS film.

4.11.3 SOA (Size, Sensitivity, Uniformity, Bandpass, Etc.)

In general, pyroelectric arrays have D* values of $< 10^8$ at room temperature, with slight cooling to –20 or –40° C providing a slight noise reduction. Commercial pyroelectric devices seem to be limited in size to 64×64 arrays. Published examples include:

- A 32×32 TGS was made with a $\pm$ 2 percent uniformity and a D* of 1×10^8 [Carlson et al., 148].

- A 64-element pyroelectric arrays has been produced with 20 micron unit cells [Pankratov, 149].
- A linear pyroelectric array has been shown with a ± 7 percent uniformity and a responsivity of 4,000 V/W [Schopf et al., 150].
- EDO/Barnes has produced arrays made of $LiTaO_3$ with D* values to 1×10^8. Slight cooling to –20 or –40° C provides noise reduction [EDO/Barnes, 151].
- 100 × 100 arrays for 8–14 mm detection have been made in the U.K.

In 1987, the Army's Center for Night Vision funded Texas Instruments to develop uncooled pyroelectric arrays [Lytle, 152]. TI developed a ferroelectric uncooled detector array on 1.9 mil centers employing a polymide bump for thermal isolation. It is temperature stabilized at 23° C and must be chopped. The detector was integrated into a camera that weighs about 2 kg and draws only 5 W of power [Miller, 153].

4.11.4 Producibility, Cost, and Power

Producibility is questionable for large arrays or for high-sensitivity specifications. Small, low-sensitivity arrays can be easily produced and purchased for a few thousand dollars. The resulting arrays have low power consumption (e.g., a few milliwatts for a 32 × 32).

4.11.5 Appropriate Applications

Pyroelectrics are most appropriate in applications that require uncooled broad spectral response without much sensitivity. They are useful for laser diagnostics, thermography of high temperatures or at very close ranges, and detection of very long wavelengths where sensitivity is not a concern. They have great promise for inexpensive LWIR sensors.

References

1. J. Vincent. 1990. *Fundamentals of Infrared Detector Operation and Testing.* New York: John Wiley & Sons, 39.
2. R. Leftwich. July 1988. "Multi-Detector Thermal Imagers." *Proc. SPIE* 510:69–80.
3. S. Botts. July 1988. "New Horizons For Focal Plane Arrays." *Photonics Spectra,* 125–7.
4. J. Wimmers. 1988. "Indium Antimonide Detector Technology Development at Cincinnati Electronics Corporation." *Proc. SPIE* 930:125–138.
5. Ibid.
6. Ibid.
7. J. Monin et al. 1990. "Applications of Infrared Bidimensional Devices in Astronomy." *Proc. SPIE* 1341:202–213.
8. Ibid.

9. J. Chatard. 1990. "SOFRADIR IR Focal Plane Array Production." *Proc. SPIE* 1341: 316–26.
10. J. Kreider et al. 1991. "Multiplexed Mid Wavelength IR Long, Linear Photoconductive Focal Plane Arrays." *Proc. SPIE* 1488:376–88.
11. R. Balcerak et al. 1987. "Evolution of a New Semiconductor Product: Mercury Cadmium Telluride Focal Plane Arrays." *Optical Engineering* 26:191–200.
12. J. Haystead. April 1991. "Thermal Imaging Technology Has Versatile and Bright Future, *Defense Electronics,* 48–52.
13. J. Rotolante. November 1991. "Changes In The IR-Detector Market." *Laser Focus,* 38–42.
14. Information courtesy Hughes SBRC, 1993.
15. S. Whicker. 1992. "New Technologies for FPA Dewars." *Proc. SPIE* 1683:102–112.
16. J. Chatard. 1990. "SOFRADIR IR Focal Plane Array Production." *Proc. SPIE* 1341: 316–326.
17. J. Rotolante. November 1991. "Changes In The IR-Detector Market." *Laser Focus,* 38–42.
18. July 1989. "Detector Mini-Magazine." *Photonics Spectra,* 100.
19. J. Rotolante. November 1991. "Changes In The IR-Detector Market." *Laser Focus,* 38–42.
20. I. Spiro and M. Schlessinger. 1989. *Infrared Technology Fundamentals.* New York: Marcel Dekker.
21. R. Schoolar. 1991. "Photconductivity in Photodiodes." *Infrared Physics* 31, 5:467–474.
22. R. Balcerak et al. 1987. "Evolution of a New Semiconductor Product: Mercury Cadmium Telluride Focal Plane Arrays." *Optical Engineering* 26:191–200.
23. M. Nelson, J. Johnson, and T. Lomheim. 1991. "General Noise Process in Hybrid Infrared Focal Plane Arrays." *Optical Engineering* 30:1682–1700.
24. J. Piotrowski, W. Galus, and M. Grudzien. 1991. "Near Room Temperature IR Photo Detectors." *Infrared Physics* 31, 1:1–48.
25. V. Gopal and V. Dhar. November 1992. "Control and Design of Nonuniformities in a Hybrid Focal Plane Array." *Optical Engineering,* 2330–2335.
26. J. Mooney et al. November 1989. "Responsivity Nonuniformity Limited Performance of Infrared Staring Cameras. *Optical Engineering* 28:1151–1161.
27. D. Scribner et al. 1987. "Measurement, Characterization and Modeling of Noise in Staring Infrared Focal Plane Arrays." *Proc. SPIE* 782:147–159.
28. M. Nelson, J. Johnson, and T. Lomheim. 1991. "General Noise Process in Hybrid Infrared Focal Plane Arrays." *Optical Engineering* 30:1682–1700.
29. Ibid.
30. D. Scribner et al. 1987. "Measurement, Characterization and Modeling of Noise in Staring Infrared Focal Plane Arrays." *Proc. SPIE* 782:147–159.
31. Hansen, Schmidt, and Cassleman. 1982. "Energy Gap vs. Alloy Composition and Temperature in $Hg_{1-X}Cd_XTe$." *Journal of Applied Physics* 53:7099–7101.
32. J. Chu, Z. Mi and D. Tang. 1991. "Intrinsic Absorption Spectroscopy and Related Physical Quantities of Narrow-Gap Semiconductors HgCdTe." *Infrared Physics* 32: 195–211.
33. A. Rogalski. 1988. "Analysis of The R_0A Product in n+ p– HgCdTe Photodiodes." *Infrared Physics* 28:139–153.

34. K. Vural. March 1987. "Mercury Cadmium Telluride Short and Medium Wavelength Infrared Staring Focal Plane Arrays." *Optical Engineering* 26:201–208.
35. R. Balcerak et al. 1987. "Evolution of a New Semiconductor Product: Mercury Cadmium Telluride Focal Plane Arrays." *Optical Engineering* 26:191–200.
36. R. Bailey et al. May 1991. "256 × 256 Hybrid HgCdTe Infrared Focal Plane Arrays." *IEEE Transactions on Electron Devices* 38, 5:1104–1109.
37. Ibid.
38. Ibid.
39. R. Balcerak et al. 1987. "Evolution of a New Semiconductor Product: Mercury Cadmium Telluride Focal Plane Arrays." *Optical Engineering* 26:191–200.
40. Ibid.
41. Ibid.
42. B. Joyce and C. Foxon. May 1987. "Molecular Beam Epitaxy of Multilayer Structures With GaAs and $Al_xGa_{1-x}As_x$." *Philips Tech Review* 43, 5:143–53.
43. R. Balcerak et al. 1987. "Evolution of a New Semiconductor Product: Mercury Cadmium Telluride Focal Plane Arrays." *Optical Engineering* 26:191–200.
44. K. Vural. March 1987. "Mercury Cadmium Telluride Short and Medium Wavelength Infrared Staring Focal Plane Arrays." *Optical Engineering* 26:201–208.
45. P. Norton. November 1991. "Infrared Imaging Sensors." *Optical Engineering* 30: 1649–1662.
46. April 1990. "HgCdTe Infrared Arrays Grow on Si-Based Substrates." *Laser Focus World,* 11, 13.
47. J. Nordwall. June 29, 1992. "Hughes Manufacturing Strides to Allow Better Performance in Low Cost Systems." *Aviation Week and Space Technology,* 53–54.
48. C. Adams. May 1990. "IR Focal Planes Efforts Exceeds DARPA Goals." *Military & Aerospace Electronics,* 3.
49. P. Norton. November 1991. "Infrared Imaging Sensors." *Optical Engineering* 30: 1649–1662.
50. R. Balcerak et al. 1987. "Evolution of a New Semiconductor Product: Mercury Cadmium Telluride Focal Plane Arrays." *Optical Engineering* 26:191–200.
51. J. Haystead, J. April 1991. "Thermal Imaging Technology Has Versatile and Bright Future." *Defense Electronics,* 48–52.
52. J. Rotolante. November 1991. "Changes In The IR-Detector Market." *Laser Focus,* 38–42.
53. C. Adams. May 1990. "IR Focal Planes Efforts Exceeds DARPA Goals." *Military & Aerospace Electronics,* 3.
54. A. Tebo. March 1990. "Arrays and Superlattices Expand Detector Options." *Laser Focus World,* 125–140.
55. M. Reine, A. Sood, and T. Treadwell. 1981. "Photovoltaic Infrared Detectors." In *Semiconductors and Semimetals,* ch. 18. New York: Academic Press, 202.
56. A. Tebo. March 1990. "Arrays and Superlattices Expand Detector Options." *Laser Focus World,* 125–140.
57. A. Davis. 1989. "UK Thermal Imaging Common Modules CLASS II." *Proc. SPIE* 1157:176–84.
58. S. Braim, A. Ford, and M. Thomas. "System Implementation of a Serial Array of Sprite Infrared Detectors." *Infrared Physics* 29, 5:907–14.

59. D. Hughes. December 12, 1988. "Honeywell Delivers IR Detector for Use in SDI Technology Tests." *Aviation Week & Space Technology,*117–121.
60. September 21, 1992. "GAO: Cost of Javelin Focal Plane Array Component Has Been Underestimated." *Inside The Pentagon's Inside the Army,* 3–4.
61. J. Chatard. 1990. "SOFRADIR IR Focal Plane Array Production." *Proc. SPIE* 1341: 316–26.
62. Ibid.
63. J. Chatard and D. Zenatti. 1992. "Analysis of Criteria Selection For IRFPA Detector Production." *Proc. SPIE* 1683:81–91.
64. Hoshino et al. "Fabrication Procedures of Photovoltaic Lead-Chalcogenide-on-Silicon Infrared Sensor Arrays For Thermal Imaging." *Microelectronic Engineering* 15: 293–296.
65. Hoshino et al. 1991. "Monolithic $Pb_{1-x}Sn_xSe$ Infrared Sensor Arrays on Si Prepared by Low Temperature Processes." *Infrared Physics,* 169–175.
66. P. Norton. November 1991. "Infrared Imaging Sensors." *Optical Engineering* 30: 1649–1662.
67. Hoshino et al. 1991. "Monolithic $Pb_{1-x}Sn_xSe$ Infrared Sensor Arrays on Si Prepared by Low Temperature Processes." *Infrared Physics,* 169–175.
68. Tetyorkin et al. 1990. "Carrier Transport Mechanisms and Photoelectrical Properties of PbSnTe/PbTeSe Heterojunctions." *Infrared Physics* 30, 6:499–504.
69. B. Halford, Y. Guan, and M. Tacke. 1991. "$Pb_{1-x}Eu_xSe$ Photodiodes For The 3–5 μm Ranges." *Infrared Physics* 3, 3:255–263.
70. A. Joshi. et al. 1992. "Popcorn Noise in Linear $In_{0.53}Ga_{0.47}As$ Detector Arrays." *Proc. SPIE* 1683:200–207.
71. V. Ban et al. 1989. "Room Temperature Detectors For 800–2600 nm Based on InGaAsP Alloys." *Proc. SPIE* 1106.
72. G. Olsen et al. 1989. "Room Temperature InGaAs Arrays for 2.5 μm." *Proc. SPIE* 1106.
73. G. Olsen. 1988. "Multiplexed 256 Element InGaAs Arrays for 0.8 to 1.7 μm Room Temperature Operation." *Proc. SPIE* 972:279–285.
74. A. Tebo. March 1990. "Arrays and Superlattices Expand Detector Options." *Laser Focus World,* 125–140.
75. V. Ban et al. 1989. "Room Temperature Detectors For 800–2600 nm Based on InGaAsP Alloys." *Proc. SPIE* 1106.
76. G. Olsen et al. 1989. "Room Temperature InGaAs Arrays for 2.5 μm." *Proc. SPIE* 1106.
77. J. Vincent. 1990. *Fundamentals of Infrared Detector Operation and Testing.* New York: John Wiley & Sons, 39.
78. Information courtesy EDO/Barnes, 1993.
79. A. Tebo. March 1990. "Arrays and Superlattices Expand Detector Options." *Laser Focus World,* 125–140.
80. Resso and Harris. April 1990. "The Hunt For Cost Effective IR Transducers." *Photonics Spectra,* 114.
81. P. Norton. November 1991. "Infrared Imaging Sensors." *Optical Engineering* 30: 1649–1662.
82. V. Khryapov et al. April 1992. "Optical Sensors." *Optical Engineering* 31:678–684.
83. Information courtesy Litton, 1992.

84. J. Kreider et al. 1991. "Multiplexed Mid Wavelength IR Long, Linear Photoconductive Focal Plane Arrays." *Proc. SPIE* 1488:376–388.
85. J. Vincent. 1990. *Fundamentals of Infrared Detector Operation and Testing.* New York: John Wiley & Sons.
86. J. Silverman, J. Mooney, and F. Shepard. March 1992. "Infrared Video Cameras." *Scientific American,* 78–83.
87. F. Shepard. 1988. "Silicide Infrared Staring Sensors." *Proc. SPIE* 930:2–7.
88. M. Kimata et al. March 1987. "256 × 256 Element Platinum Silicide Schottky-Barrier Infrared Charge Coupled Device Image Sensor." *Optical Engineering* 26:209–215.
89. F. Shepard. 1988. "Silicide Infrared Staring Sensors." *Proc. SPIE* 930:2–7.
90. D. Hughes. March 27, 1989. "Platinum Silicide Detectors Incorporated Into New Generation of Missile Seekers." *Aviation Week & Space Technology,* 51–62.
91. Tebo, A. May 1992. "Infrared Imaging: Detector Arrays." *OE Reports,* 1, 13.
92. D. Hughes. March 27, 1989. "Platinum Silicide Detectors Incorporated Into New Generation of Missile Seekers." *Aviation Week & Space Technology,* 51–62.
93. P. Norton. November 1991. "Infrared Imaging Sensors." *Optical Engineering* 30: 1649–1662.
94. A. Fowler et al. 1990. "A 256 by 256 Hybrid Array For Astronomy Applications." *Proc. SPIE* 1341:52–55.
95. D. Hughes. March 27, 1989. "Platinum Silicide Detectors Incorporated Into New Generation of Missile Seekers." *Aviation Week & Space Technology,* 57–62.
96. P. Pellegrini. December 1989. "Range Calculations For Staring Schottky Barrier Sensors." *Optical Engineering* 28:1289–1292.
97. J. Murguia and W. Ewing. 1987. "Statistical Characterization of a Large Pt:Si Focal Plane Array." *Proc. SPIE* 782:121–8.
98. F. Shallcross et al. 1992. "Development of a 640 by 480 LWIR Focal Plane Arrays. *Proc. SPIE* 1683:191–9.
99. B. Tsaur, M. McNutt, R. Bredthauer, and R. Mattson. August 1989. "128 × 128 Unit Cell Ir:Si Schottky Barrier Focal Plane Arrays For Long Wavelength Infrared Imaging." *IEEE Electron Device Letters* 10, 8:361–363.
100. B. Tsaur, M. Weeks, and P. Pellegrini. February 1988. "Pt-Ir Silicide Schottky Barrier IR Detectors." *IEEE Electron Device Letters* 9, 2:100–103.
101. D. Hughes. March 27, 1989. "Platinum Silicide Detectors Incorporated Into New Generation of Missile Seekers." *Aviation Week & Space Technology,* 57–62.
102. Ibid.
103. November 1992. "$Si_{1-x}Ge_x$/Si-Heterojunction Internal-Photoemission Detectors." *NASA Tech Briefs,* 44–48.
104. B. Tsaur, C. Chen, and S. Marino. June 1991. "Long wavelength Ge_xSi_{1-x}/Si Heterojunction Infrared Detectors and 400 by 400-Element Imager Arrays." *IEEE Electron Device Letters* 12, 6:293–296.
105. A. Tebo. May 1992. "Infrared Imaging: Detector Arrays." *OE Reports,* 1, 13.
106. E. Yamaka. 1989. "Recent Infrared Study in Japan." *Proc. SPIE* 1157:286–288.
107. Z. Djuric and J. Piotrowski. September 1992. "Infrared Photodetector With Electromagnetic Carrier Depletion." *Optical Engineering* 31:1955–1960.
108. Ibid.
109. F. Lacombe et al. 1990. "Advances in IR Technology at Paris Observatory." *Proc. SPIE* 1341:187–191.

110. J. Wimmers. 1988. "Indium Antimonide Detector Technology Development at Cincinnati Electronics Corporation." *Proc. SPIE* 930:125–138.
111. H. Fujisada, M. Nakayama, and A Tanaka. 1990. "Compact 128 InSb Focal Plane For Thermal Imaging." *Proc. SPIE* 1341:80–90.
112. D. Pocock et al. 1981. "Monolithic InSb Charge Injection Device Technology." *Proc. SPIE* 267:31–38.
113. P. Norton. November 1991. "Infrared Imaging Sensors." *Optical Engineering* 30: 1649–1662.
114. J. Wimmers. 1988. "Indium Antimonide Detector Technology Development at Cincinnati Electronics Corporation." *Proc. SPIE* 930:125–138.
115. J. Wimmers and F. Smith. 1983. "Characteristics of InSb Photovoltaic Detectors at 77 K and Below." *Proc. SPIE* 364:120–125.
116. Osbourne. "InAsSb Strained Lattice Superlattices For Long Wavelength Detector Applications." *Journal Vacuum Science Technology* 2:266–287.
117. A. Tebo. March 1990. "Arrays and Superlattices Expand Detector Options." *Laser Focus World,* 125–140.
118. September 1991. "Hole-Impeded-Doping-Superlattice LWIR Detectors." *NASA Tech Briefs,* 22–23.
119. J. Rowell. October 1986. "Photonic Materials." *Scientific American,* 146–157.
120. M. Kobayashi et al. 1990. "GaAs/AlGaAs Multiquantum Well IR Detectors." *Proc. SPIE* 1341:46–51.
121. September 1991. "Hughes Eyes Superlattice Crystals For IR." *Military and Aerospace Electronics,* 13.
122. M. Kobayashi et al. 1990. "GaAs/AlGaAs Multiquantum Well IR Detectors." Proc. *SPIE* 1341:46–51.
123. B. Levine et al. 1992. "Producibility of GaAs Quantum Well Infrared Photodetector Arrays." *Proc. SPIE* 1638:41–48.
124. R. Hughes. March 1987. "III-V Compound Semiconductor Superlattices For Infrared Photodetector Applications." *Optical Engineering* 26:249–255.
125. B. Levine et al. 1992. "Producibility of GaAs Quantum Well Infrared Photodetector Arrays." *Proc. SPIE* 1638:41–48.
126. J. Leotin and C. Meny. 1990. "Far Infrared Photoconductors." *Proc. SPIE* 1341: 193–201.
127. P. Jackson. November 1991. "Infrared Pictures From ERS1 Show Kuwait Oil Fires." *Lasers and Optronics,* 15.
128. J. Leotin and C. Meny. 1990. "Far Infrared Photoconductors." *Proc. SPIE* 1341: 193–201.
129. D. Pommerrenig, D. Enders, and T. Meinhardt. 1981. "Hybrid Silicon Focal Plane Development: An Update." *Proc. SPIE* 267:23–30.
130. E. Dereniak and D. Crowe. 1984. *Optical Radiation Detectors.* New York: John Wiley & Sons.
131. J. Monin et al. 1990. "Applications of Infrared Bidimensional Devices in Astronomy. *Proc. SPIE* 1341:202–213.
132. P. Norton. November 1991. "Infrared Imaging Sensors." *Optical Engineering* 30: 1649–1662.
133. Ibid.
134. V. Khryapov et al. April 1992. "Optical Sensors." *Optical Engineering* 31, 678–684.

135. S. Botts. July 1988. "New Horizons For Focal Plane Arrays. *Photonics Spectra*, 125–127.
136. P. Norton. November 1991. "Infrared Imaging Sensors." *Optical Engineering* 30: 1649–1662.
137. Lytle. May 1991. "Night Vision Devices Head for Civilian Market." *Photonics Spectra*, 68.
138. K. Liddiard. 1986. "Thin Film Bolometer IR Detectors—II." *Infrared Physics* 26, 1: 43–49.
139. J. Miller. 1992. "Status of Uncooled IR Imagers." *Proc. SPIE* 1689:379–395.
140. Private communications with Dr. Steve Cota, 1993.
141. G. Valco et al.1990. "Photoresponse of $YBa_2Cu_3O_{7-\delta}$©" NASA Technical memorandum 103144.
142. Ibid.
143. B. Johnson. et al. 1992. "$YBa_2Cu_3O_7$ Superconducting Microbolometer Linear Arrays." *Proc. SPIE* 1685:139–145.
144. January 1990. "World News Breaks." *Laser Focus World,* 9.
145. A. Tebo. March 1990. "Arrays and Superlattices Expand Detector Options." *Laser Focus World,* 125–140.
146. P. Kruse. "Physics and Applications of High-T_c Superconductors for Infrared Detectors." *Semiconductor Science Technology* 5:S229–S239.
147. Carlson, et al. 1981. "Solid State Pyroelectric Imaging System." *Proc. SPIE* 267:84–89.
148. Ibid.
149. Pankratov. 1992. "Nonselective Thermal Detectors of Radiation." *Optical Engineering* 31:689–695.
150. Schopf, Ruppel, and Wurfel. 1989. "A 16 Element Linear Pyroelectric Array With $NaNO_2$ Thin Films." *Infrared Physics* 29, 1:103–109.
151. Information courtesy EDO/Barnes, 1993.
152. Lytle. May 1991. "Night Vision Devices Head for Civilian Market." *Photonics Spectra,* 68.
153. J. Miller. 1992. "Status of Uncooled IR Imagers." *Proc. SPIE* 1689:379–395.

Appendix 4A

Representative Data Sheets

AEG 240

Manufacturer: AEG Germany

Qual: Any including S-level

Intended Application: Various

Availability: Custom made from previous work, delivery ≈ 12 months ARO

Description:

AEG has developed a 240-element linear array based on its common module experience.

Characteristics:

Spectral Response: 8 to 12 μm

Element size: 51 by 36 μm on 76 μm center to center spacings

D * = 1.5* 10^{10} cm-$Hz^{1/2}$/W @ 77 K

Approximate Price for a Dewar Integrated Unit: ≈$200,000 (prototype)

Information courtesy of AEG

AESC Pulsed Bias PbS Mosaic

Manufacturer: Aerojet Electronic Systems Division, Azusa, California

Qualification: Any including S-level

Intended Application: General SWIR detection

Availability: Developed and demonstrated; custom fabrication

Description:

PbS provides significant advantages over other materials where high dynamic ranges and linearity are desired at moderate cooling temperatures, with moderate sensitivity requirements in the spectral region below 3 μm. Its photoconductive properties, however, create some difficulties in multiplexing at the focal plane due to the required bias voltage requirements and resultant offset. However, two-dimensional arrays using a unique form of multiplexing of PbS area arrays has been built and extensively tested. This method used the photoconductive properties of PbS with its relatively long time constant and permitted the demonstration of imaging arrays of up to 12,000 detector elements. By using orthogonal code biasing of the rows and columns of the array, with suitable decoding, it was possible to achieve performance levels suitable for space sensor implementation of an appropriate detection system.

Characteristics:

Detector Temperature: 156 K

Signal Spectrum: 2.5–3.5 μm

Background Irradiance: 4.3×10^{12} Ph/cm^2–s

Detector Bias Potential: 10 V

Measurement Frame Rate: 5 frames/second

Mean NEP: 2.1×10^{-14} W

Total Number of Elements: 12,288

Percent of Good Elements: 99.8

Information courtesy of Aerojet Electronic Systems Division

Amber Engineering HgCdTe 480 × 6 Gen II LWIR Scanning Array

Manufacturer: Amber Engineering, Santa Barbara, California

Qualification: All levels available up to S-level

Intended Application: Various

Availability: Custom made from continuing line; delivery ≈ 4 months ARO

Description:

This is a LWIR HgCdTe "second generation" TDI scanning array. The FPA's 2,880 detectors are arranged in four staggered banks of 720, with detector pitch and size chosen to produce 100 percent oversampling in the in-scan direction and 50 percent oversampling in the cross-scan direction. The 480 channels of output are multiplexed to 8 voltage mode outputs using on-FPA electronics. The FPA is an indium bump bonded hybrid consisting of an AE167 readout integrated circuit and the HgCdTe array. Several arrays of various configurations have been subjected to 2,000 thermal cycles between ambient and 80 K with no observable failures.

Characteristics:

Cutoff: 10.5 μm @ 76 K

Pitch: 56 μm in scan, 76 μm in cross-scan

Power Dissipation: < 125 mW

Quantum Efficiency: 52 percent

Dynamic Range: 71.5 db (SNR = 3750)

D*: 4.4×10^{11} cm $Hz^{1/2}$/W

Approximate Price in Large Quantities: $20,000 @ 500

Approximate Capacity for Such Arrays per Month: 10

Information courtesy of Amber Engineering

Amber HgCdTe LWIR 128 by 128

Manufacturer: Amber, Santa Barbara, California

Intended Application: Various

Qualification: All levels available

Availability: Custom made on order; delivery 2 months ARO

Description:

The detector array is fabricated from LPE HgCdTe and bump bonded to a CMOS readout array.

Characteristics:

Array Size: 128 by 128

Cell Size: 50 μm by 50 μm

Average D*: 2×10^{11} cm $Hz^{1/2}$/Watt

Operability: 97 percent

QE: Greater than 50 percent from 8 to 11 μm

Approximate Price for One Unit: $195,000

Approximate Price for 500 Units: $25,000

Approximate Capacity for Such Arrays per Month: 10

Information courtesy of Amber Engineering

Amber Engineering InSb 256 by 256

Manufacturer: Amber Engineering, Santa Barbara, California

Qualification: All levels including S-level

Intended Application: Commercial, NASA, seekers, and other military

Availability: Custom made to order

Description:

The InSb array is bump bonded to a CMOS readout. The array contains anti-blooming circuitry. It is mounted in a 68-pin leadless chip carrier.

Characteristics:

Spectral Response: 1–5.5 μm

Element Size: 38 μm by 38 μm detectors

Dynamic Range: 70 dB

D*: 4×10^{11} Jones @ 80 K and 5×10^{12} Jones @ 60 K

Power Dissipation: < 50 mW

Well Size: 16 million electrons

Noise Floor: 400 e–

Approximate FPA Price for One Unit: $50,000

Approximate FPA Price for 500 Units: $10,000

Information courtesy of Amber Engineering

Amber Engineering InSb Linear TDI

Manufacturer: Amber Engineering, Santa Barbara, California

Qualification: Any including S-level

Intended Application: Commercial, NASA, seekers, and other military

Availability: Custom made; typical deliveries 2 months ARO

Description:

This is a MWIR version of the popular 480 by 4 format.

Characteristics:

Dynamic Range: 76 dB

Price: $150,000

Approximate Price for 500 Units: $10,000

D*: 1 to 2×10^{12} Jones @74 K

Uncorrected Nonuniformity: 2–5 percent

Approximate Capacity for Such Arrays per Month: 20

Information courtesy of Amber Engineering

EDO/Barnes PV Lead Salts

Manufacturer: EDO/Barnes, Shelton, Connecticut

Intended Application: Various

Qualification: Capable of any level of specification and qualification

Availability: Commercially available for single elements; delivery ≈1–2 months ARO

Description:

Barnes has created a technique of epitaxially growing lead sulfide and lead selenide on an lead electrode with a BaF_2 substrate, creating a Schottky diode. The result is a PV device that needs no bias supply and has a faster response than PC lead salts. It is also possible to grow lead selenide over lead sulfide, making a sandwiched two-color device.

Characteristics:

Array Size: Single element, small arrays possible

Operating Temperature: 77 K to 25° C

NEP @ 25°C: 1.5×10^{-11} W for PbS or PbSe

NEP @ 77 K: 2×10^{-13} W for PbS and 1×10^{-12} W for PbSe

Information courtesy of EDO/ Barnes

EDO/Barnes Sandwich FPA A-600

Manufacturer: EDO/Barnes, Shelton, Connecticut

Intended Application: Various

Qualification: Capable of any level of specification and qualification

Availability: Commercially available for single elements; delivery ≈ 4–6 weeks ARO, highly developmental for arrays

Description:

Barnes has sandwiched InSb and HgCdTe on top of each other to provide a detector that responds in two wavelength regions simultaneously. The detector utilizes n-type InSb and PC HgCdTe. The PV InSb provides sensitivity in the 1 to 5 μm region and also acts as a filter for the HgCdTe detector, which responds in the 5–15 μm region. This construction provides simultaneous focusing and registration of both spectral bands. The detector sandwich is available in standard design dewars—both glass and metal—as well as custom designs.

Characteristics:

Array Size: Single element, 2 by 2, 4 by 4

Operating Temperature: 77 K

Bandwidth: 6 Hz

Detector Area for Single Element: 0.1 cm^2

D* (λ_p,900,1): HgCdTe = 8.8×10^9 Jones, InSb = 2.1×10^{11} Jones

Approximate Price for Few Units: $4 K per unit

Approximate Capacity for Such Arrays per Month: 50

Information courtesy of EDO/ Barnes

Graseby Infrared: Sandwich InSb/HgCdTe Two-Color Detector

Manufacturer: Graseby Infrared, Orlando, Florida

Intended Application: Commercial and military applications

Qualification: Capable of any level of specification and qualification (given $)

Availability: Commercially available for single elements, but highly developmental for arrays or wavelengths beyond 12.5 μm.

Description:

Infrared Industries has sandwiched InSb and HgCdTe on top of each other to provide a detector that responds in two wavelength regions simultaneously. The upper element has a short wave response and transmits the long wave to the bottom element. The company also has made sandwich arrays of Ge and InSb.

Characteristics:

InSb Active Area: 2.5 mm in diameter

HgCdTe Active Area: 2.5 by 2.5 mm

InSb D* @ 1kHz: 9×10^{10} Jones

HgCdTe D* @ 10 KHz, Peak: 1×10^{10} Jones

Operating Temperature: 77 K

Array Size: Single element, possible development of 2 by 2 and 4 by 4

Approximate Price per Units: $1,000 to $3,000

Information courtesy of Graseby Infrared

HIDAD (High Density Array Development) Uncooled Microbolometer Arrays

Manufacturer: Honeywell, Minneapolis, Minnesota; further licensees under negotiation

Intended Application: Various, including commercial viewers and low-cost cameras

Qualification: None currently, but suitable for any

Availability: Developmental; delivery on special request, three months after receipt of order

Description:

Developed by Honeywell in cooperation with the US Army NVEOD uncooled IR imager program, the bolometric pixels are thin film bridges, thermally isolated from the bulk silicon by micromachining techniques. The cavity between the pixel and the silicon is designed (tuned) to peak the response at a given wavelength range. The sensor operates in the 8 to 14 micron spectral range. All pixel address and multiplexing electronics are integrated on the chip, below the pixels. This results in improved fill factors. Yields are high, and they are expected to be inexpensive in large-scale production.

Characteristics:

Array Size: 245 by 328

Operating Temperatures: –40 to 20° C

Unit cell Size: ≈50 μm center-to-center spacing

NEDT: 0.1°C

Information courtesy of Honeywell Inc.

Hughes HTC-4880 Pt:Si

Manufacturer: Hughes Technology Center, Carlsbad, California

Intended Application: Missile seekers, FLIRs, day-night driving aids, astronomy, radiometry, cameras

Qualification: Commercial, military, or space

Availability: Custom made per order; delivery 6 months ARO. Limited inventory available for "off-the-shelf" delivery

Description:

The Hughes Technology Center in Carlsbad is the corporate producer of Schottky barrier infrared imaging sensor chip assemblies. The Pt:Si arrays are indium bump bonded to an X-Y readout. Several pixel formats are available.

Characteristics:

Active Unit Cells: 488 by 640 (312,320)

Architecture: X-Y scan, non-interlaced

Unit Cell Size: 20 by 20 μm

Fill Factor: >80 percent

Noise Floor: < 200 e–

Emission Factor (C_1): > 0.3

Maximum Operating Temperature: 79 K

Maximum Frame Rate: 100/second

Uniformity: <5 percent

Approximate Price for Few Units of 640 by 480: $50,000

Approximate Capacity for Such Arrays: > 100/month

Information courtesy of Hughes Carlsbad

ISOCAM Long Wavelength Channel Detector Array

Manufacturer: CEA-LETI-LIR

Intended Application: Space-based astronomy on the ISOCAM on ISO

Availability: Highly developmental, custom made

Description:

The FPA is a gallium doped silicon photoconductive array hybridized via indium bump bonds to a direct voltage readout circuit. The device has been specially designed for the ISOCAM's requirements. The same technology has been applied to a 64 by 64 array.

Characteristics:

Useful Bandpass: 4 to 17.5 μm

Operating Temperature: ≈ 4 K or less

Fill Factor: ≈ 100 percent

Detector Pitch: 100 μm

Crosstalk: ≈ 8 percent

Responsivity: 7×10^{13} V/W

Uniformity ≈ 7 percent

NEP: Under 1×10^{-17} W

Information from 1990. P. Mottier, C. Lucas, M. Ravetto, and P. Agnese, "Recent Developments on ISOCAM Long Wavelength Channel Detector." Proc. SPIE *1341: 368–374*

Litton 200 by 200 InSb

Manufacturer: Litton, Tempe, Arizona

Intended Application: Seekers, astronomy, surveillance, industrial

Qualification: Military

Availability: 2 months after order

Description:

These are hybrids FPAs with a CMOS silicon readout.

Characteristics:

D*: 4.2×10^{11} Jones

Responsivity @ λpeak, 1,000: > 3.5 A/W

Unit Cell Spacing: 20 by 20 μm active areas on 30 by 30 μm centers

Operating Temperature: 60 to 90 K

Dynamic Range: > 10,000:1

Approximate Price for Few Units: $34,000 each

Approximate Capacity for Such Arrays per Month: 100

Information courtesy of Litton

Litton 1 × 256 PbSe

Manufacturer: Litton, Tempe, Arizona

Intended Application: Smart munitions, thermal sights, seekers

Description:

These are PC arrays on a CMOS multiplexer. The dewar package has 20 feedthroughs and can be integrated with a TEC.

Characteristics:

Array: 1 by 256 staggered with 5 percent overlap

Unit Cell Spacing: 50 μm

Detector Impedance: 2 × 107 Ω @ 214 K

Dynamic Range: 4,000:1

Average D*: 4×10^{10} cm-Hz$^{1/2}$/W at 240 K and 1,400 Hz

Approximate Price for a Few Units: $13,000 each

Approximate Production Capacity for Such Arrays per Month: 200

Information courtesy of Litton

Martin Marietta Linear Mercury Cadmium Telluride Arrays

Manufacturer: Martin Marietta, Syracuse, New York

Intended Application: Various

Qualification: Capable of any level of specification and qualification

Availability: Commercially available

Description:

This linear FPA is packaged on an alumina (Al_2O_3) substrate and is mated to a low-noise CMOS preamplifier/mux in an expandable modular configuration. The CMOS multiplexer amplifies and frequency-band limits the signal output. Improved spatial response characteristics and minimal pixel-to-pixel crosstalk are achieved through a unique multilayer design. The planar, front-illuminated FPA provides high sensitivity by circumventing the signal loss associated with transmission through a substrate. Additionally, the planar device structure is passivated to provide the additional benefit of low-noise operation, even under reverse bias conditions. The individual diode outputs are time multiplexed to keep the number of dewar penetrations to a minimum.

Characteristics:

Array Size: 1 × 256

Cutoff: 10.5 μm

Active Area: 38 × 38 μm

Pitch (LXW): 19 × 76 μm

Integration Time: 80 μs

QE: >70%

D*: >1E11 Jones

Dynamic Range: ≥72 dB

Power Dissipation: <250 mW

Operating Temperature: 77 K

Number of Dewar Inputs/Outputs: 7 (2 clocks, 3 biases, and 2 signal)

Uniformity: >90%

Operability: >90%

Price: <$10,000 in quantities exceeding 100, and in commercial qualification

Information courtesy of Martin Marietta

Mitsubishi 1040 by 1040

Manufacturer: Mitsubishi Electric Corp., Japan

Qualification: Capable of any level of specification and qualification (with sufficient $ and time)

Description:

This high density array was developed for thermal imaging in the 3–5 micron MWIR window. This is the first reported million pixel FPA chip in the IR.

Characteristics:

Minimum Feature Size: 1.5 μm

Pixel Size: 17 by 17 μm

Fill Factor: 53 percent

Saturation Level: 1.6×10^6 electrons

NEDT With 1.2 Optics: 0.1 K @ 300 K

Clock Frequencies: 9.2 MHz supporting 30 frames per second with a field rate of 60 Hz

$\mathbf{C_1} = \approx 0.2$

Cutoff = 5.9 μm

Information courtesy of Mitsubishi

Rockwell TCM 2000, 256 by 256 MWIR HgCdTe

Manufacturer: Rockwell, Anaheim, California

Intended Application: Various strategic, tactical, astronomical, and commercial

Qualification: None, able to accommodate 9858A and S-level with additional time and $

Availability: Commercially available ≈ 1 to 6 months ARO, depending on

quantity and custom packaging (e.g., dewar, cooler, and electronics)

Description:

Rockwell uses a sapphire substrate for these hybrid PV HgCdTe Arrays. The company's PACE-1 (Producible Alternative for CdTe Epitaxy) procedure provides state-of-the-art performance at an affordable cost. The FPA is available in miniature tactical dewars compatible with JT, TE, and closed-cycle coolers.

Characteristics:

Array Size: 256 by 256

Unit Cell Size: 40 by 40 μm

Fill Factor: > 85 percent

Spectral Response: 1 to 4.85 micron Jones

D* (Qb = 1×10^{15} ph/cm^2/second): 4×10^{11} Jones

Operability: > 95 percent

Response Uniformity: < 10 percent

No. of Outputs: 1

Well Size: 3×10^7 electrons

Maximum Frame Rate: 60 Hz

Total Crosstalk: 2 percent

Operating Temperature: 80 to 120 K

Information courtesy of Rockwell International

Rockwell TCM 6000 640 by 480 MWIR HgCdTe

Manufacturer: Rockwell, Anaheim, California

Intended Application: Full framing cameras, seekers, threat warning sensors

Qualification: None, can do up to S-level with additional time and $

Availability: Developmental

Description:

Rockwell uses a sapphire substrate for these hybrid HgCdTe arrays. Electroplated indium bumps connect the array to a CMOS readout.

Characteristics:

Array Size: 640 by 480

Unit Cell Pitch: 27 μm

No. Pins on Chip Package: 84

MWIR NEDT @ Qb = 3×10^{14} ph/cm^2/second: ≤ 15 millikelvins

Minimum D*: 5×10^{11} Jones

Operating Temperature: ≤ 120 K

Projected Cost: <$1 per unit cell in large-scale production

Information from 1991. Rockwell International and Kozlowski et al. "Large Staring IRFPAS of HgCdTe on Alternative Substrates." Proc. SPIE *1540.*

Sarnoff 640 by 480 IRFPA Imager

Manufacturer: The David Sarnoff Research Center, Inc., Princeton, New Jersey

Intended Application: TV-compatible IR imagers

Qualification: Any up to S-level available for extra time and $

Availability: Custom made per order; delivery ≈ 6 months ARO

Description:

The David Sarnoff Research Center has developed a CMOS Pt:Si infrared imager with full TV resolution. The IRFPA provides subframe imaging and variable integration time operating modes. The Pt:Si CMOS imager may be operated in progressive scan or interlaced modes. The IRFPA provides a 5,000:1 instantaneous dynamic range and a 2,000,000:1 commandable dynamic range.

Characteristics:

Array Sizes: 640 by 480

Unit Cell Size: 24 μm x 24 μm

Well Capacity: 1.3×10^6 electrons (5 V)

Nonuniformity: 1 to 2 percent uncorrected, 0.1 percent corrected

Information courtesy of David Sarnoff Research Center, Inc.

Santa Barbara Focalplane InSb 128 by 128, 256 by 256 and 240 by 320

Manufacturer: Santa Barbara Focalplane, Goleta, California

Qualification: Any up to space

Intended Application: Any terrestrial background: military and commercial

Availability: Custom made on order; two-month delivery ARO

Description:

The InSb is bump bonded to a silicon multiplexer. Leads and control inputs go into all four sides, so the devices are not buttable.

Characteristics:

Unit Cell Size: 50 μm for the 128 by 128, 31 μm for the 256 by 320

Well Capacity: 6×10^5 to 5×10^7 electrons

Quantum Efficiency: 85 percent AR coated and 60 percent non-AR coated

Fill Factor: 90 percent

D* @ 4.6 μm: 3.7×10^{12} Jones

R_0A: 1×10^5 Ω-cm^2

Operating Temperature: 77 K

QE at 1.7 μm: ≥ 90 percent

Detector Dark Current: ≤ 400 e–/second

Mean Readout Noise: 100 e– @ 2 Hz

Nonuniformity: 1.5 to 3.5% uncorrected, 0.2% corrected with operability > 99%

Base Price for 128 by 128 Array: $55 K

Information courtesy of Santa Barbara Focalplane

SBRC Astronomical InSb

Manufacturer: SBRC, Santa Barbara, California

Intended Application: Astronomy

Qualification: None, capable of up to S-level give time and $

Availability: Custom made from previous work

Description:

This array is an outgrowth of SBRC's 58 by 52 high-sensitivity arrays for astronomy.

Characteristics:

Unit Cell Format: 256 by 256

Center-to-Center Spacing: 30 μm

Well Capacity: 1 × 106 electrons

Total FPA Noise: ≈ 100 electrons at 50 K and 60 Hz

Information courtesy of SBRC

SBRC CRC 198/199

Manufacturer: SBRC, Santa Barbara, California

Intended Application: Various

Qualification: Ground

Availability: Custom made from previous work

Description:

SBRC specifically markets various formats of doped silicon detector arrays to the astronomical community. The original use was in small cameras attached to large infrared astronomical telescopes. The CRC 198/199 series are gallium doped silicon detector arrays, indium bump bonded to a readout chip of integrating MOSFETs and output drivers. SBRC also makes the same type of arrays with an arsenic dopant.

Characteristics:

Array Size: 10 by 64

Cutoff: ≈ 10 μm

Operating Temperature: 10 K

Uniformity: 10 percent

Minimum Integration Time: 1 millisecond

Information courtesy of Hughes SBRC

SOFRADIR 128 by 128 Staring MWIR HgCdTe

Manufacturer: SOFRADIR, Grenoble, France

Intended Application: Seekers, FLIRS, thermal sights

Qualification: Capable of any level of specification and qualification (with sufficient $ and time)

Availability: Developmental; delivery ≈ 6 months after ARO

Description:

SOFRADIR produces these via LPE and uses indium bump bonds to mate the array to a silicon CCD readout.

Characteristics:

Array Size: 128 by 128

Cutoff: 5 μm

Typical D*: 3×10^{11} Jones

NEDT: 12 millikelvins

Operating Temperature: 77 K

Approximate Capacity for Such Arrays per Month: 50

Information courtesy of SOFRADIR

SOFRADIR 128 by 128 High-Temperature MWIR HgCdTe

Manufacturer: SOFRADIR, Grenoble, France

Intended Application: Seekers, FLIRS, thermal sights

Qualification: Capable of any level of specification and qualification (with sufficient $ and time)

Availability: Developmental; delivery ≈ 6 months after ARO

Description:

SOFRADIR produces these via LPE and uses indium bump bonds to a silicon CCD readout. These devices have a high D* at high temperatures.

Characteristics:

Array Size: 128 by 128

Cutoff: 4.5 μm

Typical D*: 2×10^{10} Jones

Operating Temperature: 200 K

Unit Cell Size: 50 μm

Approximate Capacity for Such Arrays per Month: 50

Production Price (AUPC > 1,000): ≈ 11,000

Information courtesy of SOFRADIR

SOFRADIR 288 by 4 TDI Scanning LWIR HgCdTe

Manufacturer: SOFRADIR, Grenoble, France

Intended Application: Seekers, FLIRS, thermal sights

Qualification: Capable of any level of specification and qualification (with sufficient $ and time)

Availability: LRIP; delivery ≈ 6 months after ARO

Description:

SOFRADIR produces staggered arrays via LPE and uses indium bump bonds to a silicon CCD readout. Cutoff wavelength is controlled to ± 0.1 μm. The "on-focal-plane" electronics include TDI, anti-blooming, skimming, and partitioning.

Characteristics:

Array Size: 288 by 4 TDI

Cutoff: 10.5 μm

Typical D*: 1.5×10^{11} Jones

Operating Temperature: 77 K

Heat Load: 340 mW

Crosstalk: < 40 dB

Approximate Price in Large Quantities: $20,000

Approximate Capacity for Such Arrays per Month: 50

Information courtesy of SOFRADIR

Spiricon Pyroelectric Arrays

Manufacturer: Spiricon, Logan, Utah

Intended Application: Laser diagnostics

Qualification: Commercial

Availability: Delivery ≈ 2 months ARO

Description:

Spiricon makes mosaic and linear arrays out of $LiTaO_3$ and $LiNbO_3$. The 32 × 32 is the largest commercially available, although a 64 × 64 is possible with development.

Characteristics:

Array Size: 32 × 32, 16 × 16, and 32 × 16

NEP: 30 nW per $Hz^{1/2}$

Dynamic Range: 1: 10,000

Approximate Price for Few Units: $32,000

Approximate Capacity for Such Arrays per Month: 4

Information courtesy of Spiricon

Thompson THX 31800, SWIR InGaAs

Manufacturer: Thompson, CSF, France

Qualification: Capable of any level of specification and qualification (with sufficient time and $)

Availability: Made to order; delivery 1 month ARO

Description:

These are linear arrays of 300 unit cells on sapphire substrates and buttable on two sides. They use CCD readouts.

Characteristics:

Bandpass: Up to 1.7 μm

Cell Size: 30 μm by 30 μm

Operating Temperature: Room temperature; slightly better sensitivity when cooled to –20° C

Approximate ROM Price for Few Units: $7,000 each

Approximate Capacity for Such Arrays per Month: 100s

Information courtesy of Thompson CSF

Uncooled Ferroelectric Arrays

Manufacturer: Texas Instruments, Dallas, Texas

Intended Application: Various military and commercial uses

Description:

The uncooled LWIR ferroelectric arrays are the culmination of a decade-long government and industry development led by the Night Vision Electro-Optic Directorate at Ft. Belvoir.

Characteristics:

Bandpass: 8 to 14 μm

Detector Material: Barium strontium titanate

Array Size: 244 by 328

Frame Rate: 30 Hz

NEDT: 0.1° C

Information from 1992. R. Owen, S. Frank, and C. Daz, "Producibility of Uncooled IRFPA Detectors." Proc. SPIE *1683: 74–80*

5

Cryocooling Systems

"If you can't stand the heat, get out of the kitchen."

Harry S. Truman

5.1 TECHNOLOGY BASICS

The photon's journey through the sensor ends with its conversion to an electronic signal at the focal plane. However, as seen in the previous chapter, focal plane arrays (FPAs) frequently require some type of cryogenic cooling. Cryogenic cooling is an integral part of infrared sensor architecture. Such cooling is inherent to an infrared sensor and is essential for two reasons. First, the detectors may require cooling just to function, or for increased sensitivity. Second, cryogenic cooling reduces thermal noise from filters, baffles, and even the optics themselves. The extent to which this refrigeration is applied depends on the system design, detector material, bandpass, desired sensitivity, and the expected background. For example, it is not uncommon to gain a factor of two or three in sensitivity with a low background HgCdTe-based FPA by cooling it and its surroundings an additional 10 K.

Detailed sensor system analysis must be conducted to establish the optimal level of cooling for optical train components. For instance, there is no benefit to cooling an optical element if the focal plane array is sensitivity limited by its own noise. Conversely, cooling a filter or mirror could improve system sensitivity greatly in low background, highly sensitive systems. As a rule, the optics will not need to be cooled if the bandpass cutoff is approximately 11 μm or less and if the baffle design and the fields of view (FOV) of the FPA are properly designed. However, even with cutoffs of less than 11 μm, it is likely that one will still need to cool the closest baffle (often called a *cold shield*) to the FPA, and probably the filter. A possible exception might be if the optical bandpass cutoff is less than 4 μm and if the background is moderate. Then the baffle and filter do not necessarily require cooling. The exceptions to this include extremely sensitive but highly non-uniform focal planes viewing a cold background.

Table 5.1 gives probable temperatures to which one should expect to cool for various traditional focal plane materials. The reader is cautioned that the entire table, especially the final column, is subjective. The intent of the table is to give a general impression of the required temperatures and what components need to be cooled. The intent is not to provide a design. Actual temperatures and the components requiring cooling depend on system requirements and specifications, taken in conjunction with program considerations such as cost, manufacturability, producibility, and availability.

For all but space sensors, such cryogenic components must be isolated from the environment to prevent frosting, reduce background signal and reduce heat load. The insulating vessel, usually called a dewar, traditionally contains the FPA, a cold shield, and perhaps a filter. The FPA views through a warm window, and cooling is supplied via a thermal attachment to a cryocooler. Newer dewar designs often include the cooler (or part of it) and additional electronics. Figure 5.1 shows a "second generation" standard advanced dewar assembly (SADA) dewar and split Stirling cryocooler.

Five distinct alternative methods are currently used to provide cryogenic cooling for sensors: thermoelectrics, Joule-Thomson (J-T) blow-downs, closed-cycle refrigerators, radiators, and cryogen-filled dewars. In addition, advances are being made in absorption and magneto-caloric coolers.

Thermoelectrics utilize the Peltier effect to cool a small area and heat load via a semiconductor process. They are solid state refrigerators with low efficiencies. They are most applicable to spot cooling with low loads and temperature drops between their hot end and the FPA of less than 100 K.

Open cycle blow-down systems use an expanding gas from a pre-charged reservoir to cool a small area via the Joule-Thompson effect. This is the same effect one observes with the spray from an aerosol can. Rapid spot cooling can be accomplished. The main drawback is that the gas is a consumable. The gas containment bottle will eventually run out. This cooler type is useful for systems that have a short life and only operate once or a few times, such as a missile seeker.

Closed-cycle refrigerators provide cooling via a mechanically-driven thermodynamic process, as in a home refrigerator. Cryo-refrigerators are applicable to sensors that require many on/off cycles. Cool-down is not as rapid as with blow-down systems: it can take four to 20 minutes to reach operational temperatures. Traditional drawbacks have included short life, low reliability, large weight, and high vibration, but recent advances have mitigated those drawbacks.

Radiative cooling consists of orienting a large highly emissive surface to a colder environment (such as deep space). Usually a small focal plane assembly is

TABLE 5.1 Representative Temperatures of Detector Materials

Detector Material	*Configuration*	*Suggested Spectral Range (μm)*	*Representative FPA Temp (K)*	*Likely Cold Train Components*
PbS	PC	1–3.5	195–270	FPA
PbSe	PC	1–5	195–270	FPA, cold shield
InAs	PV	1–3.5	195	FPA
InGaAs	PC, PV	0.5–1.6	200–270	FPA
HgCdTe	PC	1 to 5	190	FPA, cold shield
HgCdTe	PV	1–4	150	FPA, cold shield
HgCdTe	PC	8–12	70–100	FPA, cold shield, inner baffle, filter
HgCdTe	PV	3–12	70–80	FPA, cold shield, inner baffle, filter
HgCdTe	HIT	8–20	40–60	FPA, cold shield, baffles, filters, optics
Si: As	Extrinsic	1–16	10–18	Entire telescope
Si:In	Extrinsic	1–8	15–40	Entire telescope prior to lyot stop
Si:Ga	Extrinsic	1–18	10–18	Entire telescope
Si Bi	Intrinsic	1–20	10–18	Entire telescope
Pt:Si	Mono/hybrid	1–5	60–80	FPA, cold shield
Ir:Si	Mono/hybrid	1–9	35–50	FPA, cold shield, filter, inner baffle
InSb	Hybrid	1–6	70–120	FPA, cold shield, inner baffle
InSb	Stained lattice	4–10	30–70	FPA, cold shield, inner baffle, filter
LiTaO3	Pyroelectric	1–50	250–300	Nothing
TGS	Pyroelectric	1–50	250–300	Nothing
Quantum wells	LWIR	14–18	50–100	Entire telescope
Uncooled bolometers	MWIR to LWIR	4–20	250–300	Potentially minor cooling of dewar baffles and filters

FIGURE 5.1 An External View of a SADA Dewar and Cryocooler (courtesy of Infrared Components Corporation)

attached to a large radiator that radiates the heat away. Radiative coolers are eminently dependable and have provided cooling at temperatures down to 100 K (in space) with little or no vibration for many years. Their main drawbacks are weight and applicability. There must be a suitable position for the radiator and a proper environment to radiate to. They are most applicable in large, long-life space assets and have been used on certain meteorological (e.g., NIMBUS), deep-space, and earth resources technology satellites.

The vacuum insulating vessels that house the cryogenic portions of IR sensors are frequently called dewars. In this sense, they are merely the cryogenic insulating assembly, and provide no cooling. The containers of liquid or solid stored cryogenic material (e.g., liquid nitrogen, helium, neon, solid nitrogen, or methane) that provide cooling are also called dewars. Liquid cryogen containment dewars are quick to cool, can accommodate large heat loads, and are reliable. They are also heavy and rapidly consumed. Few tactical systems use containment dewars since it is inconvenient to fill them during a military operation. Although a number of space sensors have been built employing containment dewars [e.g., infrared astronomical satellite (IRAS), Teal Ruby, cryogenic infrared radiometer in shuttle (CIRRIS), Delta Star, and infrared space observatory (ISO)], they are viewed with disfavor. This is because of their weight, complexity on the launch pad, and consumable nature. The longest life of a stored cryogen system on orbit is about one year, as on IRAS. Some space infrared telescope facility (SIRTF) concepts call for several years of cooling capability in a very high orbit. New systems and almost all currently marketed cameras use mechanical cryocoolers. The applicability of dewars is being limited primarily to labs, and therefore will not be discussed further.

Hybrid cooling systems can provide the best of two or more of the above systems. Depending on your needs, the expected weight and cost penalty may not exist, because one system may allow a lighter weight selection of the other. It is a system trade issue. One can imagine a search-and-track sensor on a fighter that needs to be used for long periods, leading the natural choice to be a Stirling cryocooler. This same system may have a requirement for rapid operation in an emergency situation. For emergency turn-on, a Joule-Thomson blow-down cooling system can be included in addition to the Stirling cryocooler. While it must be recharged after being used the designed number of times, the J-T provides the rapid cooling that is required in many applications. Although such a hybrid approach is more costly, it may have system-wide cost savings (such as saving a pilot and a $50 million dollar aircraft). Furthermore, the J-T blow-down may permit the selection of a lower capability cryocooler, possibly offsetting a portion of the cost and weight of the blow-down system. Likewise, the designers of a space surveillance sensor may want to include both a radiator and a closed-cycle refrigerator. The radiator will keep the cold train colder than otherwise, allowing the cryocooler to cool down the assembly faster and with less electrical power.

Regardless of type of cooler, the user will find an abundance of companies and research institutions that supply cryogenic systems. An international industry has developed to satisfy electro-optic cryogenic cooling needs by all the aforementioned methods. This legion of players extends from France to Israel and from Liechtenstein to New Hampshire. The industry is composed primarily of thermodynamicists and mechanical engineers, not sensor designers.

This results in lengthy community arguments about thermodynamics and Carnot efficiencies, rather than system applicability. The sensor designer is advised not to be overly concerned with these figures of merit. There may be little relationship between an efficient and pragmatic system design and Carnot efficiency.

For example, suppose one had to cool a focal plane array with a heat load of 0.5 W to 80 K for 60 seconds. Closed-cycle Stirling coolers have more efficient Carnot cycles than J-T systems. If the designer employed an integral Stirling cooler, it would take roughly 40 W of power for six to ten minutes, weigh near a kilogram, and cost from $6,000 to $20,000. However, the designer could select the less efficient J-T, which uses no electrical consumption, has a weight of only two or three hundred grams, and costs just a few thousand dollars. There is no question that Stirling coolers have better Carnot efficiency; however, there is also no question that, from the overall system perspective, the J-T has lower weight, cost, and power demand. The choice depends on the sensor and FPA design, requirements for multiple cool-downs, and logistic concerns with getting the gases for the J-T.

One generally applicable favorite figure of merit in the industry is the coefficient of performance (COP). It is defined as the dimensionless ratio of cooling effect output to electrical power input, as follows:

$$\text{COP} = \frac{W_e}{Q_c} \tag{5.1}$$

where

COP = coefficient of performance, dimensionless

Q_c = cooling capacity effect in watts

W_e = electrical input in watts

In the previous chapter, we used an example focal plane that was cooled to 70 K. If the cooling were accomplished mechanically by a Stirling that could remove 0.5 W at 70 K while requiring 30 W of input power, then its coefficient of performance at 70 K would be 0.5/30 = 60 watts per watt.

Equation (5.1) represents a useful and facile concept that could be used for making meaningful comparisons. Unfortunately, not everyone measures and defines it in the same way. To effectively apply the concept of COP, one must know if it includes control electronics losses, rejection (or compressor case) temperature, and the temperature control specification. It is difficult to compare one cooler to another unless all parameters are equal. The most useful definition to the sensor designer includes all losses from the power supply to the cold finger.

Nature has dealt the sensor designer a lamentable condition. Efficiency decreases as required cooling power is decreased. The unfortunate result is that the new, highly efficient large focal planes may need close to the same amount electrical power to cool them as their less efficient smaller predecessors.

Other important factors influencing selection of a cryo-system include reliability, vibration, weight, lifetime, cold head temperature, and cooling effect. The last is the most difficult factor to trade and determine. Intricacies of design and operating condition can greatly affect the heat load. A misplaced heat conducting strip of indium or a slightly misaligned baffle will impact the load in an unforeseen and detrimental way.

Obviously, the best way to determine heat loads is to measure them. However, in the early stage of design, the engineer rarely has hardware to measure. Instead, he must estimate the load to select a cooler. The first step in this estimation process is to define the cold train, which (if the gods of design smile on you) may only be the FPA or (if you have bad engineering karma) the entire telescope assembly. Once the bandpass, FPA, and optical concept are chosen, one should weigh component temperatures against sensitivity, keeping important attributes of weight, cost, manufacturability, and, if applicable, producibility in mind. The next step is to perform radiometric calculations on the entire optical train and focal plane assembly cavity, including baffles, filter, shields, and walls. This is done to determine what components other than the FPA need to be cooled. Typically, the FPA is the coldest component, with optics and baffles being tens to hundreds of Kelvins hotter. Regardless, anything that is to be cooled must be considered in the heat load estimate, and is, by definition, part of the cold assembly.

Calculating the precise heat load that matches the real world is difficult, but developing a crude estimate is not. Heat loads can be estimated and scaled by adding the conduction, convection, radiation, and heat produced by the devices in the cold train.

Conduction of heat happens through contacts and wires. This is usually the dominant heat load. Care must be taken to include all wires and touching surfaces. Fourier's law of conduction, expressed in Equation (5.2), can be used to estimate this for each wire and surface.

$$Q_{cd} = \frac{kA\,(T_1 - T_2)}{L} \tag{5.2}$$

where

Q_{cd} = static heat load from conduction in watts

k = material thermal conductivity in watts/m-K

A = cross sectional area in m^2

T_1 = hot end temperature in K

T_2 = cold end temperature in K

L = distance between the two temperatures in M

There is a subtlety in determining the cross-sectional area. This is the only area in thermal contact that depends on the surface finish and microstructure. On a microscopic level, conduction occurs through contact conduction, radiation in interstitial sites, and convection through the trapped gases in the microscopic vacancies. The effective thermal surface area is always less than the geometric area. For malleable metals, such as gold and indium, it approaches the geometric area, whereas it can be very low for rough surfaces. This indicates that when thermal contact is required, an indium or gold washer should be installed.

Convective heat load depends on the gases, pressure, and area exposed to the gases. It is negligible for pressures of less than 10^{-5} Torr. At standard temperatures and pressures, for an atmospheric gas mixture, it is about 0.5 W/cm^2. Unquestionably, this can be a very large heat load, requiring the cold train to be sealed in a vacuum or partial vacuum. A pressure of 10^{-2} Torr yields an order of magnitude drop. Another practical reason to encase the cold train in a vacuum is to prevent condensation and freezing.

Radiation can be another major heat load. All sides of the entire cold train surface will receive infrared thermal radiation, resulting in a static heat load.

The radiation load can be calculated by the classic thermal equation

$$Q_r = \varepsilon\sigma VA\,(T_1^4 - T_2^4) \tag{5.3}$$

where

Q_r = heat load from radiation

ε = emissivity*

σ = Stefan-Boltzmann constant

V = viewing factor (often just the cosine of the angle between the surfaces)

A = area of the surface

T_1 = temperature of surface radiating

T_2 = temperature of cold surface

Heat production is generally from the FPA input bias and the readout. However, one should be careful to include Z-planes or any thermally attached processors, moving filters, tunable filters, and other components that might be in the cold train. When each heat load mechanism has been estimated, merely add them together to estimate the total static heat load.

---

* If you don't have a measured value, get an approximate value from the *IR Handbook,* or assume it is 0.9 for a black surface, 0.2 for a shiny surface, or 0.8 for anything else.

FPA manufacturers have a good idea of the heat dissipated from their focal planes. Consulting them is usually more accurate than trying to estimate heat loads. If it is a new design for a focal plane, linear scaling on a pixel, voltage, and area basis from previous similar FPAs will usually suffice for a rough estimate. Similar FPAs mean the same basic architecture, same material, and bias control. Currently, collections of photoconductive discrete elements, such as PbS or HgCdTe, usually consume approximately 10 mW/element, and two-dimensional photovoltaic arrays, such as InSb or HgCdTe, take 0.1 to 10 μW/pixel. Most Pt:Si arrays dissipate less than 0.1 W, which is about 0.3 μW/pixel for the larger formats. Finally, a classic common module at 80 K poses a heat load of 150 mW and an interface loss of 100 mW, for a total demand of 0.25 W [Ross, et al., 1].

5.2 THERMO-ELECTRIC COOLERS

A thermo-electric cooler (TEC) has semiconductor materials with different characteristics connected electrically in series and thermally in parallel. An n type material is connected to a p type, and an electrical current of a few amps and low voltage is applied. Electrons pass from a low energy level in the p to a higher energy level in the n. The cold side (p) temperature will decrease as heat is absorbed to supply the energy. Thus, a thermo-electric cooler is basically a solid state heat pump or refrigerator. Being a heat pump, reversing the cycle by reversing the electrical current will generate heat. This has the advantage of heating the detector immediately after use to avoid contamination (or for the occasional cookout).

Each p-n junction is called a *stage*. Using multiple stages will make the final cold side increasingly colder. Seven stages is the maximum for commercial coolers, although custom ones have been made with more.

The TEC industry has defined a figure of merit Z, in which the higher the Z, the better the cooler performance will be. It can be given as

$$Z = \frac{\alpha^2}{\rho\kappa} \tag{5.4}$$

where

Z = material figure of merit

α = Seebeck coefficient

κ = thermal conductivity

ρ = electrical resistivity

The Z for combined n and p is

$$Z = \frac{(\alpha_p - \alpha_n)^2}{[(\kappa_p \rho_p)^{1/2} + (\kappa_n \rho_n)^{1/2}]^2} \tag{5.5}$$

where

Z = stage figure of merit

α_p = Seebeck coefficient for p material

α_n = Seebeck coefficient for n material

κ_p = thermal conductivity of p material

ρ_p = electrical resistivity of p material

κ_n = thermal conductivity of n material

ρ_n = electrical resistivity of n material

Equation (5.4) can be used when the n and p are the same material. Otherwise, Equation (5.5) should be used. The conundrum with this figure of merit is that α, κ, and ρ vary with temperature. Thus, different stages are made of different materials, so Z is maximized at the expected temperature. When comparing, the engineer must be sure Z is quoted at the operating temperature. The highest Zs are approximately 3×10^{-3} /K (with 0.5 being typical for ZT). Research is being done to increase this as much by as a factor of five.

The most technologically advanced TECs employ a superconductor in place of the p type material. As can be seen in Equations (5.4) and (5.5), if the electrical resistivity drops, the Z increases. The Z for a superconductor is almost infinite, so the overall Z for the junction is the same as the n material. High-Z n materials are available at temperatures <100 K, where high-T_c superconductors operate. These TECs have high performance and operate in the <140 K temperature range that many infrared detectors require. While these devices represent the state of the art and are developmental in nature, they hold great promise.

Because TECs lack moving parts, they produce no vibration or acoustic noise. In spite of their useful attributes, however, their range of application is limited. A cardinal limit in their applicability is that they can only cool from about 100 to 150 K below ambient. Their efficiencies are low—only about one percent of Carnot. Reliability is still a concern, as thermal stresses tend to cause mechanical fractures at the edges of the stages. Their choice application is to cool focal planes 100 K or less from ambient, where vibration and weight are prime concerns but power consumption is not.

TECs have amazingly high reliabilities. Units have been operating for so long that MTTFs of 10 to 20 years are indicated. For example, "With failure defined as only a five percent loss in cooling effectiveness, the results of a typical three stage TEC containing 101 couples shows that the MTBF is 82,000 hours" [Kreider, 2].

5.3 JOULE-THOMSON (BLOW-DOWN) SYSTEMS

The Joule-Thomson (J-T) effect occurs when a non-ideal gas at high pressure expands to a low pressure. The expansion causes the gas to cool until the freezing point of the gas is reached. At this point, a stable temperature will result. For cryogenic cooling, this occurs in a counter current heat exchanger, usually close to the detector and inside the dewar assembly. The cooling capacity and speed of cool-down depend on the cryostat design, gas mixture, flow rate, and pressure. Used with infrared FPAs, they ordinarily provide up to 2 W of cooling. Cooling capacity can be enhanced by exotic gas mixtures and considerably high flow rates at high pressures, with about 5 or 6 W at approximately 80 K being the brute force state of the art.

The gas used depends on the application. High purity, extra dry Mil-N-6011 nitrogen or argon is frequently used. Freons and special mixtures are sometimes are used. To efficiently cool to 80 K, a pressure ratio on the order of 150 is required, causing the tank pressure to be on the order of 3,000 to 6,000 psi. Cryostat clogging is possible and should always be a concern to the IR engineer. As a rule of thumb, clogging tends to occur when there is about 2 ppm of water or CO_2 and 3 ppm of a hydrofluorocarbon, or particles of 6 μm size, while commercial grade gases have approximately 6 ppm of water [Bonney and Longsworth, 3].

Advantages of open J-T systems are high reliability, fast cool-down, potentially low weight, low vibration, and low cost. Disadvantages are limited cooling duration, limited temperature control, and low thermodynamic efficiencies. J-T systems are most suitable for one-time use of sensors, such as a missile seeker, or for extremely rapid cool-down times, such as in a surveillance sensor that must operate within a few seconds of turning on. Also, key to obtaining rapid cool-down is a reduction in cold end thermal mass. Metal dewar components can decrease this mass substantially (i.e., in the Javelin seeker). Replacing the glass components with metal components reduced the thermal mass from 298 to 97 joules [Whicker, 4].

The counter current heat exchanger, commonly referred to as a cryostat, is customarily made of externally veined capillary tubing wrapped around a conductive post. A representative cryostat, from EG&G Judson, is shown in Figure 5.2. J-T cryostats generally weigh 20 to 50 grams and are sized to fit into a common module dewar. In the early eighties, an ingenious method of producing cryostats was developed by Stanford and MMR Inc. Figure 5.3 is a drawing of one of their units. Photolithography is used to define channels, a liquid reservoir, and expansion cap-

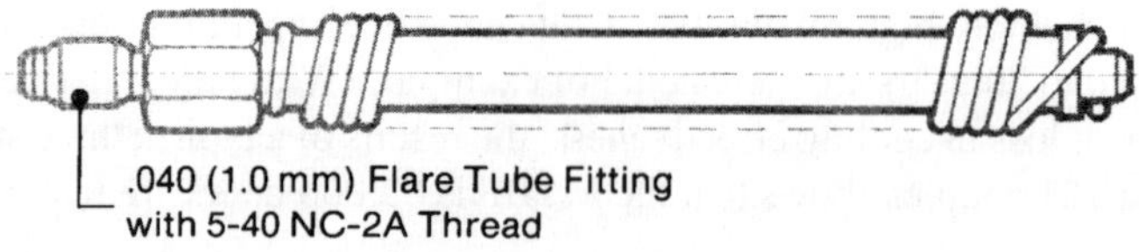

FIGURE 5.2 Joule-Thomson Cryostat (courtesy of EG&G Judson)

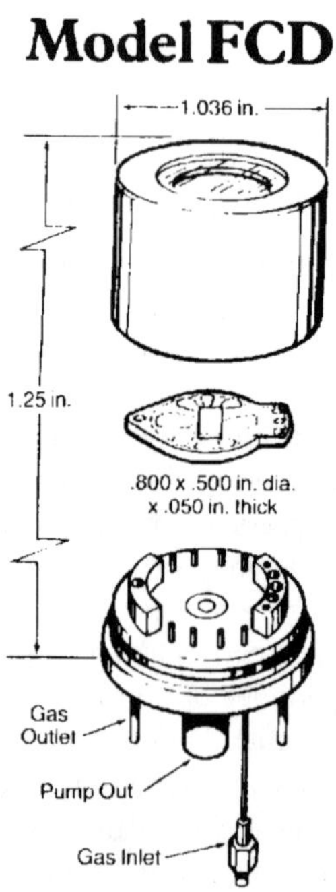

FIGURE 5.3 MMR Microchannel (courtesy of MMR)

illaries in thin slices of glass. This process draws on the methodology used to make integrated circuits. The channels are small (e.g., 200 μm wide by 30 μm deep) and numerous. Large area panels can provide more than 50 W of cooling capacity. Of more interest to the IR engineer is a fast-cool-down version (2.5 by 2.0 by 0.125 cm) that cools to 90 K in two seconds and has survived acceleration of 30,000 Gs without failure [MMR, 5].

The main system weight constituent for a J-T system is the weight of the gas bottle and the stored gas it contains. Equation (5.6) provides the weight of a cylindrical gas bottle, excluding the gas [Wolf and Zeisis, 6].

$$M_{lb} = (2\times10^{-8})\, V^{1.067} P^{1.5} FS^{1.36} \tag{5.6}$$

where

M_{lb} = bottle weight in pounds

V = volume in cubic inches

P = operating pressure in PSIa

FS = safety factor (assume approximately 1.5)

and for spherical gas tanks as [Wolf and Zeisis, 7]

$$M_{lb} = (3.7\times10^{-4})\, V^{0.86} P^{0.49} FS^{0.7} \tag{5.7}$$

where

M_{lb} = bottle weight in pounds

V = volume in cubic inches

P = operating pressure in PSIa

FS = safety factor (assume 1.5–2)

Other prominent technological advancements are occurring in tank production for propulsion systems. This same tank technology can be applied to the reservoir of high-pressure gas. For pressures required by J-T systems, the dry weight of a composite wound tank can be estimated by

$$W = \frac{P \times V}{1\times10^{6}} \tag{5.8}$$

where

W = weight in kilograms

P = pressure in pounds per square inch

V = volume in cubic inches

For example, if we wanted to cool down our notional example focal plane rapidly, a Joule-Thompson cooler could be included. If it had a pressure of 4,000 psi and a volume of 80 cubic inches, the weight of the tank would be

$$\frac{4000 \times 80}{1 \times 10^6}$$

or approximately 320 grams.

Advances in high pressure tank technology could elevate the constant in the denominator by a factor or two or three. This will lead to even lighter tanks in the future. When calculating weight, be sure to add the weight of the gas. As any scuba diver knows, the weight of nitrogen is not trivial.

5.4 STIRLING REFRIGERATORS

Closed-cycle refrigerators are frequently used in tactical/military FLIR systems. There is a historical impression that these refrigerators are large, unreliable, and power hungry—characteristics that are no longer accurate. Therefore, they are debuting in commercial cameras, environmental sensors, and space sensors.

There are a plethora of thermodynamic processes to accomplish cryogenic refrigeration with a closed-cycle mechanism. It is not the scope of this chapter to provide a comprehensive dissertation on each one. However, it is useful to managers and engineers on the periphery to have at least a sophomoric understanding of the Stirling, Vuilleumier, magneto-caloric, Sorption, and Brayton, since these are commonly used or show future promise.

5.4.1 The Stirling Cycle

The quintessential and most important thermodynamic cycle is the Stirling cycle. A Stirling cooler uses compression and expansion of gas, usually helium, to accept heat in one location and reject heat in another location. The Stirling cycle can be implemented in a system that uses either two pistons or a piston and a gas displacer to produce the required compression and expansion effects.

Stirling coolers are characterized as either "integral" or "split." Integral Stirling coolers have the two moving components located together in one housing. Split Stirling coolers typically have a piston and motor mounted in two housings. Figure 5.4 is a cutaway drawing of a typical split Stirling showing the larger compressor unit attached to the expander by a single pressure line. Figure 5.5 is an exterior drawing with dimensions for a typical split Stirling used in tactical FLIRs. Physically, most split Stirlings' compressors are cylinders 5 to 19 cm in diameter and 10 to 25 cm long, with a thin, single line connecting to the expander. Integral Stirlings are usually "boxy" with various protrusions, one being the cold finger.

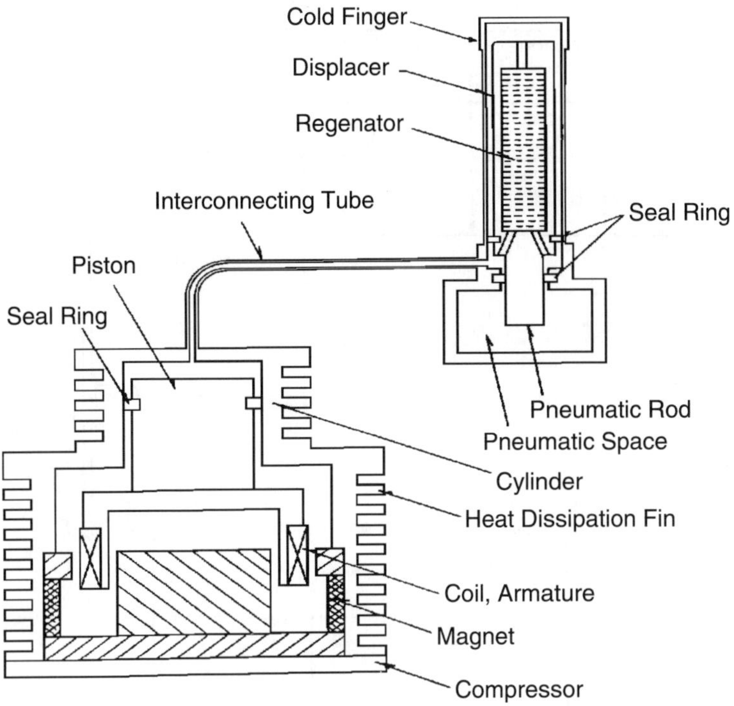

FIGURE 5.4 A Cutaway View of a Split Stirling Cryocooler (courtesy of Mitsubishi)

By convention, the assembly that houses the piston and motor is called the compressor. Also by convention, the assembly that includes the displacer is called the expander. This naming convention is confusing because compression and expansion occur in both the compressor and expander. In split Stirling systems the heat-acceptor-heat-exchanger and heat-rejector-heat-exchanger are normally located, respectively, at the cold and warm ends of the expander assembly.

Some split Stirling systems use the oscillating pressure of the Stirling cycle to drive the displacer. Only a single line is needed, as the gas moves both ways within the line, creating the aforementioned pressure change, which moves the displacer piston. An alternative approach is to drive the displacer piston with a small second motor. The expander has the smaller displacer piston moving up and down, powered by this pressure wave.

The motion of the compressor piston(s) and the displacer are phased so that when expansion is occurring, most of the gas in the cycle is located in the vicinity of the region to be cooled. When compression is occurring, most of the gas in the cycle is located in the vicinity of the rejection heat exchanger. Hence, they cool at one location and reject the heat at another. The small expander piston is really a

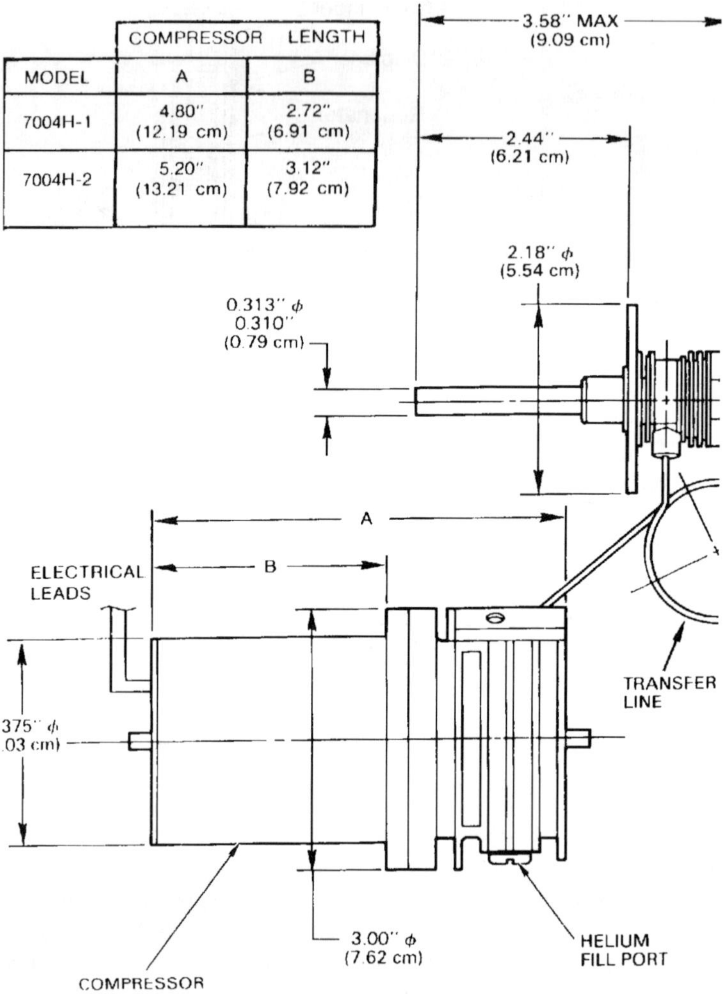

	COMPRESSOR LENGTH	
MODEL	A	B
7004H-1	4.80" (12.19 cm)	2.72" (6.91 cm)
7004H-2	5.20" (13.21 cm)	3.12" (7.92 cm)

FIGURE 5.5 The Hughes 7004H Split Stirling Cryocooler (courtesy of Hughes Aircraft Company, Electron Dynamics Division)

valve (although piston-like) working 90 degrees out of phase with the big compressor piston. In this way, it moves gas closer the detector to collect heat during the expansion process and move the hotter gas away during the compression part of the cycle. Being 90 degrees out of phase, most of the working gas is expanded at the cold end and compressed at the hot end.

Figures 5.6 and 5.7 graphically represent this Stirling cycle. In going from step one to step two, the large compressor piston compresses the gas, generating heat

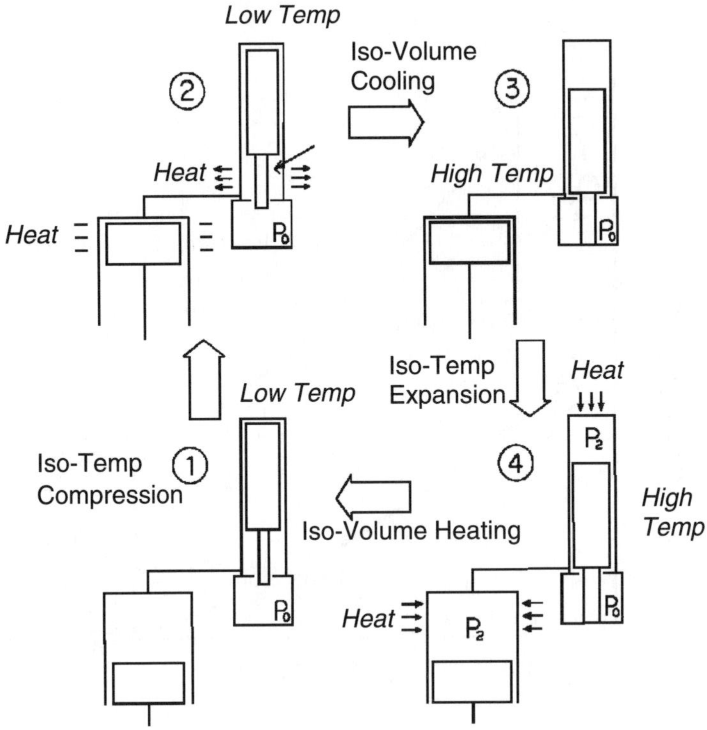

FIGURE 5.6 Operation of a Split Stirling Cryocooler (courtesy of Mitsubishi)

which is dumped to the environment in the vicinity of the compressor. From step two to step three, the displacer piston moves down from the pressure gradients resulting from the previous step. The working gas moves through the regenerator (being further cooled) to the upper part of the expander. From step three to step four, the gas moves into the expander assembly, expands, and provides cooling. Returning to step one, the displacer piston then rises due to pressure gradients, and the working gas moves to the lower part of the expander through the regenerator, cooling the regenerator [Fujino, 8].

The piston in the compressor module is typically driven by either a rotary electric motor or a linear electric motor. The rotary systems use a short-throw crank and connecting rod assembly to transform the rotary motion into linear piston motion. This includes side loads on the piston and makes long life difficult to achieve. In linear compressors, the piston is connected directly with the linear motor armature by a rod. In this case, the piston and motor armature share the same linear motion. In linear compressor systems, the side loads induced in driving the piston are usually small or negligible, making long life and high reliability much easier to achieve.

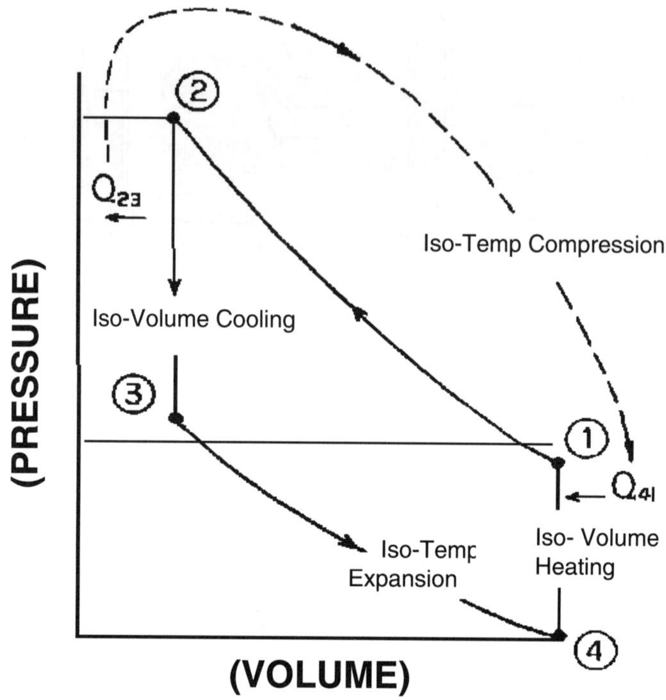

FIGURE 5.7 PV Diagram for the Stirling Cycle (courtesy of Mitsubishi)

Finally, electrically isolating the compressor and associated components from the detector is usually important. This can be accomplished simply by placing an insulating washer in the gas transfer line. This is best left to the cryocooler manufacturer to implement.

5.4.2 Vibration

Vibration is frequently a concern in FPA cooling applications. It is difficult to eliminate the vibration from rotary compressors, which generate both translational and torsional vibrations. Simple counterweights on the crank assembly can only reduce compressor translational vibration amplitude by about half.

In linear coolers, vibration is easier to reduce. Linear coolers have vibration largely contained in one axis. This allows placement to minimize vibration-induced line-of-sight errors, especially by orienting the vibration with respect to scanning mechanisms. One approach is to use opposed counter-oscillating components in the compressor and expander modules. The compressor pistons move 180 degrees out of phase to each other, canceling out most of the vibration. Figure 5.8

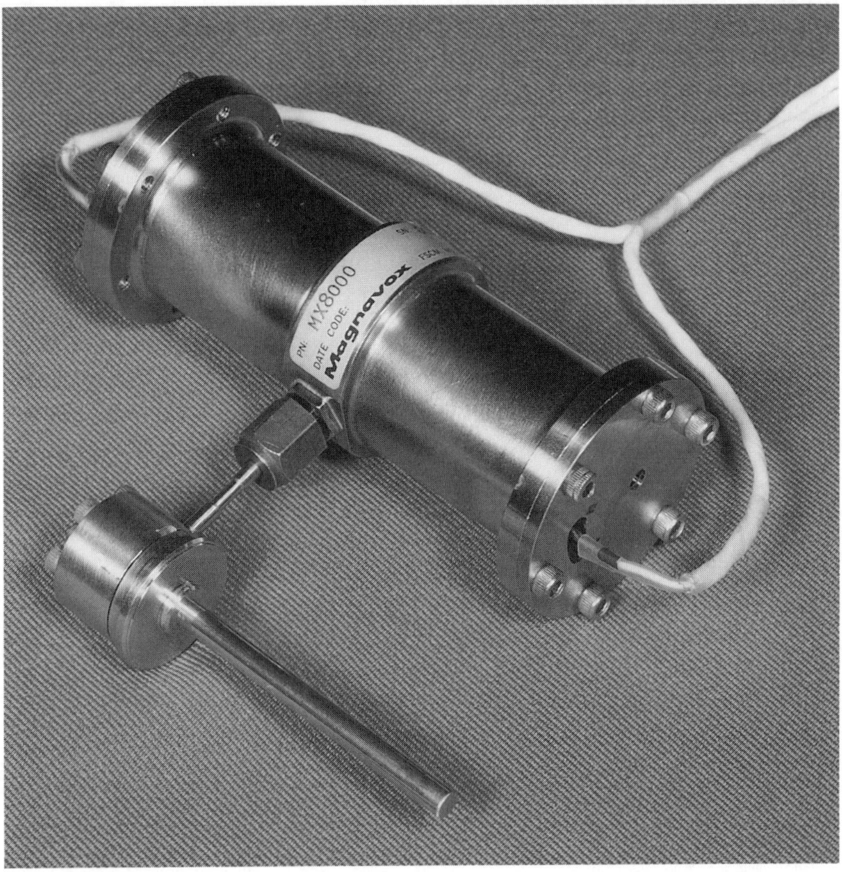

FIGURE 5.8 The Magnavox 8000 Prototype Dual Opposed Piston, Split Stirling Cryocooler (courtesy of Magnavox)

is a photograph of such a cryocooler configuration. These "opposed piston" designs exhibit the lowest vibration characteristics. Some designs employ dual opposed compressor systems with active balancing for a single displacer. The linear vibrations have numerous significant harmonics. However, the higher harmonics have small amplitudes and isolation is easy. Vibration reduction has become quite advanced for spacecraft sensor cryocoolers, which may employ electronic active vibration cancellation. Coolers are being developed to respond to requirements for vibration forces below 0.1 lb. In all cases, one should be aware of unexpected vibration components. Metal diaphragms, for instance, sometimes have a rotational vibration.

Work is also progressing in counterbalancing the pistons in Stirling cryocoolers. In Stirlings, a vibration value of 0.2 newtons (N) has gained community ac-

TABLE 5.2 Stirling Cycle Cryocoolers and Attributes

Manufacturer	*Model*	*Type*	*Cooling Capacity (W)*	*Temp. Supplied (K)*	*Weight (kg)*
AEG	HD1033	IS	1	80	1.68
	SC025	SS	0.25	80	1.1
	SC100-01	SS	1	80	1.81
Balzers	UCH 065		65	77	13
	UCH 065		6	20	14
	UCH 110		100	77	13
	UCH 110		12	20	14
	UCH 150		160	77	13
CTI-Cryogenics	VM-1	VM	1	77	4.5
	CM-2A	IS	1	80	1.77
	CM-3	SS	3.5	80	2.95
	CM-4	SS	1	80	1.72
	CM-5	SS	0.35	80	1.14
	CM-6	IS	0.35	80	0.95
	CM-7	SS	0.9	72	3.6
Hughes	7000H	SS	1	80	3.18
	7004H	SS	1	80	1.81
	7012H	IS	1	80	1.81
	7013H	SS	0.35	80	1.14
	7014H	SS	0.4	85	1.14
	7022H	SS	1	80	1.81
	7024H	SS	0.4	85	1.14
Magnavox	MX7043		0.88	77	1.8
	MX7045L	SS	0.35	80	1.13
	MX7048	SS	0.4	80	0.9
	MX7049	SS	1.75	80	1.59
	MX7058	SS	1.3	80	1.8
	MX7051	SS	0.65	77	1.37
Ricor	K506B	IS	0.3	70	0.45
	K526	SS	0.25	80	0.93
	K526S	SS	0.5	80	0.9
	K542 IDCA	IS	0.5	75	0.65
Signaal	UP7056	SS	0.5	80	1.2
	UP7057	SS	0.7	80	1.9
	UP7058	SS	1	80	1.8

Power Consumed (W)	*Reliability (MTTF or MTBF)*	*Watts per Watt*	*kg per Watt*	*Information Source*
50	1800	50	1.68	AEG
35	2000	140	4.40	
50	5000	50	1.81	
		0	0.20	Balzers
	8000	0	2.33	
		0	0.13	
	8000	0	1.17	
		0	0.08	
	2200	0	4.50	CTI-Cryogenics
55	2500	55	1.77	
180		51	0.84	
60	1000	60	1.72	
35	1000	100	3.26	
22	1000	63	2.71	
75	500	83	4.00	
90	1000	90	3.18	Hughes Aircraft Co., Electro-Dynamics Div.
40	1500	40	1.81	
35	1000	35	1.81	
17		49	3.26	
17	1500	43	2.85	
40		40	1.81	
17		43	2.85	
55	2500	63	2.05	Magnavox
30	2500	86	3.23	
55	2500	138	2.25	
90	2500	51	0.91	
55	4000	42	1.38	
40	2500	62	2.10	
12	6000	40	1.50	Cincinnati Electronics
17.5	1500	70	3.72	
40	1500	40	1.80	
20	8000	40	1.30	
30	3500	60	2.40	Signaal USFA
50	3500	71	2.71	
55	3500	55	1.80	

ceptance as a vibration goal. Other goals are for approximately 1/100 g or 0.01 N RMS at 30 to 60 cycles per second (CPS). The expander itself is expected to produce at least 0.1 N, but if counterbalanced, only 0.01 N may result from the unit. The amplitude can be expected to be only a few microns—much smaller than conventional FPA unit cells. The vibration is generated at a frequency (and its harmonics) that matches the piston movement—typically 20 to 60 cycles per second.

5.4.3 Figures of Merit

The current state of the art for power efficiency (or COP) for Stirlings is about 30 to 40 W of electrical input for 1 W of cooling output, at around 70 or 80 K. This is a good estimate if a particular cooler has not been chosen. Electrical efficiencies are expected to progress to 20 to 25 watts per watt by the mid-nineties for cold temperatures in the neighborhood of 80 K. The power required is more like a thousand watts per watt at the 20 K temperatures required by doped silicon. Not much effort is being expended with Stirlings into this low temperature range due to the maturing of HgCdTe and Pt:Si.

Weight is on the order of 2 kg/W of cooling at 70 to 80 K and 20 kg/W at 20 K. System cost per watt depends on size, reliability, cooling temperature, and the quantity of the order. System mass is on the order of kilograms per watt at 70 K and tens of kilograms per watt at 20 K.

Table 5.2 lists commercially available Stirling cycle cryocoolers and some of their attributes. Note that Hughes and Magnavox dominate the Stirling cycle cryocooler industry. In the early nineties, Magnavox was producing about 25 units per month, with a capacity for 50 units per month in a single eight-hour shift [Magnavox, 9].

Manufacturer-supplied numbers are hard to compare. The first problem is that some manufacturers quote mean time between failure (MTBF), but others quote mean time to failure (MTTF). Still others quote a 95 percent confidence level. MTTF is the best concept to use for remote systems that are difficult to repair (i.e., those in space). MTBF is more proper for lab or tactical units, where one expects a long life with suitable maintenance possible. Second, cryocoolers are not all tested in the same manner (case temperature, for instance). Third, manufacturers do not all define failure the same way. Suppose the efficiency of your system degrades 20 percent. Is this a failure or not?

Efficiency as a function of temperature and case temperature can be estimated using empirical data from various manufacturers. A Stirling's electrical efficiency as a function of cold finger temperature is roughly related to the change in temperature (from 77 K) raised to approximately the 2.5 power. Thus, a refrigerator operating at 38 K will be about 5.7 times less efficient than a similar version operating at 77 K. Conversely, one operating at 150 K will be 5.6 times more efficient than a similar one operating at 77 K.

Electrical efficiency (or COP) is also sensitive to the case temperature of the compressor for Stirlings. For small changes within the operating ranges, this is empirically related by

$$P_t \approx P_c + 0.3\Delta \tag{5.9}$$

where

P_t = power adjusted for temperature

P_c = power consumed

Δ = difference between actual and quoted case temperature

Mechanical cryocoolers lend themselves to customized figures of merit (FOMs) based on system requirements and specifications. The best way to evaluate one against another depends on your system-specific attributes, including delivery schedule, cost, and so forth. However, generalized FOMs can be developed easily and are useful for course comparisons. Equations (5.10) and (5.11) give starting points for space-based IR sensor cryocoolers and FLIR cryocoolers. Different mathematical emphasis is given for the differing importance of specific attributes.

For space basing, a figure of merit can be expressed as

$$\frac{\text{MTTF}}{(\text{Power FOM})\,(\text{Weight})^2} \tag{5.10}$$

This assumes a remarkably reliable cryocooler of 10 to 50 thousand hours MTTF and a spacecraft in which cost was not an important concern. Typically, the penalty on a spacecraft for consuming power via solar panels and recharging batteries is so large in weight (100 to 200 g/W) and cost that sometimes more emphasis is given to the power consumption.

FLIRs are usually produced in relatively large numbers, adding increased emphasis to cost. Their weight and power are less of a concern when considering jets, tanks, and ships as platforms. Because of this increased emphasis on real-world production issues, a FLIR figure of merit may be defined as

$$\frac{\text{MTBF}}{(\text{Weight})\,(\text{Power per watt})\,(\text{Dollars per watt})^{1.5}} \tag{5.11}$$

In both cases, the higher the number the better. (Managers seem to like that relationship.) Remember, only the relative differences matter.

These equations can represent a starting point in the search for the best cryocooler for a given application. In engineering reality, the particular system design will determine the figures of merit. The engineer should determine a custom figure of merit based on the individual IR sensor system. For example, cost may be of less concern than thermal control. The figure of merit should be adjusted accordingly. It is really up to the individual engineer on an individual project to determine the figure of merit. Finally, the actual selection may be determined by politics. Is one company a competitor? Does another company have an excellent record with your organization? It is also always wise to consider availability, configuration, and security issues.

5.4.4 The Quest for Long Life

Several U.S. and foreign government development efforts are currently directed toward increasing the performance and reliability of Stirling cryocoolers. Recent advances employing linear motors, flexure bearings, clearance seals, ultra pure gas, and increased efficiency have the promise of increasing reliability by orders of magnitude. The physics behind long life are well understood. First, the gas must be free from contaminants. Second, lubricants must not be present in the design. Third, wear debris must be minimized or eliminated. And fourth, changes in friction must be mitigated or eliminated. Today, the latter two considerations are usually accomplished via clearance seals. Exceptionally clean units with magnetic bearings and clearance seals have been operated for more than 20,000 hours, and MTTFs of 50,000 to 80,000 hours are likely.

Oddly enough, non-operating shelf life is important because many coolers have high-pressure pure gases that may leak. Gas contamination is a key factor in allowing demand control when the unit does not turn on and off. Gas contamination can be reduced by starting with an extra pure gas, selecting materials that do not outgas, vacuum baking parts, utilizing detailed cleaning procedures, and eliminating wear debris. Wear debris is reduced through materials, tribology, and clearance seals. Material selection must consider even the very subtle issues, such as plastic insulated wires, graphite epoxy, and ceramics—all which can be culprits responsible for uninvited outgassing. Mixing gases holds promise for higher efficiency and easier design. By mixing a monotonic and diatomic gas, better heat transfer and viscosity can be achieved.

Motors burn out and pollute the working gas. Simply locating the motor outside the working gas volume reduces contamination, thereby increasing reliability. Additionally, moving voice coil motors improve reliability over standard motors. A moving-coil motor is similar to an audio speaker. A changing magnetic field causes a relatively small linear movement. These motors have high reliability, allowing years of operation.

Magnetic bearings can increase life by supplying a touch-free, "frictionless" operation. Also, simply pre-loading the bearings and using precisely balanced shafts improves reliability. Other efforts have been directed toward compressors in an attempt to reduce (and in some cases eliminate) dynamic disturbances, giving the motor remarkably long life. One method uses a rotary compressor suspended on magnetic bearings sealed by a clearance seal. Additionally, new materials are being used to increase efficiencies. These include diamond coatings for heat transfer, composite materials for weight reduction, neodymium-boron-iron magnets for improved motor response, and titanium nitrite ion implanted surface coatings.

Seals (or piston rings) are another potential problem component. A recent trend has been the use of flexure bearings and clearance seals to support the linear motions of the pistons. Linear bearings can maintain radial clearance between a piston and bore on the order of 0.12 mm without rubbing. Eliminating the rubbing eliminates both wear and the need for lubricants. The elimination of wear also reduces contamination of the gas, which can also cause failures. Therefore, these linear bearings can lead to great reliability and long lifetimes. It is important to realize that flexure bearings will not support large acceleration loads. Thus, they do not appear to have a major role in tactical missions, where accelerations loads can be hundreds of Gs.

Using clearance seals with flex bearings provides for the longest life coolers. The staff of Stirling Technology Company developed flexure bearings in the early seventies and has used them in implantable Stirling artificial heart power sources. The company has implemented accurate computer optimization for Stirling coolers, employing flexural bearings and clearance seals to increase life and reliability. A 60,000 hour life with 98 percent reliability is considered achievable, with lab units demonstrating a life potential exceeding 60,000 hours. Flexure bearings have enormously long lifetimes if flexural stresses are kept below the endurance limit [Stirling Technology, 10]. Likewise, Creare has also developed a Stirling cryocooler in which metal diaphragms replace all seals and bearings. This cooler produces 2.6 W of cooling at 65 K, weighs 20 kg, and is expected to last ten years [Creare, 11].

5.4.5 Special Considerations For Space Applications

There is an increasing desire to use commercial Stirling coolers in non-commercial strategic search-and-track and space systems. This lure is easy to understand, given the current price and performance coupled with future promise. However, there are some subtleties that must be understood.

The space environment is a vacuum. There is simply no convection cooling. The cryocooler must be properly heat sunk and designed to allow operation by heat sink only. Space basing usually implies replacing the rubber seals and o-rings with crushed metal, such as aluminum, copper, indium, or gold. Casting leaks are also a problem with aluminum casting, especially when He is used as a working gas.

Spacecraft are electrically self-contained, raising issues about charging, electromotive forces, and electromagnetic induction. Commercial cryocoolers typically have a multiple-point grounds, poor electromagnetic interference (EMI) and electromagnetic susceptibility characteristics that require extensive shielding, and substantial ground loop control. Electronics are not convection cooled in space, so their thermal design must be scrutinized, often leading to redesign. The electronics are not made with S-level parts, and some parts are not even available at S-level. Many manufacturers simply are not capable of full space qualification or of meeting Mil-STD-1540B (testing of space vehicles). Vendors that can meet Mil-STD-1540B usually increase their price by approximately 8 to 12 times for this qualification.

However, electronics have not contributed to many space-based failures. The main reasons for failures of tactical coolers used in space can be traced to unknown or little-known considerations. In substandard clean rooms and with low gas purity, the contaminants freeze and throw the system off balance, resulting in increased part wear, increased vibration, and decreased cooling capacity. Cryocoolers should be assembled in class 1,000- to 10,000-level rooms, with materials strictly controlled and the parts vacuum baked. With rare exceptions, tactical producers usually don't understand these levels.

Standard tactical testing includes power-off periods every 20 to 24 hours; a useful and important attribute for FLIRS, but not applicable if one wishes to operate continuously. When operating continuously, problems arise from contamination in the working gas. Small amounts of water or CO_2 gradually freeze on the cold surfaces (e.g., the expander piston) when operating at 70 or 80 K. This may deposit enough mass to cause excessive vibration. If stopped every 20 hours and allowed to heat to room temperature, the contaminants will evaporate and redistribute themselves throughout the working gas. If these must run continuously, excessive vibration will result well after 20 hours of operation. Therefore, special tests and consideration to gas purity must be given. A unit should be tested in a manner simulating operating conditions. If a cooler is to run continuously, it should be tested continuously.

5.4.6 Pulse Tubes

Recently, a variant of the Stirling has been devised. Called the *pulse tube,* it doesn't employ a mechanical piston in the expander. The pulse tube is related to the Stirling cooler, but it flows gas through an orifice to replace the motion of the displacer or second piston. Thus, the expander module has no moving parts. The pulse tube is simpler and more reliable than standard Stirling coolers, offering advantages in cost, packaging, and vibration control. The primary drawback of the basic pulse tube cryocooler is an efficiency of about half that of comparable Stirling cryocoolers. Researchers have been busy finding new ways to configure the pulse tube to reduce its efficiency penalty relative to Stirlings. However, the thermodynamic irreversibility of the orifice implies that orifice pulse tubes will never match the efficiency of Stirling coolers.

The pulse tube works on the phase lag of the gas flow through an orifice. The mechanical motion of the expander's piston is replaced with a pneumatic equivalent. In essence, a "gas slug" replaces the displacer piston. This has the advantage of no moving parts near the cold end. The elimination of the moving mechanical displacer (at the cold end) improves reliability. The orifice pulse tube has high efficiency and simplicity and is capable of low operating temperatures [Radebaugh, 12]. Pulse tubes have a slight reduction in thermal efficiency, but they allow an easier compressor design, increased expander producibility, improved reliability, and reduce vibration concerns. While mechanical displacers are about 70 percent efficient, pulse tubes are more closer to 60 percent efficient. This slightly reduced efficiency requires a greater mass flow (approximately twice) and, therefore, a larger compressor to meet cooling requirements. However, the offset to this increased mass is that pulse tubes are expected to cost 60 to 80 percent of a standard Stirling, and they should be slightly more reliable. Additionally, the drive electronics are slightly less complicated. Because pulse tubes are highly developmental, commercial units are not yet available.

5.4.7 Integrated Dewar Cooler Assemblies

Originally, Stirling coolers were split to reduce vibration of the FPA, allowing one to place the heavy expander away from the detector for easier thermal management and lighter gimbals. This made perfect sense in the seventies, when the compressor weighed many kilograms, was as efficient as a personnel office, and vibrated like a jumping bean. However, as previously explained, this is no longer the case. Examining the data sheets in Appendix 5A, one finds many complete units weighing under two kilograms. Vibration is low, and efficiency is becoming higher and higher (requiring less heat to be rejected). In fact, as the technology advances, there are fewer and fewer reasons to split the compressor and the expander. The likely trend indicates that split Stirlings will disappear in the next few decades. There is a penalty for splitting the Stirling cycle. Split Stirlings are inherently less efficient and, as total systems, weigh more than integral Stirlings.

It is simply more thermodynamically efficient to place the entire cryocooler assembly within the sensor head and integrate the focal plane assembly with the cryocooler. The industry has loosely termed this concept the integrated dewar cooler assembly (IDCA). What does the resulting increased thermodynamic efficiency mean? First, a smaller cooler can do the same job. Second, less power is consumed. Third, reliability is increased, because the job is easier and requires less piston travel over its life. Again, the penalties of vibration, heat rejection, and weight are being mitigated.

Realizing this, several manufacturers have specially produced both integral and split Stirling coolers designed to be used as IDCAs. Most have dynamic clearance seals, dc brushless motors, preloaded bearings, and high-endurance lubricants to achieve high MTTF. One implementation has a cold finger so rigid that a 5 G, 5–5,000 Hz vibration causes less than a 2 μm deflection—far less than conventional pixel sizes [Owen, 13]. The entire weight is less than half a kilogram, and it can provide a third of a watt of cooling at 70 K for 12 W input [Owen, 14]. The ICDA concept must be considered and traded for any project. Even a gas blow-down on a missile seeker may weigh and cost as much as a modern IDCA.

5.5 OTHER CLOSED-CYCLE CRYOCOOLERS

As mentioned previously, there is a plethora of thermodynamic processes to accomplish cryogenic refrigeration with a closed-cycle mechanism. The Stirling was just discussed. Following are discussions of Vuilleumier, closed-cycle J-T, Magnetic-caloric, Sorption, and Byatron systems.

5.5.1 Vuilleumier Systems

Vuilleumier (VM) systems are heat driven and have constant volumes. The working gas is moved from one section to another by easily sealed displacers. The VM process, named after its inventor, is a heat-driven, constant-volume derivative of the Stirling. Piston-like displacers move gas 90 degrees out of phase through three working constant volumes. The cold and hot end volumes act like expanders, and the intermediate volume acts as a compressor. Displacers move the gas without compression. It was once believed that the VM had the theoretical advantage of longer life than Stirlings. Hughes Aircraft ran a VM cooler successfully for 2.5 years at double normal speed to simulate a five-year operation in space ["New Chapters in the Science of the Icy Cold," 15]. However, the extension of life and reliability for the Stirlings has kept pace. While much effort has been expended by government and industry on development of VM coolers for space and tactical applications, their success has not been sufficient to replace Stirlings. VMs are not commercially available and are unlikely to provide benefit over Stirling in the nineties. It is commonly felt that VMs are heavy, not easily space qualifiable, prone to leakage, and subject to rapid wear. It is doubtful if the VM design can compete with new generation Stirling coolers.

5.5.2 Closed-Cycle J-Ts

J-T coolers may be open or closed. The gas may escape and be lost or recaptured, repressurized, and used again. Closed systems are typically heavy, as they require a multi-stage, ultra-clean compressor to repressurize the gas to thousands of pounds per square inch. These high pressures require compressors that are large, inefficient, and unreliable in field systems. Currently available tactical units require maintenance every 200 hours and must be overhauled at 2,500 hours. Their use is becoming limited to laboratory systems. The key to their problems is the very high pressures to which the expanded gas must be repressurized. This is orders of magnitude higher than the maximum pressure variations encountered in a Stirling, making the valves, seals, and motors less reliable.

5.5.3 Magneto-Caloric

Magneto-caloric coolers operate by moving a field relative to a paramagnetic material. The materials experience a drop in temperature when they are demagnetized. Typical designs have a magnetic material formed into a wheel that rotates near a material with the proper temperature entropy characteristics. As the field changes, temperature drops at one end and heat is rejected at the other. The major advantage of these systems is their capability to cool below a single K with a complete lack of pressurized gases. They tend to weigh a lot and have reliability driven by the motor, which either moves the field or the material though the field. An example developed by Hughes used a gadolinium gallium garnet disk moving at 10 rpm to cool below liquid helium temperatures [Nordwall, 16].

5.5.4 Sorption

Sorption refrigerators employ a chemical and a mechanically and/or thermally-driven process in tandem. A sorption compressor is usually combined with an expander. The fundamental process of gases being absorbed onto a surface or vacant interstitial sites is exothermic. Chemical van de Waal bonds (usually covalent) are formed between the sorbent and sorbate [Wade, 17]. Heat is then required to desorb (emit) the gas. This process can be configured to provide cooling. For cryogenic cooling, a cycle is initiated with mechanical compressors and/or electrical heaters that cause heating, desorption, cooling, and, finally, absorption—often in several stages of each. Some designs use charcoal, while others use the outgassing of O_2 from a metal hydride at high pressure. Sorption compressors are more efficient when producing high pressure ratios and low flow rates, as favored by J-T expanders [Wade, 18]. Thus, many designs can send gas through a Joule-Thompson cryostat to provide cryogenic cooling.

Most current sorption coolers require hundreds of watts to produce a watt of cooling, although efficiencies are projected to approach 30 or 40 W/W at 65 K. Cascading multiple stages can result in very low temperatures. For instance, Aerojet was awarded a $2.7 million, 36 month development contract to build a space-suitable 10 K cryocooler ["Aerojet to Make Cryocooler for Missile Sensors in Orbit," 19]. In addition to cascading, hybrid designs with magnetic coolers have advantages for cooling below 20 K.

There exist unanswered reliability questions due to the required heater filament wires, contamination control, and mechanical pressure pumps. However, there are reasons to suspect that system reliability can approach levels required for ten-year lifetimes. These include [Wade, 20]

- minimal strain on the plumbing
- moderate temperatures and pressures for the compressor elements
- in some designs, J-T expanders that eliminate moving parts at the cold end
- compressors located 23 meters or more from the cold end
- compressors that have been tested to over 17,000 hours and 37,000 cycles

NASA's Jet Propulsion Laboratory (along with several private companies) has been perusing sorption technology. This promises high efficiency and high cooling capability at a moderate weight. Today, it is applicable only to large sensor systems with large periodic cooling needs.

5.5.5 Brayton and Claude

Brayton, reversed Brayton, and Claude cycles are mechanical systems providing cooling based on gas expansion through a turbine. Basically, a compressor pressurizes the working gas at ambient temperature. Obviously, the gas heats up, so it is sent to a heat exchanger to be cooled. Thus far, the process is analogous to an intercooled automobile turbocharger. Finally, the gas expands across a turbine and provides cooling and energy for the mechanical process. The gas then goes to the compressor.

5.6 RADIATORS

Radiators are a wonderful way to cool a focal plane on a spacecraft. There are few or no moving parts or vibration, reliability is high, and achievable temperatures can be impressive with tiny heat loads. Unfortunately, they are not usable on airplanes or ground systems for direct FPA cooling because they cannot radiate to cryogenically cold environments. They are especially useful for interplanetary probes or orbiters for the outer planets. Radiators have been used to cool focal planes to 110 K on such missions. Earth orbiting spacecraft can employ radiators, but proper consideration for the level of solar synchronization and beta angles must be thoroughly examined. Often, moving shades or slightly gimbaled radiators are required to eliminate direct Earth, sun, or moon viewing.

A radiator simply radiates heat to a cold surface or the vacuum of space by

$$Q_r = \varepsilon\sigma V\,(T_1^4 - T_2^4) \tag{5.12}$$

where

Q_r = heat load from radiation in W/m^2

ε = emissivity*

σ = Stefan-Boltzmann constant of 5.67×10^{-8} W/m^2/K^{-4}

V = viewing fudge factor, depending on placement on spacecraft†

T_1 = temperature of radiator surface

T_2 = temperature of cold end

Equation (5.12) allows one to rapidly size a radiator for steady-state, constantly radiating conditions. System design requires that the size be kept to a minimum to reduce cost, weight, and packaging problems. In light of these requirements and Equation (5.12), the radiator should have high emissivity, high temperature, and a view of nothing but deep, cold space. The latter is rarely the case 100 percent of the time. Sizing radiators for complex duty cycles based on orbits and spacecraft configuration is better left to computer simulations. However, if the above equation yields an area that is even borderline in acceptability, a radiator will not be acceptable.

There are plenty of broadband black coatings in the infrared that are reasonably reflective in the visible (to mitigate heat input from the sun). Black coatings are a three-edged sword. The blacker the radiator, the better it is at conducting heat away from the focal plane—exactly what is desired. However, it will also be better at absorbing heat from other sources (e.g., the sun or solar panels) and conducting it to the focal plane. The third edge comes with time: a major concern with such coatings is their ability to withstand wear and aging. As spacecraft contaminants are deposited on them, as they are exposed to UV and other radiation, and as they chip, flake, and age, they tend to become less black in the IR and less reflective in the visible. Finally, one must consider the emissivity of the entire radiator, not just

---

*If you don't have a measured value, get an approximate value from the *IR Handbook* or assume it is 0.9 for a black surface, 0.2 for a shiny surface, or 0.8 for anything else. A good radiator coating, even when aged, should be near 0.9.

†If you don't know this value, assume 0.7.

the coating. The radiator can be constructed like a blackbody with a rough surface and deep groves to maximize radiating area while minimizing heat input area.

The higher the operating temperature of the radiator, the better. Unfortunately, physics requires the radiator to be at a colder temperature than the focal plane to remove heat from the focal plane. For a radiator cooling only a focal plane, a refined design can keep this temperature difference to 5 to 20 Ks. Thus, cooling a focal plane to 120 K requires a 100 to 115 K radiator. Keeping a radiator this cold is difficult, because other parts of the spacecraft will tend to radiate onto the radiator and heat it up. Additionally, astronomical sources of radiation such as the sun, moon, and Earth (or any planet and moon system) will also tend to heat up the radiator. Shades are often applied during times of direct (or near direct) solar viewing. Unfortunately, motor-driven mechanical shades raise reliability questions. However, shaped memory shades and materials that change emissivity based on temperature or flux are potential solutions to shading with high reliability.

When shaded, radiators cannot provide cooling. A thermal mass must provide the needed ballast or inertia to keep the FPA cool during shaded periods. Phase change materials are often employed to provide cooling during such radiator outages These challenge engineers in a zero-gravity environment, because things tend to float around and not separate by weight as they do in a gravity environment. The result is the loss of efficient thermal contacts of the various phases

However, there have been significant improvements in coatings, materials, heat pipes, thermal straps, and thermal diodes. If weight is a critical parameter, the radiator may weigh as little as a gram per square centimeter. The entire system, including shades, heat pipes, and so on, can weigh a few grams per square centimeter. A rough, empirical gauge (good within a factor of three between 100 and 250 K) of their weight can be gained by knowing the temperature to which one wishes to cool and the cooling capacity. An empirical relationship based on data in the *IR Handbook* with the constant updated for modern designs indicates

$$W \approx \frac{(9\times 10^{10})\ (Q)}{T^{3.5}} \tag{5.13}$$

where

W = weight in grams

Q = cooling capacity in watts

T = temperature desired after cooling

Examining Equation (5.13), one can see that radiators are only competitive with other methods when the load is a few tens of milliwatts, and for temperatures above about 130 K for Earth orbiters. However, they can be especially applicable to long-life missions to the outer planets that can more easily view deepest, darkest space. The few systems that still employ this technique do so to avoid the questions of reliability and vibration associated with mechanical cryocoolers.

Radiators must be considered for use in any spacecraft, regardless of cooling method. This is because the heat from any reusable cooler (be it a TEC, VM, or Stirling) eventually must be rejected into space. The benefit of using a TEC or mechanical cooler between a cold FPA and warm radiator is that it allows heat to be removed at a very low temperature (i.e., 80 K) and rejected at the radiator at a high temperature (i.e., 300 K). From Equation (5.12), one can see that the higher the temperature of the radiator, the more heat can be dumped via radiation per unit area. As a result, one must consider a tradeoff using a mechanical or TEC cooler for any FPA temperatures. In this case, we trade a smaller radiator for the additional cooler.

References

1. R. Ross, D. Johnson, and R. 1991. Sugimura "Characterization Of Miniature Stirling Cycle Cryocoolers For Space Application." *Proc. Sixth International Cryocoolers Conference (DTRC-91/002),* 27–38.
2. J. Kreider, et al. 1991. "Multiplexed Mid-Wavelength IR Long Linear Photoconductive Focal Plane Arrays." *Proc. SPIE,* 376–388.
3. G. Bonney and R. Longsworth. 1991. "Considerations in Using Joule Thompson Coolers." *Proceedings of The Sixth International Cryocoolers Conference (DTRC-91/002),* 231–44.
4. S. Whicker. 1992. "New Technologies For FPA Dewars. *Proc. SPIE* 1683:102–112.
5. 1992. Information courtesy of MMR Inc.
6. W. Wolf and G. Zeisis, eds. 1985. *The IR Handbook* (Ann Arbor, Michigan: The Infrared Information and Analysis Center of ERIM) 14–15.
7. Ibid.
8. S. Fujino. 1989. "Mitsubishi Thermal Imager Using the 512 by 512 Pt:Si Focal Plane Arrays." *Proc. SPIE* 1157:136–143.
9. 1993. Information courtesy of Magnavox.
10. 1993. Information courtesy of Stirling Technology Company.
11. Information courtesy of Creare and from B. Henderson. April 6, 1992. "US Industry to Produce Long Life Space Cooling System." *Aviation Week and Space Technology,* 41–43.
12. R. Radebaugh. 1987. "Pulse Tube Refrigeration—A New Type of Cryocooler." *Japanese Journal of Applied Physics,* 26:2,076–2,081.
13. A. Owen. 1990. "Miniature Integrated Dewar/Cooler Assembly." *Proc. SPIE* 1308: 303–311.
14. Ibid.
15. November 3, 1987. "New Chapters in the Science of the Icy Cold." *Vectors* 24:14–17.
16. B. Nordwall. January 18, 1988. "Hughes Develops Magnetic Refrigerator For Space Based Sensors, Signal Processors." *Aviation Week and Space Technology,* 53.
17. L. Wade. 1992. "An Overview of The Development of Sorption Refrigeration." *Advances in Cryogenic Engineering* 37, part B:1,095–1,105.

18. Ibid.
19. November 23–29, 1992. "Aerojet to Make Cryocooler for Missile Sensors in Orbit." *Space News,* 12.
20. L. Wade. 1992. "An Overview of The Development of Sorption Refrigeration." *Advances in Cryogenic Engineering* 37, part B:1,095–1,105.

Appendix 5A

Representative Equipment Data Sheets

AEG Fast Cool-Down J-T Cooler Detector Unit

Manufacturer: AEG, Germany

Intended Application: Missile seeker, smart ammunition

Description:

The entire unit is a compact cylindrical assembly. The gas is supplied to a spiral J-T heat exchanger with the exhaust in the center.

Characteristics:

Operation Time: 15 seconds

Cool-Down Time: 0.4 seconds

Cooler Temperature: <100 K

Shock Survival: 40,000 g

Gas: 20 cc of argon @ 800 bar

Information courtesy of AEG

AEG HD 1033

Manufacturer: AEG, Germany

Qualification: DLSM-C-971500

Intended Application: Military products

Availability: approximately 3 to 6 months ARO from continuing production line

Description:

AEG coolers have long-life operation and low noise and vibration. This integral Stirling houses the compressor in a cylinder and has a common module applicable displacer. The motor is a rotary type.

Characteristics:

Cooling Capacity: 1 W @ 80 K and 23° C ambient

Cool-Down Time: 11 minutes to 100 K and 13 minutes to 80 K

Power Consumption: 150 W for the first 5 seconds and < 50 W steady state

Vibration Output: ≤ 0.35 lb in all axes

Weight: 1.7 kg

MTBF: >1,800 hours

Price: approximately $4,000

Information courtesy of AEG

AEG Linear Split Stirling

Manufacturer: AEG, Germany

Intended Application: Military products

Availability: approximately 6 months ARO from continuing production line

Description:

AEG coolers have long-life operation and low noise and vibration. This split Stirling houses the compressor in a cylinder and has a common module applicable displacer. The motor is a linear dc type.

Characteristics:

Power In: <30 W for 0.35 W load @ 77 K to <55 W removing 1 W @ 77 K

MTBF: >7,000 hours

MTTF: >4,000 hours

Information courtesy of AEG

AEG SC 100

Manufacturer: AEG, Germany

Intended Application: Military products

Qualification: DLSM-C-971500

Availability: approximately 6 months ARO from continuing production line

Description:

This split Stirling houses the compressor in a cylinder and has a common module applicable displacer. The motor is a linear dc type.

Characteristics:

Cooling Power: 1 W @ 80 K and 23° C ambient

Typical Cool-Down Time: 10 minutes to 100 K and 13 minutes to 80 K

Power Consumption: <50 W

Compressor Dimensions: 6 cm diameter and 12 cm long

Operational Temperature Limits: –54° C to +71° C

MTBF: >2,200 hours without DRC, >3,000 hours with DRC

Vibration: 0.8 N max

Weight: 1.8 kg

Price: approximately $10,000

Information courtesy of AEG

AESC Sorption Refrigerator

Manufacturer: AeroJet, Electronic Systems Division, Azusa, California

Intended Application: Space-based NASA, USAF

Qualification: Any available, including space

Availability: Under development

Description:

The sorption refrigerator operates by the adsorption of the coolant gas by an appropriate medium, such as metal hydride. The coolant gas is expanded at high pressure through a Joule-Thomson cryostat to liquefy the gas and produce cooling. The return gas is adsorbed in the sorbent bed. This bed is subsequently heated to drive off the adsorbed gas, at high pressure, hence repressurizing the supply vessel for the cryostat. As valving for this process operates at a very low rate, there is little wear, and, hence, intrinsic long life. Cascading of the refrigeration processes provides additional capabilities and temperature ranges. For example, the required precooling of the hydrogen for a 20 K system can be achieved by an upper stage consisting of another sorption cooler operating at 65 K. Likewise, the achieving of 10 K cooling is facilitated by the solidification of the liquid hydrogen generated by a 20 K cooler. Such solidification is achieved by "vacuum pumping" the liquid using still another sorption bed. Both periodic and continuous sorption refrigerators are under development. The periodic device uses a single (or redundant, as desired) set of sorption beds and reservoirs. Cooling is followed by regeneration in a sequential manner. The continuous cooler typically uses multiple beds and reservoirs, such that a continual process may be implemented, wherein one set of reservoirs is being used for cooling while another set of beds is being heated to repressurize the related reservoir. The refrigerator is capable of achieving specific efficiencies comparable to those of mechanical refrigerators. Programs are currently underway to demonstrate space performance of such refrigerators in both continuous and periodic modes of operation.

Characteristics:

Cooling Capacity: <1 W

Temperature: 130 K

Refrigerant: Krypton

Sorbent: Saran charcoal

Volume: <0.03 m3

Weight: < 18.2 Kg

Efficiency: < 40 W/W

Information courtesy of Aerojet ElectroSystems

Ball Stirling Space Cryocooler

Manufacturer: Ball Corporation, Boulder, Colorado

Intended Application: Space applications for NASA and surveillance

Qualification: Space qualification

Availability: Delivery approximately 8 months ARO

Description:

This is a long-life, low-vibration space cryocooler. It utilizes the Stirling cycle with duel opposed compressor pistons and a momentum compensated displacer. Diaphragm springs and clearance seals are used to eliminate all wearing surfaces. The unit contains ultra pure gas and few nonmetallic parts. All parts are processed by vacuum baking to prevent gas contamination.

Characteristics:

Minimum Cold Temperature: 17 K

Cooling Capacity:

Temperature	*Heat lift*	*Efficiency*
30 K	0.44 W	132 W/W
55 K	1.06 W	55 W/W
65 K	1.25 W	46 W/W
80 K	1.55 W	37 W/W

Lifetime: 10 years

Vibration: <0.05 lbf

Weight: approximately 14 kg

Information courtesy of Ball Corporation

CM-2A

Manufacturer: CTI-Cryogenics, Waltham, Massachusetts

Intended Application: Various sensor systems

Qualification: US Army spec. standard, all levels on a custom request basis

Availability: Custom made based on previous work

Description:

The CM-2A is an integral Stirling closed-cycle refrigerator with the regenerator at right angles to the main compressor units.

Characteristics:

Cooling Capacity: 2 W @ 13 K

Input Power: 6,100 W

Weight: 104.5 kg

Information courtesy of CTI- Cryogenics

CM-4

Manufacturer: CTI-Cryogenics, Waltham, Massachusetts

Intended Application: Various infrared sensors

Qualification: All levels on a custom request basis

Availability: Custom made based on previous work

Description:

The CM-3 is a split Stirling cryocooler that has been in low-volume production since 1980.

Characteristics:

Length: 23 cm

Diameter: 10.2 cm

Cooling Capacity: 3.5 W @ 80 K

Input Power: 180 W

Weight: 2.95 kg

Information courtesy of CTI- Cryogenics

CM-5

Manufacturer: CTI-Cryogenics, Waltham, Massachusetts

Intended Application: Various infrared sensors

Qualification: All levels on a custom request basis

Availability: Custom made based on previous work

Description:

The CM-3 is a split Stirling cryocooler.

Characteristics:

Length: 22.9 cm

Diameter: 10.2 cm

Cooling Capacity: 3.5 W @ 80 K

Input Power: 180 W

Weight: 2.95 kg

Information courtesy of CTI-Cryogenics

Creare 65 K Stirling

Manufacturer: Creare, Hanover, New Hampshire

Intended Application: Space

Qualification: Space—all levels on a custom request basis

Availability: Custom made based on previous work

Basing: Space

Description:

Double-acting diaphragm Stirling cooler with dual opposed compressors.

Characteristics:

Cooling Load: 2 W @ 65 K

Survive Loads: Launch

MTTF: 90,000 hours with a 95 percent reliability

Start/Stop Cycles: 1,000

Weight: Approximately 20 kg including electronics

Information courtesy of Creare

Creare Reverse Brayton

Manufacturer: Creare, Hanover, New Hampshire

Qualification: None (in brassboard development)

Availability: Highly developmental

Basing: Various, including space

Description:

This effort is directed to the next generation of high-performance, closed-cycle cryocoolers. They are developing a reverse Brayton suitable for long life. The key components (turbo compressor, turbo expander, and heat exchanger) have been designed and exist as a functioning prototype. Vibration is undetectable as the turbine and compressor run in self-acting gas bearings.

Characteristics:

Turboexpander Speed: 500,000 rpm

Compressor Speed: 450,000 rpm

MTTF Goal: >5 years. Currently, the bearings have operated for >8 years without degradation.

Cooling Load: Various from 30 to 250 K (5 W @ 70 K and 0.5 W @ 40 K)

Gas: Neon or helium

Efficiency Goal: 10 percent of Carnot for 5 W @ 70 K

Electrical Efficiency: Consumes 200 W to produce 5 W @ 70 K

Information courtesy of Creare

EE.&G Open Cycle Cryostats

Manufacturer: EG&G Judson, Montgomeryville, Pennsylvania

Intended Application: Various

Qualification: Any specification or qualification available on custom basis

Availability: Custom made from continuing line, delivery approximately two months ARO

Basing: Various

Description:

These are fixed-orifice cryostats suitable for use with common module and advanced dewar configurations. They are miniature, self-regulating devices designed for rapid cooling with minimal gas consumption. The flexibility of their small diameter tubing is essential in reducing the stress imposed on glass dewars by high-pressure plumbing. The tubing and fittings are rated for 7,500 PSI service; burst pressures are in excess of 20,000 psi.

Characteristics:

Gas: Argon or nitrogen, depending on cryostat selection

Diameter: 5.3 or 8.3 mm

Total Length: 67.3 or 68.3 mm

Cooling Capability: Up to 2 W

Price: approximately $2,000

Information courtesy of EG&G Judson

Hughes 7014H Split Stirling Cooler

Manufacturer: Hughes Electron Dynamics, Torrance, California

Intended Application: FPA cooling

Qualification: Mil-Spec

Availability: Custom made from continuing line; delivery approx. 3 months ARO

Description:

The 7014H is a single-stage, split Stirling cooler with a rotary drive. It has a cylindrical shape, about 1.75 inches in diameter and 5.6 inches long, with a flexible 18 inch line leading to the cold finger. The unit was made to cool U.S. and British common module detectors and is often used in FLIR applications. Special attention to vibration was given during the design. Hughes has built over 5,000 of these units, and current production is approximately 150 per month.

Characteristics:

Cooling Capacity: 0.4 W @ 85 K

Power Consumption: 30 W

Cool-Down Time from 300 K to 80 K: 6 minutes

Weight: 1.1 kg

Information courtesy of Hughes Electron Dynamics Division

Hughes 7050H Split Stirling Linear Drive Cooler

Manufacturer: Hughes Electron Dynamics, Torrance, California

Qualification: Mil-Spec

Availability: Custom made from continuing line; delivery approximately 3 months ARO

Description:

The 7050H is a single-stage split Stirling linear drive cooler. The compressor is cylindrical, with the line to the regenerator extending from the center. There is an additional small electronics box to control the cooler. The cooler is specifically designed for commercial and industrial applications requiring cooling to temperatures from 45 to 100 K in ambient temperatures of –50° C to 71° C. This closed-cycle cooler provides the user with high reliability at low cost. The cooler operates at low acoustic noise levels and very low vibration forces.

Characteristics:

Cooling Capacity: 0.8 W @ 80 K

Input Voltage: 18 to 32 Vdc

Power Consumption: 50 W maximum

Cool-Down Time from 300 K to 80 K: <3.5 minutes

Weight: 1.6 kg

Acoustic Noise: 15 dB @ 1 kHz

Expander and Compressor Vibration Forces: <0.15 lbf peak

Compressor Size: 12 cm by 6.4 cm in diameter

Electronics Box Size: 9.4 by 12 cm by 3.5 cm

Information courtesy of Hughes Electron-Dynamics Division

Inframetrics Microcooler

Manufacturer: Inframetrics, No. Billerica, Massachusetts

Intended Application: Thermal imaging

Qualification: Any available, including space

Availability: Custom made from continuing line

Description:

The microcooler was designed for IDCAs and has been used in NASA's microgravity laboratory on the shuttle. The cooler is a tiny integral Stirling with the compressor at right angles to the displacer.

Characteristics:

MTTF: 2,000 hours

Weight: 300 g

Power Required: <3 W

Cooling Capacity: 150 milliwatts @ 77 K, 100 milliwatts @ 65 K

Information courtesy of Inframetrics

Magnavox MX 7045L

Manufacturer: Magnavox, Mahwah, New Jersey

Intended Application: Various

Qualification: Flight qualification

Availability: From Continuing production line

Basing: Ground, flight, space potential

Description:

The 7045L is representative of the Magnavox line. The units rely on convection cooling of the compressor but can be adapted for space use. The coolers run from dc input. The circuitry supports a demand cycle where, after cool-down, it only operates enough to maintain the static heat load.

Characteristics:

Cooling: 1 W @ 80 K, 0.4 W at 70 K

Weight: <2 kg

Price: $15,000 (1990)

MTTF: >2,500 hours.

Electronics Price: $4,500 (1990)

Input Power: 30–55 W @ 32 volts

Size: approximately 15.2 cm long and 7.6 cm in diameter

Cool-Down Time: 7.5 minutes or less to 100 K with a 1.8 gram copper mass

Information courtesy of Magnavox

Magnavox MX 8000 Integrated Microcooler/Dewar

Manufacturer: Magnavox, Mahwah, New Jersey

Intended Application: Various commercial and government

Qualification: Mil-spec/flight qualification

Availability: approximately two months ARO

Basing: Ground, flight, space potential

Description:

The MX 8000 is Magnavox's entry into the small IDCA product area. The compressor employs dual opposed piston design with clearance seals. It is a linear resonance design with low acoustic noise and vibration.

Characteristics:

Weight: <450 grams

Cooling Capacity at 80 K and 23° C Ambient Case Temp: 0.15 W nom, 0.4 W max @120 K 0.70 W max

Operating Ambient Temp: –41° C to 71° C

Cool-Down Time: 5 minutes from 300 to 80 K with a 250 J thermal mass

AC Power Input: 8 W nominal, 18 W peak

Input Voltage: 6 Vdc or will customize

MTTF: >4,000 hours

Information courtesy of Magnavox

MMR FCD OPEN CYCLE CRYOSTATS

Manufacturer: MMR Technologies, Mountain View, CA

Intended Application: Various

Qualification: Specifications available on custom basis

Availability: Custom made from continuing line; delivery approximately 1–2 months ARO

Basing: Various

Description:

These cryostats use photolithographically-defined channels to provide numerous channels in a small area. These basic designs have been around since 1984.

Characteristics (FDC-3-1):

Cool-Down Time: 5 seconds

Final Temperature: 90 K

Capacity: 0.15 W

Stability: ± 0.1 K

Gas: Argon

Pressure: 4,500 psig

Weight: 0.5 grams

Price: approximately $5,000

Information courtesy of MMR

Oxford 80 K Cooler

Manufacturer: British Aerospace, Oxford, RAL

Intended Application: Space

Qualification: Space

Availability: Industrial off-the shelf

Description:

The Oxford cooler was a breakthrough in reliability and applicability to spacecraft. It was based on coolers made for the shuttle-launched ISAMS atmospheric instrument. The cooler is a split Stirling using metal diaphragm springs, gas clearance seals, a conservative thermal design, and extremely clean parts and gases, coupled with extremely low leak rate containers.

Characteristics:

Cooling Capacity: 0.8 W @ 80 K

Lifetime: 10 years with 93 percent reliability

Vibration: 0.1 Newton seconds peak uncompensated, 0.001 Newton seconds compensated

Weight: compressor: 3 kg, displacer: 0.9 kg and 3.5 kg for the electronics

Power Consumption: compressor: 30 W, displacer: 1 W, electronics: 9 W

Operating Frequency: 40 Hz

Compressor Size Envelope: 20 cm by 12 cm in diameter

Displacer Size Envelope: 21 cm by 7.5 cm in diameter

Electronics Size Envelope: 25 cm by 25 cm by 12 cm

Lowest Temp Achievable: approximately 55 K

Information from C. Jewell, C. and B. Jones, "Mechanical Coolers: An Option for Space Cryogenic Cooling Applications," ESA Bulletin *(May 1990): 79–85*

Phillips Prototype 5-Year Stirling

Manufacturer: Phillips Labs, Briarcliff Mannor, NY

Intended Application: Lab demo for eventual space deployment

Qualification: None

Availability: Highly developmental prototype

Description:

The prototype is a linear integral Stirling with a single compressor and single expander. System vibration is kept extremely low with a counter-balancer. The refrigerator uses closed loop controlled moving magnet linear motors. Moving elements are supported by active magnetic bearings with clearance seals of 20 μm.

Characteristics:

Cooling Capacity: 3 W @65 K with 20° C rejection

Lifetime: 3 years min, 5 year goal with 1,000 on/off cycles

Launch Load: 3 g steady, 5.4 g at 7 Hz

Working Gas: He

Displacer Diameter: 3.155 cm

Regenerator: phosphor bronze wire mesh with 53 μm diameter wires

Piston Diameter: 4.445 cm

Power Consumption: 250 W operating, 50 W standby

Weight: 84 kg, excluding electronics module

Information from C. Keung. et al., "Performance of a Prototype, 5 Year Lifetime, Stirling Cycle Refrigerator for Space Applications." Proceedings of The Sixth International Cryocooler Conference *(DTRC-91/002) (1991): 53–73*

Ricor K506 B Micro IDCA

Manufacturer: Ricor, Israel, distributed by Cincinnati Electronics, Mason, Ohio

Intended Application: Various

Qualification: Various per application

Availability: Made to order

Description:

The K506B is an integral (one-piece) mini-Stirling cryocooler. It has an integral cold finger/dewar for mounting a detector, cold shield, cold filter, and window. There is a standby mode that consumes 3 W and provides for quick cool-down. CE is the exclusive marketing representative for Ricor Ltd., the original designers and manufacturers of these minicoolers.

Characteristics:

Dimensions: 7 by 5.4 by 12 cm

Cooling Capacity: 100 to 400 milliwatts @ 90 K

Cool-Down Time from 300 K to 85 K: 4 min or less for thermal inertia of 50 J

Cool-Down from Standby Mode: 30 seconds

Input Power: 10 to 16 W

Input Voltage: 18–28 Vdc

Weight: 450–480 grams

MTTF: >5,000 hours

Information courtesy of Cincinnati Electronics

Ricor K526 Cryocooler

Manufacturer: Ricor, Israel, distributed by Cincinnati Electronics, Mason, Ohio

Intended Application: US Army, 60 element common module type thermal imagers

Qualification: Internal specification: Ricor IH0094

Availability: Made to order

Description:

The K526 is a split Stirling cryocooler. A package contains the compressor, which is a patented brushless dc motor. A transfer line connects this to the cold finger assembly, which supplies 1/4 W cooling capacity at 80 K.

Characteristics:

Unit Operating Temperature: –40 to 72 C

Cooling Capacity: 250 mW @ 80 K

Input Power: 30 W

Input Voltage: 17.5 ± 0.5 Vdc

Weight: 850 grams

MTBF: 1,500 Hours

Information courtesy of Cincinnati Electronics

SADA: Standard Advanced Dewar Assembly (SADA)

Manufacturer: Various planned; Santa Barbara Research Center currently

Intended Application: Replacement for Common Module, initial uses for the Comanche, LOSAT, ITAS and Javelin systems

Qualification: US Military BZ-A3165824 (SADA I) and BZ-A31658ZZ (SADA II)

Availability: Developmental with transition to production planned

Description:

SADA is a family of standard advanced dewar/cooler modules meeting the advanced infrared sensor deployment requirements of the 1990s and beyond. The SADA families include the high-performance SADA I, the medium/high performance SADA II, the compact SADA III, and the emerging SADA IV family. The SADA I interfaces with a 1.75 W linear cooler and the SADA II with a 1.0 W.

Characteristics:

FPA Temp Stability: ± 0.5 K

SADA I Dewar Heat Load: 225 mW active; 175 mW passive

SADA II Dewar Heat Load: 225 mW active; 175 mW passive

SADA I Cool-Cown Time from 300 K @ 23 ° C ambient: 8 minutes

SADA II Cool-Down Time from 300 K @ 23 ° C ambient: 10 minutes

SADA I Weight: dewar, 0.73 kg; cooler, 2.3 kg

SADA II Weight: dewar, 0.73 kg; cooler, 0.45 kg

MTTF: approximately 4,000 hours

Information courtesy of NVEOD

TADOPTR (Thermo acoustically Driven Orifice Pulse Tube Refrigerator)

Developer: NIST, Boulder, Colorado; Los Alamos Lab, Los Alamos, New Mexico

Intended Application: Various

Description:

This is a wonderfully creative concept for a cryocooler that requires no moving parts. Heat is applied to create a temperature gradient. When this gradient exceeds a critical value, oscillations generate an acoustic wave in a tube that is formed into a hoop. This acoustic wave acts like a compressor and provides a pressure oscillation that drives the refrigerator. This is simple and inexpensive, with easy mechanical tolerances. It was suggested that solar concentrators in orbit can be the heat source. Unfortunately, the design requires a long resonant tube and is inefficient compared to a mechanically driven pulse tube refrigerator.

Characteristics:

Hoop Hot End Temperature: 1,000 K

Pulse Tube Cold End Temperature: 90 K

Power Required: 2,000 W

Hoop Diameter: 3 meters

Information courtesy of NIST

UCH 110 Cryo-Refrigerator

Manufacturer: Balzers, Hudson, New Hampshire

Intended Application: Lab and test facilities

Qualification: None from production line, any available

Availability: Custom made per order or contract from line

Description:

The UCH 110 is an integral unit providing a cold surface from which a line may provide thermal connection to a remote surface. This unit is very large, made for massive cooling capacity applicable to cooling optics, test equipment (e.g., collimators), and sensor suites. It has two stages. The final can go down to 10 K, providing the necessary cooling for doped silicon detectors and semiconductor photo multipliers. The UCH 110 and 65 uses a patented ceramic sliding helium flow control valve for increased reliability.

Characteristics:

Cooling Capacity: 85 to 100 W @ 77 K

Cooling Capacity: 10 to 12 W @ 20 K

Cool-Down Time from 300 K: 20 minutes

Minimum Temperature: 10 K

Weight: 14 kg

MTBF: approximately 8,000 hours

Information courtesy of Balzers

UP 7056

Manufacturer: Signaal USFA

Intended Application: Various FPA cooling

Qualification: Commercial

Availability: In production, delivery approximately 4–6 months ARO

Description:

The UP 7056 is one of a line of high-reliability, split Stirling coolers made by Signaal USFA. The cooler is a sealed, split Stirling device with the compressor and cold finger constructed as separate components, connected by a gas-filled malleable metal tube. The Signaal line of coolers requires no maintenance and is in current use throughout the world in military applications.

Characteristics:

Cooling Power: 0.5 W

Cool-Down Time: <5 min.

Cooling temp: 80 K

Power Required During Cool-down: 30 W

Length of Split Tube: 350 mm max

MTTF: 3,500 hours

Storage Life: 10 years

Weight of DC/AC Converter Module: 550 grams

Weight of Cooler: 1.2 kg

Information courtesy of Signaal USFA

VM-1

Manufacturer: CTI-Cryogenics, Waltham, Massachusetts

Intended Application:

Qualification: All levels on a custom request basis

Availability: Custom made based on previous work

Description:

The VM-1 employs a Vuilleumier cooler in a split mode. The cryo-engine is separated from the cold finger by a transfer line.

Characteristics:

Length: 25.4 cm

Diameter: 10.2 cm

Cooling Capacity: 1 W @ 77 K @ 25° C ambient

Cool-Down Time to 85 K: 10 minutes

Weight: 4.5 kg

Scheduled Maintenance: 2,200 hours

Information courtesy of CTI-Cryogenics

6

Image and Signal Processors

"The real definition of a supercomputer is a machine that is just one generation behind the problems it is asked to solve."

Neil Lincoln

6.1 TECHNOLOGY BASICS

6.1.1 Introduction

The photon's journey through the sensor ends with its conversion to an electronic signal at the focal plane array. However, electronic signals per se are of little use to people. It is the function of the signal and image processor to extract and form useful information from these photon-generated signals. In other words, the signal and image processor's function is to form useful information from the FPA output. To illustrate the importance of these functions, currently about 30 percent of an IR system's weight is devoted to signal and image processing.

Providing useful information to the IR system operator while avoiding overload is important for contemporary high data rate systems. This data reduction will be accomplished largely via signal and image processing. The more autonomous and advanced a sensor system, the more signal and image processing is needed and, therefore, the more these functions will grow as a system performance driver.

Processing plays a determining role in sensor performance as the need for information outgrows the need for pretty pictures. Signal and image processing is a vast, varied, and rapidly changing discipline that must be customized for each application and sensor. To dwell on the performance of various algorithms, heuristics, and machines is outside the scope of this treatise. Instead, we will briefly examine the state of the art of emerging technologies and discuss only some of the basic principles.

Before diving into the technology, a little context setting and history are in order. Since the advent of imaging sensors, electro-optical designs have tended to overload the processing and communications systems supporting them. Current state-of-the-art (SOA) processors barely meet sensor bandwidth requirements. Historically, neither computer nor communication technologies have been effective in handling the real-time data rates generated by sensors. One bottleneck always exists where image processing must occur, and another occurs if this data has to be transmitted. Because of this, the image processor is often a weight and power driver for an IR system. Although great improvements continually occur in miniaturization and processing speed, the disparity remains. As focal planes grow (and grow), so do the megabits per second generated by them. The raw data rate from an FPA is expressed as

$$\text{Raw data rate} = \#\text{ of pixels} \times \text{frame rate} \times \#\text{ of bits of digitization} \qquad (6.1)$$

For an illustration, consider a single Mitsubishi 1,040 by 1,040 Pt:Si array with a 1/60 second frame rate and 12 bits output

$$1040 \times 1040 \times 60 \times 12 = 779 \text{ Mb/s}$$

For this example, to accomplishment of simple thresholding, clutter rejection, and temporal filtering might require tens of gigaflops (i.e., tens of billions of floating point operations per second). Performing this with conventional electrical circuits could result in the electronics weighing hundreds of kilograms, consuming thousants of watts, and potentially causing a reliability problem.

Fortunately, a number of ways exist to reduce this processing load without losing the indispensable sensed information. Three generic approaches can satisfy the need for reduced weight and power without information loss. One is the use of advanced and powerful processors—a brute-force approach. A second is to reduce the data rate from the focal plane through an intermediate processing step called "preprocessing," often performed in the analog domain. The third approach is to reapportion the processing requirement by closely examining overall system performance requirements.

A basic definition of signal processing versus image processing is useful here. Signal processing usually refers to the time stream of data coming from a single detector, much like the line trace of a heartbeat on an oscilloscope. Signal processing works on, and looks at, the rise and fall of the trace to detect targets against the background noise signal.

Image processing goes the next step. It works on, and looks at, the difference between neighboring scene pixels in a two-dimensional arrangement of detectors, and it usually includes the time changing component of signals. Generally, a level of signal processing (such as nonuniformity correction) occurs before the higher-order image processing functions such as edge detection. The main thing to remember is that signal processing is single detector (or pixel) related, and image processing is multiple pixel related.

6.1.2 Fundamental Architectures

The signal/image processing architecture is highly dependent on functionality, state of the art employed (read cost and schedule), packaging constraints, and focal plane architecture. Like other sensor functional areas, the clear division between hardware components that was obvious in past systems is disappearing quickly. Signal and image processing is fusing with, or becoming embodied in, focal planes, optics, control loops, and even structure.

"On-focal-plane" processing is maturing, being enabled by the microlens (see Section 3.4) to regain the fill factor. On-focal-plane processing places signal processing circuits on each unit cell for great reduction in the weight and power of downstream electronics. Optics can be used to provide spatial and spectral filtering and even some of the target identification function (see Section 6.6). Control loops can dither the sensor around a target to smear out some clutter, and they provide information for subpixel processing. Even support structures can function as the electronics board by placing the ICs directly on the support structure, baffle, or dewar sidewalls. The most attractive alternative architectures include performing some or all these functions on the focal plane or in the Z-plane.

Typical architectures usually include an analog or mixed analog/digital preprocessor physically located near the focal plane as shown in Figure 6.1. Its function is to collect the data stream, tag, digitize, and perhaps do some simple uniformity, thresholding, or data compression operations. This compressed, digitized data stream is then sent off of the gimbal or away from the dewar to an electronics box (at a more thermally convenient location) for more complicated digital image processing and object-based processing. This segmentation and division of function avoids long, high-rate analog lines, allows easier thermal design, alleviates packaging concerns. and can reduce the mass on the gimbal.

Typically, functions to be performed by the processing electronics include mitigating unwanted noise from the detectors, identifying potential targets, enhancing the target, and higher order functions such as tracking, data compression, and fusion. A generic camera and FLIR system usually concentrates more on producing a visually pleasing image, whereas seekers, IRST, and surveillance systems concentrate on detecting and tracking both point source and imaged targets.

Some of these system processing functions have components that are sensor driven, and others that are scene driven. These present a nice dividing point for the hardware and software architectures. Sensor-driven functions are those that de-

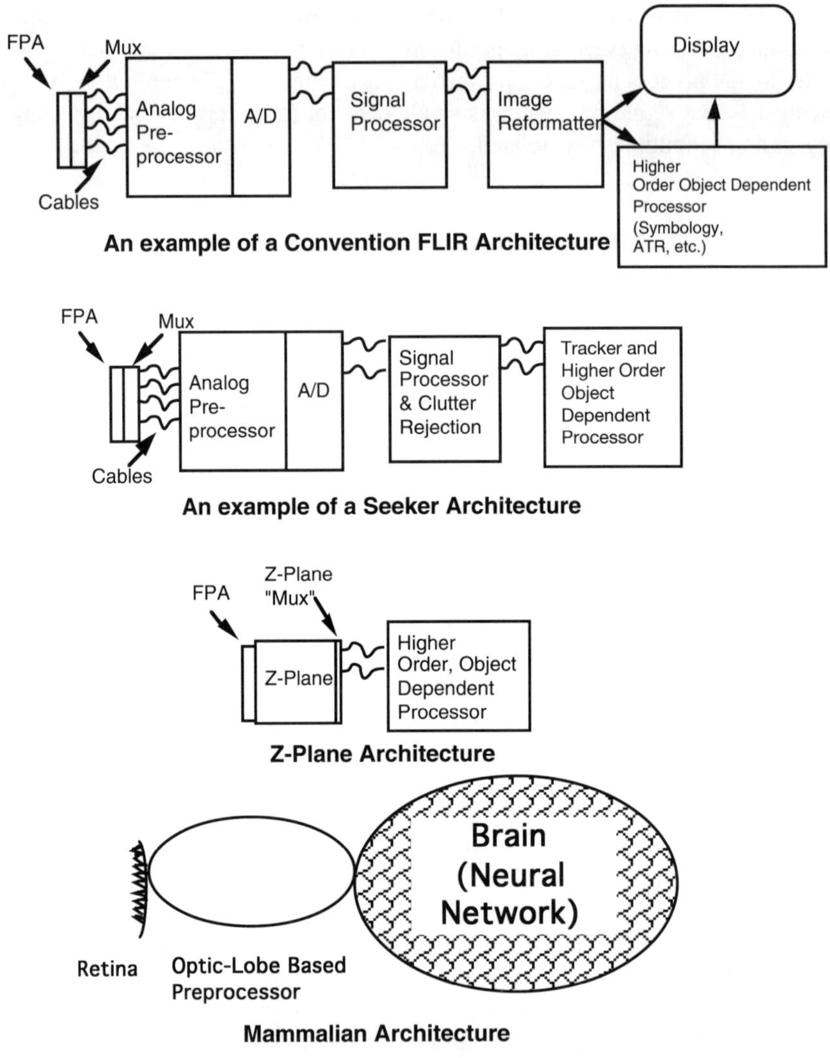

FIGURE 6.1 Examples of Signal Processor Architectures

pend on the number of pixels and their readout rates. These include functions such as nonuniformity corrections, A/D conversion, and various forms of filtering and data compression. They are included in the preprocessor and signal processor box of Figure 6.1. Other processing functions are more dependent on the given scene and the number of targets and false alarms. These functions include tracking, range estimation, and range rate estimation, and would be included in the general purpose processor of Figure 6.1. These usually are not lumped in with the signal and image processor but are separate, object dependent, and used more as general system computers.

A modern controversy concerns sensor head design for ease of signal processing verses signal processor design for ease of sensing. The "tail wagging the dog" metaphor is often used in the debate. Should the sensor head be designed to support the processor, or should the processor be designed to support the sensor head? The problem with this controversy lies in that it is often difficult to match the sensor head and signal processor to either the tail or the dog. In the rare cases where the design is driven clearly by one, the application is either so basic or specific that the same philosophy does not apply to any other system. Regardless, the sensor head is usually designed first and given precedence because it is the cost and performance driver. However, with modern IR systems that don't display images per se but identify materials and perform search and track in highly cluttered backgrounds, the system performance is a balance between the sensor head and the signal processor. In some of these applications, the signal processor is of such great size and power consumption that it behooves the engineers to reconsider the sensor head design to reduce the processing load and lower the system-wide weight, power, and cost. Oddly enough, in some highly cluttered background/target combinations this may mean more pixels and greater resolution rather than fewer.

The architecture of the processor itself also represents an array of choices and can be a large driver in system trade-offs. The architecture of the processor depends on the mission, available funds, available time, and sensor configuration. It can be a single, fast (pipeline) processor or several processors connected together in a semi-independent, nodal style. An extreme version of connected processors are the massively parallel processors containing tens to thousands of processing elements. A Z-plane can physically place a processor on every unit cell of an FPA, and neural nets begin to emulate the design of the human brain.

Algorithm development, software coding, testing, and implementation can be a large part of an IR system's cost and one of the largest schedule drivers for the system. Algorithms must be developed early in the program and matched to the architecture; otherwise, the hardware may not be suited to them. Moreover, algorithm design and coding must be in concert with hardware development. The end result in performance is highly dependent on synergism between the hardware and software.

Before software code can be written, algorithm and heuristic development must take place. This consists of deciding what sequential manipulations and processes the computer should perform to achieve the desired results. Algorithm development always costs too much, and the sky is the limit on the cost. The IR engineer should control costs by seeking out algorithms that have already been developed under another program, or ones that can be modified from another EO or IR program, to control costs. Once the algorithms/heuristics have been developed, writing the software typically costs about $100 per line of code.

6.2 SIGNAL AND IMAGE PROCESSING FUNCTIONS AND TECHNIQUES

6.2.1 Listing of Processing Techniques and Processor Applicability

There exists a plethora of desired functions and techniques to perform signal and image processing. A brief overview of those most commonly used in IR sensors is provided in Sections 6.2.2 and 6.2.3, but this is not meant to be a comprehensive discourse on every technique.

Table 6.1 lists some commonplace desired actions one might wish from the signal and/or image processor, plus the generic techniques used to accomplish them. Although not exhaustive, the list covers the most popular (like a current list of top-40 song hits). Any given sensor design will probably employ only a few of these. Moreover, the order and purpose might be modified.

The division between signal processing and image processing is blurred by the applied architecture and end application. Therefore, there is significant mixing of functions and techniques between Sections 6.2.2 and 6.2.3.

6.2.2 Signal Processing Functions and Techniques

The operations to which these varied processor types are applied are varied and diverse. We begin immediately after the sensor has converted real photons to representative signals. The outputs of the focal plane's multiplexer (mux) are analog electrical signals. These signals must be amplified, converted into digital signals, conditioned to correct for nonuniformities, and processed to form images, targeting information, or track commands.

Special consideration must be given to processing and reformatting for a pleasing and effective visual display. There are several issues to be addressed for a display. Among them are that most displays do not have the dynamic range that the FPA has, the FPA has a high noise level that washes out contrast, the spatial resolution is usually poor, nonuniformities can cause speckle or blotches, and update rates should be visually pleasing.

Correction of each pixel on an FPA to smooth out its response and make it more uniform is one of the first and most critical functions accomplished by the signal processor component. This can be done in the digital or analog domain. With very nonuniform detectors such as HgCdTe, however, correction should be done in the analog domain if one wishes to avoid reliance on enormous A/Ds. This is because most A/Ds have limited dynamic range relative to detectors and because of the detectors' magnitude of nonuniformity. Analog correction is also usually more weight and power effective. Using analog techniques can save an order of magnitude in volume, weight, and power. Usually, most FPAs require two-

TABLE 6.1 Typical Sensor Processor Functions and Techniques

Desired Action	*Signal Processing Technique*
Signal conditioning (prepares the signal for other processing, transmission, and storage)	One-point connection Two-point connection Three-point connection N-point connection Calibration adjustment Analog-to-digital conversion Dead or noisy pixel substitution Amplification
Image formation (produces a two-dimensional image convenient for additional processing or for visual display)	Reformatting Deskew Timing control Electronic deroll Warping
Background subtraction (reduces dc and clutter in background to enhance target detection routines)	Threshold AC coupling DC coupling Spatial filtering Temporal filtering Morphological filtering Frame-to-frame differencing Various order differencing Pseudoregistration
Signal enhancement (provides contrast adjustment, gain control for saturation control, and various integration times)	Frame coaddition Integrator or decoder Negative contrast enhancement Time delay and integration Automatic gain control
Target detection (identifies a potential target, whether a star, missile, or tank)	Threshold Adaptive thresholding Correlation Streak detection Temporal filtering Frame-to-frame differencing Various order differencing Soft detection (maybe it is a target) emissivity mapping Temperature mapping Material mapping
Target tracking (temporally and spatially follows target and allows prediction)	Kalman filtering Adaptive pole bandpass Recursive filtering Multi-hypothesis tracking
High-order processing	Feature extraction Homing commands Intensity centroiding Silhouette centroiding Subpixel accuracy Hadamard transform Discrimination Automatic target recognition Emissivity mapping Material identification/mapping Circumvention Cottrell-Zipser identity mapping Sobel operator Discrete cosine transform Vector quantization Complex data compression (e.g., fractals)

point correction that adjusts each pixel for both gain and offset by having them view two known radiance sources. These radiance sources are most appropriately blackbodies of two different temperatures, but a source may be an aperture cover or another uniform source such as deep, dark space. The offset correction is done to normalize the outputs of the pixels when viewing a single given radiance. This is simply multiplication or addition. The gain is adjusted by comparing the ratio in output between two levels of radiance and involves solving a curve fit and applying correction. Gain correction is usually more complicated than offset.

Where *calibration* is needed, some radiometric applications employ three- and four-point correction. These adjust the linearity of the gain. If the gain is adjusted from two points (the radiance levels) of data, only a straight-line gain correction can be applied. Detectors are all somewhat nonlinear, so adding more radiance levels allows application of a gain correction curve. Calibration adjustment can be thought of as a formal correction of sorts. The difference is that it adjusts to provide an "absolute" calibration to a known source that is formally traceable in some way to NIST. This is usually accomplished during the integration, assembly, and test phase of a program. Calibration entails extensive repetitive testing based on test set-up and sensor repeatability, the pixel correction, radiance flux, integration time, AGC setting, spectral content, bias, and any other parameter that might effect the output. Without a formal error trail back to a primary standard, any blackbody comparison cannot be correctly called "calibration." Merely comparing a FLIR's output to a commercial blackbody is "characterization," not "calibration." In-flight calibration usually entails performing a correction to a known source (again traceable to NIST) carried as part of the sensor. Any calibration/characterization is best accomplished through the complete sensor system, including all windows, optics and processing electronics, as they might effect the results.

AC coupling removes a large "pedestal" of noise, allowing a smaller temporal "blip" to be more easily identified when on top of a large, constant signal platform. It removes the constant amplitude component of the scene and, in the temporal domain, passes only changes. For a scanning system, it will eliminate most of the signal emitted by the detector until an object brighter (or dimmer) than the average background is scanned. In the time domain, this will appear as a change and be allowed to pass through the electronic filter. AC coupling is commonly used with scanning systems. *DC coupling* does the corollary for constant radiation and staring sensors. A dc coupled signal, with a lowpass filter, will remove large fluctuations and pass the constant background. This approach is sometimes used with staring sensors.

Thresholding is simply the method of comparing a pixel to a value whereby the pixel will be ignored if it is lower than that value. If it is higher than the value of the threshold, it will be identified as a possible target, tracked, and have other, higher-order techniques applied to it. A *fixed threshold* compares a pixel's value to a constant standard. A *variable threshold* compares the pixel's value to another value which is determined by the values of several (or all) pixels in the scene.

Thus, the latter can change from frame to frame and more accurately falls into the category of *image processing.*

Soft detection assigns a qualifier to the detection process. Traditionally, detection is viewed as a binary process; either detection occurs or it does not. Clearly, the human brain does not function this way. In difficult situations, one may "suspect" that an object has been detected long before its presence becomes obvious. So-called "soft detection" implies the application of complex processing without the constraint that the algorithm/heuristic must assign only a yes or no. In the simplest form, a third level of "maybe" can be added to the algorithm/heuristics repertoire. In more complicated versions, any number of levels can be added. Alternatively, the algorithm/heuristic may simply assign a probability that the object is some given target. This single probability assignment is most suitable with automatic target identification and detection in highly cluttered environments, such as those frequently encountered with IRST applications.

6.2.3 Image Processing Functions and Techniques

Spatial filtering removes false targets (reduces clutter) by using a weighted average of the neighboring pixels to determine a "localized" mean. This mean can be used to set an adaptive threshold to determine if the pixel has a potential target in it. *Temporal filtering* does the same in the time domain. This manifests itself as a type of spatial filter for a scanning system and a type of frame-to-frame subtraction for a staring sensor. *Morphological filtering* is a type of spatial filtering. It employs a specific one-dimensional (rolling ball) or two-dimensional (rolling pin) operator that detects objects of a given spatial frequency.

In *frame-to-frame subtraction,* one frame (two-dimensional array of pixels) is simply subtracted from another taken at a different time. It is a form of a moving target identifier. This subtraction is usually accomplished from one sequential frame to the next. If the frame-to-frame footprints are perfectly registered, then the only phenomenon that will pass is that of an object's motion across a pixel boundary. Any residual signal that "leaks" through this process is usually caused by a slight misregistration of the pixel positions from one frame to the next. This is the result of jitter, sensor movement, or platform movement. One must carefully design the IFOV such that the footprint is the proper size to accomplish this for a given target and background. If the footprint is too large, the target will take too much time to cross a pixel (more than multiple frame times) and the sensor will drift excessively in this time period. If the footprint is too small, the target will cross too many pixels, making identification and tracking difficult. When properly designed and applied, this is an extremely powerful technique for a clutter filter.

Both the atmosphere and optics tend to cause a Gaussian blur of the image. Much work is underway to "de-blur" the image via processing. This provides crisper images, particularly in the infrared, where diffraction caused by the long wavelengths can be a major cause of blur. One technique to is use a Gaussian deblurring kernel, while another uses an analytically derived, highly structured Toeplitz matrix [Kimia and Zucker, 1].

Streak detection is a higher-order form of frame subtraction. Again, it is a moving target indicator. A pixel and its immediate (and then not immediate) neighbors are examined from frame to frame in the time domain. This process looks for a similar change in pixel level moving in a straight line ("streak") or in some predicable pattern.

Image formation (e.g., reformatting, warping, electronic de-roll) consists of taking the FPA output and forming a visually pleasing image. Depending on the circumstances, this can be a trivial task, requiring little electronics, for highly uniform imaging arrays. On the other hand, this can be anything but "simple" for a common module detector being scanned by a nonlinear scanner that is undergoing image roll caused by platform or gimbal motion. In this latter case, it can require several electronic boards because the outputs from a detector must be memorized and held until its neighbors in both dimensions are acquired and corrected. In addition, large scanning arrays often have every other row of pixels offset to allow dense (no dead space) packing. This requires a time shift to be applied to half of the detectors to align them on an image. This process sometimes is called *deskew*. Special-purpose large-scale integrated (LSI) electronics have been produced to perform such "warping" of images.

The time needed to test, evaluate, and refine complex automatic target recognition (ATR) algorithms is the longest part of the image processing subsystem development cycle, and it may take several years [Nasr and Sadjadi, 2]. For instance, a common target recognition scenario involves a 64×64 pixel reference templates searching through a 512×512 image. ATR correlation problems may require a billion operations per frame to be performed. Since the object orientation is not known, the template must be scaled and rotated. At a frame rate of 30 per second, this can result in 10^{12} operations per second [Molley, 3].

6.2.4 Estimating the Number of Operations

Many advanced image processing techniques have not yet been successfully implemented on a given processor architecture. Since electro-optics have such high data rates, a driver of the algorithms is often the number of operations that the hardware can support.

A "word" is the digitized stream that can be operated on as a unit. For the case in electro-optics, it is usually just the digitizing of the pixels (be it 8 or 12 bits). Estimating the number of operations per word is trickier, because it really depends upon the individual sensor's requirements and the way in which requirements were divided between the image processor and other parts of the system (such as the FPA, people, or processor). For simple techniques like thresholding or correction, the number of operations might be as few as 1 to 20 operations per pixel. For complicated functions, it can be hundreds or thousands of operations, and the

TABLE 6.2 Expected Number of Operations for Popular Techniques

Technique	*Approximate Typical Number of Operations*
Gain correction	1 multiply or add per pixel
DC offset correction	1 multiply or add per pixel
3 × 3 pixel spatial filter	9 multiplies and 8 adds per pixel
5 × 5 pixel spatial filter	25 adds and 24 multiplies per pixel
Larger matched filters	Tens to hundreds per pixel
Simple threshold	1 compare per pixel
Mean of frame	(No. of pixels in frame) + 1 per pixel
Sigma of frame	3 × (No. of pixels in frame) + 3
Adaptive threshold	5 × (No. of pixels in frame)
Intensity centroid	5 × (No. of pixels in shape) + 2
Shape centroid	3 × (No. of pixels in shape) + 2
Frame-to-frame subtraction	2 × (No. of pixels) + 1
Streak detection	(No. of pixels in array) × [(Pixels crossed – 1) × (No. of potential directions) + No. of potential directions)]
Frame registration	Can be several hundred per pixel
Automatic target recognizer	Can be several thousand

number of operations may even vary in real time. Table 6.2 lists some techniques and their typical number of operations. The best advice is to discuss the requirements with an algorithm developer. Failing that, assume 20 operations per word per technique. The number of instructions also depends on the processor's instruction set, so one should assume more instructions per technique for low instruction set machines, and fewer for specialized ASIC-based processors.

For a notional example of the use of a mix of these image processing techniques, one can imagine that a staring IRST or automatic target recognizer would require a minimum of

- Gain correction
- Bias correction
- Five-by-five convolution
- Scene sigma
- Adaptive threshold
- Threshold violation (or the exceeding thereof)

For this example, the processing load then can be estimated based on the number of pixels (n), the addition of all the operations divided by the time allowed (t), or

$$\frac{\text{Math steps}}{\text{Seconds}} = \frac{n + n + 50n + (3n + 3) + 5n + n}{t}$$

$$\approx \frac{61n}{t}$$

Therefore, for a 640 × 480 array operating at 30 Hz, the processing load is

$$\frac{61 \times 480 \times 640}{0.033} = 570 \text{ million operations per second (MOPS)}$$

One can imagine that a scanner would require a minimum of

- Gain correction
- Bias correction
- One-dimensional scene sigma (along the line of the array)
- Adaptive threshold
- Analog matched filter
- Threshold violation

The processing load then would be the addition of all the math steps required to accomplish the above divided by the time allowed, or

$$\frac{\text{Math steps}}{\text{Seconds}} = \frac{n + n + (5n + 1) + 5n + 3n + n}{t}$$

$$\approx \frac{17n}{t}$$

The scanner may require less processing because of the low load of the natural techniques to be employed. However, these techniques are employed in only one dimension, so they are less effective than that of the starer. These techniques can be used in two dimensions (even with a scanner) for increased performance, but the processing load penalties apply.

Regardless of sensor style, the total processing load is a complex function of the number of pixels (increasing load) and the algorithms needed to achieve required clutter rejection levels. Clutter rejection is a system-level issue that is a strong function of the bandpass, target, range, resolution, slant angle, and so forth. It is important to keep digital throughput low to minimize cost, weight, and power. This can be accomplished through limiting pixels, limiting field of regard, processing in the analog domain, and allowing as much time as possible to cover the field of view.

Converting the number of operations to the number of instructions is difficult to accomplish in a universal manner. It depends on the size of the instruction set of the processor—the smaller the instruction set, the larger the instructions per operation. It is about 1 to 4 instructions per operation for conventional machines, 3 to 10 instructions per operation for reduced instruction set chip (RISC) machines, and 6 to 600 for super RISC machines. Moreover, the reader is cautioned that all machines add in one instruction, but none can calculate a square root in less than several operations!

Benchmark instruction tests must be taken with more than a grain of salt. Clock speed, operations per second, and the like don't tell the whole story. To begin with, software (and even algorithms) can be written differently to enhance the performance of a given machine. Moreover, the access times and data shifting times may drive the image processor speed and never show up in a standard benchmark test. Better benchmark tests for image processing are the so-called "image benchmark" tests such as the "Abingdon Cross." For instance, Thinking Machine's CM and Wavetracer's Zephyr-8 took about 6.3 milliseconds to process a 512 × 512 cross [Preston, 4].

6.3 FUSION

A recent development in image processing is to combine IR data with data from other sources in what is called *image fusion, multispectral fusion,* or *sensor fusion.* Fusion is a process of combining images of different wavelengths (colors), or even from different sensors of the same scene, at the same times to form a composite image. The composite image is formed in a way that increases image content and makes it easier for the user to detect, recognize, and identify targets.

Image fusion is the generic combining of images taken from different sensors or with different bandpasses. Multispectral fusion implies multiple bandpasses within the same sensor package such as an LWIR with an MWIR image. In contrast, sensor fusion implies different sensors and/or widely different parts of the electromagnetic spectrum such as fusing IR with UV, visible, millimeter wave, or radar data. Sensor fusion is not limited to two sensors, and multispectral is not limited to two bands; in fact, the more independent sensors, the better the information. (However, more processing load is incurred.) Sensor fusion can also occur using range data from radar or laser rangers. This is sometimes called *range fusion.*

To maximize information content, different areas or objects in a scene must be combined differently. The IR image tends to provide more content than a visible image when an object has a higher contrast in the IR than in the visible part of the spectrum. This usually only occurs at night, in heavily shadowed areas, or where a large thermal gradient exists. Otherwise, the IR usually has less contrast than a visible image. If one merely averages (or adds or subtracts) the two bandpasses, the result will often be a net reduction in contrast and image content! This is the opposite of what is desired. The two images must be combined with a weighted process where the weights are adjusted on a per-pixel basis to maximize image content. For example, one may take a pixel and compare it to its neighbors to get an estimate of contrast, then compare it to the same pixel in another bandpass and "fuse" the two, giving more weight to the one with the most contrast.

Figures 6.2 and 6.3 (both courtesy of the David Sarnoff Research Center, Inc.) illustrate this technique and some of the benefits of fusion. In Figure 6.2, the small airplane shown in the upper left photo is pictured in a visible bandpass. Selected features are apparent in the visible, such as the tail numbers and the generally high resolution of the image. The photo to the right is in an LWIR bandpass. Notice how other features are apparent, such as the heat of the engine and tires, indicating a recent flight. If the pixels are properly registered and averaged, the image on the lower left results. Although increased image content exists, a general decrease in resolution and de-emphasis of the IR intensity results. A properly fused image is displayed at the lower right. This image uses a pyramid processor technique that compares the relative contrast of a pixel to its neighbors. Here the resultant image retains the resolution of the visible image yet displays the increased image information derived by IR observation. Note how specific features stand out more in the fused image than with either the IR only or the visible image only. Also note that while some objects stand out in one unfused scene, they still maintain high contrast in the fused scene.

Figure 6.3 is another example, showing a small plane flying toward the sensors above a tree canopy, and a deer grazing near the trees. Again, the visible image is in the upper left and the LWIR IR image is in the upper right. In the visible image, the plane is easy to see, yet the deer is invisible. In the IR, the deer is easily visible, but the plane is difficult to see. Simple pixel averaging in the lower left blurs the resolution but tends to present both the deer and plane. The best image, however, is again the lower right picture, in which the fusion process compares the contrast of neighboring pixels. Here, both the plane and deer are visible with no loss in resolution. Maintaining contrast for as many objects as possible and registering the scenes are the tricks of the image fusion trade.

There are three key advantages to multisensor fusion. First is an increase in the probability of detection/probability of false alarm ratio due to the increased image content, as demonstrated by Figures 6.2 and 6.3. The second advantage is a graceful degradation: if one sensor fails, the system performance is decreased but not eliminated. Third is the synergistic combination of information that yields a better human–machine interface, resulting in improved target discrimination, recognition, identification, and typing. When performance is optimized, multiple-sensor approaches with fused images are found to provide improved performance in all comparative performance measures. Qualitative advantages and disadvantages are tabulated in Table 6.3 for various missions.

FIGURE 6.2 Example of Fused Visible and IR Images (photos courtesy of David Sarnoff Research Center, Inc.)

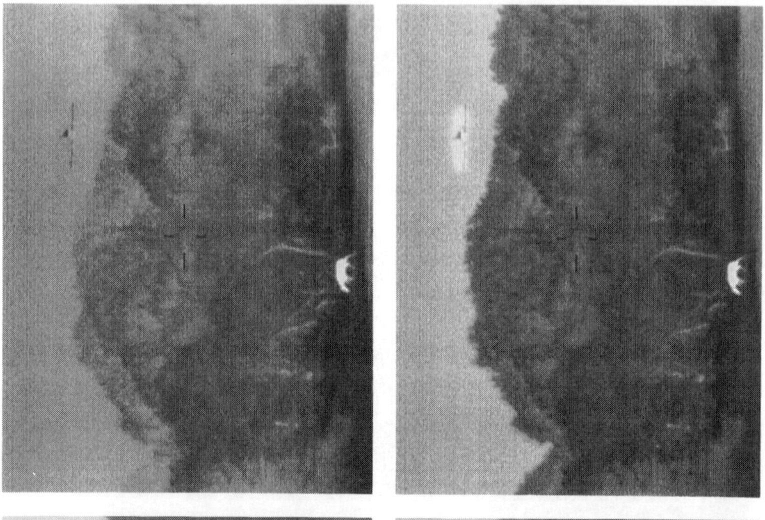

FIGURE 6.3 Example of Fused Visible and IR Images (photos courtesy of David Sarnoff Research Center, Inc.)

TABLE 6.3 Advantages and Disadvantages of Multispectral/Multisensor Function

Function	*Advantages*	*Disadvantages*
Military target ID	Exploits multiphenomenology Graceful degradation Increased P_d/P_{fa}	Weight, cost, and power increase Large signal processing increase
Police evidence gathering	Better all-weather capability Increased confidence of identification Exploits multiphenomenology Increased information to human operator	Weight, cost, and power increase Large signal processing increase May not be admissible in court Difficult to perform in real time Difficult to optimize for all possible scenarios
Rescue	Exploits multiphenomenology Increased P_d/P_{fa} Graceful degradation Increased information to human operator	Weight, cost, and power increase Large signal processing increase Difficult to perform in real time Difficult to optimize for all possible scenarios
Search and track	Increased information to human operator Increased P_d/P_{fa} Graceful degradation	Weight, cost, and power increase Large signal processing increase Difficult to perform in real time
Target detection	Increased P_d/P_{fa} Graceful degradation	Weight, cost, and power increase Large signal processing increase Requires extra development to set bandpasses and fusion techniques

As mentioned previously, multisensor and multispectral fusion can result in a synergistic increase in the ratio of detection to false alarms. This is because the clutter is not well correlated across the bands, while the target has predictable correlated characteristics. In actual tests, the number of a false alarm per detection has been greatly reduced by fusion.

The fusion architecture must be chosen based on individual requirements and systems. Fusion can occur at any of the levels of field of regard, pixel, target, feature, classification, or report. The pixel level is the most basic, requiring the greatest registration and processing. It also provides the greatest payoffs. The classification level combines the independent classifications of the individual sensors and can function with any individual sensor not functioning [Selzer and Gutfinger, 5]. Either pixel- or classification-level fusion can be used for any application and are required for imaging applications such as police evidence gathering. Feature- or symbol-level fusion usually is appropriate with vastly different sensors, such as range radar and IR, while pixel-level fusion is preferred with sensors that are only of different bandpasses [Duane, 6].

Regardless of the above architectural considerations, registration is a concern for multisensor fusion, especially across large portions of the electromagnetic spectrum. Obviously, the field of regard is easiest to register, and the pixel level is the most difficult. Fusing IR with radar entails mixing different resolutions, fields of view, fields of regard, and update rates, and it presents even more difficulty. This registration often is implemented in steps where a gross registration (*macroregistration*) aligns the different sensors' fields of view and is followed by a fine registration (*microregistration*) that aligns on a pixel basis. The microregistration aims to superimpose the target from one sensor on top of the same target from another sensor. Fine registration is needed for target identification and feature extraction, and it is most difficult to achieve.

When registering, differences in range must also be considered. For example, a FLIR typically will have a longer range than laser intensity direction and ranging (LIDAR), so adjustments must be made. Radar may also have a different range from that of the FLIR. Lastly, registration within an IR sensor suite can always be accomplished—often most appropriately by "brute force" methods. These involve keeping the optical bench excessively stiff and having all focal planes of the same IFOV viewed through as much of the same optics as possible.

Several developmental multisensor systems have been made that usually include separate sensors with a common processor and display. This provides a convenient architecture because, while some processing is unique to each sensor, significant processing (such as registration and final target identification and typing) requires input from all fused sensors. Some integrated systems exist, including several developmental multicolor sensors and integrated FLIRs and lasers. Common sensors have been developed for different parts of the spectrum, such as an integrated lightweight receiver telescope that collects MWIR and 94 GHz radiation and relays it to a dichroic that efficiently splits the two into separate detectors ["Westinghouse Develops Brassboard Aperture for Simultaneous Radar, IR Applications," 7].

Fusion with a laser or radar ranger gives the user an image superimposed with ranges. One technique uses the difference in ranges from one pixel to another to enhance object edges. The small changes within an area measured by a laser radar point to surfaces that can be part of a target. The large changes in range point to possible boundaries. The large infrared changes characterize the relative temperature signature of objects. Thus, one could "use small range changes in laser radar data for segmentation and large changes in infrared data for enhancement" [Tong et al., 8].

The display of fused scenes is important because the end goal is to provide the user with increased effectiveness. Much effort is being expended to determine the best way to present humans with the data, but no universal consensus has emerged at the time of this writing. It seems that the optimum way to present the data depends upon the fusion technique and the mission. One method is a contrast-based black-and-white display such as shown in Figures 6.2 and 6.3. Another method is to present a false-color display of fused images. This can be implemented simply using the near IR band, MWIR, and LWIR to provide substantially optimized target detection and recognition [Miller, 9]. For a case like this, the near IR provides reflectivity differences, while the MWIR and LWIR can provide emissivity differences between target and background.

Whether employing image fusion, multispectral fusion, or sensor fusion, most users require the fusion to be presented on the display (in real time) while they are using it. This requires large amounts of real-time processing, exact registration, and coordinate transformations of several images—all simultaneously. Fusion is demanding on the processor and dependent on the algorithms used. Great processing weight increases can occur if generalized processors are employed for fusion. This is because a large amount of processing is needed to select the way in which the image is mixed—usually on a per pixel basis. ASICS, heuristics, and artificial intelligence work well with multiple sensor fusion and tend to reduce the processing load, hardware size, weight, and power.

6.4 SIGNAL PROCESSING WITH CONVENTIONAL INTEGRATED CIRCUITS

6.4.1 Technology Overview

Processing functions can be accomplished satisfactorily using conventional large scale integrated (LSI) circuits on circuit boards. Conventionally, the digitized signals from the focal plane are fed into general processors on standard circuit boards and operated. Unfortunately, the weight, volume, and power consumption may be prohibitively large for complicated functions (e.g., multispectral fusion) or for large focal planes.

For those designing for the U.S. military, consideration must be given to military standards and military specifications, which often do not focus on state-of-the-art technology. These effectively act like a time machine to move the available technology about five years back while ensuring high reliability. Electronic components usually do not become available for military use until their commercial market has peaked. This condition is worse for space-qualified hardware; S-level and fully qualified parts usually lag a few more years. In many cases, this is sadly ironic given that the military may have initiated the initial technological advancements. Yet, in the fast product cycle world of electronics, sometimes the electronic technology is found in children's toys before it is incorporated into an operational satellite or military device.

Electronic technology does advance, being perhaps the only technology that makes clear advancements in short time spans. When an appropriate and focused effort is applied to specific image processing hardware, amazing decreases in weight, power, and cost can be obtained. For instance, an image processor that weighed about 20 kg and performed 1,250 MIPS was shrunk to 540 grams and performed 2,652 MIPS within about three years. This corresponds to increasing the processing power by 467 MIPS per year while reducing the weight by over a third per year. Additionally, by 1992, even a PC could obtain 10 times the image processing power that was available in 1990, at essentially the same cost [Preston, 10].

Work is also progressing on very large scale integrated (VLSI) circuits and wafer-scale integrated circuits that will push the current technology. "Wafer-scale" simply means that most of the wafer is used as a chip, so the available real estate is about 10 cm (4 inches) in diameter. With wafer-scale integration, the number of interchip connections is reduced and those remaining are more efficient. In addition, the chip can be physically large and therefore more powerful overall. A wafer-scale vector processor with 800 MFLOPS capacity is expected to consume 65 W and provide 50 MFLOPS per cubic inch [from UCLA Short Course 823.29, 11]. Approximate reductions through the use of wafer-scale technology are by factors of almost 100 in volume, 20 in weight, and 5 in power.

The Rome Air Development Center (RADC) is sponsoring work to produce processors that are applicable specifically to signal processing. The brassboard can perform over 100 MFLOPS using less than 300 W, with evolution to an enhanced version in the works. RADC is also developing in house a wafer-scale vector processor. The objective is to accommodate GFLOPS of power onto a single wafer with low power consumption. Additionally, the Phillips Lab is funding work on an image processor containing 10 to 20 MIPS of radiation-hard electronics ["Electronic Intelligence," 12].

Application specific integrated circuits (ASICs) and hybrid circuits can greatly reduce the weight and power of a processor. These integrated circuits are custom made for a particular user and application. Weight and power can be reduced by an order of magnitude compared to conventional electronics performing the same functions. When properly structured, designed, and configured, even rudimentary technology, such as a 386 chip, can be combined with ASIC designs for impressive weight and power. A 170 gram, 100 W electronics package has been developed to support a seeker with a 128 × 128 FPA using the Intel 80386 and 11 custom LSI circuits [Scott, 13]. The Strategic Defense Initiative Organization (SDIO) has developed tiny, powerful missile computers by investing in advanced and creative packaging. They are about the size of a deck of cards (seven times smaller) and have 25 to 100 times the processing capability of other advanced computers [from *Technology Applications Program,* 14].

ASICs represent a critical and exciting development because their cost has rapidly declined. Recently, ASICs have come down in costs to levels ≈ $50,000 or less per foundry run. To avoid unnecessary additional runs, breadboarding the integrated circuit with discrete components is recommended before going to "print" with the photolithography. Even then, the first attempt at an integrated circuit rarely works, so one should usually allow another $20,000 to $50,000 for reworking the custom chip. This expenditure is high for a single chip for a single sensor. If, however, the sensor were to go into large-scale production, additional copies of the ASIC would be extremely low in cost—not more than a few dollars to few tens of dollars per copy. Almost all of the expense is in designing and troubleshooting the ASIC and chip rework. An example of a commercially available image processing ASIC is Sarnoff's "... 1 μm CMOS ASIC containing algorithms that pay selective attention to details in images, such as shape, motion and texture. Called the Pyramid Integrated Circuit (PRY-1), the ASIC is being fabricated by VLSI Technologies Inc." [Keller, 15].

6.4.2 Estimation of Weight, Power, and Volume

The best way to estimate the weight, power, and volume of a processor is to design and lay out the circuit boards and estimate them from each component on the board. However, this is not possible during a proposal or when attempting to trade various architectures and/or technologies in the concept development stage. As a second technique, the engineer should scale the processor by the following relationships, using a similar system with which he is familiar. If sufficient information does not exist, the following procedures, based on empirical data, are somewhat useful for roughly estimating weight and power. These rules of thumb assume VHSIC chips, and special concern is given for light weight and low power consumption. For some generic analog preprocessors, the weight can be guessed by the following steps:

1. Assume each focal plane chip will probably need a controller, clock, and driver. Assume 30 to 100 grams of electronic control support per FPA chip.
2. Calculate the A/D needs and the number of A/D chips. Each A/D chip weighs about 20 grams with board overhead.
3. Double the weight for packaging overhead.

For a generic signal processor, the weight can be crudely estimated this procedure:

1. Estimate the number of memory chips necessary by assuming a 256×1 byte memory chip, and double for overhead.
2. Estimate the number of processing chips needed by assuming 10 MIPS per CHIP with VHSIC 1. If dealing with partial chips, round the number up to the next nearest integer. The 10 MIPS per chip number is likely to increase to 80 or 100 MIPS per chip by the mid nineties.
3. Calculate the weight of the memory and processing chips at ≈ 20 grams each.
4. Assume the interface circuitry weighs the same as the processing chips.
5. Assume the residual discrete circuitry weighs about one-third of the processing weight.
6. Estimate the weight of cabling and wiring as 60 percent of the weight of the memory chips plus 80 percent of the weight of the processing chips.
7. Add the results of Steps 6, 5, 4, and 3.

8. Double the above for margin, packaging inefficiencies, and cooling.

To estimate the power consumed, use the following steps:

1. Estimate the power of the memory and processing chips by assuming they consume 0.5 W per chip (or look them up in manufacturers' data books).
2. Assume the residual circuits will consume as much power as one-fourth of the processors plus one-eighth of the power that the memory chips use.
3. Estimate power for cooling (if any). This includes fans, thermoelectric coolers, and refrigerators.
4. Add the results of Steps 1, 2, 3, and 4.

Now, if the situation does not offer enough detail to perform the above, then one can very crudely guess at weight and power by scaling, using "rough order of magnitude" techniques.

Preprocessing power, P, can be crudely approximated or scaled from the following equation:

$$P = A + (4\times10^{-6}\ \text{watts})\ (\text{bits per second}) \tag{6.2}$$

where

P = power in watts

A = overhead amount of power in watts (usually 1 to 5 W)

A "ballpark" estimation of the processor's weight in kilograms, W_{kg}, can be calculated by

$$W_{kg} = (5\times10^{7})\ (\text{operations/word/second})\ \left(\frac{\text{number of words}}{\text{bits/word}}\right) \tag{6.3}$$

and it will consume ≈ 50 W/kg. These relationships were derived by fitting a curve to a highly scattered plot of existing systems. Severe caution must be exercised when using the above methodologies and relationships. First of all, electronics and image processors advance rapidly, with throughput relative to size, power, and cost doubling every three years or so. Therefore, it is almost impossible to use absolute numbers for weight, power, and cost relationships for more than a year. Moreover, this hardware can be made much more efficient by employing a custom design for a given task. Finally, the technology may be pushed by the project. The level at which the technology is pushed usually depends on the basing of a system. FLIRs and cameras usually do not employ expensive and advanced electronic packaging techniques unless weight, volume, or power is at a premium. Conversely, space-based systems almost always employ advanced techniques, as do IRSTs.

A simple, small board can take two to four man-weeks to design, check, and lay out, while a complicated image processing board can take six months for the same processes. Usually, three or four man-months per board is a good rule of thumb. It often takes about six months to year from concept to delivery of a usable signal processing board using conventional ICs. This time can be reduced slightly by building on similar boards from another program, or the time can be increased dramatically by requiring new algorithms and processes. A Mil-Spec image processing board usually costs about $10,000 in production, and a commercial-grade board about $1,000.

Additionally, when estimating weight, engineers and managers should not overlook the mundane parts of electronics such as wiring harnesses and connectors. Surprisingly, these can add substantial weight and create very long lead times. For example, vendor delivery of an S-level electrical connector can take more than a year.

6.5 PARALLEL MACHINES

Single instruction multiple data (SIM-D, pronounced "sim-dee") is an architecture that performs the same instruction on each parallel processor and each bit of data. SIM-D works well with image processing, which is usually a number of the same simple operations performed on every pixel and frame. Multiple instruction multiple data (MIM-D, pronounced "mim-dee") performs multiple instructions on the data using the parallel processors and is more applicable to generalized processing or higher-order image processing.

A *nodal machine* has several processors with their own memory connected to various other processors. This distributed processing increases fault tolerance and hardness. Theoretically, one could shoot the computer with a shotgun, and the pellets might destroy half the nodes. If so the machine would hiccup, redistribute the processing load itself, and continue (albeit at a slower rate).

Processor architectures can be *coarse grained,* with a single high-speed super processor (e.g., a Cray). The same high throughput can also be realized by a *fine grained* architecture, or a series of small, simple SIM-D processors connected in parallel. Examples of parallel processors are the Connection Machine and the *digital array processor* [Jacobi et al., 16]. An intermediate architecture with tens of MIM-D processors has been implemented in the NCUBE® and Intel Touchstone® machines [Jacobi et al, 17].

Parallel machines have a prolific number of processors (perhaps several thousand) working in parallel, every one executing an operation on a different data point. Parallel machines are the devices of choice for doing the same operation a multitude of times on different data. According to Zyszkowski,

> These massively parallel computer engines achieve computational efficiency at a low cost by applying thousands of simple and inexpensive processors to divide a complex problem in smaller, more manageable data components. The price/performance advantages of massively parallel computing systems are being demonstrated on a broad range of complex problems [Zyszkowski, 18].

Nodal machines have a few (perhaps ten) processors among which the processing is distributed. At first, they sound similar to parallel processors, but significant architecture differences exist. Parallel machines are array processors that require the data inputs and outputs to be in an array, whereas a nodal machine commonly requires serial inputs and outputs. In general, parallel machines offer a factor of five improvement in cost to performance [Dornheim, 19] over vector machines for a large number of simple operations.

DEC, IBM, and HP have processing chips for parallel machines [Dornheim, 20], and Thinking Machines Corporation offers a four gigaflop machine starting at $750,000 [Lieberman, 21]. Applied Intelligence Systems' AIS50000 has 1,024 processors in a linear array and leads the currently available commercial parallel processors in speed and price performance, with complete systems for a few tens of thousands of dollars [Preston, 22]. In addition, special-purpose processors developed on VME cards have good cost effectiveness. Examples are sold by Datacube, VITec, and Vicom [Preston, 23]. But parallel signal or image processors frequently are expensive—about $5 thousand to $20 thousand per megabit per second of image processing. The cost is due to the low number made and because each is essentially a custom design (although the chips are not custom designed).

In Orlando, Martin Marietta has developed a C++ or Ada programmable, two-dimensional, fine grained, parallel SIM-D architecture processor specifically for image processing. Called the Geometric Arithmetic Parallel Processor (GAPP), it provides a relevant example of this architecture. GAPP is oriented toward applications that benefit from maximum image processing algorithm throughput support within minimal size, weight, and power constraints. The basic idea is to match the computer's architecture with the unique characteristics of image processing applications. Thus, GAPP is a two-dimensional grid of interconnected processors that can be scaled to match the image, or a portion of the image, generated by the FPA. Images are usually digitized with 8 to 12 bits of precision per pixel sample, and some image processing algorithms are typically preformed with relatively low precision (e.g., edge operations may be accomplished with only a single bit of precision). Therefore, the GAPP processors use a simple, single-bit processor.

GAPP can be considered an extreme example of the RISC concept. It uses a single-bit arithmetic logic unit (ALU) that can be used to perform multibit operations, including floating-point. Each GAPP cell is small and simple, containing a one-bit full adder/subtracter, 128 bits of memory and four one-bit registers. Each cell is connected to its four nearest neighbor processors.

Several versions of the GAPP chips have been produced to date. A single GAPP II chip contains 72 processing elements organized into a 6 × 12 array and has an input bandwidth of 7.5 MBps along with a concurrent 7.5 MBps output bandwidth. Using a 10 MHz instruction clock and consuming 0.6 W, a single GAPP II chip provides 28.8 million 8-bit additions per second. The most current GAPP III chip version is implemented with 1.0 micron feature sizes, allowing it to operate at a 25 mHz clock rate. GAPP III chips contain an 8 × 16 processing element array and 128 bits of RAM per cell, and they have a capacity of 177.8 million 8-bit adds per second.

Matching of the computer architecture to the characteristics of an image's data structure allows the GAPP to achieve high execution speeds. A complete GAPP-based image processor may contain anywhere from 16 to over 1,000 chips, depending on the algorithm requirements and data rate. A typical GAPP III processor uses 540 chips to form a 144 × 480 array and provides throughput capacity of 96 billion 8-bit additions per second [Martin Marietta, 24].

6.6 Z-PLANES

Processing on the focal plane, right next to the FPA, offers tremendous payoffs in reduction of weight, power, and (potentially) cost. However, crowding the pixel with electrical circuits tends to make the complicated focal plane more elaborate and harder to manufacturer. Moreover, it can greatly reduce the fill factor. From a configuration perspective, the obvious solution is to make the focal plane a three-dimensional unit and use the real estate behind the optical surface for the processing electronics (analogous to the human eye). A typical Z-plane employing this concept is depicted in Figure 6.4 (courtesy of Irvine Sensors Corporation). It is an 1.3 cm cube containing 206 square centimeters of electronic real estate for integrated circuits, all located directly behind the FPA.

Opticians call the optical axis in a telescope, where the elements are spaced, the Z-axis. Traditionally, it ends with the focal plane as photons are transduced into electrical signals. However, it is philosophically and engineeringly pleasing to extend the direction of the electrical signal along the same axis that the photon travels: the Z-axis. Processors that deliberately take the electrical signal and process it with electronics extending in this direction are termed *Z-plane processors.* In an ideal world, a Z-plane would put a series of electronic circuits on every pixel, such as a processor, an A/D, or whatever is needed. Since there is a processor for every pixel, these units are massive parallel processors—but more than just processors, as they can accomplish A/D conversion. If the focal plane is analogous to the retina, Z-planes are the optic lobes. Sometimes called *smart focal planes,* these detector/processor combinations emulate the eye–brain architecture and offer the potential to emulate the eye's characteristics.

These electronic extensions of the focal plane can perform at staggering rates because there is a processing element allocated to each unit cell. They are physi-

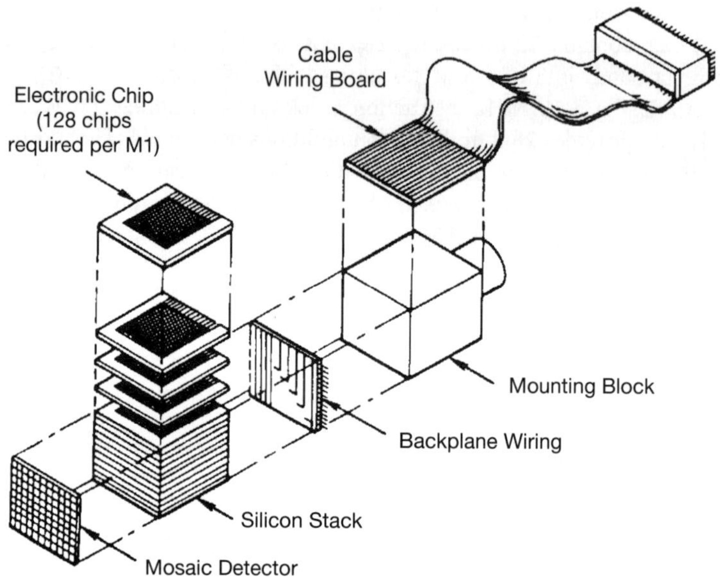

FIGURE 6.4 HYMOSS Z-Plane Concept (courtesy of Irvine Sensors Corporation)

cally attached close to the FPA (eliminating data rate transmission and bus timing concerns) and can perform operations in the analog and/or digital domain. The actual functions that they can perform are limited by the Z-plane methodology, the circuit design employed, and human ingenuity. The Z-plane lends itself to a wide range of analog processing. Common functions using the Z-plane include preamplification, gain control, offset control, TDI, multiplexing, image motion compensation, temporal filtering, and gamma circumvention. However, it is possible for high-order processes to be implemented. The bottom line is that Z-planes can perform just about any function.

As mentioned before, the impetus for using Z-plane processing are speed, reduced power consumption, and reduced weight. It is possible to perform any analog signal processing techniques in Z-plane configurations (including data compression). This reduces the digital processing and can reduce the A/D rate. Reducing the A/D rate offers great weight and power savings while lightening the entire downstream processing load.

Irvine Sensors Corporation is one of the leaders with its Hybrid Mosaic On Stacked Silicon (HYMOSS) sandwiches of silicon. The process is to design a thin (usually 4 mil) integrated circuit with a series of electrical channels, one for each pixel in a row, fabricated in standard CMOS. Each circuit chip is then cut to the proper thickness (usually 4 mil) and stacked (glued) to a number of identical slices. Then the focal plane is then attached to the edge via indium bump bonding. So, if one would use a 128 × 128 focal plane, the Z-plane would have a stack of 128 slices of an integrated circuit with 128 analog channels on each slice.

Present HYMOSS Z-planes include a 128 × 128 buttable developed under an U.S. Air Force research contract. Dynamic ranges are typically 10^4. The cubes are buttable with only a pixel or two loss between cubes. This allows building large format Z-planes. Z-plane cubes weigh about 6 grams and consume less than a watt. A 13 bit A/D Z-plane was built that requires only 3 μW per channel at 1,000 Hz. Efforts for the near future will concentrate on producing up to a 512 × 512 Z-plane, including neural nets on chip, and enhancing capability at a 2 mil thickness [Irvine Sensors Corporation, 25].

Although somewhat programmable, Z-planes are more "hard wired" than general processors made using conventional electronics. Each Z-plane contains specific circuits to perform specified functions. However, if properly designed, the function can be switched off or modified. The Irvine Sensor Corporation's key to the real-time tunability of the circuits is the use of switched capacitors as resistors in the circuit. The clock frequencies establish equivalent resistance values, allowing the user to change values [Irvine Sensors Corporation, 26].

An alternative technology is provided by ceramic-based Z-plane processors. The detectors are mounted on the edge of a multilayer circuit board, and the boards are stacked via lamination to provide a two-dimensional array. This approach is to bond together layers of ceramic substrate electronic boards with epoxy. Each layer is metalized and delineated to provide a place for bump bonding HgCdTe arrays. Again, the signal from each HgCdTe unit cell is fed directly into the electronics through indium bump bonds.

The fact than Z-planes can perform A/D conversion beckons an architecture yet to be implemented. This would be one that uses a Z-plane to perform the A/D only, while a processor conveniently located elsewhere conducts signal and image processing.

Availability, producibility, and cost are major concerns for all forms of Z-planes, since none has ever been produced in large quantities. No production FLIR, seeker, or camera system utilizes them; however, several advanced seekers and space sensors are planned that will use Z-planes. Costly Z-planes can easily pay for themselves by lowering the need for other electronics, reducing weight, and providing increased performance. With complicated sensor systems, a few high component costs may often lead to system-wide savings. It is not intelligent to be penny wise and pound foolish, or component cheap and system expensive.

Reliability is another concern. The Z-plane's reliability problems occur due to thermal stress. If the FPA is not of the same material as the Z-plane, there will be thermal coefficient of expansion mismatch between the Z-plane and the FPA. Under rapid cooling, the FPA can shear off. Likewise, the larger the Z-planes is, the hotter the temperatures internal to the Z-plane become. Improving producibility and reliability calls for extra dicing and lapping process control for edge roughness minimization, especially for unit cell dimensions of less than 2 mils. Improved thermal conduction is also advantageous for increased temperature uniformity across the Z-plane.

The costs of implementing a Z-plane are minimal. To estimate weight, simply calculate the volume by assuming the Z-plane is the entire size of the focal plane and extends several centimeters in the Z-plane, then multiply by the density of silicon (or the ceramic) and add 3 or 4 grams for additional ceramic or Teflon® supports. The weight is trivial.

Power is also inconsequential (usually less than 0.5 W) compared to conventional processors. Since the power is dissipated right next to the focal plane, the Z-plane must be cooled to the focal plane temperature, which calls for a larger cryocooler. If one is using a Stirling cryocooler, add an additional 20 W to the Stirling to supply the additional 0.5 W of cooling at 75 K.

6.7 NEURAL NET APPLICATIONS

Neural networks are computers that form an interconnected branching network like the neurons of the mammalian brain. Theses are low-power, massively parallel, autonomous, nodal processors. Neural nets are the engineer's attempt to recreate the biological brain in style and context in electronics. Typically, electronics are characterized by rapid speed and low connectivity, while biological systems are characterized by relatively slow speed but high connectivity (50,000 connections per cell). Hence, the challenge to the engineer is the programming and implementation of high connectivity in electronics.

A neural network is set up to propagate data through a series of interconnected branches and nodes with various "weights" applied to these interconnects. Figure 6.5 is a notional example of a neural network that might be applied to image processing. The first layer of the neural net is the input of data and is represented in the figure by the top row of shaded circles. In this example, two attributes may be the input for a pixel (e.g., intensity and contrast to neighboring pixels). The next two layers are called hidden layers, and they consist of the real neural processing (represented by the circles with the checkerboard fill). The notional Figure 6.5 shows two hidden layers, which tends to be the maximum number of hidden layers required for any class separation problem. These hidden layers finally feed the output layer.

For a modern image processing application such as infrared search and track, the goal is to identify a target among clutter. There is system-wide value to providing slightly more information, which a neural net is particularly well suited to do. Instead of merely forcing the processor to classify an object as a target or not, a third level of "maybe" can be assigned. This then can be further investigated by the system or the operator. Therefore, the notional architecture of Figure 6.5 contains three possible outputs. As Van Camp points out [Van Camp, 27], "In real

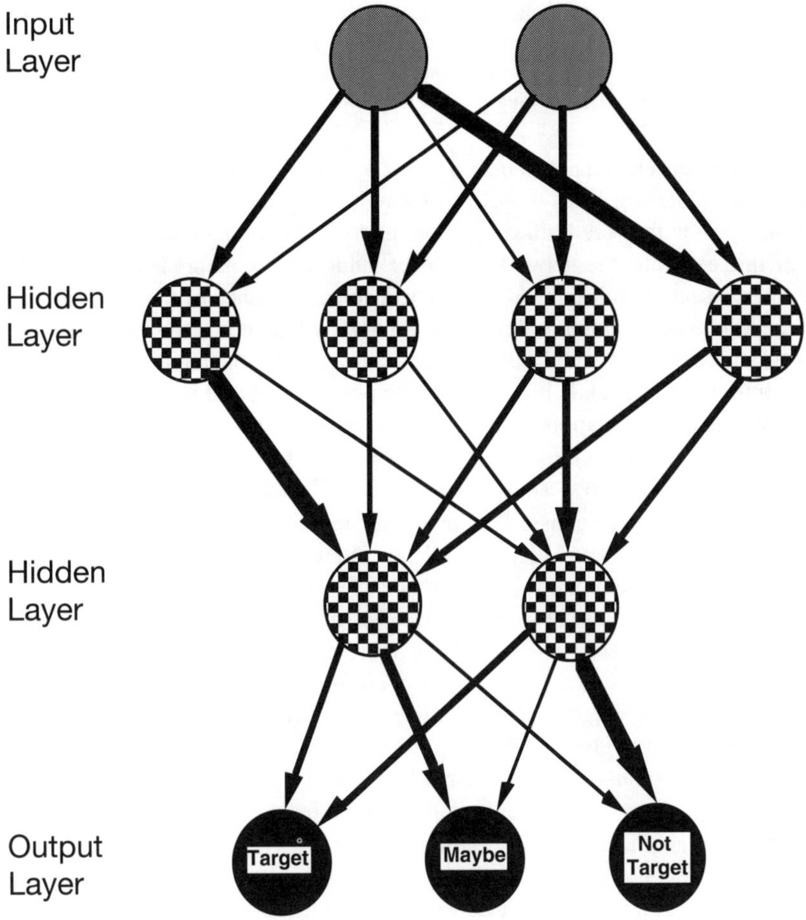

FIGURE 6.5 Representative Neural Network Architecture

neurons, learning is thought to occur in the synapses: when the connection strengths between synapses change, the firing behavior of the network changes. In artificial networks, the learning occurs when the weighting factors on the links change." So the learning occurs within an artificial neural net by modifying the weight between the nodes.

In Figure 6.5, this is represented by the thickness of the connecting lines. Thickness indicates the relative strength of the connectivity, which can be thought of as the likelihood of information flow. The value may be altered in real time, and this adjustability allows neural networks to "learn" by converging on the output of the net with the training data definition of target and non-target.

> It performs this conversion in two stages. First, it multiplies each incoming activity by the weight on the connection and adds together all these weighted inputs to get a quantity called the total input. Second, a unit uses an input-output function that transforms the total input into the outgoing activity [Hinton, 28].

Neural nets are not programmed per se but are trained. They are set up to solve a given problem such as detection of a point source in a highly cluttered background. Then they are trained by allowing the "net" to run cases of the problem. For this example, the network would be subjected to hundreds or thousands of cluttered scenes with imbedded targets. The performance of the net will tend to peak after a given amount of training. Following that, it might actually decline as it becomes "overtrained." Just as with a dog or child, the amount and quality of the training is critical. If it is trained with poor data, it will perform poorly when given real-world situations. The training set must be realistic; i.e., free of artifacts that are not present in the real world infused with real-world clutter. The latter is imperative e for those working with "simulated" data.

Neural nets can be considered "mapping" functions that take object descriptors (e.g., information about object size, shape, intensity, etc.) and "map" or match up these data descriptors with the output node for each object class. Classes of "target" or "non-target" is the simplest case. The most popular type of neural net, the so-called *back propagation neural net,* achieves this mapping by minimizing the error in separating one set of class objects from another set. This is accomplished by solving a high-order, greater than linear or quadratic, curve fit between the set of input descriptors for the two classes. As such, this technique often requires a tremendous amount of data to accurately produce a good separation curve and accurate target versus non-target declaration [Duvoisin, 29].

Less popular, but often very useful, are the neural nets that discretely represent the input data descriptors. These nets represent the data as discrete mathematical vectors. Rather than develop a curve separating the data that represents the target versus non-target classes, these nets use the training vectors themselves to compare with the unknown input data. They declare targets versus non-targets by whether the descriptor data for the unknown object comes closer to the stored target vectors or non-target vectors [Duvoisin, 30].

Neural nets function heuristically and as such do not necessarily give the same answer each time. They work well with fuzzy logic in adaptable, fault-tolerant situations. When given feedback, some types are capable of adapting in real time. Their most promising use in electro-optics is with complex pattern recognition such as human faces, automatic target recognizers, separating sub-pixel closely spaced objects [Forsell et al., 31], soft detection, and target detection in highly cluttered environments. Another promising use is on binary morphological operations, which are extremely sensitive to spatial noise [Sinha, 32].

A neural net can be implemented in software on a parallel or serial pipeline machine, but they operate most efficiently when the hardware is specifically configured as a network. Herein lies the problem. Hardware neural network machines are highly developmental, temperamental, and immature. However, several government agencies and corporate R&D labs around the world have implemented simple hardware neural networks. Siemens, Intel, Adaptive Solutions, HNC, and Ricoh are the current leaders in commercially available neural chips [Johnson, 33; Yoshida, 34]. Also, in Europe, the Esprit program is developing a neural net computer with 1,024 neurons and 512 synaptic connections per neuron under a three-year, Ecu 17 million program [Johnson, 35]. "Meanwhile," according to Johnson [Johnson, 36], "Hitachi Ltd., Hughes Technology Center, IBM Corp., and even the National Aeronautics and Space Administration have revealed intentions to field commercial devices over the next couple of years."

Examples of the current state of the art can be described by the following accomplishments:

- The Ricoh RN-200 neural net chip realizes 256 synapses in a 16×16 neural architecture providing a processing of 3 billion connections per second and learning of 1.5 billion connections updates per second [Yoshida, 37].
- Irvine Sensors is applying its HYMOSS technology to achieve massive amounts of interconnects for neural networks. The neural network computer is a 128×128 Z-plane stack mated with another, giving 10^{14} interconnects per second (a factor of 10 less than the human brain) running at 16 MHz [Irvine Sensor Corporation, 38].
- The HNC chips offer from 320 million to 1.28 billion connections per second of operation and learning rates of 100 million connections per second [Johnson, 39].
- The Wavetracer Data Transport Computer System is composed of 16,384 single-bit processors interconnected in a 3-D network of $32 \times 32 \times 16$ with nearest neighbor connectivity [Zyszkowski, 40].

6.8 MISCELLANEOUS HARDWARE ADVANCEMENTS

Perhaps the most exciting advance in signal processing technology is the potential of making image processors utilizing all three dimensions. Several organizations are attempting to produce 3-D computers. The circuits boards are built in all three dimensions to save weight and power and to increase processing speed. One concept being pursued is to stack and connect wafer scale processors vertically. Currently, there are difficulties with alignment during manufacture and problems with structural strength. Additionally, thermal overload is a concern for practical applications. This technology is highly developmental, but it holds great promise for production within the nineties. The expected benefits may include factors of hundreds in volume reduction, up to 100 in weight reduction, and up to 50 in power reduction as compared with conventional LSI computers. For instance, a programmable three-dimensional signal processor that could perform ten billion operations per second has been condensed into a volume of less than 164 cm^3 [Shepnal, 41].

Another, more advanced, concept is to package the individual integrated circuits in three dimensions and just mount them on a planar board. These are similar to the Z-plane electronic packaging but are not located on the sensor's Z-plane and are not directly connected to the FPA. Hence, they are "off focal plane" and do not require cooling to FPA temperatures. Both processing and memory chips can be made that exploit the third dimension in this way. Organizations pursuing 3-D chip packaging technology include Irvine Sensors Corporation, Texas Instruments, and Thomson CSF [Seur, 42], while whole 3-D computers are also being pursued by General Electric, Phillips Labs, Aerospace Corporation, Mission Research Corporation, and Hughes [Shepnal, 43]. By making chips three dimensional, we can achieve a great decrease in weight and power while increasing performance. Line drivers, which consume much power, are eliminated because the lengths are shortened from centimeters to mils [Suer, 44]. The three dimensional chips populating a board are one-half to a few centimeters high, and the board made from them looks like an architect's tiny model of a downtown city.

Irvine Sensors has bootlegged its Z-plane HYMOSS technology to include high-density, three-dimensional memory cubes. As an example, they have stacked 64 layers of 1 Mb RAMs. In a cube of approximately 0.5 inch dimensions, 64 Mb of RAM can be contained with a weight of only 10 g. This type of dense memory can greatly reduce weight and cost for higher-order processing, especially for applications that require multi-frame comparison [Irvine Sensors Corporation, 45].

Also within the subject of memory are solid state recorders. These are arrays of memory chips and their control circuits encased in a black box and designed to replace analog experiment tape recorders. The weight efficiency is high at > 40 Mb/kg. Texas Instruments has demonstrated a 1.2 gigabit (Gb) solid state recorder that weights 227 grams, is 1 cm thick, 10 cm wide, and 15 cm long ["Texas Instruments Demonstrates Solid State Recorder," 46].

In the arena of general processing, rapid improvements are occurring with optical (or photonic) computers. These computers accomplish the processing with photons, not electrical signals. Their architecture is typically parallel in that they process an array of photons at one time rather than a single serial stream. Optical computers allow computing up to three orders of magnitude faster than ordinary electronic computers ["Optical Computing," 47], with little or no power required or heat generated. An architecture has been suggested that mimics the human eye–brain system with a photonic computer in the front end backed up by a neural net.

In optical computing, the image is often presented in Fourier space and optically filtered to find targets. The key to practical optical computing lies with the devices that form photons from the data and that optically and Fourier filter the photon stream. This optical modulator in an image plane represents an array of optical switches; it transmits or blocks light. A Fourier pattern is presented (e.g., a wheel), and any round object is passed. Now the pattern must be scanned throughout the scene and scaled to pass all possible "wheels." These spatial light modulators (SLMs) must be accurate, fast, lightweight, and low power. The SLM performs a spatial scan of the presented image looking for a programmed target or feature of a target. It passes representative light to the target. There are also research efforts to replace the focal plane with a lens that converts the information into a Fourier transform. The light will be collected by the optics, Fourier transformed, and smart filtered by the modulator without the need for a detector. Beyond SLMs are electron trapping devices that store data optically. It is predicted that electron trapping devices could cost less than SLMs and respond almost a million times faster ["Optical Computing," 48]. Texas Instruments has developed a micromachine mirror array with 50 μm mirrors on a 97 mm pitch which has been used as an SLM ["Optical Computing," 49].

In the realm of image processing ASICs, Harris has developed image processing ICs that are specially made to perform spatial filtering convolutions. They employ a parallel array of nine multipliers and address, forming an 8 bit or 12 bit 3 × 3 convolution in a single clock cycle. They are made as 0.7 high-speed CMOS process and can support real time spatial filtering form large arrays [Hoq et al., 50].

Returning to "on-focal-plane processing," several groups are exploring this technique in a manner that doesn't really qualify as Z-plane technology. This is because the electronics are physically located on the X- and Y-axes, alongside the detector rather than behind it. Simple circuits are made on HgCdTe, InSb, and the silicon of a Schottky barrier and placed next to the photo-active area of a pixel. Making electrical circuits out of HgCdTe and InSb is tricky but possible. Additionally, this technique has an undesirable effect on fill factor. Fill factors are typically reduced to less than 50 percent to make room for the circuits. However, to compensate for this lenslets can be deposited upon the FPA (see Chapter 3) to boost the effective fill factor to 80 percent or more.

References

1. B. Kimia and S. Zucker. January 1993. "Analytic Inverse of Discrete Gaussian Blur." *Optical Engineering,* 166–176.
2. H. Nasr and F. Sadjadi. 1989. "Automatic Target Recognition Algorithm Performance Evaluation: The Bottleneck in The Development Life Cycle." *Proc. SPIE* 1098: 156–160.
3. P. Molley. 1989. "Implementing The Difference Squared Error Algorithm Using an Acousto-Optic Processor." *Proc. SPIE* 1098:232–239.
4. K. Preston. September 1992. "Sci/Industrial Image Processing: New System Benchmarks Results." *Advanced Imaging,* 46, 48,49, 82.
5. Selzer and Gutfinger. 1988. "LADAR and FLIR Sensor Fusion For Automatic Target Classification." *Proc. SPIE* 1003:236–246.
6. G. Duane. 1988. "Pixel-Level Sensor Fusion For Improved Object Recognition." *Proc. SPIE* 931:180–185.

7. May 15, 1989. "Westinghouse Develops Brassboard Aperture For Simultaneous Radar, IR Applications." *Aviation Week and Space Technology,* 65.
8. S. Tong et al. 1987. "Multisensor Data Fusion of Laser Radar And Forward Looking Infrared (FLIR) For Target Segmentation And Enhancement." *Proc. SPIE* 782:9–14.
9. W. Miller. 1991. "Sensor Fusion Approach to Optimization For Human Performance." *Proc. SPIE* 1482:224–233.
10. K. Preston. September 1992. "Sci/Industrial Image Processing: New System Benchmarks Results." *Advanced Imaging,* 46, 48,49, 82.
11. Presented at The UCLA Short Course 823.29.
12. January 1, 1990. "Electronic Intelligence." *Aviation Week and Space Technology,* 47.
13. W. Scott. June 17, 1991. "Leap Begins Flight Tests to Demonstrate Kinetic Kill Missile Defense Capability." *Aviation Week and Space Technology,* 207.
14. August 1991. From SDIO Public Document, "Technology Applications Program."
15. J. Keller. January 18, 1993. "Sarnoff Spins Off New Venture." *Military & Aerospace Electronics,* 4, 14.
16. W. Jacobi et al. 1991. "Miniature Signal Processor For Surveillance Sensor Applications." *Proc. SPIE* 1479:111–119.
17. Ibid.
18. E. Zyszkowski. August 1992. "Molecular Modeling on 3-D Massively Parallel Processing System." *Scientific Computing & Automation,* 21–34.
19. M. Dornheim. June 22, 1992. "Parallel Processing Gains In Supercomputer Market." *Aviation Week and Space Technology,* 62–64.
20. Ibid.
21. D. Lieberman. October 19, 1992. "Massive Parallelism For Less." *Electronic Engineering Times,* 14.
22. K. Preston. September 1992. "Sci/Industrial Image Processing: New System Benchmarks Results." *Advanced Imaging,* 46, 48,49, 82.
23. Ibid.
24. 1993. Information courtesy of Martin Marietta Corporation.
25. 1992. Information courtesy of Irvine Sensors Corporation.
26. Ibid.
27. D. Van Camp. September 1992. "Neurons For Computers." *Scientific American,* 170–172.
28. G. Hinton. September 1992. "How Neural networks Learn From Experience." *Scientific American,* 145–151.
29. 1993. Private discussions with Herbert Duvoisin III.
30. Ibid.
31. R. Forsell, A. Thronbrugh, and C. Preyer. 1990. "Applications of Smart Neuromorphic Focal Planes." *Proc. SPIE* 1339:1330–1334.
32. D. Sinha. February 1993. "New Paradigms Emerge For Morphological Processing of Noisy Binary Images." *OE Reports,* 1.
33. R. Johnson. July 6, 1992. "E-Metrics May Market Neural-Net Chip." *Electrical Engineering Times,* 34.
34. J. Yoshida. July 6, 1992. "Ricoh Announces Neural Microchip." *Electrical Engineering Times,* 34; and R. Johnson. October 5, 1992. "Neural Chips Pour 100M 'cups,' *Electrical Engineering Times,* 35.
35. C. Johnson. December 21, 1992. "Siemens Fields Big, Fast Neural IC." *Electronic Engineering Times,* 17.

36. R. Johnson. July 6, 1992. "E-Metrics May Market Neural-Net Chip." *Electrical Engineering Times,* 34.
37. J. Yoshida. July 6, 1992. "Ricoh Announces Neural Microchip." *Electrical Engineering Times,* 34.
38. 1993. Information courtesy Irvine Sensors Corporation.
39. R. Johnson. October 5, 1992. "Neural Chips Pour 100M 'cups,' *Electrical Engineering Times,* 35.
40. E. Zyszkowski. August 1992. "Molecular Modeling on 3-D Massively Parallel Processing System." *Scientific Computing & Automation,* 21–34.
41. J. Shepnal. December 16, 1992. "Powerful 3-D Computer Unveiled." *Military & Aerospace Electronics,* 1, 10.
42. M. Seur. January 21, 1991. "Pursuing 3-D Packages." *Electrical Engineering Times,* 66.
43. J. Shepnal. December 16, 1992. "Powerful 3-D Computer Unveiled." *Military & Aerospace Electronics,* 1, 10.
44. M. Seur. January 21, 1991. "Pursuing 3-D Packages." *Electrical Engineering Times,* 66.
45. 1993. Information courtesy of Irvine Sensors Corporation.
46. February 24, 1992. "Texas Instruments Demonstrates Solid State Recorder." *Aviation Week And Space Technology,* 129.
47. 1990. "Optical Computing." *SDI High Technology Update* 1, no. 3.
48. Ibid.
49. M. Hoq, T. Kirle, and J. Walkup. November 1992. "Optical Logic Function Implementation Using a One Dimensional Deformable Mirror Device." *Optical Engineering,* 2413–2421.
50. D. Landeta and C. Malinowski. 1989. "A 2-Deconvolver Architecture For Real Time Image Processing." *Proc. SPIE* 1098:261–271.

Appendix 6A

Representative Data Sheets for Typical Systems and Components

ALLADIN (Algorithm Adaptive and Diminished Dimension)

Manufacturer: Several companies

Intended Application: DARPA, NVEOD

Availability: Highly developmental

Description:

ALLADIN is a multipurpose digital image processor. It is specifically made to reduce non-recurring costs and fit within weapon systems. It is a versatile and modular automatic target recognizer image processor being developed by DARPA and NVEOD for various potential users including LHX. ALLADIN is to provide digital processing for real-time target identification, tracking and aimpoint selection. ALLADIN is to have a growth capability to a 512 × 512 FPA and multisensor fusion, especially between IR and MMW radar.

Characteristics:

MIPS: 100 million

MFLOPS: 400 million

Size: "Hockey puck size"

Information from S. Chow and T. Jones. "Army's FLIR/ATR Evolution Path," Proc. SPIE *1050(1989): 42–49; and "Microsize High Speed Parallel Processor,"* Aviation Week and Space Technology *(June 8, 1992): 67*

ASCM (Advanced Spaceborne Computer Module)

Manufacturer: IBM, Manassas, Virginia

Intended Application: Space systems

Qualification: MIL-STD 1750A, 1541A, 461-C

Availability: Custom made from continuing line

Description:

ASCM offers the application user a cost-effective architecture and associated product set. Space qualified parts, materials, and processes combine with the autonomous, fault-tolerant, real-time operating system to enable high levels of reliability and availability.

Characteristics:

Technology: Radiation—hard one micron CMOS

Total Dose: 10^6 rads (Si)

Prompt Dose Upset: 10^9 rads (Si)/second

Survivability: 10^{12} rads (Si)/second

Single Event Upset: 10^{-10} fails/bit/day

Neutron Fluence: 10^{14} N/cm^2

Pryotechnics Shock: 100 Gs/100 Hz to 2,000 Gs/1K–10 kHz

Processor Subassemblies: Two dual lockstep 3.7 MIPS GVSC

Local Memory per Processor: 356 kilobytes

Global Memory: 1 megabyte

Dimensions: 20.3 × 20.3 × 31.8 cm

Information courtesy of IBM

AST (Airborne Surveillance Testbed Signal and Image Processor)

Manufacturer: Boeing, Seattle and Hughes, El Segundo

Intended Application: U.S. Military

Qualification: U.S. Mil Flight

Availability: Developmental, one made

Description:

The AST signal and image processor is a large nodal rack-mounted system. It performs simple and higher-order signal and image processing functions equivalent to 13 times every volume of the *Encyclopedia Britannica* every second. The AST processors are divided into three stages, beginning with the analog signal processor, which performs basic signal conditioning. The second stage is a highly parallel time-dependant processor. It performs at rates of 15 billion operations per second, accomplished by 86 logic boards composed of 25,000 integrated circuits. The final stage is the object dependent processor which conducts processing on specific objects of interest. It consists of a parallel arrangement of 31 single-board computers performing more than 20 million operations per second. Approximately 250,000 lines of code are used.

Characteristics:

Data Input Rate: 38,400 detectors @ 3000 Hz

A/D Rate: More than 387 million A/D conversions per second

Information courtesy of U.S. Army Strategic Defence Command

GAPP III

Manufacturer: Martin Marietta, Orlando, Florida

Intended Application: Various

Qualification: Various

Availability: Custom made; delivery ≈ 9 months ARO

Description:

The Geometric Arithmetic Parallel Processor (GAPP) is a two-dimensional, fine-grained, SIM-D architectured image processor chip. It can be implemented in arrays to form a massively parallel processor.

Characteristics:

Programmable in: C++ or Ada

Cells per Chip: 8 × 16

Memory per Cell: 128 bits

Speed: 25 MHz

Type: CMOS 1.0 micron

No. of Pins: 128

Power Consumption: 0.85 W/chip

Information courtesy of Martin Marietta Corporation

GVSC (Generic VHSIC Spaceborne Computer)

Manufacturer: IBM, Manassas, Virginia

Intended Application: Various

Qualification: MIL-STD 1750A

Availability: Custom made from continuing line

Description:

GVSC is the CPU used in a family of IBM space processor products that includes the ASCM, Zeus, Cassini, and communication satellite applications. The GVSC is a high-throughput processor capable of supporting a wide range of spaceborne, avionics, or ground-based applications, including image processing. The CPU design exploits concurrent processing and branch prediction techniques. The computer can be built using 1.0 μm or 0.5 μm ruling technology. It is a 16 bit Ada programmable processor.

Characteristics:

Memory Interface: 32 bit

Memory Addressability: 8 megaword

MFLOPS: 150 Million

MIPS: 6.4 @ 27°C DAIS nominal throughput

Information courtesy of IBM

HYMOSS Z Plane

Manufacturer: Irvine Sensors, Costa Mesa, California

Intended Application: Various

Qualification: Any level available

Availability: Highly developmental in large arrays, somewhat developmental in small sizes

Description:

Irvine Sensors makes a HYMOSS stacked silicon Z plane that is indium bump bonded to the detector array. Essentially, their Z-plane is like a three-dimensional unit composed of other chips glued together so closely that an array can be bonded to it. The circuitry can support preamplification, temporal filtering, gamma circumvention, and nonunifomity correction. The circuits are tunable, so they can be programmed (to a certain level, usually just values changed) after the Z-plane is made.

Characteristics:

Array Size: Up to 128 × 128 demonstrated

Spacing: 2 to 10 mils, 4 mils preferred

Dynamic Range: Depends on power, typically 2×10^4:1

Present Functions: Preamplification, tunable filters, mux

Power Dissipation: Typically 0.7 μW per channel

Frame Rate: 15–30 Hz

Size: 0.5 × 0.5 × 0.5 inch

Information courtesy of Irvine Sensors Corporation

MA 16 Neural Network Computer Chip

Manufacturer: Siemens, Munich, Germany

Intended Application: Various

Qualification: TBD

Availability: New product

Description:

The MA16 employs four parallel processors that can be configured into either a one-dimensional vector or two-dimensional arrays. "Designed as a special-purpose DSP, the MA16 takes just 100 ms, for instance, to perform a two-dimensional convolution on an image of 512 × 512 pixels with a 16 × 16 pixel convolution window."

Characteristics:

Connections per Second: 800 million

Technology: 1 micron CMOS

Die Size: 13.7 × 13.6 mm

Cost: About $2,000 each

Information from C. Johnson, "Siemens Fields Big, Fast Neural IC." Electronic Engineering Times *(Dec 21, 1992), 17*

RHVP (Radiation Hardened Vector Processor)

Manufacturer: IBM, Manassas, Virginia

Intended Application: Various

Qualification: Any up to S-level

Availability: Custom made from continuing line; delivery depends on customer requirements

Description:

RHVP is a versatile, high-throughput, programmable processor for spaceborne digital signal processing. It is applicable to image processing as well as other uses. The RHVP utilized VHSIC technology and can be easily programmed in Ada. The architecture combines the high-performance scalar processing capabilities with the high-throughput vector processing capabilities of IBM's floating point VHISIC chips.

Characteristics:

MFLOPS: 150 Million

MIPS: 4

Volume: < 2,800 cc

Power Required: < 50 W

MTBF: 250,000 hours

Weight: < 2.3 kg

Information courtesy of IBM

SCC-100

Manufacturer: Space Computer Corporation

Intended Application: Various

Qualification: None currently, any level available

Availability: Highly developmental

Description:

This is a miniaturized, low-power parallel processor for electro-optical surveillance image and signal processing. It employs low-power VLSI devices with three-dimensional hybrid wafer-scale integration and packaging. High-speed, low-power ASICs are used for interfacing and inter-node communication.

Characteristics:

Number of Nodes: 12

Peak Throughput: > 1.0 GFLOP

Volume: 245 cc excluding enclosure

Peak Power Consumption: 60 W

Weight Excluding Enclosure: 365 g

Information courtesy of Space Computer Corporation

Zeus Space Computer

Manufacturer: IBM, Manassas, Virginia

Intended Application: Space systems

Qualification: Space, including MIL-STD 1750A

Availability: Custom made from continuing line

Description:

IBM began delivery of the ZEUS computer, test set, and software in 1991. The Zeus product line consists of four configurations based on the 16 bit GVSC processor.

Characteristics:

Throughput: 2.4 MIPS

Reliability: >0.94 for eight-year missions

Radiation Survivability: Greater than 10^6 rad (Si) total ionizing dose and greater than 10^{12} protons/cm^2 (20 MeV)

Power Required: < 23 W

Dimensions: 17.8 × 20.3 × 22.9 cm

Weight: 8.2 kg

Information courtesy of IBM

7

Pointing, Scanning, and Stabilization Mechanisms

"The trouble with nude dancing is that
not everything stops when the music stops."

Sir Robert Helpmann

7.1 TECHNOLOGY BASICS

7.1.1 Introduction

The photon's journey through the sensor ends with its conversion to an electronic signal at the focal plane array. During the sampling (exposure, or integration) time, where photons are collected, the line of sight must be maintained within specified limits. That is, the light gathered from a given point in the scene must be registered. Otherwise, the photons from a given point in the scene will fall on and be recorded by pixels that are supposed to represent different parts of the scene. Similarly, in many applications, the stability of the scene with respect to the focal plane must also be maintained between frames. It is frequently required that both these criteria be met while the sensor is attached to a vibrating platform. The function of controlling and stabilizing the optical line of sight (LOS) falls to the servo/gimbal subsystem, which plays a significant part in determining overall infrared sensor performance, cost, and schedule.

The servo subsystem may provide some combination of three basic functions in an IR sensor. The first is to provide the capability of pointing a limited field of view (FOV) throughout a much larger field of regard (FOR). The notional telescope example of Chapter 3, Section 3.1.2, might have a $2° \times 2°$ FOV at any given instant. It could be mounted on a two-axis gimbal, allowing positioning of the $2° \times 2°$ field of view over a near spherical range. Another important function of a sensor's servo system might be to scan a linear focal plane array across the scene to create a two-dimensional image frame. The final major potential function provided by the servo subsystem is to correct for wavefront inaccuracies. While the mechanisms used to correct for this are somewhat different from those used for pointing and stabilization, the automatic control theory and techniques are similar. Infrared sensor performance, cost, and schedule can rely substantially on the attributes of the servo system.

When scanning or pointing, a key sensor attribute is to have accurate knowledge of the LOS with respect to a known reference. This is necessary to accurately reconstruct a scanned scene or, as in the case of a fire control system, to boresight the LOS of a viewed target to the expected trajectory of the weapon. LOS jitter, quantifiable in units of angular displacement, is usually held between 1/10 and 1/2 of a pixel's subtense throughout the integration time to prevent blurring. The sources of jitter that the servo system tries to eliminate are typically stochastic processes stemming from the motion of the base platform and unavoidable attributes (e.g., friction, sensor noise) that are inherent to the mechanisms and sensors used in the servo system itself.

These functions can occur in object space (before the rays are condensed by the telescope) or in image space (after the rays are bent). Object space gimbaling has the advantage of large fields of pointing with the disadvantages of higher mass and power than image space gimbaling. Image space pointing provides low weight and power because the steering element deflects a condensed beam within the telescope. Image space pointing also suffers from a field of regard limited by the telescope's FOV, fore of the steering mirror, and the acceptance angle aft of the steering element. Table 7.1 compares various gimbaling functions in object and image space.

When a design is mature enough, it is important to accurately model the entire dynamic system, including structural flexures and modes. Following this model, the design should be validated with real hardware. This validation can occur early in the program by using a brassboard, qualification unit, or even just some critical assemblies of the actual sensor.

A key paradox to be aware of is that increasing the acceleration, if improperly applied, may not decrease settling time. This is because higher accelerations, when applied faster, tend to excite more structural vibration modes. This may cause the system to take longer to settle to within jitter limits or possibility even drive the system to instability. In some cases, a design with increased acceleration capability may only add weight to a system and produce no corresponding increase in performance [Neal and Overy, 1].

7.1.2 Control Theory Overview

The controls designer's task involves trade-offs between control loop bandwidth, noise magnification, and stability. The classic method used by control engineers to model, analyze, and design automatic control systems makes use of block dia-

TABLE 7.1 Object Space vs. Image Space

Function	*Object Space*	*Object Space Techniques*	*Image Space*	*Image Space Techniques*
Fine pointing	Appropriate, but heavy weight penalty	Inner yoke gimbals Adaptive optics Fast steering mirrors for small apertures Rotating binary	Appropriate, but may drive optical design	Tilting primary Tilting secondary Tilting fold mirror Rotating binary
Gross pointing	Most appropriate	Yoke gimbals Pointing mirror Rotating binary	Appropriate for limited angles & small wide-field optics	Tilting primary Tilting secondary Tilting fold mirror Rotating binary
Image stabilization	Appropriate, but requires severe weight and power increases	Small-aperture adaptive optics Inner yoke gimbals for small sensors	Appropriate	Adaptive optic fold mirror Tilting/dithering Fold mirror Electronic motion compensation
Scanning	Appropriate	Pan/tilt assemblies Rotary mirrors	Appropriate but limited by optics' FOV	Oscillating mirrors Rotary polygons
Wave front correction	Appropriate for first surface	Adaptive optical element	Appropriate	Adaptive optical elements Phase conjugate cells

grams. A block diagram consists of several blocks, each representing the dynamics of different parts of a system in terms of their Laplace transforms and the flow of information between them.

A simple illustration is shown in Figure 7.1 for an object space pointing mirror. In this case, the LOS is tilted by a mirror before the photons enter the telescope. The mirror is driven by a motor. When a voltage, V, is applied to the motor, a torque, T, is created that accelerates the mirror. The mirror moves through an angle, θ_m. Due to Snell's law of reflection, the LOS moves through an angle (θ_{LOS}), which is exactly twice θ_m.

The reduced block diagram shown in Figure 7.1 represents a very simple "open-loop" control system. "Open loop" pointing is controlling the pointing of the sensor's line of sight without comparing the sensors output for correction. This method is adequate for crude pointing of large fields of view. Given the desire to move the LOS to a given position with absolute knowledge of all servo parameters (e.g., motor constant, mirror inertia, initial mirror position, and so on), a voltage can be calculated and applied to adequately position the beam. The punch line, of course, is that no engineer ever has absolute knowledge of anything. Uncertainties in each of the system parameters will combine so that the final LOS position will

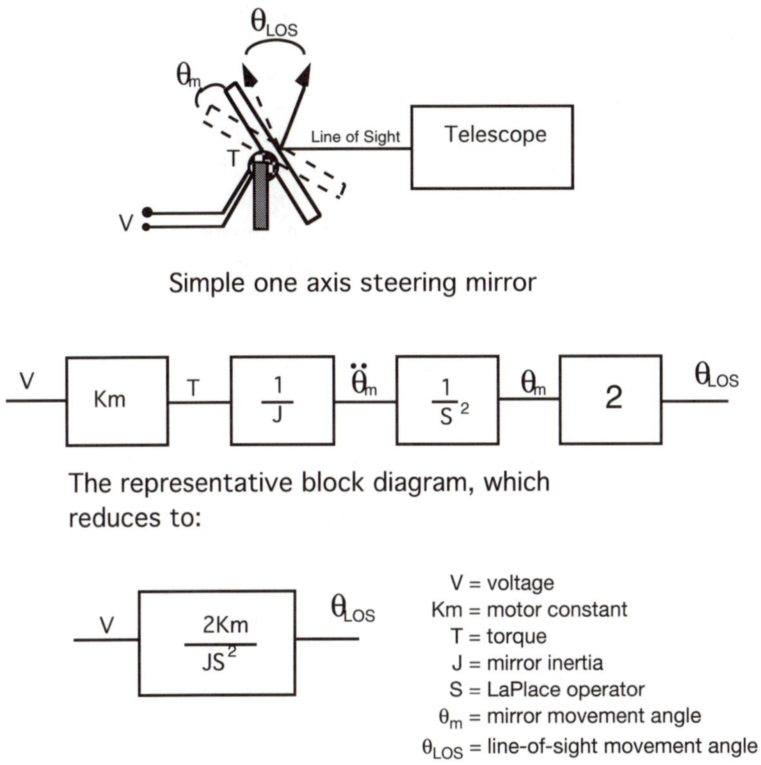

FIGURE 7.1 Block Diagram of a Simple One-Axis Steering Mirror

only approximate the desired position. The error between the two will be roughly proportional to the uncertainty of the system parameters.

The solution to this problem consists of providing a measurement device, such as an optical encoder, that directly measures the mirror position and "feeds" the measurement back to the control system to be accommodated. This process is called *closing the loop* with feedback. Closing the loop allows for real-time shaping of the motion (called *command shaping*), accurate position knowledge, and accurate results. If the mirror overshoots the desired position, a negative force is created that drives the mirror and LOS back. When the desired position is reached, the combined command and position feedback exactly cancel, and the system will stay in position. A pointing control system that uses the sensor's output to adjust pointing via a feedback loop is considered a closed loop. Closed-loop pointing or scanning is required for fine positioning of the LOS. For high performance in pointing accuracy, high slew speeds, or accurate retargeting, a closed-loop system is required.

However, if not carefully designed, closed-loop control systems tend to be unstable because the feedback loop will tend to overcompensate or oscillate—just like with audio feedback. To avoid this, the control system must respond rapidly and exhibit reasonable damping [Ogata, 2]. However, closed-loop designs have the advantage of compensating in real time for LOS changes due to thermal and random (such as bearing hash) effects. As such, they are always superior in performance to open-loop systems. But when tracking an exceptionally weak target (i.e., weak compared to noise and clutter), it is not realistic to expect superior performance from a closed loop. This is because the loop may get closed on the noise.

A block diagram of such a closed loop system is shown in Figure 7.2. Here, the block labeled "plant" is equivalent to the reduced block shown in Figure 7.1. "Plant" is "controlese" for the subsystem that represents the dynamics of the device that will be used to control the beam. The "feedback" block contains the dynamics of the sensor used to measure the output parameter (LOS angle) and convert it into a voltage that can be used by the control system electronics. Typically, this is considered constant over the dynamic range of the control system if the measurement sensor is correctly chosen, and it represents a conversion factor associated with the sensor (e.g., degrees/volt). The remaining block labeled "compensation" is the object of the control system designer's task. This consists of items such as electronics that differentiate the signal to calculate velocity and provide damping to the system, integrators for integrating the error signal over time, and electronic filters to reduce the sensitivity of the system dynamics to various resonances in the system.

Again, closing the control loop allows shaping of the velocity, acceleration, and position profile in real time during the slew. The result of command shaping during slews provides less settling time and more accurate initial positioning. The mechanism will cover less ground (move slower) than an average of 1/2 at^2, but reduced settling time will often offset the slower average movement. Additionally, this method tends to be more gentle on the telescope structure by not changing

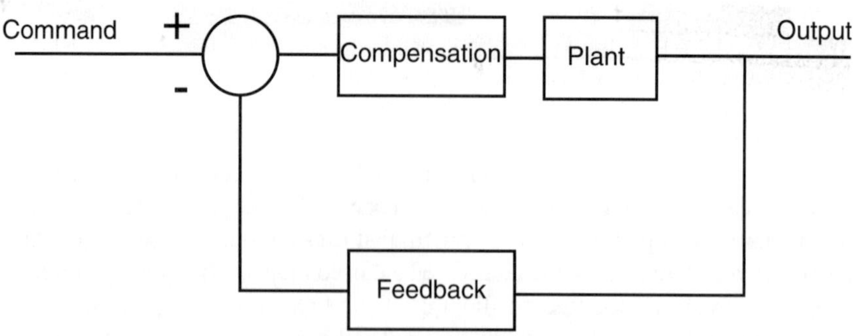

FIGURE 7.2 Simple Block Diagram of a Closed-Loop System

acceleration as severely as a *bang-bang control system* (i.e., one that uses a square wave command that only commands to a full amplitude).

Properly shaping the acceleration can reduce the settling time to near zero. However, this requires an accurate and quick motion feedback sensor, increased acceleration, and rapid processing. There are several techniques and methodologies for shaping the acceleration profile for both open- and closed-loop systems, including proximate time optimal servo (PTOS), modal avoidance, and slew profile following. For example, Workman came up with a PTOS controller that has produced the quickest retargeting in some hardware [Neal and Overy, 3]. This control philosophy commands a bang-bang (square wave) in the beginning of the slew but employs a shaped deceleration that shortens settle time. This shaped slew profile avoids the unwanted excitation of structural modes and large bias errors [Neal and Overy, 4].

To perform such closed-loop corrections, the position (or velocity) must be independently sampled, and corrections must be calculated and applied. The servo system must be able to accurately follow a given command. In general, a command such as a sine wave or a unit step function can be represented in terms of its spectral content by means of a Fourier transform. The former would appear as a sharp spike at the characteristic frequency on the frequency axis. The bandwidth of the control system defines that portion of the frequency domain to which it can accurately respond. Consequently, a sine wave with a frequency within the control loop bandwidth will be tracked accurately, but if it is significantly outside the bandwidth, it essentially will be ignored.

While it would seem that unlimited expansion of the system bandwidth would be desirable, the ugly realities of life intrude on the control engineer's dream. The first limitation on bandwidth comes from stability requirements placed on the system. While a detailed discussion is outside the scope of this work, a well known example of instability is the screeching sound generated by audio feedback. This results from the feedback signal being combined additively with the normal command signal. In a beam control system, the stability limitations are usually encountered when the system bandwidth approaches the primary structural resonance. Another limitation to large bandwidths arises from excessively noisy measurements. In these situations, the control system will tend to follow the noise, resulting in increased jitter.

For digital controls, the frequency of the measurements should be twice the Nyquist frequency, which implies five to ten times the bandwidth. The analog equivalent would be to choose sensors and actuators with capabilities of ten times the system bandwidth. The applied changes to control the slew should not be applied at a frequency close to any natural structural mode of the plant. A good practice is to keep the control bandwidth frequency a factor of five or ten below the natural frequency—especially the first structural mode. A notched filter is sometimes employed to prevent changing torque in a natural vibration harmonic. Keeping control corrections below the natural frequencies has serious implications for the stiffness of a telescope or mirror to be slewed. If the slew is to be updated 50 times during a 1 second slew, this implies a ≥ 100 Hz measurement and a 250 to 500 Hz natural frequency.

Typically mirrors and telescopes are required to have fundamental frequencies above 100 Hz. For example, the fundamental frequency on the Hubble primary is above 200 Hz [How et al., 5], which seems to be a good number for mirrors in active control systems. Additionally, a 17.8 cm steering mirror has been reported having a 5,000 rad/s^2 acceleration at control bandwidths of 4,000 Hz and supporting 10 nanoradian LOS stability with a 2 percent settling time for small angles [Medbery and Germann, 6]. Additionally, the Relay Mirror Experiment steering mirror was a 61 cm beryllium nonreactionless system with an 80 Hz servo bandwidth [Germann and Braccio, 7].

Closed-loop control systems can be underdamped, overdamped, or critically damped, as depicted in Figure 7.3. In general, an underdamped system overshoots the desired value then tends to oscillate about the desired value with decreasing amplitude. Slightly underdamped systems are frequently used for pointing mirrors because they reach the desired position the quickest, but only if they can be designed so that the oscillation is within the stabilized jitter requirement (such as 1/10 of a pixel). An overdamped system will always be sluggish in response but will eventually reach the desired state with no overshoot. A critically damped system approaches the desired position in the least time from any non-oscillating approach.

This discussion so far has been an extraordinarily brief overview of the control system designer's art. In reality, a beam control system could consist of several

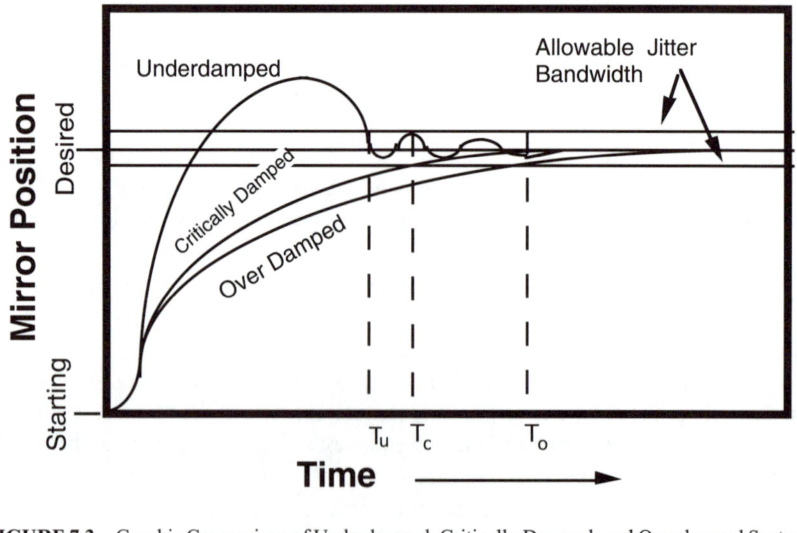

FIGURE 7.3 Graphic Comparison of Underdamped, Critically Damped, and Overdamped Systems

loops wrapped around each other and feeding back position, velocity, or acceleration. Noise sources from measurement devices and approximation of base motion disturbances would need to be quantified and added to the model. Unlike the simple plant model described here, structural resonances of the mirror, hysteresis, and nonlinearities must be accommodated in a real system.

7.1.3 Measurement Devices

The designer should select measurement devices that provide sufficient accuracy, resolution and bandwidth to describe the observed state in a timely fashion. Similarly, actuators should be selected to provide low noise commands over a sufficient bandwidth yielding direct and timely control. Structural design should be evaluated in light of controllability and observability by designing actuators and sensors that are rigidly coupled to the control surfaces.

The control loop can be controlled using the focal plane's output, a gyroscope, a tachometer (rate sensor), and a magneto-hydro-dynamic (MHD) sensor or encoder pick-off. Gyroscopes mounted on the focal plane and/or mirrors are an excellent but expensive way of controlling the slew. High-tech gyroscopes can weigh less than 100 grams but cost over $50,000. More common ones weigh more but cost less (e.g., $10,000). All of these sensing devices may be long lead-time items, taking 4 to 12 months for delivery. A gyroscope must have high bandwidth plus good resolution and linearity to be used to close the control loop. Gyro-stabilized gimbaled mirrors have been reliably and effectively used on several airborne platforms. MHDs are another type of position sensor, and they work via a deflection of an electron beam. They have a rapid and accurate response but need a rapid change to measure; they respond poorly to measurements of less than about 1 Hz. Eddy current position sensors "utilize a small coil to emit a fluctuating magnetic field, generating eddy currents in a conducting medium. The strength of the eddy current is used to measure position" and transduces this into a readout voltage [Medbery and Germann, 8]. Optical encoders optically count dark "ticks" on a transparent disk. Optical encoders can resolve a full rotation with up to about 22-bit accuracy (corresponding to 1.5 μrad), with a knee in the cost and weight curves at an accuracy of 16 bits (about 100 μrad). Optical encoders can weigh several hundred grams and be a weight driver for lightweight systems.

The feedback information, regardless of where it comes from, must be of a finer resolution than the closed-loop tracking error requirement. Experiments with the multi-mirrored telescope have verified the theoretical tracking error per axis of [Ulich, 9]

$$\sigma = \frac{\theta_{LSB}}{\sqrt{12}} \tag{7.1}$$

where

σ = tracking error

θ_{LSB} = amplitude of the least significant bit

Equation (7.1) indicates that the tracking error is directly proportional to the resolution of the measuring device.

Although desirable, it is often difficult to close the loop off of FPA coordinates due to limitations in bandwidth. Sampling at video rates (1/30 s) to stabilize a video image is rarely successful in a dynamic environment. Often, significantly higher FPA sampling rates are required to achieve a reasonable bandwidth. If the focal plane can be sampled faster, it should be. Although image stabilization at video rates is a challenge, many successful target trackers have been implemented in FLIRS using video update rates.

If the position information is derived from the focal plane, subpixel algorithms may be used. These yield pointing information to an accuracy greater than a pixel's angular subtense. If the spot size allows even rudimentary centroiding, these usually work to an accuracy in FPA coordinates as a function of the signal strength as

$$\theta_{LSB} = \frac{\text{Pixel IFOV}}{\text{SNR}} \tag{7.2}$$

where

θ_{LSB} = least significant bit of greatest useful resolution

Pixel IFOV = instantaneous field of view of a pixel

SNR = signal to noise ratio

Equation (7.2) has been experimentally validated from SNRs of about 5 to 100 and is most appropriately applied to starers with low-crosstalk square pixels. Star trackers often do better due to advanced algorithms and blurring techniques. For example, the ASTROS1 has demonstrated a 1/250 of a pixel centroiding [Stanton et al., 10].

Scanners usually have different algorithms for deriving subpixel accuracy in different directions. This is because of the temporal nature of signals in the scan direction, as opposed to spatial and temporal nature in the cross-scan direction. In the scan direction the accuracy is a function of the time it takes the sensor to generate a signal pulse, the SNR and sampling rate while in the cross scan direction it is a function of SNR, center-to-center unit cell spacing, SNR and optical blur.

The control electronics is an important weight and power consideration as it can weigh 20 or 30 percent of the total servo system. It is important to select the correct motor and feedback device to minimize the control loop electronics. Also, modularity and autonomous serviceability can greatly reduce life cycle costs if they are part of the original servo system design [Medbery and Germann, 11].

7.1.4 Actuators and Motors

The actuators are the mechanical mechanisms that provide movement in response to an electrical command. These include conventional magnetic motors for large travel (in a yoke gimbal) and several types of "solid state" actuators for minute travel (as with adaptive optics).

Piezoelectrics (or piezos, or PZTs) are one type of solid state actuator. PZT require high voltage (e.g., 1,000 V) to be applied for their mechanical properties to change. They expand about 0.1 to 1 percent of their linear dimension. This expansion is a sold state phenomenon that occurs rapidly and reliably and can bear loads of several tons. There are no conventional moving parts, so reliability is high—usually several hundred million cycles before failure. Piezos are applicable to deformable mirrors, micropositioning, and stabilization. They are usually implemented in a stack of several piezo wafers to give 0.5 mm or less of total travel. Electrodes are placed between the piezo material. The electrodes separate the wafers and provide the voltage change. The entire assembly is then encapsulated in a passivation layer and a mechanical protection sleeve. Piezoelectric systems suffer hysteresis on the order of 10 percent and are fairly weak mechanically [Galvagni, 12]. The hysteresis tends to give rise to a slow drift in the direction of travel.

Another solid state actuator class is ceramic electrostrictive actuators (ESAs). These have lower hysteresis (3 percent), almost linear temperature effects, and the ability to exert a force of hundreds of kilograms per square centimeter [Galvagni, 13]. Lead magnesium niobate (PMN) ESAs have been used in adaptive optics [Cross, 14]. Ferroelectric actuators are another breed, and they have been tested to 3.6×10^9 cycles and used to deflect a beryllium mirror 24 milliradians in 100 μs [O'Neill and O'Neill, 15]. Magnetostrictive materials are similar to ESAs, except they respond to a magnetic field instead of an electrical one.

Classic motors are usually employed for large travel and slow response. Depending on the design, they can be steppers, brush or brushless, and ac or dc. All of these motors consist of a rotor and stator rings acting against each other. One of the two will consist of a number of permanent magnetic poles spaced around a ring such that the magnetic fields alternate radially inward and outward. The outer ring consists of electrical windings that will produce similar radial magnetic fields to react against the fixed fields. The outer ring can be changed in both magnitude and direction by controlling the current in the coils to produce a desired torque.

Stepper motors are frequently used because of their availability and accurate positioning over a wide range of speeds. Steppers operate by applying a large electrical pulse to the coils, producing a large torque. This causes the motor to step to the next pole. The step size is fixed by the number of poles in the device but can be geared to alter the final output. Steppers work well for both slow tracking and fast slewing. They are also easily controlled to within their step accuracy with simple electronics. Telescope drive systems providing three rates have been built with a total of four chips [Sinnott, 16].

The simplicity of stepper motors is a compelling reason for their consideration by the designer. However, the pulsed nature of their command induces high-frequency torque noise into the surrounding structure. Additionally, the limited resolution step size often makes them unfeasible for precision stabilization. Such applications call for the use of brush or brushless dc motors.

Requiring somewhat more complicated electronics, the torque in dc motors can be precisely controlled by changing the direction and magnitude of the current in the coils. However, because the control system must be aware of the direction of the fixed magnetic field it is driving against, these devices must be commutated. Brush motors accomplish this by physically closing a circuit between the rotor and stator. The circuit directly switches the direction of the coil as the rotor moves to a new pole position. However, brushes have limited reliability, induce EMI discharge, and generate undesirable debris. An alternative approach is to employ a position-sensing device to keep track of the rotor position and switch the direction of the command current in the electronics. This type of motor, called a *brushless dc,* uses the control loop's position sensors or small Hall effect sensors. For small-angle devices, where a given coil always acts against a given pole, commutation is not necessary.

For relatively small angles, linear and rotary voice coil actuators represent an attractive solution for cost, weight, and performance. Voice coils are similar to torque motors in theory, having a coil acting against a fixed magnetic field. They can be configured in a wide variety of shapes and forms to produce a desired performance. For instance, they are frequently used in two-axis steering mirrors that are mounted on a single pivot. Voice coil actuators can easily provide torque in one axis but will accommodate motion in the cross-axis, whereas a standard motor could not.

7.1.5 Bearings and Structural Elements

Motors usually work against bearings, making bearings key components in servo systems. Gimbals and large-travel steering mirrors are mounted on bearings. The bearings may limit reliability due to lubrication breakdown, and friction may limit pointing accuracy and maximum motor size. Mechanical bearings exhibit three types of friction: (1) *stiction,* a static friction in which the bearings bind and need to be broken loose; (2) *coulomb,* a constant friction torque that depends on the direction of rotation; and (3) *viscous,* which is proportional to rate in both magnitude and direction. While the bearings are in operation, random and uncalibratable variations in viscous and coulomb friction can induce self-jitter, which can only be taken out with a high-bandwidth, closed-loop controller. The bearings must be preloaded to limit radial play, but preloading should not be restrictive enough to cause jitter or stalls. Preloading removes radial and axial slop by ensuring intimate contact of all moving parts and provides stiffness to the bearing joint. Proper design must include suitable thermal paths to ensure that all motors and bearings are properly heat sunk.

Random and periodic shaft wobble can occur due to friction variances, flexing in the shaft, and/or a slight imbalance in the mirror. Shaft stiffness depends on the material and the fourth power of the shaft diameter, so a larger-diameter shaft tends to mitigate the wobble. Beryllium and composites provide for stiffer shafts. Shaft wobble can also be reduced by placing the bearings on each end of the shaft and making the shaft as long as stiffness and packaging allow.

Another type of actuator/bearing combination is the "flex pivot." This is essentially a stiff connection with a built-in actuator that can bend (flex) the joint. It accommodates rotary motion by including bending elements instead of rotating elements, such as balls. The bending action alleviates the friction issues associated with bearings but is inherently limited to small angle of a few degrees. Flex pivots can provide rapid and accurate pointing, but only within that small angle. Thus, they have limited travel, no stiction, little or no friction, and a spring-rate response instead of a friction type response.

7.1.6 Mathematical Representations of Slew Times

All gimbal/sensor systems require time to accelerate, decelerate, and settle before data can be collected. This often poses a drawback for step-staring or retargeting systems—most of the time is wasted moving and settling the sensor rather than taking data. The time it takes the sensor to step from one field to another is its retargeting time and can be defined as:

$$\text{Time to retarget} = (\text{Acceleration time}) + (\text{Deceleration time}) + (\text{Settle time}) + (\text{Integration time}) \tag{7.3}$$

At first glance, it would seem the fastest way to move from one point to another is by using a square-wave acceleration profile yielding a triangular velocity profile, as in Figure 7.4. For a first approximation, this assertion is true. Unfortunately, the structural modes of the system may end up vibrating, or ringing, so much from this violent slew that it takes a relatively long period to settle. Unless the entire structure is ultra stiff or highly damped, the fastest way to slew is almost always a smoother, controlled, closed-loop movement. As a result, square acceleration profiles are rarely used because of their deleterious interaction with the system's structural modes.

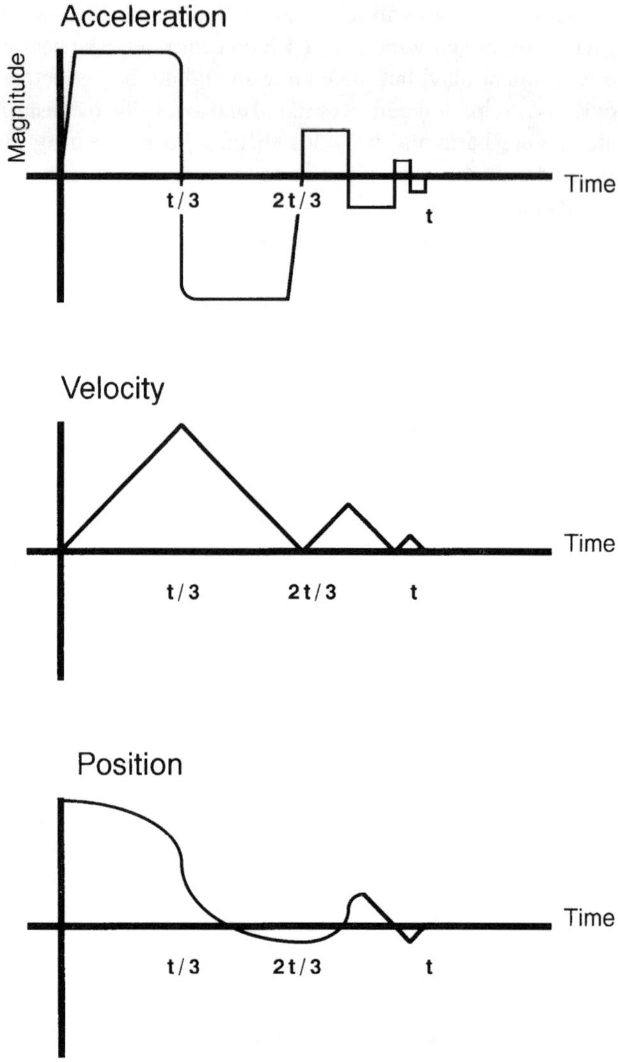

FIGURE 7.4 Graphic Representation of a "Bang-Bang" System

Nevertheless, one can still get a simplistic feel for a system's slew times based on a square acceleration profile, which produces a triangular rate profile as illustrated in Figure 7.4. A square acceleration profile is sometimes called a "bang-bang" profile, as the acceleration is commanded to its full amplitude, and only the direction is changed. Examining a bang-bang profile is especially useful in the early stages of product development, for making viewgraphs to sell an idea, or if little is known about the system's dynamic characteristics. However, a detailed understanding of the system's dynamics is critical for a comprehensive analysis.

Lacking the details, one can postulate that, for real systems, the acceleration time is the same as the deceleration time, and the settle time is on the order of 5 to 33 percent of the total time. The time it takes to retarget can be approximated as

$$T_r = \left(2\frac{\theta}{\alpha}\right)^{1/2} + t_i + T_s \tag{7.4}$$

where

T_r = time to move and take data from one position to another

θ = angular distance to be slewed

α = angular gimbal acceleration

t_i = focal plane integration time

T_s = settle time

It is assumed above that the settle time is the same as the acceleration and deceleration time. If integration time is insignificant or there is none, then Equation (7.4) can be reduced to

$$T_r = 3\left(2\frac{\theta}{\alpha}\right)^{1/2} \tag{7.5}$$

where

T_r = time to move and take data from one position to another

θ = angular distance to be slewed

α = angular gimbal acceleration

For example, if we wish to crudely estimate the time it takes to slew and settle our notional example sensor system one radian (57.3°), and it has a gimbal acceleration of 5 rad/s^2, then $T = 3(2 \times 1/5)^{1/2}$ which equals 1.9, or about 2 seconds.

The minimum angular acceleration to move a mirror or telescope from one position to another is achieved with such a triangular rate profile. Again, this does not necessarily mean it is the minimum time. As pointed out earlier, the settling time is usually large with such a "bang-bang" control. The minimum angular acceleration can be derived by integrating the second derivative of θ and solving for the angular acceleration at one half the transition time which yields [Kline-Schoder and Wright, 17]

$$\alpha_{min} = \frac{2(\theta/2)}{(0.5t)^2} \tag{7.6}$$

where

α_{min} = minimum acceleration required to conduct slew

θ = angle to slew through

t = time to slew

From Equation (7.6), one can get the minimum required acceleration, which (along with the slewing inertia) sizes the motor's volume, weight, and power. None of the previous equations accounts for the time needed for the motor to reposition itself at a starting point, which may or may not be required. As always, it is good practice to include substantial design margins.

When more is known about the dynamic behavior of a system, then more mathematics may be substituted for the above hand waving. Ulich gives a more rigorous methodology of estimating slew times, considering three cases of a second-order pointing system. The first case is for tiny slew distances where the dominant time consumer is the damping of the system. Another case covers slews of moderate positional magnitude in which the acceleration is the limiter. Finally, he cites one in which the slew is limited by the velocity [Ulich, 18].

In the first case, system damping is the driver for the time to move a small distance. This can occur in fast, fine steering mirrors where the step size is microradians. The step time depends on the natural log of the step size and the damping ratio, ζ. This damping ratio of the plant can be defined as [Ulich, 19]

$$\zeta = \frac{\beta}{(JK)^{1/2}} \tag{7.7}$$

where

z = damping ratio

β = viscous damping coefficient

J = plant's moment of inertia

K = plant's spring constant

The time to slew to a position and remain within an allowable positional error is

$$\tau_1(\Delta) = \frac{\text{In}\{\Delta/[\delta(1-\xi^2)^{1/2}]\}}{\xi\omega_n} \tag{7.8}$$

where

$\tau_1(\Delta)$ = time to move a step for case one

Δ = step size

δ = maximum allowable overshoot (or jitter amplitude)

ζ = damping ratio defined in Equation (7.7)

ω_n = natural frequency = $(K/J)^{1/2}$

The acceleration may be limited because of actuator torque capabilities, feedback sensor limitations, or allowable reaction torque on the base. The time to accomplish a larger slew, which is limited by acceleration for the second case, can be estimated by adding the coast period to the acceleration time.

$$\tau_2(\Delta) = 2\left[\frac{(\Delta - \Delta_a)}{\alpha_{max}}\right]^{1/2} \tag{7.9}$$

where

$\tau_2(\Delta)$ = time to move the step for case two

Δ = step size

Δ_a = size in which the step response is that of a linear second-order system

$$= \frac{\alpha_{max}}{\omega_n^2(1-\xi^2)}$$

α_{max} = maximum allowable acceleration

The allowable velocity may be limited by inertia effects, back EMF, the gyroscope, or some other feedback sensor's limitations. The time to accomplish an even larger slew, limited by velocity for the third case, can be estimated by adding the acceleration limited step time to the excess time caused by velocity limitations.

$$\tau_3(\Delta) = \tau_2(\Delta_v) + \left[\frac{(\Delta - \Delta_v)}{\varpi_{max}}\right]^{1/2} \tag{7.10}$$

where

$\tau_3(\Delta)$ = time to move the step for case three

Δ = step size

Δ_v = size in which the step response is linear when the velocity dominates

$$\Delta_v = \frac{(\varpi_{max})^2}{\alpha_{max1}}$$

α_{max1} = maximum allowable acceleration used to limit maximum velocity

Although pointing, retargeting, and stabilization are different functions controlled at different bandwidths and with different angular amplitudes, the above discussions and equations apply to any moving plant.

Random jitter is a result of external vibrations coupled into the sensor system as base motion disturbances. It can also be caused by bearing wobble (hash), coulomb friction, and torque ripple in the servo system. Often, it is easier to correct for base motion disturbances and jitter than it is to remove these disturbances. This is especially true with FLIRs that are mounted on tanks or helicopters. The jitter from many aircraft and motorized platforms is frequently sinusoidal in nature, and the degradation in image quality depends on the ratio of exposure time to the period of the sinusoidal jitter [Wulich and Kopeika, 20]. If sinusoidal, the resulting MTF (degradation thereof) for such a system can be described by a Bessel function given by [Wulich and Kopeika, 21]

$$M_s(f) = J_0(\pi f F \theta_{pp}) \tag{7.11}$$

where

$M_s(f)$ = jitter-dominated MTF

J_0 = zero-order Bessel function

π = pi = 3.141 …

f = spatial frequency

F = effective focal length of the system

θ_{pp} = peak-to-peak angular movement

Equation (7.11) is valid for jitter of high frequency with respect to the integration time. The blur for low-frequency movement is more akin to a random process. There is a general problem with modeling MTF degradation with Bessel functions. The Bessel function is independent of the time varying components of the vibration (which tend to be so irritating to a person's eye–brain as well as computer trackers).

To achieve optimal resolution and prevent blurring, the image must be stabilized to ≈1/10 of a pixel over the integration time. To present a comfortable image to a person on a display, the image should not jitter more than about 1/10 of a pixel over the integration time and no more than a pixel every second or so. Fawcett describes a FLIR stabilization method, which is as follows. The basic inertial stabilization is achieved by feedback from a dual-axis rate transducer mounted on a stable body. This, in turn, is modified by synthetic inertia and friction compensation techniques in all axis. A LOS stabilization of 230 microradians is accomplished at azimuth and elevation positions of zero degrees, with degradation at the ends. To accomplish this, the LOS acceleration rate required was 5 rad/s^2, with a LOS slew rate of 1.5 rad/$s^{2,}$ a gyro bandwidth of 20 Hz, and a synthetic inertia loop bandwidth of 90 Hz [Fawcett, 22]. Sub-microradian stabilization is achievable on space platforms with a combination of star trackers, gyros, and lateral accelerometers.

7.2 FAST STEERING/STABILIZATION MIRRORS

A high-bandwidth mirror can be utilized in object or image space to scan over small angles, stabilize, dither, "toggle" between FPAs of a different color, chop the beam, or provide general beam steering and low-amplitude, low-frequency drift correction. It is advantageous for fast steering mirrors to be stiff, lightweight, and highly damped. Beryllium and silicon carbide lend themselves best to this application. Typical FSM mirror closed-loop bandwidths range from 80 to 300 Hz (with over 400 Hz being achieved) and the diameter is dictated by the needed optical surface—typically from 5 to 25 cm [Cochran and Vassar, 23]. Frequently, FSMs are made reactionless to prevent exciting surrounding optical structures. The forces generated to move the mirror are cancelled by other elements in the FSM instead of the mounting structure.

Dither mirrors are usually implemented by a controlled jitter of a flat mirror element in the telescope. By dithering the scene with proper image processing, variations and focal plane nonuniformities can be mitigated. The dither serves as a bandpass spatial filter that reduces the effects of low-frequency noise components [Kline-Schoder and Wright, 24].

Thermal considerations of steering mirrors and gimbals must be carefully considered for infrared systems. The motors and actuators can generate much waste wattage, which can heat up the mirror or telescope structure and create an extra parasitic effect on the focal plane cooler.

The secondary mirrors of a reflective telescope may be employed as a steering or dither mirror over small angles. A piezoelectric actuator has been used on a Cassegrain's secondary to dither at 50 to 1,000 Hz to fill in gaps (due to poor fill factor) in a two-dimensional array [Liddaird, 25]. In astronomical uses, the secondary can be employed to chop between an astronomical source and deep space. Several ground-based astronomical telescopes use this technique with their secondary, including the 10 m Keck astronomical telescopes and NASA's Infrared Telescope Facility.

High-bandwidth mirrors are still very expensive. Units of only a few centimeters in aperture can cost tens to hundreds of thousands of dollars, with little or no environmental qualifications.

7.3 OBJECT SPACE GIMBALS

Object space gimbals are yokes surrounding a sensor and attached at a balance point, often near the primary mirror. Figure 7.5 (courtesy of Ball Corporation, Aerospace Systems Group) shows a cutaway drawing of typical yoke style, two-axis, object-space gimbal for a sensor. Figure 7.5 also shows a method of directing the beam through the gimbal. That is, the photons are directed through the elevation and azimuth supports and out the bottom, where they can be directed to a focal plane. Other designs have the focal plane, and sometimes cryocoolers and electronics, within the innermost yoke.

A one-axis gimbal usually just tilts the sensor in one dimension, with the platform providing the other degree of freedom. A two-axis gimbal is frequently configured as an altitude-azimuth (or alt-az) by rotating the base of the yoke for one axis and tilting the sensor at the yoke for the other axis. A three-axis gimbal adds another angular dimension, either a roll axis or a cross-elevation axis. The cross elevation provides an extra level of stabilization and prevents gimbal lock-up when pointed near nadir or zenith. A four-axis gimbal has four angular directions of freedom. Usually, they are composed of two azimuth and two elevation gimbals.

These gimbals are usually bearing gimbals, with the telescope held in a yoke that provides large angular travel and continuous movement. They are usually of low bandwidth and have high jitter characteristics on the order of 40 to 250 μrad [Medbery and Germann, 26]. Object space gimbals usually can provide stabiliza-

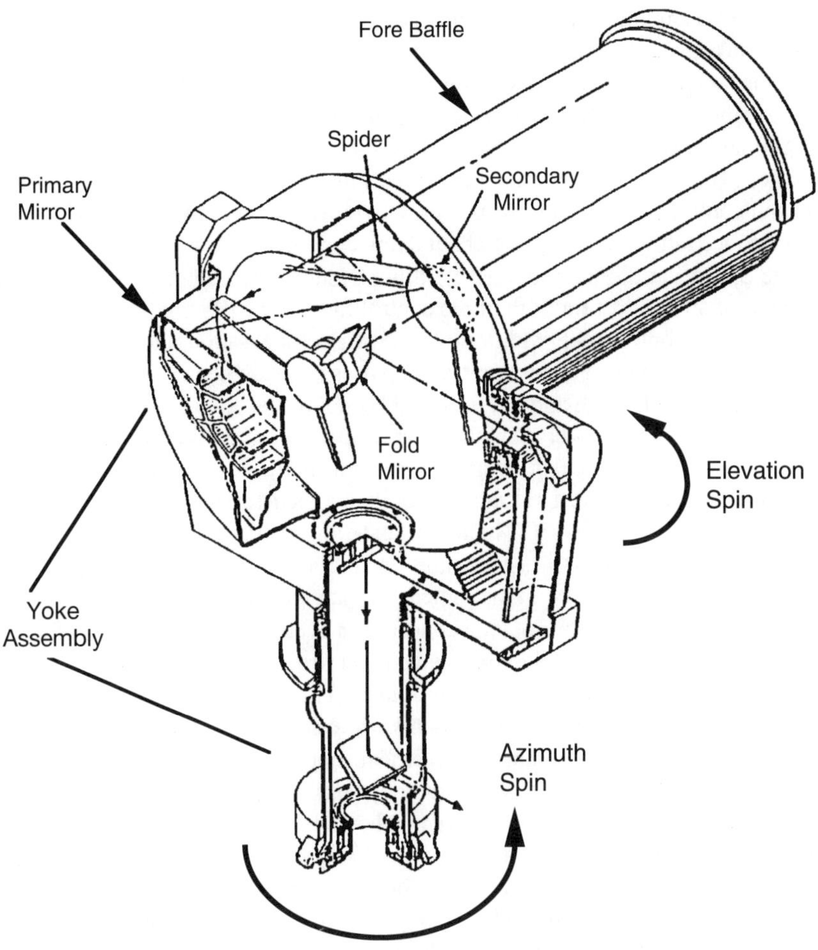

FIGURE 7.5 An On-Axis Telescope Supported by an Object Space, Yoke Style Gimbal (courtesy of Ball Corporation, Aerospace Systems Group)

tion in the range of 1 to 10 milliradian when the sensor is slaved to follow a master reference [Wolfe and Zissis, 27]. The stabilization accuracy can be increased to the neighborhood of 1 to 500 μrad by implementing inertia sensing components directly on the telescope [Wolfe and Zissis, 28] with a substantial weight penalty.

Large fields of regard, implying large travel coupled with the desire for high base motion rejection, are usually obtained by using both a coarse and fine steering mechanism. Sensors requiring both usually employ a hierarchical architecture of increasing stabilization accuracy and bandwidth with decreasing angular travel. This is achieved through nested gimbals and control loops, one completely contained within another. The fine steering control loop and hardware is nested inside that of the coarse gimbal. The coarse, on the outside, provides the large, relatively imprecise travel, while the inner loop accomplishes high bandwidth stabilization. Image stabilization with object space gimbals often employs this "inner" and "outer" gimbal for each axis. The inner is more suited for this, as it will have mechanical stops (banging against the outer gimbal) that limit travel. It also weighs less and can control faster using less power. Obviously, nested gimbals provide high performance but with a large weight and power penalty.

The Space Shuttle International Pointing System (IPS) control loop is a good example of an application of multiple control loops. The IPS control system consists of a fast control loop with long-term drift compensation and short-term vibration compensation. The fast control loop uses measurements taken from a gyro package at a 100 Hz rate and fed into a 25 Hz controller for actuation of the platform via redundant torquers. The long-term updates are derived from an inertial reference using measurements from either a dedicated star/sun sensor or an instrument on the platform at a 1 Hz rate. In parallel, a feed-forward loop uses accelerometer outputs taken at 50 Hz to mitigate the vibration effects caused by astronauts' movements or thruster firings. In addition, the fast control loop accepts offset commands issued by an instrument on the platform [Dornier, 29].

This inner and outer control loop concept can even be applied to loops of a segmented mirror. The Keck telescopes have two mirror support subsystems operating at different bandwidths and amplitudes. This prevents a segment of the primary from banging into other segments while still allowing it to tilt.

A gimbal's interface comprises electrical, optical, mechanical, and reaction forces. On an aircraft, these are the aerodynamic effects. On a spacecraft, the blocking of solar panels or radiators must be considered. Electrical connections across the gimbal are critical, as the wiring harness can be the limiter in gimbal travel. Slip rings are often too noisy for data transmission; however, several FLIRS now employ them for video analog or digital data transmission. Nevertheless, it is still desirable to digitize the focal plane's output on the gimbal near the focal plane. This allows the transmission of digital information, which is less likely to be corrupted. On the negative side, this adds the weight of the A/D to the slewed weight and increases the gimbal weight.

When implemented, most control bandwidths are limited to 1 to 50 Hz, based on typical gimbal and telescope bending modes between 10 and 500 Hz [Gupta et al., 30]. This isn't as bad as it sounds: hardware emulated retargeting results indicate that a 7 Hz bandwidth gimbal can support a two-second retarget slew [Pettit, 31]. Residual motion of a yoke-gimbaled telescope is usually between 35 and 150 microradians rms due to rotational base motion disturbances and bearing friction effects [Germann and Gupta, 32]. Additionally, empirical data from Ball Aerospace indicate that pointing errors are constant for slow slew rates (e.g., 100 μrad/s or less) and increase approximately linearly with slew rates for faster slews [Medbery and Germann, 33].

Gimbal manufacture scheduling is adversely impacted by poor simulation of the control loops. The schedule is also often adversely lengthened by lack of validated models causing a redesign after the hardware is built. It should also be noted that many gimbal components can have extensively long lead times for delivery. Off-the-shelf motors and encoders can require 12 to 16 weeks for delivery, and custom motors take significantly longer.

The weight of a gimbal is usually driven by the "plant" stiffness and required torque. The torque, in turn, is driven by the slewing inertia and required acceleration. The gimbal weight is often about 30 to 50 percent of the slewed weight if an attempt at weight reduction is made; otherwise, it is 100 percent to 500 percent of the slew weight.

The cost obviously depends on the level of testing and environmental qualification. However, it is also driven by accuracy, size, and the up-front engineering required to design the servo control loop. Requirement-driven weight, power, and cost considerations are shown in Table 7.2.

TABLE 7.2 Requirement Impact on Weight, Power, and Cost for Gimbals

Requirement	*Expected Effect on Weight*	*Expected Effect on Power*	*Expected Effect on Cost*
Acceleration	Increases with square	Increases with square	Greater than linear increase
Accuracy	Moderate until specification requires additional freedoms and control loops	Moderate until specification requires additional freedoms and control loops	Greatly increases cost as jitter decreases
Base motion rejection	Moderate until specification requires additional freedoms and control loops	Little until specification requires additional freedoms and control loops	Little until specification requires additional freedoms and control loops
Cryogenic operation	Great	No direct impact unless active cooling or heating is required	Great, due to exotic materials, lubricants, and testing
Jitter	Moderate until specification requires additional freedoms and control loops	Moderate until specification requires additional freedoms and control loops	Greatly increases cost as jitter decreases
Payload size	Increases linearly	Increases linearly	Greater than linear increase
Range of motion	Slight dependence, but may require repackaging	Slight to moderate	Slight to great, depending on packaging and interfaces
Reliability	Great, if dual strings are used	Slight	Great, if dual strings are required
Velocity	Moderate	Large	Moderate

One can see that it is advantageous to slew the smallest moment of inertia possible. This leads to the advantage of object space pointing mirrors or using a periscope in front of the telescope. The latter can afford excellent field of regard for a sensor. Periscopes also have the advantage that fewer wires and less power need to be routed across a moving interface. As with pointing, weight and power limitations dictate that periscopes work best for narrow field of view telescopes of small apertures.

Reaction cancellation on a gimbal is important for spacecraft applications. The smaller the spacecraft, and the more frequent the movement, the more important it is. Reactionless movement may also be important for ground- or flight-based systems. It has been pointed out that reaction cancelation "is needed to achieve high-bandwidth operation, good following accuracy, and low residual LOS jitter, and to prevent the forces and torques from exciting bending modes of the support structure" (and/or telescope) [Germann and Braccio, 34].

Another gimbal technique allows the sensor to be fixed and utilizes binary (diffractive) optics to tilt the incoming beam. This tilting can be rotated to steer a beam by spinning the element. This simple rotation can be accomplished at rapid rates. Moreover, it can be combined with other tilting binary elements to provide more versatile steering. Lincoln Labs has demonstrated agile beam steering over a 11.5° FOV with a ± 100 μm travel and 35 Hz sweep rate. The small mechanical displacement allowed the use of piezoelectric actuators and had a LOS speed limit of 400°/s [Goltos and Holtz, 35].

7.4 SCANNING SYSTEMS

If a two-dimensional scene is to be covered by less than a two-dimensional array (see Figure 4.2), a scanning mechanism must be implemented. The implementation of the scan depends on system requirements. There are several patterns and methods of moving an array across a field. Scanning is not simple to accomplish. For serial scans with 30 Hz updates, this can necessitate a rotary scan speed in excess of 50,000 rpm. Also, one must be careful of image rotation, skewing, and tilting, especially with a TDI focal plane. It is best to keep all surfaces perpendicular to eliminate the imaging problems that can be caused by scanning.

Scanning is sometimes done in object space, especially for wide angular coverage. The AST scans the entire telescope, although most applications use a scanning mirror. The scanning mirror must accommodate the full aperture at an angle half that of the maximum scan (usually half, due to Snell's law of reflection). This means that the scan mirror is always larger than the aperture (as can be seen in Figure 7.1). For large field angles, this easily can be a sensor system weight driver.

There are three popular object space scan methods and patterns as listed in Table 7.3. The first points the telescope to a scan wheel that scans the beam as it rotates in one direction. An additional axis can be gained by having an intermediate

TABLE 7.3 Advantages and Disadvantages of Scan Methodologies

Type of Scan	*Advantages*	*Disadvantages*	*Expected Scan Efficiency (%)*
Race track scan	High efficiency Zero mean disturbances	High complexity	95–99
Oscillating flyback mirror	Zero mean disturbance Capable of high efficiencies	Large torquers High-frequency disturbances for mirror size	98–98
Scan wheel	Smooth motion DC torque Simple controls Low acceleration Low torques Low power	Low scan efficiency	50–75

tilt mirror or by tilting some of the segments on the scan wheel. However, an undesirable image rotation may result. The second method employs a mirror that tilts in a desired fashion to scan the beam into a telescope. When its limit is reached, it rapidly snaps back to the starting position (the *flyback*). The flyback of the second scan pattern can account for 5 to 25 percent of the total time of the cycle [Germann and Braccio, 36]. The third method is a scan in which a mirror moves in one direction until its limits are reached, then reverses and scans back. Tilts of the mirror can occur in the axis perpendicular to the scan, allowing an increase in field in the perpendicular axis.

A scanning mechanism is usually more dynamically benign than a step-stare. This is because it does not require acceleration and settling between integration times—merely constant velocity.

Most FLIRs employ an oscillating flyback scanner after the afocal telescope. The size of the flyback mirror is kept small by placing it at a location in the optical train where the light bundle is 1 cm or less in diameter. By keeping the scanning mirror small, many of the drawbacks listed in Table 7.3 are mitigated. Flyback scanning designs usually have about a 50 percent efficiency, ± 1 to 3 percent nonlinearity, a few watts of power consumption and a few hundred grams weight.

It is common with serial and parallel scan FLIRs to employ different types of scanners for elevation and azimuth. Typically, scanning in the azimuth direction must be much faster than in the vertical. Several sensor designs employ a rotating polygon mirror to scan the azimuth and an oscillating mirror to provide the elevation.

Galvanometer scanners are frequently employed as the small flyback scanning devices. When operated in their "resonant" mode, they provide high linearity at fast scan rates suitable for television-like displays. Galvonometer scanners operate from a few degrees to an amplitude of tens of degrees.

Velocity stability (scan linearity) is critical to forming a good image. Although it is possible to continuously read out the scan mirror position, it is more common to just read it off once or twice per scan and interpolate between points to determine position. The "open loop" part of the scan must be linear for proper interpolation, as a slight change will cause a misplacement of the image pixel and/or seriously degrade TDI. Linearity is enhanced by just using the center part of an oscillating scan or running a rotating motor as fast as possible. Linearity is based on the entire subsystem, including control electronics; thermal effects; and mechanical effects of the mirrors, shafts, and motors.

The scan efficiency for a flyback mirror can be calculated as

$$\mathrm{SE} = 100\% - 2\,(\mathrm{T_W}) \tag{7.12}$$

where

SE = scan efficiency = useful percentage of the scan

T_W = wasted portion of one edge of the scan

T_W is the part of the scan not used because of linearity concerns. Such concerns often are driven by the time it takes the mirror to slow down and change direction. Equation (7.12) assumes that T_W is equal on both ends of the scan. The efficiency can be increased if the "flyback" portion can also be used for scanning, as implied in Eq. (7.12).

Rotary scan mechanisms usually fulfill the desire for smooth rotary motion at a high velocity. They can be easily made so that their velocity varies by only 0.1 percent or less, unless they are operated at slow speeds (e.g. <50 rpm). Rapidly rotating polygons are usually made in a clean room and come as a sealed unit. The unit may be in a vacuum or pressurized in helium to reduce windage power losses and reduce the chances of the mirror surface colliding with a dust particle. Direct drive is usually preferred for high-bandwidth devices because of the lack of hysteresis and gear clatter. One disadvantage to a rotating polygon is that when a corner swings around, it will cut the aperture, directing a portion of the aperture in one direction and part of it to another. When a facet face crosses the aperture, the FPA output is not useful during this portion of the scan. The scan efficiency for a rotating polygon can be mathematically defined as

$$\mathrm{SE} = 100\% - \frac{\text{beam diameter}}{\text{facet length}} \tag{7.13}$$

When using Equation (7.13), it is important to determine the beam diameter on the reflecting surface of the rotating polygon. Even a displacement of a few centimeters in object space will cause the beam footprint to be larger than the aperture.

Brushless dc motors are usually the choice for scanners because of their lack of debris, friction, and brush-generated electromagnetic interference (EMI). They can be purchased for as little as $500. These motors are usually three-phase, and their drawbacks involve reduced reliability and increased system cost for highly accurate optics. Several commercial FLIR manufactures have staring models for the same price as scanning models. The elimination of these rapidly rotating scanning devices can allow costs to be shifted to the purchase of a two-dimensional array.

Bearing precision, preloads, and lubricants are critical. Ferranti makes conical self-acting gas bearings for rapid rotary motions. These allow speeds from 2,000 to over 90,000 rpm, with a wobble of less than 5 μrad and extremely long life (rated at many thousands of stops and starts). Other commercial scanners using hydrostatic air bearings can revolve at 100,000 rpm. Additionally, magnetic bearings perform well but are typically heavy and bulky, which limits their use in production systems such as FLIRs.

Scanning systems for FLIRS can be as inexpensive as $10,000 or so, while for space-based systems they can run a full 25 or 30 percent of the total payload cost. High-torque, small, mil-specification motors can be purchased for about $500 each. In some applications, this cost can be transferred into buying larger focal planes that eliminate the need to scan. This provides smaller sensors that offer lower vibration and increased sensitivity at the same or a lower price.

Linearity requirements can drive the cost of the scanner, especially for TDI focal planes. If linearity is an issue (and it usually is), it is not uncommon for scanning assemblies to cost $50,000 with a man-year or so of labor to get it functioning optimally.

Electro-optic modulators that change their index of refraction have been used to steer a beam rapidly. By applying a voltage, we can change their index of refraction and cause a tilt of the incoming beam. Such devices have fantastic retargeting rates because there are no moving parts—it is all a solid state effect. Their drawbacks include extremely small steering angles, fragility, high expense, and (generally) radiometric sensitivity loss resulting from throughput losses.

7.5 ACTIVE/ADAPTIVE OPTICS

7.5.1 Adaptive Optics for Hardware Error Compensation

Deformable or moving optics can be used to compensate for static or dynamic alignment errors or jitter. *Adaptive optics* is the technique of changing a light beam's wavefront by intentionally altering an optical surface both spatially and temporally. The altered surface induces changes to the wavefront to correct for other induced errors. Typical phase conjugation in this application can be described as a feedback control loop with a wavefront sensor. The incoming wavefront is measured and gradients are calculated for each segment. The tilt of the segment is adjusted to the opposite of the incoming error via pistons [Gleckler et al., 37], and the reflected beam has an improved wavefront. In addition to mechanically altering an element, an array of electro-optical modulators that can change their index of refraction can be employed to accomplish this correction. It even has been suggested that an electrostatic bulk cooling technique for optics can produce an effect analogous to phase conjugation [William, 38].

As mentioned in Chapter 4, a segmented mirror can provide weight savings and producibility enhancements for large apertures. If the segments are small enough, they can be moved to provide a degree of phase conjugation, which can be used to compensate for minor thermal drifts, stabilize the image, and/or clean up the telescope's inherent aberrations and reduce LOS vibration. An alternative to a segmented mirror is a deformable mirror in which the reflecting surface is somewhat plastic in nature and can be tilted and displaced.

These techniques also hold great promise for low-cost, large telescopes. This is because the large primary can be made without the traditional high figure accuracy and because a ground-based telescope can rival (and maybe surpass) the resolution of space-based assets by mitigating atmospheric effects. "Rubber mirror technology could save research dollars. For example, an Earth-based telescope using adaptive optics costs about $12 to $20 million, compared to the Hubble, which cost $1.5 billion" ["Rubber Mirrors Are Reshaping the Universe," 39]. Adaptive optics can be used to correct for static aberrations (inherent flaws) in a mirror. It has been suggested that this can greatly reduce the fabrication cost of large mirrors. As Meinel and Meinel point out,

> The recently completed 3.5 meter New Technology Telescope of the European Southern Observatory has demonstrated the power of tuning a thin meniscus mirror to control the total wave front of the telescope. This step has contributed to lower cost of the optics by reducing the optical requirement from being one of total precision of the surface to one where smoothness is the dominant requirement with considerable relaxation of the large scale surface figure requirement [Meinel and Meinel, 40].

Using adaptive optics adds weight for the actuators, control electronics, and wavefront sensors. Excluding electronics, a 43.2 cm 50 Hz bandwidth deformable mirror has been reported with a 50 kg/m^2 areal density, which is about triple the mirror's areal weight without the actuators and controls [Janosky, 41]. However, this added weight often can be offset by reduced structure, gimbal, and mounting weight, yielding a lower sensor system weight. This is especially true if a smaller fold mirror is used internal to the optical train, where a tripling of a mirror's weight has little system impact.

Adaptive optics used to correct for wavefront usually have high bandwidths that range from a few hertz to several kilohertz [How et al., 42]. It is easier to implement such high bandwidths with small mirrors or segmented mirrors (smaller pieces of a mirror), as on the Kecks. The smaller size allows a higher bandwidth because the mirrors weigh less and are stiffer than a single large piece. To accomplish high-bandwidth tuning, piezoelectric actuators can be used to change the shape of a deformable mirror in hundreds of microseconds, and entire control systems have been built that change and stabilize the mirror's shape in less than a millisecond. For example, the Japanese National Large Telescope has 390 actuators that recompute the shape of the primary 10 times per second [Magrath, 43]. This telescope must also point to within 1 arcsec and track to 0.1 arcsec ["Japan's 8 Meter Telescope," 44]. Likewise, the VLT has a required effective surface accuracy of 100 nm rms, which will be maintained by hundreds of active actuators compensating via a reflected star's wavefront [Martin et al., 45]. The Japanese telescope single-element primary is a mere 200 mm thick, supported by 264 actuators [Martin et al., 46].

Again, the engineer should pay attention to the weight and power impacts of the processor and control electronics to accommodate adaptive optics. Global control of a large aperture segmented or deformable mirror can require a computational capability exceeding current large mainframes. This can be circumvented by a hierarchical control that clusters segments into super-segments to speed up an iterative convergence process [Gleckler et al., 47].

Adaptive optical techniques can be employed in stages analogous to nested gimbals. A coarse correction can be done on a primary with the high frequency and more exact correction being done on a smaller flat optical element placed further down the optical train. The adaptive optical elements need to be operated at ambient temperatures. It is difficult with current technologies to cryogenically cool adaptive mirrors and their actuators. This can impose severe constraints on high-sensitivity, low-background sensors.

Real-time active compensation applies to the metering structure as well as the optical elements. Much research is progressing on adaptive structures for electro-optical systems. These structures have actuators used to cancel out vibration modes in an active sense. That is to say, they have imbedded transducers like piezoelectrics imbedded that vibrate the structural member in such as way as to cancel out system-level vibrations. Active structures act in image space since they effectively change the relative spacing and tilt of the optical elements. Active structures for image stabilization have been pioneered for space telescopes. Actuators are placed on and/or in the telescope's metering structure and are controlled such that a force is applied to counter a vibration and stabilize the optics. The drawbacks are increased processing and difficult engineering and manufacture.

7.5.2 Atmospheric Compensation with Adaptive Optics

Another form of image stabilization that blurs the boundary between optics and mechanics is *atmospheric compensation.* Adaptive optics can be employed to compensate for atmospheric turbulence. It is possible to subtract the blurred and scintillated images via phase conjunction using real-time adaptive optics. Much effort has been expended to adapt this technique from ground- to space-based applications in astronomy and satellite operations. These techniques grew out of research at Lincoln Labs, Phillips Labs, AVCO Everett Labs, Western Research, and UTOS for ground-based laser weapons. It was declassified in May 1991. Although little has been reported on other use, it is theoretically possible to use the same basic principles to mitigate deleterious atmospheric effects from space to ground, air to ground, ground to airborne platforms, and at ground level across the atmosphere to ground-level targets.

Convection, turbulence, and varying index of refraction of the atmosphere distort and blur an image, limiting its resolution. The best "seeing" astronomers can obtain on good nights, at premier high altitude observatories such as Mauna Kea, is on the order of 0.5 to 1.5 μrad. This limit is the same regardless of aperture size. It is not a question of diffraction limit but of being "atmospheric-seeing-limited." The seeing tends to improve with increasing altitude and wavelength. For a given time at a given place, it can be described as

$$R_p = \frac{\lambda}{r_o} \tag{7.14}$$

where

R_p = practical resolution of a ground based telescope when limited by the atmosphere

λ = wavelength

r_o = Fried parameter

The Fried parameter is the radius in which the incoming wavefront is approximately planar. In the visible, it ranges from about 3 to 30 cm [Tebo, 48]. The Fried parameter is strongly spatially and temporally dependent on the very localized weather at a given location and varies with the air mass (or the telescope slant angle). It can also be affected by such localized effects as the telescope dome and air flow within the telescope. Moreover, the Fried parameter can vary across the aperture of a large telescope. The parameter (hence, seeing) increases with wavelength so that, on good nights, moderate size telescopes (say less than 5 meters in aperture) operating at 10 μm or longer can be diffraction limited [Tebo, 49].

One method to correct for this atmospheric distortion is to employ a wavefront sensor to measure the spatial and temporal phase change on the incoming light—in essence, to measure the Fried parameter in real time. The wavefront sensor can be a Hartman sensor, which is a series of lenslets (or subapertures) that "sample" the incoming wavefront at the size of (or smaller than) the Fried parameter. The size of the wavefront sensor subaperture and the spacing of the actuators can be expected to be less than the Fried coherence cell size. The diameter of the telescope divided by the Fried parameter indicates the minimal number of sub-apertures needed. The optimal size of the wavefront spacing and correction actuators seems to be between 0.69 and 1.0 times the Fried cell size [Welsh, 50].

The information is fed into the computer section of a control loop that commands a deformable mirror to adjust its front surface such that it represents the negative (or conjugate) of the incoming wavefront. The result is phase conjugation that adds to the distortion-produced valleys and subtracts from the distortion-produced peaks. Often, a fast steering mirror is employed to remove large one- or two-axis overall wavefront tilt, while the higher-order distortions are delegates to the deformable mirror [Tebo, 51]. The deformable mirror can be any in the optical system, but it is usually a small, planar fold mirror aft of the primary and secondary, controlled by piezoelectric actuators. It can be of two types. One is a segmented surface consisting of several pieces of the mirror that roughly correspond to the expected Fried radius, with each segment individually controlled and actuated. In the other type, the reflecting surface can be a single sheet of a deformable nature. In either case, the highest spatial resolution that can be accommodated is limited by the discrete spacing of the actuators, and they do not allow correction for intensity distributions [Reintjes, 52], the latter not usually a strong concern for passive systems but a large concern for active systems.

Researchers at the University of Hawaii have applied atmospheric correction techniques by using real stars rather than generated ones. This allows a great reduction in complexity and cost but places the astronomer at the mercy of the heavens for a star. The HRcam used at the Canada France Hawaii telescope on Mauna Kea uses a bright natural star (not a laser-generated one) to adjust the beam 500 times per second via a three-piezo-actuated flat mirror ["Active-Optics Camera Tested," 53]. Results from HRcam on Mauna Kea are reported to increase the resolution by almost a factor of two from 3.4 to 1.9 μrad [Powell, 54], while others estimate that a factor of 25 improvement is attainable. Lastly, the French have developed the "Come-On" adaptive optics image correction system. This system employs a 5×5 array Hartman sensor controlling 19 piezo-electric actuators of a deformable mirror. In addition, they employ a separate tip-tilt mirror in the control loop to accommodate large-scale phase gradients or overall wavefront tilt.

The problem with the above approach is that it requires a high brightness guide star within the field of view of the telescope. This blessing from nature is a convenience that can be rarely counted on. The Strategic Defense Initiative Office realized this early in their high-power ground-based laser program and began research to create a "false star" to allow the use of phase conjugation wherever the telescope was pointed.

The most promising techniques include using the Raleigh scattering to excite a mesospheric sodium layer at 90 kilometer altitude (this metal layer is courtesy of decaying meteorites). The resulting point-like beacon is called a "false star"

[Tebo, 55; Powell, 56]. The pumping occurs by using a laser at the sodium D line at 0.589 μm. This sets up a resonance in the sodium whereby it absorbs and re-emits, not reflects. The latter is a subtle but important point. A reflection wouldn't work as well because it would act like a "double-pass" system where the distortions occurring on the way up to the sodium layer would be corrected on the way down, for no net deviation. Atmospheric refractive indexes, and therefore the phase distortions, are to a first order independent of wavelength, allowing IR sensing to be corrected with the visible sodium technique [Tebo, 58]. The laser power need only be on the order of 80 to 100 W ["Livermore Technology May Boost Detail of Earth Bound Astronomy," 59] and possibility as low as 5 W [Welsh, 60]. Telescopes such as the Japanese eight-meter telescope or the Keck may require more than one laser beacon to correct across their large apertures. These techniques do not completely eliminate the atmospheric distortion but only mitigate it. However, they do so to a level that challenges the utility of space-based telescopes.

The frequency to which the deformable mirror must be adjusted depends on seeing conditions (which depend on altitude, air mass, weather, and luck). It is postulated that a few hertz will compensate for much atmosphere blurring at high altitudes, while in other circumstances corrections must be applied at rates of 1,000 times per second.

The costs for false star generation and the resulting atmospheric compensation are usually estimated at about $15 million, although some believe that less complicated systems for the IR can run as low as a few million dollars [Powell, 61]. A national lab has been funded for a sodium star experiment for about $4 million, and estimates are that a similar system at an observatory would cost about $3 million ["Livermore Technology May Boost Detail of Earth Bound Astronomy," 62]. Additionally, Robert Fugate has predicted that it would only cost a few hundred thousand dollars to add an infrared adaptive system to a telescope, whereas it might cost millions for a visible system [Kirkhart, 63].

References

1. D. Neal. and C. Overy. 1991. "Rapid Retargeting and Precision Pointing." *Martin Marietta Astronautics Group Journal* 2:36–44.
2. K. Ogata. 1970. *Modern Control Engineering.* Englewood Cliffs: Prentice Hall, Inc.
3. D. Neal and C. Overy. 1991. "Rapid Retargeting and Precision Pointing." *Martin Marietta Astronautics Group Journal* 2:36–44.
4. Ibid.
5. J. How, E. Anderson, D. Miller, and S. Hall. 1991. "High Bandwidth Control For Low Area Density Deformable Mirrors." *Proc. SPIE* 1489:148–161.
6. R. Medbery and L. Germann. 1991. "Specifications of Precision Optical Pointing Systems." *Proc. SPIE* 1489:163–176.
7. L. Germann and J. Braccio. November 1990. "Fine Steering Mirror Technology Supports 10 Nanoradian Systems." *Optical Engineering,* 1351–1359.

8. R. Medbery and L. Germann. 1991. “Specifications of Precision Optical Pointing Systems.” *Proc. SPIE* 1489:163–176.
9. B. Ulich. 1988. “Overview of Acquisition, Tracking and Pointing System Technologies.” *Proc. SPIE* 887:40–63.
10. C. Stanton et al. 1987. “Optical Tracking Using Charge Coupled Devices.” *Optical Engineering* 29:930–938.
11. R. Medbery and L. Germann. 1991. “Specifications of Precision Optical Pointing Systems.” *Proc. SPIE* 1489:163–176.
12. J. Galvagni. November 1990. “Electrostrictive Actuators and Their Use in Optical Applications.” *Optical Engineering* 29:1389–1389.
13. Ibid.
14. S. Cross. 1989. “Piezoelectric and Electrostrictive Sensors and Actuators for Adaptive Structures and Smart Materials.” *Proc. ASME* Ad-Vol 15.
15. C. O’Neill and C. O’Neill. November 1990. “Ferroelectric Actuators in the Electromechanical Interface.” *Optical Engineering* 29:1383–1387.
16. R. Sinnott. May 1989. “The Amazing Stepper Motor.” *Sky & Telescope,* 554–557.
17. R. Kline-Schoder and M. Wright. 1991. “Design and Analysis of a Dither Mirror Control System,” *Proc. SPIE* 1489:189–200.
18. B. Ulich. 1988. “Overview of Acquisition, Tracking and Pointing System Technologies.” *Proc. SPIE* 887:40–63.
19. Ibid.
20. D. Wulich and N. Kopeika. June 1987. “Image Resolution Limits Resulting From Mechanical Vibrations.” *Optical Engineering,* 529–533.
21. Ibid.
22. J. Fawcett. 1991. “Lightweight Surveillance FLIR.” *Proc. SPIE* 1498:82–91.
23. R. Cochran and R. Vassar. 1990. “Fast Steering Mirrors in Optical Control Systems.” *Proc. SPIE* 1303:245–51.
24. R. Kline-Schoder and M. Wright. 1991. “Design and Analysis of a Dither Mirror Control System,” *Proc. SPIE* 1489:189–200.
25. K. Liddaird. 1988. “Technical Note: A Novel Focal Plane Scanning Technique.” *Infrared Physics* 28:195–197.
26. R. Medbery and L. Germann. 1991. “Specifications of Precision Optical Pointing Systems.” *Proc. SPIE* 1489:163–176.
27. W. Wolfe and G. Zissis, eds. 1985. *The Infrared Handbook.* Ann Arbor: ERIM, 22–11.
28. Ibid.
29. 1993. Information courtesy of Dornier.
30. A. Gupta, C. Van Houten, and L. Germann. 1990. “Attitude Determination for High Accuracy Submicron Jitter Pointing on Space Based Platforms.” *Proc. SPIE* 1303: 336–349.
31. C. Pettit. 1991. “Line of Sight Stabilization, Sensor Blending.” *Proc. SPIE* 1498: 278–287.
32. L. Germann and A. Gupta. April 1990. “Inertially Referenced Pointing for Body Fixed Payloads.” Technical Symposium on Optical Engineering and Photonics in Aerospace Sensing.
33. R. Medbery and L. Germann. 1991. “Specifications of Precision Optical Pointing Systems.” *Proc. SPIE* 1489:163–176.

34. L. Germann and J. Braccio. November 1990. "Fine Steering Mirror Technology Supports 10 Nanoradian Systems." *Optical Engineering,* 1351–1359.
35. W. Goltos and M. Holz. November 1990. "Agile Beam Steering Using Binary Optics Microlens Arrays." *Optical Engineering,* 29:1392139–7.
36. L. Germann and J. Braccio. November 1990. "Fine Steering Mirror Technology Supports 10 Nanoradian Systems." *Optical Engineering,* 1351–1359.
37. S. Gleckler, B. Ulich, S. Shepard, and J. Conklin. 1990. "Surface Control Techniques for Large Segmented Mirrors." *Proc. SPIE* 1303:288–298.
38. S. William. March 9, 1992. "Black World Engineers, Scientists Encourage Using Highly Classified Technology for Civil Applications." *Aviation Week and Space Technology,* 66–7.
39. "Rubber Mirrors Are Reshaping The Universe." *SDI High Technology Update.* Vol. 1, no. 3.
40. A. Meinel and M. Meinel. November 1992. "Two Stage Optics: High-Acuity Performance From Low-Acuity Optical Systems." *Optical Engineering,* 2271–2281.
41. M. Janosky. November 1990. "Development of a Lightweight Active Optic System for a Spaceborne Relay Mirror Application." *Optical Engineering* 29:1328–1332.
42. J. How, E. Anderson, D. Miller, and S. Hall. 1991. "High Bandwidth Control For Low Area Density Deformable Mirrors." *Proc. SPIE* 1489:148–162.
43. B. Magrath. November 1990. "Optical Astronomy Looks to the Future." *Astronomy,* 35–43.
44. November 1991. "Japan's 8 Meter Telescope." *Sky & Telescope,* 456.
45. B. Martin, J. Hill, and R. Angel. March 1991. "The New Ground Based Optical Telescopes." *Physics Today,* 22–30.
46. Ibid.
47. S. Gleckler, B. Ulich, S. Shepard, and J. Conklin. 1990. "Surface Control Techniques for Large Segmented Mirrors." *Proc. SPIE* 1303:288–298.
48. A. Tebo. December 1991. "Adaptive Optics: The Promise For High Resolution Ground Based Astronomy." *OE Reports,* 14.
49. Ibid.
50. B. Welsh. December 1, 1991. "Image Performance Analysis of Adaptive Optical Telescopes Using Laser Guide Stars." *Applied Optics,* 5021–5030.
51. A. Tebo. December 1991. "Adaptive Optics: The Promise For High Resolution Ground Based Astronomy." *OE Reports,* 14.
52. R. Reintjes. December 1988. "Nonlinear and Adaptive Techniques Control Laser Wavefronts." *Laser Focus,* 63–78.
53. July 1989. "Active-Optics Camera Tested." *Sky & Telescope,* 12–13.
54. C. Powell. November 1991. "Mirroring The Cosmos." *Scientific American,* 112–123.
55. A. Tebo. December 1991. "Adaptive Optics: The Promise For High Resolution Ground Based Astronomy." *OE Reports,* 14.
56. C. Powell. November 1991. "Mirroring The Cosmos." *Scientific American,* 112–123.
57. A. Tebo. December 1991. "Adaptive Optics: The Promise For High Resolution Ground Based Astronomy." *OE Reports,* 14.
58. January 6, 1992. "Livermore Technology May Boost Detail of Earth Bound Astronomy." *Aviation Week and Space Technology,* 57.
59. B. Welsh. December 1, 1991. "Image Performance Analysis of Adaptive Optical Telescopes Using Laser Guide Stars." *Applied Optics,* 5021–5030.

60. C. Powell. November 1991. "Mirroring The Cosmos." *Scientific American,* 112–123.
61. January 6, 1992. "Livermore Technology May Boost Detail of Earth Bound Astronomy." *Aviation Week and Space Technology,* 57.
62. J. Kirkhart. November 1991. "Adaptive Optics." *Ad Astra,* 40–42.

Appendix 7A

Data Sheets for Typical Systems and Components

AAI Graphite Composite Gimbal System

Manufacturer: AAI, Hunt Valley, Maryland

Intended Application: Turret gimbal for air vehicle (fixed wing and rotary), shipboard, and ground vehicle applications

Qualification: Commercial

Availability: Limited stock maintained; delivery ≈ 3–6 months ARO

Description:

AAI's composite gimbal development is based on high-strength, readily producible, graphite designs. Stabilization specifications and methods are adaptable to most mission requirements. Rate or rate integrating gyros may be incorporated, depending on application. High-output, rare-earth torquers are used, providing angular accelerations consistent with requirements.

Characteristics:

Size: 33 cm sphere supported by a 46 cm high turret

Gimbal Order: Azimuth over elevation

Field of Regard: 360° azimuth continuous, +82° to –82° in elevation

Stabilization Accuracy: 80 μrad

Angular Acceleration: >5 rad/s^2 in both axes

Position Pickoffs: Resolvers accurate to ± 3 minutes of arc

Payload Volume: 32 cm sphere limited by a 23 cm width

Payload Weight: 11.4 kg

Gimbal Weight: 13.2 kg

Gimbal Electronics Weight: 5.9 kg

Power Required: 500 W @ max acceleration both axes (28 Vdc per Mil-STD-704)

Information courtesy of AAI

Aerotech AOM360D

Manufacturer: Aerotech, Pittsburgh, Pennsylvania

Intended Application: Steering mirror, video camera pointing, star tracking

Qualification: Commercial

Availability: Commercially

Description:

A general-purpose two-axis gimbal and controller.

Characteristics:

Accuracy: 5 μrad

Velocity: 60°/second

Resolution and Repeatability: 0.5 μrad

Wobble: 5 μrad

Information courtesy of Aerotech

Cambridge Technology Model 6400

Manufacturer: Cambridge Technology, Waterton, Massachusetts

Intended Application: Laser scanners, FLIR systems

Qualification: None

Availability: ≈ 2 months ARO

Description:

The scanner is a moving coil galvanometer which moves the mirror in a single-axis flyback. The coil is supported by aluminum struts attached to a center shaft; preloaded ball bearings support the rotor. Magnetic flux density exceeding 8,200 Gauss is achieved by using neodinium-iron-boron magnets. The position encoder is located as close as possible to the mirror, at the end of the shaft enclosure. A control board weighing 300 g (model 650X) can be used to control the 6400.

Characteristics:

Total Weight: 8.2 kg

Angular Excursion: 40 degrees mechanical with 0.025% nonlinearity

Rotor Inertia: 550 gm-cm^2

Wobble: 5 μrad peak-to-peak max

Jitter: 5 μrad peak to Peak max

Max Mirror Load: 50,000 g-cm^2

Price: $4,650 for one unit, $3,210 for 25 or more units

Unit Size: ≈ 10.2 × 12.7 × 26.7 cm

Information courtesy of Cambridge Technologies

Electro Optical Director Model 066-3

Manufacturer: Contraves, Pittsburgh, Pennsylvania

Intended Application: Naval gimbal for EO sensors

Qualification: Military, including MIL-STD-901C and MIL-STD-167

Availability: Custom made per contract; delivery ≈ 9 months ARO

Description:

The gimbal system is designed for shipborne applications, with each axis being gyro-stabilized. The azimuth axis corrects a ship's yaw motion, the elevation provides pitch correction, and the optional cross-elevation axis eliminates the ship's roll effects. The modular design of the servo system and control electronics allow the system to be fabricated as a two- or three-axis system.

Characteristics:

Payload Capacity: 61 kg in three axes and 91 kg in two axes

Stabilization: 25 to 250 μrad

Max Slew Velocity: 200°/second in all axes

Acceleration: 265°/s^2 in az, 235°/s^2 in el and 300°/s^2 in cross-elevation

Bandwidth (3 db) (Hz): 25 in az and el, 30 in cross-elevation

Gimbal Weight: 445 kg

Operational Temperature: 0 to +50° C

Field of Regard: Continuous in az, –40° to +90° in el and +30° to –35° in cross-elevation

Power Required: 900 W

Information courtesy of Contraves

General Scanning M3

Manufacturer: General Scanning

Intended Application: LOS and laser scanning

Qualification: Commercial

Availability: Off the shelf

Description:

A general-purpose one-dimensional scanning mirror using rare earth magnets. the system can accommodate mirrors up to 5 cm in length.

Characteristics:

Scanner Weight: 350 g

Scan Angle: 60 ° standard, 140° special

Typical Linearity Error: 0.01% at 30°, 0.03% at 60°

Gain Drift: 140 ppm/°C

Offset Drift: 5 μrad/°C

Wobble: 4 μrad

Jitter: 20 μrad

Price: $500 and up, depending on availability, mirror, and electronics selection

Information courtesy of General Scanning

Instrument Pointing System (IPS)

Manufacturer: Team lead by Dornier, Friedrichshafen, Germany

Intended Application: Space shuttle

Qualification: Space

Availability: One of a kind

Description:

A general-purpose gimbal system used to point electro-optical payloads from the space shuttle's cargo bay. The IPS control system consists of a fast control loop with long-term drift compensation and short-term vibration compensation.

Characteristics:

Total IPS Weight: 1,184 kg

Field of View: π Sr, 2π radians in roll

Slew Rate: 1°/second for inertial pointing missions, 3°/second for target tracking

Power Consumption: 500 W mean for inertial pointing, 1,280 W full torque slewing

Payload: 2,000 to 7,000 kg

Pointing Accuracy: 2 μrad 1 sigma

Stability: 3.4 μrad

Information courtesy of Dornier

ISTEC 32D Gimbal

Manufacturer: ISTEC, Hamilton, Canada

Intended Application: Stabilized platform for electro-optical sensors

Qualification: High-grade commercial

Availability: ≈ 3 months ARO

Description:

ISTEC produces a line of gimbaled sensor balls suitable for ground or flight vehicles. They employ a hybrid stabilization system. Gyroscopic forces are used directly to augment the inherent stability of the mass, plus separate active actuators are applied to steer the line of sight. Continuous panning (>360°) is enabled by use of slip rings. The gimbaled sensor balls can be configured with a user's sensors.

Characteristics:

Stabilization: Three axis

Max LOS Jitter: < 5 μrad

Ball Diameter: 0.8 m

Slew Rates: 0 to 60 °/second in each axis

Tilt Range: 30 to 90 degrees

Weight: 70 kg

Price: ≈ $250,000 to 350,000 for one unit

Information courtesy of ISTEC

Lincoln Lab High Bandwidth Steering Mirror

Manufacturer: Lincoln Labs, Cambridge, Massachusetts

Intended Application: Ground based

Qualification: None

Description:

Reportedly, this is a general-purpose, small, high-bandwidth beryllium steering mirror activated by voice coils.

Characteristics:

Total Weight: 700 g

Aperture: 16 mm

Typical Power: 1 W

Range: 26 milliradian peak to peak

Closed Loop Bandwidth: 10 kHz

Peak Acceleration: 13,000 rad/s^2

Resolution, Accuracy, and Repeatability: 0.2 μrad

Unit Size: ≈ 10.2 cm diameter, not including control electronics

Information from Loney, "Design of a Small Aperture Steering Mirror For High Bandwidth Acquisition And Tracking." Optical Engineering *29 (Nov 1990): 1,360-1,365*

Physik Instrumente S-380

Manufacturer: Physik Instrumente Gmbh & Co, Waldbronn, Germany

Intended Application: Various

Qualification: High grade commercial

Description:

The S-380 is a compact, high-speed, piezoelectric-driven tilt mirror with two orthogonal axes for a tilting range of 0.8 × 0.8 millirads. A servo loop with high-resolution eddy current sensors provide an absolute tilt angle resolution of 1 microradian. The built-in piezo-drive momentum compensation allows highest possible mechanical stability for the mirror.

Characteristics:

Tilting Range: 0.8 × 0.8 milliradians

Response Time: 1 millisecond

Resolution: 1 μrad absolute

Resonance Frequency: 1.3 kHz (with the mirror)

Weight: 4.5 kg

Mirror Quality: 1/8 wave

Mirror Weight: 0.3 kg

Information courtesy of Physik Instrumente

Schaeffer Type 55 BiAxial Drive System

Manufacturer: Schaeffer Magnetics, Chatsworth, California

Intended Application: Space flight motion control

Qualification: Qualified for multiple programs

Availability: 9 months or Mil Spec and 16 months for S-level electronics

Description:

The Type 55 is one of a family of standard biaxial systems ranging from 1 kg to 18 kg. It is a cantilevered design consisting of two orthogonally oriented actuators.

Characteristics:

Weight: Drive, 4.3 kg; electronics box, 1.8 kg

Step Angle: 0.0075 degrees in each axis

Steps per Revolution: 48,000

Slew Rate: 2.5 deg/second; higher with frequency ramping

Gimbal Size: ≈ 15 cm in diameter and 26 cm tall

Information courtesy of Schaeffer Magnetics

III

Systems

8

General-Purpose/Ground-Based IR Cameras

"In differentiation, not in uniformity, lies the path of progress."

Louis Dembitz Brandeis

8.1 TECHNOLOGY BASICS

8.1.1 Introduction

There is a plethora of cameras that have the form, feel, and function of a conventional television (video) camera except that they sense the infrared portion of the spectrum. Figure 8.1 (courtesy of Mitsubishi) is a photograph of a high-resolution, full-framing Pt:Si infrared camera that represents the state of the art. Infrared camera systems supplying quantitative temperature data are called *thermographic imagers,* while those providing qualitative images are called *thermal viewers. Radiometers* provide quantitative radiometric data on the scene as radiance or irradiance, or process this to yield information about temperatures. The latter must have some form of calibration, hopefully traceable in some manner to NIST standards.

Historically, an electro-optical "camera" includes neither the storage medium nor the display, while "camera systems" usually include the complete package. It is much easier to sell complete camera systems, so most "cameras" are now in fact "camera systems," or the manufacturer offers an optional recording medium (usually diskette or tape), display, and formatter electronics for the display. Possible exceptions to this are scientific cameras. Often, the scientific user supplies the recording device and wants only raw data to allow the greatest analysis breadth and flexibility.

It is sometimes difficult to explain the difference between a FLIR and camera. No sentence that follows in this paragraph is true all the time—just most of the time. In general, cameras are designed for generic purposes, without much con-

FIGURE 8.1 A State-of-the-Art Infrared Camera, the Mitsubishi IR-M5000 (photo courtesy of Mitsubishi Electronics America)

sideration for form and fit, whereas FLIRs are designed for specific applications and specific platforms, or class of platforms. Cameras are used by computers and machines (not just people), but FLIRs are used mostly by people (although weapon systems can be controlled by FLIRs). Cameras can be used with many different fore optics (e.g., like commercial 35 mm cameras), but FLIRs are custom designed with all optics integrated into the package. Current thermography has a highly processed image with the display being in false color, while FLIR images usually are still displayed in black and white. The archaic term "FLIR" usually implies military or paramilitary use, air-based units, and scanners. In summary, some FLIRs are definitely cameras, but others are not. Some cameras are FLIRs, but others are not. Some instruments are cameras, but others are not. Some space sensors are cameras, but others are not, et cetera, et cetera.

Finally, cameras also usually rely on "imaging" the "target." When imaging, the item of interest (target) must completely fill at least one pixel, and preferably several. For most common situations, this helps greatly in reducing noise sources and subtracting out unknowns such as reflection components of the scene.

There exists a class of night vision "cameras" not covered in this text. This class is composed of devices that operate the in very near IR using image intensifiers to amplify the very low light levels that exist in dark conditions. An example was depicted in the movie, "The Silence of the Lambs." Conversely, the movie "The Predator" depicts the alien using a miniature IR camera.

A representative (but notional) camera architecture is shown in Figure 8.2. It has three distinct hardware pieces: a camera head, and electronics/control/processing box, and the display. For user convenience, the camera head often contains the minimum hardware needed to keep volume, weight, and power to a minimum. The camera head must contain all the optics, including collecting, imaging, zoom, focusing, and spectral filtering assemblies. Electronics and motors

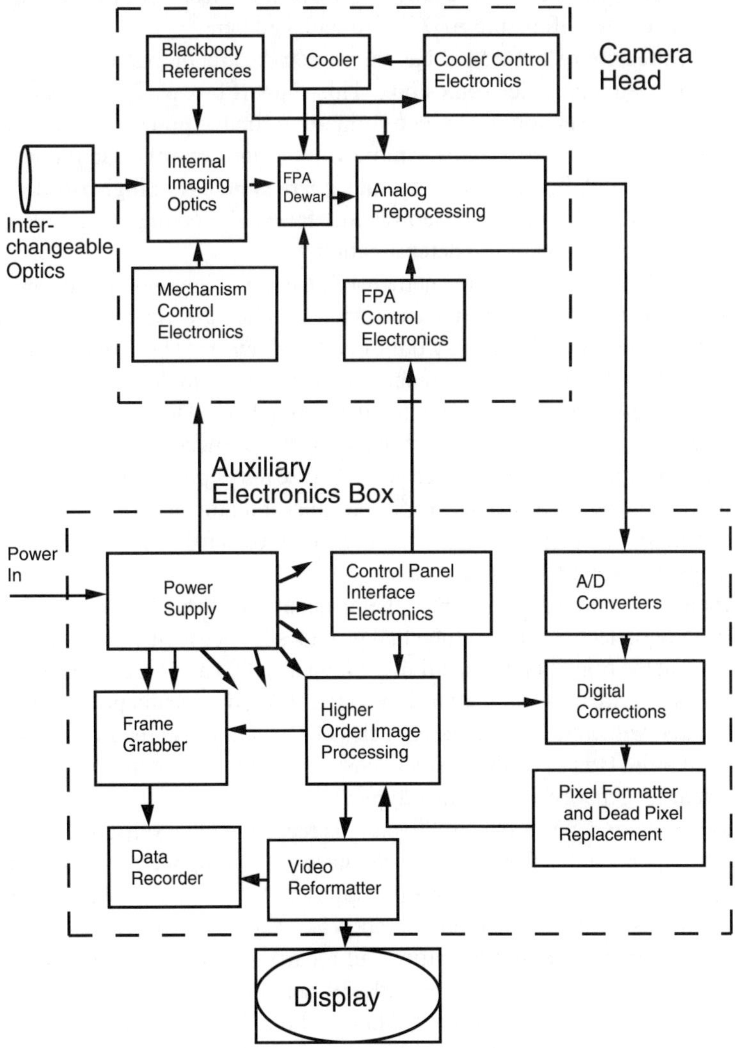

FIGURE 8.2 Representative Camera System Architecture

to control and drive moving parts must be included. Some form of thermal reference is often added for self-calibration or uniformity correction. The focal plane and its control electronics are in the camera head. The control electronics usually consist of communication circuits, bias generators, and clocks. If the focal plane needs cooling, some form of cooler must be included, along with its closed-loop cooling control electronics. A focal plane's output is of low voltage and amperage, making it a delicate signal. Therefore, it usually requires analog preprocessing, including amplification, control, and correction, before it can be sent through a long line. The basic analog electronics must be physically near the FPA and included in the camera head. Often, the A/D is also included here.

The architecture of Figure 8.2 has the camera head's output as a video signal running to an auxiliary electronics box. This support box performs control, processing, power conversion, and video formatting. Image processing is becoming increasingly sophisticated, requiring more volume and controls than is convenient for the camera head. For instance, several camera systems use higher-order image processing to provide emissivity subtraction, false color coding, target identification, tracking, and temperature determination.

Figure 8.2 does not depict the actual architecture of any individual camera. In fact, no camera is likely to have the exact same architecture. Several commercially available systems have the display built into the electronics box, but not a recorder. Despite the weight and power penalty, it is often wise to digitize on the camera head to avoid the noise associated with long analog lines. There are still many systems on the market that employ serial scanning, but the needed electronics, timers, pick-offs, and optics for complicated scanning are not depicted in the figure. Some recently developed focal planes (Pt:Si and microbolometer) are so uniform and stable in time that thermal references are included outside the camera head—sometimes just using the aperture cover.

Pyroelectric vidicons represent a major departure from Figure 8.2. They are based on a pyroelectric vidicon tube that provides low-cost, uncooled detection to 50 μm, but at the price of low sensitivity. These usually have NEDTs of 0.5 to 1° C (although some are now reported as low as 0.2° C) and typically poor image quality compared with a full-frame starer such as Pt:Si. They also require a chopper in the optical train. However, they usually cost a few thousand dollars and can detect radiation at longer wavelengths than typical detectors.

Another major departure in architecture is represented by self-contained viewers and weapon sights. They have all the necessary components nestled into the camera head, including battery power supplies and miniature displays. A key implementation is to mount them on a weapon as a sight. They must be produced in very large quantities, so producibility and recurring cost are of paramount concern. Hughes won a contract to provide 39 demonstration units to the Army's NVEOD at a cost of 15.8 million ["Hughes Puts Thermal Sights in Infantry's Hands," 1]. This produced a $405,000 average unit cost for the first run, including the nonrecurring costs. This thermal weapon sight uses a TEC cooled HgCdTe

3–5 micron array ["Hughes Lands Sight Device Contact," 2]. Lightweight, self-contained thermal viewers are also considered for other military and commercial products. Examples include night driving aids for cars and a 2 kg IR sight for the TOW anti-tank weapon [Lytle, 3].

No IR detector (and therefore no camera) directly measures temperature. As explained in Chapter 4, it senses some version of radiant excitance. This is an important distinction that must be understood by anyone associated with the IR industry, but one that is often neglected by the casual observer. First of all, an object's temperature is a result of random motion of its constituent molecules. If a sensor directly measured temperature, it would not work through insulating media such as the vacuum of space. Clearly, IR sensors do work though a vacuum. This is because they detect electromagnetic radiation emitted as a result of molecular translation, vibration, rotation, and random Brownian motion. Like a visible camera, they detect electromagnetic radiation—just of a different color. The amount of radiation from an object is a function of its surface temperature, emissivity, refection, projected area, and molecular resonance. The latter can cause selective jumps or dips such as the renstralung phenomena of silica or the resonant vibration of $C0_2$ at 4.2 μm.

The amount of radiance that a camera receives from an object is affected by the environmental situation via range, projected areas, reflection, absorption, and scatter from any medium between the camera and object. Finally, the amount of radiance that a camera identifies from the object is also a function of the camera hardware including optics, spectral bandpass, integration (or exposure) time, and focal plane sensitivity.

To produce a temperature measurement from a radiance measurement, processing and calibration must be employed. A temperature can be interpreted only after corrections are made for emissivity, reflection, and selective emission/reflection factors. Most cameras either ignore selective features by assuming everything is a gray body or a black body of emissivity less than one, or by reducing selective features by scene-to-scene processing. Since cameras usually operate in a wide spectral bandpass and these features are often spectrally narrow, this is not a bad assumption. However, in pathological cases, it may lead to slight or even gross misrepresentations of the image. Reflection is more difficult to subtract out, especially solar reflection for cameras operating in the SWIR and MWIR. Unless the camera is multiband or uses polarization modes, reflection is usually not considered. Reflection generally has the effect of making an object appear hot on one side as it reflects IR radiation from another hot object some distance away. An example is the classic phenomenon of the road underneath the engine of a car appearing hot in an IR image. The road is not being heated; it is reflecting the electromagnetic radiation (light) from the hot engine. With some training, humans usually become proficient at identifying reflection in a scene.

The emissivity component of a scene can cause a great amount of radiance variation—easily an order of magnitude in an otherwise low-contrast scene. For some applications, such as thermography, emissivity variations must be subtracted out. For other applications, such variations provide an excellent clutter rejection methodology. The temperature component of most terrestrial scenes is the dominant clutter component, and an emissivity map usually is of lower clutter, allowing man-made objects to stand out better. Emissivity correction is accomplished by viewing in multiple colors or spectral bandpasses. This provides a large increase in temperature accuracy and measurement reliability. Subtracting out the emissivity can be accomplished with a few as two bands by assuming that everything is a grey body with constant emissivity over the bands. It is then processed out. In the early eighties, EDO/Barnes was the first company to offer spatial emissivity correction with its CompuTherm system [Kaplan, 4].

Emissivity and reflection correction is especially important for medical applications of IR cameras. This is because the human skin is of a temperature (300 to 310 K) that makes accurate thermal difference measurements difficult due to background radiation and reflection of nearly the same temperature. Additionally, skin tends to have an average emissivity of about 0.975 ± 0.005, with emissivity uniform over the MWIR and LWIR regions. However, skin damage, healing, and problematic conditions can represent themselves as a difference in emissivity rather than temperature [Hejazi et al., 5].

Experienced camera builders have observed that, for electrical engineers, noise joins death and taxes in a triad of inevitabilities [Silverman et al., 6]. Noise sources arise throughout an IR camera system and must be mitigated at every step. If not painstakingly designed, the optics can induce narcissus, shadowing, or fixed pattern noise (like looking through dirty glasses). The optics and dewar FPA cavity can induce unwanted thermal noise (i.e., the FPA "seeing" the optics that it should be looking through). The focal plane itself is perhaps the largest noise source, as discussed in Chapter 4. Mechanical cryocoolers can induce microphonics, electronic noise into the analog circuits, and a noise caused by the uncompensated thermal drift of the focal plane. The latter is usually reduced to a negligible consequence by operating the focal plane a few Kelvins below any significant deltas in the noise as a function of temperature (i.e., operating it at a temperature where significant sensitivity changes sensitivity do not occur if the cryocooler varies by a one-half or one Kelvin). This is easier said than done, since every Kelvin counts in camera design, and the design cannot accommodate this for some detectors. Utilizing a closed-loop control system to maintain the FPA within $\pm$ 0.1 to 0.5 degrees, and allowing sufficient time for the focal plane dewar to stabilize after cooling, helps reduce noise induced by thermal drift. The electronics can also induce substantial noise, especially through ground loops and long analog video lines.

8.1.2 Applications

Law enforcement agencies are looking to infrared cameras to serve as night driving aids and provide improved surveillance methods, including covert videotapes of crimes to use in court as evidence. Night driving aids permit operations and approaches with no visible light. To be admissible in court, images on IR videotape must be clean, high quality, and of high spatial resolution. Otherwise, lawyers may discredit them.

The security industry has finally recognized infrared cameras for stealthy observation and perimeter control. The key requirements for a security camera are low-cost detection of possible intruders, a large field of regard, and high reliability. Few security installations can afford placing two or three cameras to cover the same scene. The benefits of passive IR include covert viewing without alerting those being viewed, better all-weather capability, and lower costs for light sources and power as compared with conventional visible spectrum surveillance.

Use of infrared cameras for pure and applied science abounds. Generally, a scientific camera will offer high sensitivity, an exact user-defined bandpass, and calibration traceable to a national standard. Astronomical cameras tend to have low noise, high sensitivity, and low well capacity. They can be cooled substantially lower than LN_2 temperatures. It is not uncommon to "pump a vacuum" on a liquid helium dewar to reduce the temperature an extra Kelvin or two lower than the 4.2 K liquid He temperature. The extra cooling provides a reduction in noise, resulting in an increase in sensitivity.

An excellent example of an astronomical camera is one that was developed by the Paris Observatory. It is optimized to fit the f/36 focus of the Canada France Hawaii telescope on Mauna Kea. It employs a 32×32 InSb CID device operated at liquid helium temperatures and manufactured by SAT [Lacombe et al., 7]. Another widely published camera was made by the University of California at Berkeley for attachment to large telescopes based on doped silicon. The FPA and pupil are cooled to about 10 K with liquid helium, and the signal is digitized to 15 bits [Arens et al., 8]. These are true scientific instruments with emphasis on calibrated and understandable data, not on weight, power, and volume.

Infrared cameras are the most commercially driven part of the IR industry. Anytime it is important to sense the temperature, radiance, or emissivity differential between surfaces, commercial infrared cameras can help. Cameras also can provide remote temperature information for convenience, efficiency, or increased human safety. Cameras are good to diagnose electrical power lines and components and plant failures. Cameras also can be used to detect impending equipment failure, implement maintenance checks, inspect for hot or cold fluid leaks, perform geologic prospecting, ensure plant/crop health, spot fires, check for gas leaks, and verify the structural health of buildings. Other examples of market-driven uses are identification of hot parts in assemblies, remote and noncontact temperature measurement, and material identification.

Market expansion is likely in the nineties as military technology is injected into the cost-conscious commercial market. Investment is required to develop solutions to industrial/commercial problems. This differs from the past, when the main challenge was in selling a technological wonder.

One way to adapt to new markets is to develop software that transforms a generic camera into a user-specific instrument. This provides an end user with an answer that is important to him, not the general IR industry. One generic camera can support many applications by changing the software. Another way is to infuse military technology economically to reduce weight, size, and power requirements. For example, Hughes has built a tiny IR camera (only a few centimeters across) using binary optics and a 3–5 μm 128 × 128 HgCdTe array [Nordwall, 9]. We can also exploit the advent of uncooled FPAs or the advent of cryocoolers with 10,000-hour lifetime, thereby producing cameras that could operate continuously at night, with years between service calls.

Commercially available quantum well cameras will be making their debut soon, perhaps before the publication of this book. Their strengths will be in applications demanding high-resolution, low-sensitivity LWIR imaging. The cost is expected to be lower than HgCdTe, and their camera costs may be only 5 to 20 thousand dollars higher than a similar InSb or Pt:Si array.

8.2 ENVIRONMENTS

The environments experienced by ground-based cameras run the gamut and depend on the user and application. As such, these cameras are usually built either for one specific environment or rugged enough to survive and operate in a variety of conditions. They are meant to be operated by humans, so acceleration and vibration are minimal. Standard off-the-shelf models are not designed for severe vibration, acceleration, or shock. However, most are designed to survive a drop onto a hard surface from 10 to 30 centimeters, with minimal or no damage.

Most cameras are built to operate in the temperature range of outside weather. Low-end temperatures typically are bounded at –20° or –25° C by mechanism lubricants. At the high end, the electronics tend to overheat above 30° or 40 ° C. With the exception of a few rainproof models, getting cameras wet is usually a very bad idea.

8.3 TARGETS AND BACKGROUNDS

The classic application for these cameras is to display slight differences in temperature. Detecting, displaying, or recording this variation allows a human to identify a target. Table 8.1 gives some typical required NEDTs for sufficient signal to noise and imagery of a few targets.

In addition to NEDT, several other issues affect target detection. Contrast (not just signal to noise) is a critical issue. Contrast enhancement is a concern in infrared camera design. Cameras operating in the visible can have contrast variations of a factor of 1,000, but an image in the IR varies by only a factor of two [Silver-

TABLE 8.1 Representative Temperature Differences

Function	*Targets*	*Typical Temperature Difference (°C or K)*	*Typical NEDT for (SNR-5) (°C or K)*
Surveillance	People—skin	8	1.5
Surveillance	People—clothed	2	0.4
Surveillance	Aircraft	10	2
Surveillance	Vehicles	5	1
Surveillance	Boats	2	0.4
Medical thermography	Skin temperatures, vein/circulation imagery	0.2–0.5	0.05–0.1
Astronomy	Planets, galactic core, dust clouds, gas clouds, IR stars	Varies	As low as possible!

man et al., 10]. If the target temperature, size, and emissivity just happen to produce an in-band radiance that balances with the average background, there will be no contrast and no change in the display. Although this balance is difficult to maintain for a period of time, any object heating up or cooling down to a large degree with respect to the background will go through a transition of near zero contrast and be invisible in a single band for a period of time. The inherently low contrast of a typical IR scene implies a reliance and emphasis on hardware that maintains or enhances the contrast. This is especially true of the optics, correction electronics, and display.

Most systems set a threshold or perform background subtraction to eliminate the large dc pedestal. A background subtraction across a scene (or localized portion) will subtract out some of the target signal. "A high-contrast image may have a low signal-to-noise ratio, while a low-contrast image must have a high signal-to-noise ratio for the same threshold resolution condition" [Johnson, 11]. An effective way to analytically account for contrast is to reduce the signal used in any performance calculation based on the signal processing, background and target characteristics. One simple procedure is to subtract the radiance in watts per square centimeter per steradian per second (W/cm^2/sr) or photons per square centimeter per steradian per second (Ph/cm^2/sr/sec) from the target that the background would contribute to the same area. This can result in positive or negative contrast, with negative contrast being a dark object against a hot background. Targets of negative contrast can be detected easily given proper signal/image processing.

Clutter presents another problem for target identification in infrared scenes. Clutter is essentially a noise source that cannot be subtracted via root-sum-square techniques and usually "leaks" through traditional noise reduction algorithms. Scene clutter is based on real, physical attributes of the scene that present a contrast of the same general level as the target. In almost every scene, clutter is nonlinear, and changing parameters even slightly will cause great changes in clutter. Generally, the lower the threshold, the greater the clutter, and the finer the resolution, the lower the clutter.

The picture on the display can be shown as *white-hot, black-hot,* or *false color*. In a white-hot picture, the intensity of the display is coupled to the intensity of the focal plane pixel so that high-radiance objects are brighter than low-radiance objects. Black-hot is the reverse; therefore, high-radiance objects are displayed as dark areas. False color display couples several colors to incremental radiance (or temperatures) within the scene. Most systems have a switch for the user to choose the display mode in real time. There is no universal best choice; it is largely a matter of user preference. Generally, the black-hot gives a cleaner picture that more closely resembles what your eye sees. White-hot gives an image that is what your brain thinks; that is, a hot object appears brighter. False color gives a lot more quantitative data and can facilitate detail analysis, but it requires a longer adjustment period for the user.

8.4 FIGURES OF MERIT

A camera's performance is usually driven by the size of the optics, field of view, update rates, background, detector sensitivity, and range. The typical system figure of merit for cameras and FLIRs is the noise equivalent delta temperature (NEΔT, or NEDT). It is also known as the noise equivalent temperature difference (NETD). Unfortunately, sometimes it is called the noise equivalent temperature (NET). The latter is regrettable because it causes confusion with another figure of merit, the noise equivalent target, described in Chapter 9. Although one may find any electro-optical figure of merit applied to a given sensor system and situation, no figure of merit is correct for every combination of hardware and application. In this chapter, we will discuss NEΔT simply because it seems to be the most quoted radiometric figure of merit for cameras, and it makes the most sense.

NEΔT is equal to the temperature difference, properly stated in degrees centigrade or Kelvins, that must be present in a scene to produce an output signal from the focal plane equal to the resultant noise of the entire camera (SNR = 1). One pixel can be distinguished from its neighbor when the average temperature of the scene composing the pixel differs by this amount. NEΔT is an appropriate figure of merit for thermal imaging systems, such as most camera applications and FLIR sensor heads. It assumes an imaging system with the target extending beyond one pixel. It can be idealized and defined as

$$\mathrm{NE\Delta T_t} = \frac{(A_d \Delta_f)^{1/2}}{A_o D^* K_o K_b K_e \omega \Delta L_t} \tag{8.1}$$

where

$NE\Delta T_t$ = increment of temperature (about temperature "t") that produces a signal equivalent to the system noise. The value of "t" is usually 300 K, although it is sometimes quoted at 500 or 1,000 K. It should be a temperature representative of the scene. Assume 300 K unless otherwise stipulated.

A_d = area of the detector in cm^2

D_f = noise bandwidth in hertz, as approximated by Equation (8.5)

A_o = clear aperture area in cm^2, as defined in Equation (8.6)

D^* = a detector parameter (see Chapter 4) typically measured by the FPA vendor as

$$\frac{(A_d\,\Delta_f)^{1/2}}{NEP}$$

K_e = electronic efficiency (usually 0.5 to 0.9)

K_o = in-band average optical train transmission including all elements and filters (usually 0.5 to 0.7)

K_b = optical blur efficiency as described below

ω = detector solid FOV in sr (elevation multiplied by azimuth, both in radians)

ΔL_t = change in radiance caused by a temperature change per Equation (8.7)

Several people in the IR community wish to express figures of merit as a function of focal length (F#). It can be convenient to do this, depending on the specific design trade or operational conditions of the equipment. Equation 8.1 is not inconsistent with this, but it uses inputs of the areas of the detector and the entrance pupil. Doing so with a fixed detector size makes optical trades easier to judge in light of system effects. Equation (8.1) can be rewritten by substituting a value of A_o calculated per Equations (8.2) and (8.3). Assuming there is no central obscuration and knowing that F# = f/d, Equation (8.1) can be reconstituted as Equation (8.4).

$$A_o = \frac{\pi D_o^2}{4} \tag{8.2}$$

where

A_o = area of the clear aperture in cm^2

π = pi = 3.14…

D_o = diameter of the entrance aperture

$$A_d = \omega f^2 \tag{8.3}$$

where

A_d = area of the detector in cm^2

ω = IFOV in steradians

f = focal length in cm

$$NE\Delta T_t = \frac{4F\#\ (\omega \Delta f)^{1/2}}{\pi A_o D^* K_o K_b K_e \Delta L_t} \tag{8.4}$$

where

$NEDT_t$ = increment in temperature (from temperature “t”) that produces a signal equivalent to the system noise

F# = F# of the optical train

A_d = area of the detector in cm^2

Δ_f = noise bandwidth in hertz, as approximated by Equation (8.5)

π = pi = 3.14…

A_o = clear aperture area in cm^2, as defined in Equation (8.6)

D* = detector D* evaluated with as many noise sources as possible.

K_e = electronic efficiency (usually 0.5 to 0.9)

K_o = in-band average optical train transmission including all elements and filter (usually 0.5 to 0.7)

K_b = optical blur efficiency as described below

ω = detector solid FOV in sr (elevation times azimuth)

ΔL_t = change in radiance caused by change in temperature per Equation (8.7)

Extra caution is required with this format to include unwanted effects from the central obscuration and spiders, if they exist. Usually, users of this form have refractive collection optics. The user of this (or any equation) is cautioned to double check the units to verify that they are consistent. It is easy and common to mix microns with centimeters or meters.

The noise bandwidth (Δf) is an effect of the coupling of electronics to detectors. It depends on the electronics and dwell time and can be defined as

$$\Delta f = \frac{a}{t_d} \tag{8.5}$$

where

Δ_f = equivalent noise bandwidth in hertz

a = yet another efficiency constant that depends on the dwell time and electronics. If you don't know it, assume 1/2.

t_d = dwell time. If you are not sure, assume it is the same as the integration time in seconds.

Seyrafi indicates that most systems would have an "a" between 1/2 and 2, with detection systems designed for a small value and imaging systems designed for a larger value [Seyrafi, 12]. Hudson points out that the optimum occurs at approximately 1/2 [Hudson, 13].

The area of the entrance aperture must be calculated. Equations (8.1) and (8.4) assume that all of the energy falling on the primary makes it to the detector, with the exception of what is accounted for in the transmission factor. This is a reasonable assumption for refractive telescopes. This requires those using on-axis reflective telescopes to subtract out any central obscurations and spiders. With any telescope, vignetting by stops or mechanisms may occur, especially at far-field angles. These must be properly treated.

$$A_o = (\pi r^2) - (\pi r_o^2) - A_s \tag{8.6}$$

where

A_o = area of the clear aperture in cm^2

π = pi = 3.14...

r = radius of the primary mirror or primary lens in cm

r_o = radius of the central obscuration in cm^2 (usually 10 to 20 percent more than the radius of the secondary)

A_s = projected area in cm^2 of the spider struts that support the secondary

Lost energy from a mismatch of the optical blur and the physical layout of the detector must be considered. Some engineers include this in the optical efficiency figure, but it is often more convenient with modern systems to break it out separately. Therefore, K_b should be included as an adjustment to account for fill factor, Airy disks split between more than one detector, and the general energy distribution from a point spread function. Under typical conceptual designs, the Airy disk is sized to match the center-to-center pixel spacing, so for a 100 percent fill factor focal plane, K_b should be about 0.8. For lower fill factor systems, this should be multiplied by the fill factor. A more accurate parameter can be obtained by estimating the optical train's point spread function to determine the percentage of the encircled energy in a circle that fits within the detector. When considering point sources, the K_b should also take into account the division of energy by more than one detector because the Airy disk might fall between two or four pixels, even with 100 percent fill factor. Some specialized IR systems deliberately over-sample the blur by many pixels to achieve subpixel accuracy of the position of a point source. Results have routinely exceeded the Rayleigh criteria. This "super-resolution" frequently is used with star sensors for satellite attitude determination. It is an effective method but requires signal to noise to burn.

The ΔL is the change in in-band radiance ($W/cm^2/sr$) as a function of temperature. This is simply the change in in-band radiance for the given spectral band and starting temperature. This is caused by a change in temperature as calculated by Planck's law and can be written as

$$\Delta L_t = \int_{\lambda s}^{\lambda 1} \left(\frac{\partial L}{\partial T}\right) d\lambda \tag{8.7}$$

where

ΔL_t = change in radiance for an incremental change in temperature

λl = long-wavelength cut-off of the camera spectral bandpass

λs = short-wavelength cut-on of the camera spectral bandpass

$\partial L/\partial T$ = partial derivative of the radiance as a function of temperature

Do not be intimidated by Equation (8.7); it has been solved many times for typical temperatures, and more details can be obtained from radiometry books listed in the bibliography. This integral can be easily evaluated with a spreadsheet or radiation slide rule, especially about a given temperature (t). Hudson gives some average and typical values for the differential radiance as 6×10^{-9} W/cm^2/sr/K for SWIR, 5.2×10^{-6} W/cm^2/sr/K for MWIR, and 7.4×10^{-5} W/cm^2/sr/K for the LWIR [Hudson, 14].

ΔL is determined by bandpass and target temperature rather than the focal plane, optics, and so forth. This part of the figure of merit is target/background dependent in that the target's emissivity and temperature affect the result. Since it occurs in the denominator of Equations (8.1) and (8.4), it is beneficial to maximize this value to reduce the NEDT. The only way the engineer has to maximize this is with bandpass selection, because the target and background will be what it will be (que sera sera). Thus, the 8 to 12 micron atmospheric transmission bandpass has greater target detection capability (lower NEDT) than the MWIR 3–5 μm for targets and backgrounds of about 500 K or less. The typical range of interest for cameras and FLIRs is around 300 K (sort of room temperature). The derivative of in-band radiance with respect to temperature reaches maximum in the 8–12 band for typical room temperature objects. Thus, the same differential in temperature gives a larger change in radiance (ΔL_t) in the LWIR than MWIR or SWIR. Additionally, the 8 to 12 bandpass is less affected by colder winter temperatures of man-made objects than the 3 to 5 μm MWIR [Haystead, 15]. Although physics has blessed the LWIR, technology development has not. The MWIR can be imaged using producible Pt:Si arrays with high uniformity. The LWIR requires expensive arrays that have relatively poor uniformity and are difficult to produce and use. These pragmatic considerations can defeat the physics argument for LWIR versus MWIR for some applications.

As in most figures of merit, there are several subtleties to be considered. First is that the NEDT depends upon the optics and cameras, which often have different fields of view and f numbers. Thus, a given camera will have a different NEDT for different fields of view. Second, the perceived performance could be background limited and depend totally on the scene and not the sensor. Third, the clutter may produce a signal that is passed by the signal processor to the screen, causing false alarms and missed detections which cannot be predicted by the NEDT figure of merit. Fourth, for point sources, the NEDT is a function of the range and background. Equations (8.1) and (8.4) assume the target is larger than a pixel. This is not necessarily true. Practical NEDT is very sensitive to resolution if the target is not fully imaged. This is a common camera and FLIR sensitivity driver. Fifth, the NEDT must be quoted for a temperature to have meaning. Standard temperatures are 300 and 500 K. The temperature that has meaning to your application is the temperature of the background. Sixth, the NEDT is very sensitive to bandpass, so don't expect to have an identical NEDT if you plan to modify the bandpass. Table 8.2 gives other cautions for using this figure of merit.

The SNR can be derived using NEDT by a simple comparison of the difference in temperature between the target and background with the NEDT. Unfortunately, targets often have emissivities different from the background. This must also be considered, and it can be the driver.

8.5 COMPONENT CONSIDERATIONS

The key to camera component implementation is effectiveness at a low cost and minimal weight. Since there is usually only one bandpass, the optics are usually refractive to provide large fields of view at low cost.

One of the major component advances for cameras has been low-cost, small, reliable cryocoolers. In the past, cooling has been accomplished via dewars that held a liquid cryogen (usually, nitrogen). This irritated users and hampered widespread acceptance of IR cameras. Currently, most system offer integral or split Stirling cryocoolers. Not requiring the user to fill dewars or keep the camera upright greatly facilitates use in the field.

Another enabling technology is the use of staring arrays in the cameras. When used in cameras, large-format Pt:Si FPAs offer lower weight and temporal uniformity by eliminating the scan mechanisms. Staring HgCdTe offers high sensitivity, but at the cost of reduced resolution, because affordable arrays are small compared to Pt:SI and standard TV. Quantum well cameras should be making a commercial debut soon—perhaps before the publication of this book. Their forte will be high-resolution, low-sensitivity LWIR imagery in excess of 11 microns.

General purpose image processors that pack high processing horsepower have been a boon to the camera systems. They allow vendors to merely change their software and have a camera tailored for different applications and customer bases.

8.6 SYSTEM CONSIDERATIONS

Cameras usually can produce high-quality images with NEDTs of 0.05 to 0.2° C. Detail and resolution vary by optics and focal planes, but several serial scanners

TABLE 8.2 Concerns and Issues with NEDT

Concern	*How to Handle It*	*Comments*
1/f noise	Include in D* or as a separate noise term	Fine for a given integration and readout rate, but may not be constant or linearly scalable if either varies
Aging effects	Degrade everything	
Atmospheric absorption/ scattering/emission	Include a visibility fudge factor	Not sufficiently handled by this equation for NEDT
Clutter effects	Can be included only as a fudge factor in this figure of merit	Really should not be considered part of NEDT. NEDT should be augmented by SCR to describe a camera in an application where clutter is a driver.
DC background effects	Properly addressed by this FOM	Detector must not be near saturation
Fill factor	Should be included in K_b	Especially critical for monolithic FPAs
Filter corner effects	Best considered in ΔL_t	
Flicker or display effects	One could attempt to degrade the result by a fudge factor	Could be a system driver, but not properly addressed by this version of NEDT
Not square detectors	Addressed in A_d and K_b	
Nuclear radiation noise	Included in D*	Not specifically addressed by NEDT
Optics noise effects	May be included in D* or fudge factor	Not properly addressed by NEDT
Other than circular aperture	Include as a fudge factor or in K_b	Not properly addressed by NEDT
Point source of less than a projected pixel (or IFOV or footprint)	Could attempt to degrade the NEDT by some weighted average	Not properly addressed by NEDT
Spectrally sensitive detectors	Must integrate ΔL_t and D* by the detector's sensitivity function. This is easily done on a spreadsheet.	Equations (8.1) and (8.4) using a D* do not properly represent Schottky barriers or quantum wells with bandpasses of more than approx. 0.5 microns.
Stray light	Extra noise can be included in effective D* or as a fudge factor	Not properly addressed by NEDT
Target contrast	NEDT virtually considers and defines contrast for imaged objects	
Target crossing pixel in less than dwell time	Can be included as a fudge factor	
Target spectral effects	Use weighted average over bandpass	
TDI	Should decrease the NEDT by slightly less than the square root of the number of TDI stages	

and Pt:Si starers produce a full television frame of data with excellent resolution. Gone are the days of streaks, dead pixels, and uncontrolled blooming. A good camera will produce an image akin to that of a black and white television.

The Night Vision and Electro-Optics Directorate (NVEOD), designated the lead institution for this technology within the U.S. Department of Defense, has pioneered several thermal imaging devices in the United States. These devices and modern upgrades are the main focus of U.S. government activity in the camera area. Table 8.3 gives some system technical data on currently fielded systems.

When compared to many other forms of infrared sensors, ground-based IR cameras are often characterized by higher weight and power consumption than seekers and space sensors. For many applications, there simply is no need to minimize weight and power. Camera heads usually weigh several kilograms, with 10 to 20 kg auxiliary boxes. The weight, power, and cost are very application specific. For example, it has been suggested that a law enforcement camera should be small (< 16,400 cc), lightweight (< 9 kg), have a wide (30°) and narrow (5 to 8 °) FOV, and low cost ($100,000 to $150,000) [Aikiens and Young, 17]. Also, handheld/man-portable units must have minimal weight and power consumption. This poses a special problem with the infrared, which requires cooling and larger optics than the visible. The weight issue can be mitigated by using uncooled focal planes, lightweight optics, and ASIC-based electronics.

Cost is a basic system driver for cameras. This is because there is so much competition and a rather limited demand. The customers do not have deep pockets, and the users are not in love with the technology. The field is so competitive that each manufacturer tries to find a "niche" for his system and then prices it in line with other systems. As such, there is little price difference between an InSb array, Pt:Si array, or serial scan HgCdTe camera system. Cameras based on staring HgCdTe still generally cost about twice as much as the others and have poor resolution. A non-HgCdTe basic camera head is around $50,000, with a complete system ranging from $75,000 to $200,000, depending on the software and special features.

Thermal viewers must be low in cost as they are generally chosen to upgrade to a low-cost weapon system such as the M-16 rifle, which is priced at less than $500. However, if they are to be employed on more sophisticated and costly platforms, additional cost for performance can be justified. For example, the Commanders Independent Thermal Viewer runs $300,000 per system, as made by Texas Instruments for the M1A2 tank. A gunner's primary sight manufactured by General Dynamics costs $150,000 per unit, and a TOW missile sight improvement will cost about $100,000 per unit ["Future Looks Bright for Night Vision Technology," 18].

Producibility is a driver. Most manufacturers want to be able to produce at least a few system per month, with ten or so per month being the usual maximum achievable without facilitization. This leads to designs that incorporate readily available components and low amounts of touch labor by engineers, scientists, and highly skilled technicians.

TABLE 8.3 Some Older NVEOD Sponsored IR Cameras (data courtesy of NVEOD)

Device	*Nomenclature*	*Purpose*	*Detector Technology*	*Range*	*FOV (°)*	*Weight (kg)*	*Power*
Hand-held thermal viewer	AN/PAS-7	Ambush detection, perimeter defense	Passive IR (first thermal device ever fielded)	1,000 m (vehicles), 400 m (humans)	6° by 12°, 2.5 X mag.	2.7	6 V batteries
TOW night sight	AN/TAS-4C	Provide day/night capability to detect, recognize, and identify armored vehicles	FLIR	3,000 m	1.13° by 2.26° and 3.4° by 6.8°	8.5	6.5 W @ 4.8 Vdc
Dragon night tracker	AN/TAS-5	Provide night capability to detect, recognize, and identify armored vehicles	FLIR (common modules)	1,000 m	3.6° by 6.8°	10	6.5 W @ 4.8 Vdc
Night observation device, long-range	AN/TAS-6	Observation or surveillance	FLIR (common modules)	3,000 m	1.13° by 2.26° and 3.4° by 6.8°	10.6	6.5 W @ 4.8 Vdc
Tank thermal sight	AN/VSG-2	Fire control system component	FLIR (common modules)	N/A	2.58° by 5° and 7.74° by 15°	41	18 to 30 Vdc

References

1. April 1991. "Hughes Puts Thermal Sights In Infantry's Hands." *Military and Aerospace Electronics,* 9.
2. May 1991. "Hughes Lands Sight Device Contract." *Lasers & Optronics,* 11, 12.
3. D. Lytle. May 1991. "Night Vision Devices Head for Civilian Market." *Photonics Spectra,* 68.
4. H. Kaplan. July 1989. "Marketing Through Innovation." *Photonics Spectra,* 66–68.
5. S. Hejazi et al. November 1992. "Scope and Limitations of Thermal Imaging Using Multiwavelength Infrared Detection." *Optical Engineering,* 283–288.
6. J. Silverman, J. Mooney, and F. Shepard. March 1992. "Infrared Video Cameras." *Scientific American,* 78–83.
7. F. Lacombe et al. 1990. "Advances in IR Technology at Paris Observatory." *Proc. SPIE* 1341:187–191.
8. J. Arens, et al. September 1987. "A 10 μm Infrared Camera." *Applied Optics* 26, 18: 3846–3852.
9. B. Nordwall. June 19, 1992. "Hughes Manufacturing Strides to Allow Better Performance in Low Cost Systems." *Aviation Week and Space Technology,* 53–54.
10. J. Silverman, J. Mooney, and F. Shepard. March 1992. "Infrared Video Cameras." *Scientific American,* 78–83.
11. J. Johnson. 1985. "Analysis of Image Forming Systems." Selected Papers on Infrared Design. *Proc. SPIE* 513, part two:761–781.
12. K. Seyrafi. 1973. *Electro Optical Systems Analysis.* Los Angeles: Electro-Optic Research Corporation, 238.
13. R. Hudson. 1969. *Infrared Systems Engineering.* New York: John Wiley & Sons, 311–313.
14. Ibid.
15. J. Haystead. April 1991. "Thermal Imaging Technology Has Versatile And Bright Future." *Defense Electronics,* 48–52.
16. M. Lloyd. 1973. *Thermal Imaging Systems.* New York: Plenum Press, 177.
17. D. Aikiens and W. Young. 1991. "Airborne Infrared and Visible Sensors Used For Law Enforcement and Drug Interdiction." *Proc. SPIE* 1479:435–444.
18. December 1991. "Future Looks Bright for Night Vision Technology," *Military & Aerospace Electronics,* 26.

Appendix 8A

Data Sheets for Representative Camera Systems

Amber 4256

Manufacturer: Amber Engineering, Goleta, California

Intended Application: Various

Qualification: None

Availability: Custom made per contract; delivery ≈ 3 months ARO

Description:

The camera is built around Amber's 256 × 256 FPAs. It offers high sensitivity in the SWIR and MWIR. The support electronics provides two-point correction and automatic replacement of non-responsive pixels.

Characteristics:

FPA: 256 × 256 InSb with 38 μm element pitch

D*: 4×10^{11} cm-Hz$^{1/2}$/W

Operability: > 98%

Spectral Band: 1 to 5.5 μm

Gain Correction: 0.0 to 15.9375 in 0.0625 increments

Offset Correction: 0.0 to 32767.5 in 0.5 increments

Frame Rate: Variable with a maximum of 60/second, options to 250/second

Cooling Method: Pour filled LN_2 dewar with options for Stirlings

Video Outputs: RS-170, RS-343

Power Requirements: 110 Vac, 60 Hz

Operating Environment: 0 to 40° C, to 95% relative humidity.

Price: ≈ $69,000

Information Courtesy Amber Engineering

AVIO TVS-2000 ST

Manufacturer: Nippon Avionics, Tokyo, Japan, marketed in the USA by Cincinnati Electronics, Mason, Ohio

Intended Application: Various thermography

Qualification: None

Availability: Production

Description:

The TVS-2000ST series thermal video system consists of the camera head and processor. The tripod-mounted camera head contains the scanner optics, infrared detectors, and amplification circuits. The processor box with built-in high-resolution RGB monitor provides the image processing features and screen graphics, along with image storage and a variety of video output formats. The TVS- 2000ST series has optional camera lens and hardware accessories to benefit most applications.

Characteristics:

Minimum Resolvable Temperature: 0.1° C (0.05° C S/N improvement by averaging)

FOV: 15° horizontal and 10° vertical

IFOV: 0.125°

Frame Time: 30 frames/second

Detector Material: InSb—10 elements

Spectral Range: 3 to approximately 5.4 μm

Coolant: Stirling cooling

Camera Operating Temperature: –10° C to +45° C

Camera Head Weight: 3 kg

Processor Weight: 10.9 kg

Camera Head Dimensions: 205 W × 183 H × 258 D millimeters

Information courtesy of Cincinnati Electronics

AVIO TVS-2000 TE

Manufacturer: Nippon Avionics, Tokyo, Japan, marketed in the USA by Cincinnati Electronics, Mason, Ohio

Intended Application: Various thermography

Qualification: None

Availability: Production

Description:

The TVS-2000TE series thermal video system consists of the camera head and processor. The tripod-mounted camera head contains the scanner optics, infrared detectors, and amplification circuits. The processor box with built-in high-resolution RGB monitor provides the image processing features and screen graphics, along with image storage and a variety of video output formats. The TVS- 2000TE series has optional camera lens and hardware accessories to benefit most applications.

Characteristics:

Minimum Resolvable Temperature: 0.5° C (0.25° C S/N improvement by averaging)

FOV: 15° horizontal and 10° vertical

IFOV: 0.125°

Frame Time: 15 frames/second

Detector Material: SPRITE MCT

Spectral Range: 3 to approximately 5.4 μm

Coolant: TE cooling

Camera Operating Temperature: 0° C to 40° C

Camera Head Weight: 2.5 kg

Processor Weight: 12 kg

Camera Head Dimensions: 173 W × 106 H × 250 D millimeters

Information courtesy of Cincinnati Electronics

CE IRC-160ST

Manufacturer: Cincinnati Electronics, Mason, Ohio

Intended Application: Various

Qualification: No qualification is standard, other levels available

Availability: In production; delivery ≈ 3 months ARO

Description:

The IRC-160ST is a camcorder-like IR camera. It is designed for easy operation and portability and uses a Ricor K506 Stirling cooler. It has a built-in B&W eyepiece display for viewing while in use. It can be operated from a rechargeable battery.

Characteristics:

FPA: 160 × 120 PV InSb with 50 × 50 micron unit cells.

Weight: 4.1 kg

NEDT: 0.025 K

Well Storage: 30 million electrons

Frame rate: 51.44 frames/second

Spatial Resolution: 1 milliradian with std. 50 mm EFL lens.

NETD: < 0.02 K

Focus: ≈ 1 meter to infinity

Outputs: Video and a 12-bit digital port

Size: 37 cm long, 12 cm wide and 13.3 cm high

Price: ≈ $39,500

Information courtesy of Cincinnati Electronics

David Sarnoff Labs IRC640

Manufacturer: David Sarnoff Research Center, Princeton, New Jersey

Intended Application: Various

Qualification: Commercial

Availability: Custom made on order; delivery ≈ 6 months ARO

Description:

The David Sarnoff Research Center is marketing a full-TV-resolution camera using their full-framing, monolithic Pt:Si IRCCD. The camera system is packaged as a compact camera head and a separate video processor unit. The camera is a general-purpose, tripod-mounted thermal imager. Physically, the camera head is 14 cm (W) × 15 cm (H) × 24 cm (L) with the lens protruding. It has a baffled cold shield with a cold IR filter. The camera head includes CDS signal processing, commandable integration time control, and selectable high/low gain front amplification. The FPA is cooled via a Stirling cooler contained within the camera head. Obtained images are of excellent quality and wonderful resolution. The camera has no blooming, lag, or transfer smear. The video processor unit provides full 12-bit digital video processing including two-point gain and offset uniformity compensation. The camera is supported by a family of interchangeable lens having different FOV, F#, and apertures.

Characteristics:

Array Size: 640 × 480

Array Radiometric Uniformity: < 1% RMS

Pixel Size: 24 × 24 μm

Fill Factor: 38%

Output: Std RS-170A

Integration Time: Variable from 70 μs to 1 second

NEDT: <0.08 K with f/1 lens and 300 K background

Price: ≈ $150 K

Information courtesy of David Sarnoff Research Center Inc.

ELITE

Manufacturer: THORN EMI, England, and SAT, France

Intended Application: Military views and gunsights

Qualification: Meets UK Mil specifications

Availability: In production, made when ordered; delivery ≈ 6 months ARO

Description:

The ELITE is a battery-powered, compact, camcorder-like unit providing the operation with an LWIR view of the scene. It can be hand held, weapon mounted as a sight, or tripod mounted. It comes with several cooling and optics options, which affects cost, weight, and size.

Characteristics:

Bandpass: 8–12 μm

FPA: HgCdTe

FOV Options: 12.4° × 6.2°, 6.2° × 3.1°, or dual 6.2° × 3.1° and 18.6° × 9.3°

Cooling Duration: 1.5 hours with a 0.2 liter bottle at 20° C

Information courtesy of THORN/EMI

FICAM 60

Manufacturer: Signaal USFA

Intended Application: Various commercial and industrial

Qualification: Commercial

Availability: In production, delivery ≈ 1–2 months ARO

Description:

The MWIR, battery-powered FICAM 60 (formerly the UA 9060), is a small camera system composed of two units. The camera head has a viewing display built into it. The FICAM is made for durability against intensive conditions while being lightweight to enable flexibility of movement and easy operation for the user.

Characteristics:

Bandpass: 3 to 5 μm

Field of View: 48° × 24°

Video Output: CCIR

Depth of View: 0.5 m to infinity

NETD: 0.25° C

Camera Weight: 3.5 kg

Battery Capacity: 60 minutes

Information courtesy of Signaal USFA

FSI C-FLIR

Manufacturer: FLIR Systems Inc., (FSI) Portland, Oregon

Intended Application: Paramilitary

Qualification: MIL-STD-810D

Availability: Custom made per order; delivery ≈ 4 months ARO

Description:

The C-FLIR is a ground-based FLIR that comes in two pieces: the imager head combined with an electronics box, and a monitor. The C-FLIR imager head and electronics are housed in a sealed enclosure. Items such as a hand controller or control panel and monitor can be added as desired to provide a complete system. The C-FLIR combines the design characteristics of FSI's series 2000 systems with advanced digital processing and high resolution.

Characteristics:

FOV: $10.4° \times 6.9°$ and $3.2° \times 2.1°$

Resolution: 360 μrad in the WFOV and 110 μrad in the NFOV

FPA: 4×4 HgCdTe

Bandpass: 8 to 12 μm

Cooling: Rotary split Stirling

Power Required: 80 W

Imager Weight: 5.7 kg

System Weight: 16.8 kg

Information courtesy of FLIR Systems Inc.

FSI IQ 325

Manufacturer: FLIR Systems, Portland, Oregon

Intended Application: Industrial thermography

Qualification: Commercial

Availability: Custom made to order; delivery ≈ 2 months ARO (or less)

Description:

FLIR Systems acquired the Hughes' line of thermal imagers for industrial applications that has the registered name of "Probeye." The recently included IQ 325 includes the camera head, a computer processing module with a keyboard for control, and a color CRT display. It operates in a parallel scan mode using an eight-sided rotating polygon mirror. Each facet of the polygon is a mirror that reflects the incoming IR energy to the detector array. Around the circumference of the polygon, each successive mirror is tilted at a slightly greater angle than the previous mirror. The result is that each facet creates 30 lines of the picture, with all eight facets creating the complete picture of 240 lines. The IQ 325 is the highest resolution thermo-electrically cooled camera currently on the market. The IQ line is supported by a large selection of software, printers, lenses, and displays.

Characteristics:

Array Size: 30-element HgCdTe

Bandpass: 2.0 to 5.6 μm

Cooling: Solid state thermo-electric

Useful Temperature Measurement Range: –20° to 600° C (up to 1,500° C with special filters)

Spatial Resolution: 1.8 mrad (50% SRF)

IR Resolution: 195 resolvable elements per line

MDT: 0.1° C @ 30° C

Display Resolution: 480 lines × 512 pixels (NTSC)

Update Rate: 30 frames per second

Camera Head Weight: 3 kg

Processor and Keyboard Weight: 12 kg

FOV: 20° × 27°

Dynamic Range: 8 bits

Information courtesy of FLIR Systems Inc.

Hand Held Thermal Imager

Manufacturer: Defense Research Agency (DRA), Malvern, Worchester, UK

Intended Application: Various

Qualification: None

Availability: Custom made

Description:

This compact camera uses refractive optics to focus IR radiation on an uncooled array. Three electronic boards process this and display an image on a mini CRT viewed through an eyepiece. It is battery operated and completely self-contained.

Characteristics:

Detector: Uncooled ferroelectric array of 10,000 elements made by GEC-MMT

NEDT: 0.2 K

Volume: < 2000 cc (target for production)

Weight: < 1.5 kg (target for production)

Information courtesy of DRA

Hand Held Thermal Imager (HHTI)

Manufacturer: THORN EMI, England

Intended Application: Military

Qualification: UK Mil Spec

Availability: In Production, made when ordered; delivery ≈ 6 months ARO

Description:

The HHTI is a compact, single-piece, hand-held LWIR sight. The user holds the unit like binoculars and views through an eyepiece. The bottle can be changed in a few seconds.

Characteristics:

FPA: Serial/parallel HgCdTe

Focus Range: 5 m to infinity

MRTD: 0.1 K typical

Bottle Life: 4.5 hours @ 10° C

Cool-down Time: Typically 15 seconds

Operating Temperature: –36° to + 70° C

Storage Temperature: –46° to + 70° C

Narrow FOV: 8° × 3.4° with X 5 magnification

Wide FOV: 20° × 8.6° with X 2 magnification

Dimensions: 47 cm × 21 cm × 16 cm

Weight: 5 kg, including 0.6 liter bottle and 0.55 Ah battery

Power Consumption: 4 W @ 12 Vdc

Information courtesy of THORN/EMI

ImagIR™ 128

Manufacturer: Santa Barbara Focalplane, Goleta, California

Intended Application: General-purpose IR imaging platform

Qualification: Commercial

Availability: 2 months ARO

Description:

The system consists of a camera head and a real-time image processing workstation. The workstation consists of a high-speed pipeline processor hosted by an 80386 AT-compatible computer. Interfaces to the hardware are provided through Windows™ for user-friendly system control. Camera heads are available in pour-fill and in closed cycles, integral Stirling and split-Stirling coolers. A variety of options are available including a 1,000 MIPS image processor and real-time image storage capabilities. The ImagIR™ Camera system provides high sensitivity, real-time imaging performance for a variety of applications including radiometry and nondestructive testing. The standard ImagIR™ is provided with InSb detectors for MWIR imaging, but options are available for LWIR imaging with both HgCdTe and quantum well detector arrays.

Characteristics:

FPA: 132 × 132 PV InSb with 128 × 128 active cells

Detector Pitch: 50 μm

Spectral Response: 1.0 to 5.5 μm

Array Operability: > 99%

Array Nonuniformity: < 3.5% one sigma

Storage Capacity: > 50 million electrons

Standard Optics: f/1.8 50 mm, f/3 100 mm, or f/3 50–200 mm zoom

Information courtesy of Santa Barbara Focalplane

ImagIR™ 320

Manufacturer: Santa Barbara Focalplane, Goleta, California

Intended Application: General-purpose IR imaging platform

Qualification: Commercial

Availability: 2 months ARO

Description:

The system consists of the same high-performance imaging workstation and dewar options as the ImagIR™ 128 with the exception of upgraded memory capacity to run the digitally configurable large-area FPA. The large-area FPA can be configured to run in the 320 x 256, 320 x 240, and 256 x 56 formats. The number of FPA output channels is selectable providing 1, 2, and 4 output modes. The FPA is also configurable between "Snap-Shot" (simultaneous) integration and staggered (rolling) integration modes. The Snap-Shot integration provides simultaneous integration of all pixels to eliminate downstream reconstruction hardware requirements in high relative motion applications. InSb is standard, but quantum well and HgCdTe detector array options are available for LWIR applications.

Characteristics:

FPA: 328 × 328 PV InSb with 320 × 256 active cells

Special Features: 320 × 256, 320 × 240 and 256 × 256 formats; 1, 2, or 4 output channels; Snap-Shot or staggered integration modes.

Detector Pitch: 31 μm

Spectral Response: 1.0 to 5.5 μm

Array Operability: > 99%

Array Nonuniformity: < 3.5% one sigma

Storage Capacity: > 30 million electrons

Standard Optics: f/1.8 50 mm, f/3 100 mm, or f/3 50–200 mm zoom.

Information courtesy of Santa Barbara Focalplane

Inframetrics 600L Series Thermal Imaging Radiometers

Manufacturer: Inframetrics, Bedford, Massachusetts

Intended Application: Industrial Thermography

Qualification: Ground

Availability: Custom made from continuing line

Description:

The 600 series of infrared imaging measurement systems are microprocessor-based, complete systems. The camera head is a camera and control module unit that is tripod mounted or man-portable. It scans the scene simultaneously in the MWIR, LWIR, or broadband range. The data is digitized and fed to a central control unit which processes the data and displays the imaging and temperature data on a standard TV monitor. The detector is cooled with the Inframetrics Stirling microcooler.

Characteristics:

MWIR Bandpass: 3 to 5 μm

LWIR Bandpass: 8 to 12 μm

Broadband Bandpass: 3 to 14μm

MRTD: ≈< 0.1 K

Price: ≈ $40,000

Information courtesy of Inframetrics

Inframetrics Model 760 Imaging Radiometer

Manufacturer: Inframetrics, Billerica, Massachusetts

Intended Application: General purpose thermography

Qualification: None

Availability: Delivery ≈ 1 month ARO.

Description:

The 760 is a "market driven" thermal imager. It comes as a camcorder-like sensor head and a portable control/display panel. The portable console can store and record images and data and perform some image processing on the spot. The system has six standard modes: black and white, false color, point temperature measurement (a temperature is given for any point on the display), line scan temperature graph, isotherm, and area average temperature. It employs a serial scan of a single HgCdTe detector which is cooled using the company's microcooler.

Characteristics:

Spectral Range: 8–12 μm standard with optional 3–12 μm and 3–5 μm

FOV: 15 × 20°

Resolution: 1.8 millirad

Image Rate: 60 Hz RS-170 or 50 Hz CCIR

E-O Zoom: 4:1

Dynamic Range: 8 bits

MRDT: <0.1 K

No. of Filters: 4 optional

Temperature Measurement: –20 to 400° C normal, 20 to 1,500° C extended

Vibration: Operating 5 to 22 Hz, @ 0.76 mm amplitude

Sensor Head Dimensions: 21.6 × 12.7 × 18 cm

Control Electronics Dimensions: 26.4 × 23.6 × 17 cm

Sensor Head Weight: 3 kg without lens

Control Electronics Weight: 5.6 kg

Power Supply Weight: 0.7 kg

Price: ≈ $54,000

Information courtesy of Inframetrics

Infrared Laboratories HgCdTe Array Cameras

Manufacturer: Infrared Industries, Tucson, Arizona

Intended Application: Various

Qualification: None

Availability: Custom made per contract; delivery in 3–6 months ARO

Description:

Infrared Industries Inc. is marketing a camera head employing a Rockwell sapphire substrate array. This is one of the first HgCdTe array cameras to be marketed as a commercial product. Infrared Laboratories (IRL) is willing to modify and customize their basic design as needed for the customer. The standard system uses a liquid nitrogen cryostat to cool the HgCdTe to 77 K. By correlated sampling, KCT noise can be reduced to lower than the readout noise. A large and varied line of optional hardware to enhance functionality is available. IRL also supports cameras based on 256 × 256 In:Sb arrays.

Characteristics:

No. of Pixels: 256 × 256

Bandpass: Selectable from 1 to 2.5 μm or 1 to 5 μm

Readout Rate: Selectable once per second to 30 frames per second

Integration Time: Selectable from 33 millisecond to 5000 second

Saturation Level: 1×10^6

Readout Noise: <100 e–, maybe as low as 50 e– for custom applications

Quantum Efficiency: Typically 70% at peak response

Pixel Size: 40 μm × 40 μm

Price: $115k+ for 256 × 256 HgCdTe, $175k+ for 256 × 256 In:Sb. Prices may drop as array production becomes more efficient.

Information courtesy of Infrared Laboratories

IRCAM

Manufacturer: Royal Observatory, Edinburgh, UK

Intended Application: Astronomy on the 3.8 meter United Kingdom Infrared Telescope

Qualification: None

Availability: Not for commercial production. This camera is custom made for specific astronomical uses.

Description:

The IRCAM has a "side looking cryostat mounted on a general-purpose Instrument Mounting Platform attached to the UKIRT Instrument Support Unit at the f/36 Cassegrain focus." The camera has a digital LSI preprocessor and signal processor which condition and process the data so that it can be recorded on a VAX 11/730 computer. The IRCAM can be easily configured to function as a starer, chopped camera, imaging polarimeter, or an imaging spectrometer. The chopping mode is useful during daylight operations. Being in the SWIR and MWIR, it is especially useful for observations through galactic dust. The array is supercooled to 35 K for added sensitivity. The IRCAM uses gold reflective mirrors for high optical throughput across a wide IR band.

Characteristics:

Image Scales: 0.6, 1.2 or 2.4 arcsec per pixel.

Weight: A few kilograms

FPA: SBRC 62 × 58 DRO InSb

Uniformity: Better than 25% across the array

Readout Noise: About 600 electrons

Capacitance: 0.65 pF

Information from I. McLean, 1987, "Results With the UKIRT Infrared Camera." Proc SPIE *72: 138–141; and 1989, T. Less, I. McLean, and R. Wade. "A Camera for Infrared Astronomy and Its Performance on the 3.8M UKIRT Telescope,"* Infrared Physics *29, 2-4: 175-84*

Mitsubishi Thermal Imager IR-M-500

Manufacturer: Mitsubishi Electric, Tokyo, Japan

Intended Application: Various commercial interests

Qualification: None

Availability: Standard model is mass produced; delivery ≈10 days ARO

Description:

Mitsubishi sponsored development of its 512 × 512 Pt:Si Infrared camera with a 3–5 micron bandpass. The device employs a non-transparent readout at the same physical level as the detector material, operating as a charge swept device (CSD). This facilitates production of large scale arrays, yielding a 39 percent fill factor. The focal plane is cooled by a Mitsubishi Stirling cycle cooler that takes about 10 minutes to cool down the focal plane. No vibration is present in the image. The cooler is expected to last 6,000 hours. There is a remote controller for their camera based on LSI circuitry. They have successfully employed it for nighttime ship navigation, security, surveillance, remote sensing, aviation landing aid, fire detection, quality control, and nondestructive testing. The camera exhibits excellent spatial and thermal resolution. There is little to no thermal blooming or distortion.

Characteristics:

Camera Head Weight: 8 kg

FPA: 512 × 512 Pt:Si

Spectral Band: 3–5 μm

FOV: 11° (V) × 14° (H) with a 50 mm lens

Frame Time: 1/60 second

NETD: 0.15° C (black body at 27° C with a 50 mm f/1.2 lens)

Price: ≈$75,000

Information courtesy of Mitsubishi Electronics America

MIVS (Magnavox Infrared Video System)

Manufacturer: Magnavox, Mahwah, New Jersey

Intended Application: Various general purpose thermal imaging

Qualification: Military, including: MIL-STD-810D

Availability: Custom made per contract

Description:

MIVS is a rugged, self-contained LWIR thermal imaging system. It comes in two pieces: a box-like camera and a display/control panel. It contains a continuous zoom telescope, linear resonant cryocooler, and a high-resolution display that is readable in direct sunlight.

Characteristics:

FPA: Multiple-element HgCdTe SPRITE

FOV: 21 × 13 and 5.3 × 3.3

NEDT: 0.15 K

MRTD: 0.3 K @ 1.8 cy/mr

Bandpass: 8 to 12 μm

Sensor Weight: 21 kg, display/control box: 5.5 kg

Power: 120 W average

Sensor Head Dimensions: 23 × 34.5 × 62.2 cm

Display/Control Box Dimensions: 19 × 18.5 × 34.7 cm

Information courtesy of Magnavox

Multi-Role Thermal Imager

Manufacturer: THORN EMI, England

Intended Application: Various

Qualification: UK Mil specification

Availability: In production, made when ordered; delivery ≈ 6 months ARO

Description:

The MRTI is a tripod-mounted, self-contained unit developed for troops to use on the front lines of battle at night. It employs UK Class One common modules and either a 0.6 liter gas bottle or a self-contained, closed-cycle cryocooler. On the back of the unit is aN LED display, which is its primary display device. Outputs are provided for recording the data.

Characteristics:

FPA: HgCdTe SPRITE

Bandpass: 8 to 13 μm

Temperature Sensitivity: 0.1 K

Narrow FOV: 4.9 × 3.2°

Wide FOV: 12.9 × 8°

Dimensions: 48 × 28 × 19 cm

Weight: 10.7 kg including bottle and battery

Information courtesy of THORN/EMI

Panther Thermal Observation Device

Manufacturer: Spar, Kanata, Canada

Intended Application: Military generic thermal imaging

Qualification: Mil-std 810D, 461, 1472, 470, 756, 2000 and AQAP-1

Availability: In production, made when ordered

Description:

The Panther is a self-contained observation device. It is an upgrade to the Tiger Eye with extended range. It is designed to be man-portable with optional mounting kits for masts, tripods, and vehicles.

Characteristics:

NFOV: 1.3 × 1.9°

WFOV: 4.2 × 6.3°

MRT: Typically better than 0.1° C

IFOV: < 0.1 mrad

Aperture: 18.4 cm

Spectral Band: 8 to 12 μm

Operating Temperature: –40° C to +52° C

Storage Temperature: –46° C to +71° C

Weight: 15 kg

Information courtesy of Spar Aerospace

Probeye 699

Manufacturer: FLIR Systems, Portland, Oregon

Intended Application: Industrial thermography

Qualification: None

Availability: Custom made to order; delivery ≈ 2 months ARO (or less)

Description:

FLIR Systems acquired the Hughes' line of thermal imagers for industrial applications with the registered name of "Probeye." The 699 is supported with a large line of standard filters and is designed to be user friendly. The viewer is box-like, with the operator peering in an eyepiece display. The 669 combines powerful problem solving capability with convenience and economy of all electric operation and operates off a battery.

Characteristics:

FOV: 18° (H) × 7.5° (V)

Useful Thermography Range: 0° C to 1,000° C

Display Resolution: 2.12 mrad

Operating Time @ 22° C: 4 hours standard

Temperature Resolution: 0.4° C

Detector: Thermal—electrically cooled HgCdTe

Viewer Unit Weight: 3.1 kg

Auxiliary Unit Weight: 5.1 kg

Information courtesy of FLIR Systems Inc.

Probeye 7300

Manufacturer: FLIR Systems, Portland, Oregon

Intended Application: Industrial thermography

Qualification: None

Availability: Custom made to order; delivery ≈ 2 months ARO (or less)

Description:

FLIR Systems acquired the Hughes' line of thermal imagers for industrial applications that has the registered name of "Probeye." They come in series 3000, 4000, and 7000. They are designed to be user friendly. The systems are augmented with a large line of standard filters, lens, video accessories, and image analysis software. The system includes the camera head, a processing box with a keyboard for control, and a color CRT display. It operates in a parallel scan mode using an eight-sided rotating mirror. It can be powered by household current or a NiCad battery pack that provides an hour and a half of operation.

Characteristics:

Array Size: 30-element HgCdTe

Bandpass: 2 to 5.6 μm

Camera Head Weight: 3 kg

Processor and Keyboard Weight: 5.0 kg

FOV: 20 × 27°

Display Resolution: 480 lines × 512 pixels (NTSC)

Resolvable Elements per Line: 260

Dynamic Range: 8 bits

Information courtesy of FLIR Systems Inc.

SAT and SAGEM ALIS Thermal Imager

Manufacturer: SAT, Paris, France

Intended Application: Various military

Qualification: Fully applicable to French military specs

Availability: In production

Description:

ALIS is a light and compact infrared camera designed for numerous operational applications. ALIS uses the 8–12 μm infrared waveband using a HgCdTe detector cooled by an integrated Stirling cycle microcooler.

Characteristics:

Display: Integrated bi-ocular or standard CCIR TV monitor

Weight: approximately 3 kg

Input Voltage: 11.5 to 32 Vdc

Cooling Time: <3 minutes

MTBF: >1,500 hours

Operating Temperature: –40° to +60° C

Information courtesy of SAT

SAT ATHOS Thermal Imager

Manufacturer: SAT, Paris, France

Intended Application: Various military

Qualification: Fully applicable to French military specs

Availability: In production

Description:

The ATHOS thermal imager is a version of the Modular Thermal System (developed jointly by TTD and SAT), specifically designed for integration in tank sights and for various thermal operational applications. Several hundred of these systems have been ordered and are now coming into service. ATHOS has been selected for the French LECLERC battle tank and for a Middle Eastern country's infantry carrier.

Characteristics:

Cooling Time: < 10 minutes

Spectral Bandwidth: 8 to 12 μm

Image Rate: 25 Hz

Number of TV Lines: 510

Basic Resolution: 500 points per line

Wide FOV: 5.7° × 8.6°

Narrow FOV: 1.9° × 2.8°

Power Consumption: >200 W

Power Supply: 28 Vdc

Weight (approximate): 25 kg

Dimensions: 445 × 200 × 316 mm

Information courtesy of SAT

SAT MURENE Thermal Imager

Manufacturer: SAT, Paris, France

Intended Application: Various military

Qualification: Fully applicable to French military specs

Availability: In production

Description:

MURENE utilizes the French modular thermal system (SMT). This imager was designed for naval thermal imaging applications and is particularly suited for observation and long-distance reconnaissance missions. MURENE has been widely deployed within the French Navy as well as several foreign navies. The most notable applications of MURENE thermal imagers to date include the Mistral missile firing positions (Sandral and Najir Turrets), tracking and identification of targets using lightweight stabilized turrets, and deck landing systems on aircraft carriers.

Characteristics:

Narrow FOV: 1.9° × 2.8°

Wide FOV: 5.7° × 8.6°

Display: CCIR TV standard

Focusing: 30 m to infinity

Power Consumption: 180 W

Spectral Bandwidth: 8 to 12 μm

Cooling System: Split Stirling

Cooling Time: approximately ≤10 minutes

Weight: 62 kg

Dimensions: 972 × 370 × 222 mm

Information courtesy of SAT

SAT SIRENE

Intended Application: Various military

Qualification: Fully applicable to French military specs

Availability: In production

Description:

SIRENE is a modular surveillance and tracking system capable of operating independently or as part of an air base defense or battlefield C^3I system. SIRENE is designed to support medium-, short-, and very short-range air defense weapon systems. It provides tracking, surveillance, and fire control for ground weapons. The basic SIRENE system is composed of three portable units.

Characteristics:

Azimuth Coverage: 360°

Elevation Coverage: 6° positioned between –10° and +20°

Tracking Capability: 64 tracks at 2 Hz

Number of Independent Target Designation Outputs: 6

Typical Designation Range for Fixed-Wing Aircraft: 20 km

Typical Designation Range for Helicopters: 10 km

Typical Designation Range for Missiles: 8 km

Information courtesy of SAT

Thermal Weapon Sight (TWS) An/PAS-13

Manufacturer: Hughes, El Segundo, California

Intended Application: Passive IR gun sight

Qualification: U.S. military qualification

Availability: Developmental

Description:

The TWS is a lightweight, battery-powered, infrared weapons sight that will enable U.S. Army troops to locate targets in total darkness and obscured battlefield conditions. The TWS user can see at significant distances with absolutely no light, using an MWIR detector. It can be applied to a wide variety of weapons, including rifles, machine guns, grenade launchers and shoulder-launched missiles. Physically, it looks like a camcorder with a large lens. It employs a staring HgCdTe array which is thermo-electrically cooled. Hughes has been awarded a 36-month developmental production prove-out contract calling for delivery of 30 sights with options for initial production. The TWS employs binary optics to reduce the number of optical elements.

Characteristics:

Bandpass: 3 to 5 μm

FPA: PV HgCdTe

Cooling: TEC

FOV: 15° for light weapons, 9° and 15° for medium weapons, and 3° and 9° for heavy weapons (with interchangeable telescopes).

Weight: ≈1.8 kg for light weapons, 1.9 kg for medium weapons, and 2.0 kg for heavy weapons.

Power Consumption: ≈ 9.9 W

Information courtesy of Hughes Aircraft

Thermovision® 210

Manufacturer: AGEMA Infrared Systems, Sweden, marketed through regional sales representative and direct sales office in Secaucus, New Jersey

Intended Application: Surveillance, search and seizure, and rescue

Qualification: Industrial

Availability: Regular production

Description:

The Thermovision 210 is a portable, self contained, one-piece surveillance camera. It produces TV-quality images and incorporates a freeze frame function.

Characteristics:

Cooling Method: Thermoelectric

FOV: 8° × 16° (9° × 4° with optional telelens)

Spectral Bandwidth: 2 to 5.6 μm

Spectral Bandwidth for the 900 LW and 900: 8 to 12 μm

Weight: 1.36 kg

Price: $20,000 to $25,000

Information courtesy of AGEMA

Thermovision® 400 Series

Manufacturer: AGEMA Infrared Systems, Sweden, marketed through regional sales representative and direct sales office in Secaucus, New Jersey

Intended Application: Industrial predictive/preventative maintenance

Qualification: Industrial

Availability: Regular production

Description:

The Thermovision 400 series is a family of versatile infrared imagers specifically designed for predictive/preventative maintenance condition monitoring programs. The 450 and 460 combine scanner and control unit in one integrated 5.9 kg unit. Automatic level and range functions speed site measurement and simplify operator training. The Thermovision 470 includes an integral disk drive. Up to 70 images can be stored on disk, allowing in field measurement and comparison with previously recorded information or detailed analysis on a PC using AGEMA's proprietary CM-SOFT software. Images and measurement data can also be recorded on a standard VCR. The 480 series, with the same ease of use and recording capabilities as the 470, is intended for use in predictive/preventative maintenance programs which include mobile land, airborne use, and/or on-foot inspection.

Characteristics:

Cooling Method for 450, 470 And 487: Thermoelectric

Cooling Method for 489: Stirling micro-cooler

FOV: Various, depending on choice of lens

Spectral Bandwidth for the 450, 470 and 487: 2 to 5.6 μm

Spectral Bandwidth for the 489: 8 to 12 μm

Price: $30,000 to $60,000

Information courtesy of AGEMA

Thermovision® 900 Series

Manufacturer: AGEMA Infrared Systems, Sweden, marketed through regional sales representative and direct sales office in Secaucus, New Jersey

Intended Application: Research and development, quality control and assurance

Qualification: Industrial

Availability: Regular production

Description:

The Thermovision 900 series is a family of thermal measurement and analysis systems combining high image quality and data acquisition rates with ease of use and sophisticated data manipulation and analytical capabilities. These effectively allow IR thermal data collection and analysis at the advanced environment of the desktop workstation. The new IR scanner and fast A/Ds that digitize data prior to transmission, employ a number of totally new components and concepts, and eliminate the need for a separate computer.

Characteristics:

Cooling Method for 900 SW/TE: Thermoelectric

Cooling Method for 900 SW and 900 LW: Cryogenic dewar

Cooling Method for 900: Stirling micro-cooler

FOV: Various, depending on choice of lens

Spectral Bandwidth for the 900 SW/TE and 900 SW: 2 to 5.6 μm

Spectral Bandwidth for the 900 LW and 900: 8 to 12 μm

Price: $60,000 to $250,000

Information courtesy of AGEMA

Tiger Eye (AN/VAR-501)

Manufacturer: Spar Applied Optics Systems Group, Kanata, Canada

Intended Application: Thermal viewing by infantry/artillery

Qualification: Military, including MIL-STD-2000, 461, 810D, 1472, 470, and 756

Availability: Custom made per contract; delivery ≈ 12 months ARO

Description:

The Tiger Eye is a tripod-mounted, LWIR thermal observation device for the military. It is man-portable and can be configured with a gimbal, extra electronics box for use on ships and motorized vehicles. In its standard configuration it is self-contained and operates from a pack.

Characteristics:

FOV: 7 × 11 and 2 × 3 degrees

MRT: Typically better than 0.1 K

IFOV: < 200 μradians

Aperture: 11.4 cm

Bandpass: 7.5 to 12 μm

Display: 525 or 625 line standard video

Weight: 11 kg

Power: 55 W average

Operating Temperature: –40° C to 52° C

MTBF: > 1800 Hours

Information courtesy of Spar Aerospace Limited

TI Short Range Thermal Sight (SRTS)

Manufacturer: Texas Instruments, Dallas, Texas

Intended Application: Military gunsight

Qualification: U.S. military

Availability: Highly developmental

Description:

The SRTS is a passive, uncooled thermal sight for use on a M16A2 gun or M203 grenade launcher. It is rugged, portable, runs off a mil-std battery, and is completely self-contained. It employes surface-mounted circuit cards and ASICs. It uses a chopped, uncooled detector array of barium strontium titanate. It is of modular construction, making piece parts easily replaceable.

Characteristics:

Spectral Range: 8–12 standard with an optional 3–12 μm

Size: 32.5 × 9.6 × 9.4 cm

Weight: 1.6 kg

Power: 4 W

Operation: 8 hours at 25° C

Startup Time: < 10 seconds

Information courtesy of Texas Instruments

UP 1043/02 Thermal Imaging Tracking Camera

Manufacturer: Signaal USFA

Intended Application: Various

Qualification: NATO AQAP-1, STANAG-4108

Availability: In production, delivery ≈ 7 months ARO

Description:

The UP 1043 is part of a line of MWIR and LWIR cameras. It is a self-contained unit designed for passive day and night tracking applications. The reliable, high-performance camera may be ship, vehicle, or hard mounted.

Characteristics:

NEI: 5 nW/m^2

NETD: 0.2 K @ 22° C

NFOV: 35 × 18 milliradians

WFOV: 260 × 130 milliradians

IFOV: 140 × 140 μrad in narrow and 1 × 1 milliradian in wide

MTBF: 1,500 hours

FPA: 24-element CMT

Bandpass: 8 to 12 μm

Power Consumption: 50 W

Weight: 18 kg

Information courtesy of Signaal USFA

9

Smart Weapon Seekers

"Give us the tools and we will finish the job."

Winston Churchill

9.1 TECHNOLOGY BASICS

Infrared guided smart weapons have proven their utility in modern warfare. The first introduction was the WALLEYE, used against the Thank Tika bridge in Vietnam. Countless missions, resulting in lost lives and aircraft, failed to destroy the bridge until the WALLEYE accomplished the mission in one strike. The Afghan war demonstrated that a simple Sidewinder could wreak havoc on the Soviet Hind helicopter. Desert Storm illustrated the military and humane usefulness of smart weapons. Robot weapons allow the surgical destruction of a target without risking human lives to pilot them. Both sides in Desert Storm used IR missiles effectively, as "most of the 27 U.S. aircraft lost were victims of IR-guided missiles" [Hughes, 1].

Smart weapons include missiles, bombs, artillery projectiles, and standoff cruise missiles. Most of the following discussion applies to missiles, but the principles are largely applicable to all types of seekers.

The seeker is the primary homing instrument for the weapon. Seekers are categorized as:

- *active*—emits a signal and homes off of the returned signal
- *semi-active*—homes off a return signal emitted by another platform (not the seeker itself)
- *passive*—homes on a signal by the target

The grandfather of passive IR seekers is the Sidewinder seeker, developed in the 1950s. The original Sidewinder was simplistic in design and employed old technologies such as vacuum tubes. A rotating gyroscope provided a stabilized line of sight and rotated a simple "spoke" reticule in front of a single-element detector. The reticule induced an amplitude modulation to the detector that was used to drive precession coils to turn the gyroscope-package toward the target. The same current was used to move the missile's control fins. Since the gyroscope's line of sight (LOS) rate was proportional to the drive current, and the missile's turning rate is a function of the LOS rate, the resulting guidance law is known as "proportional navigation."

Modern versions of the Sidewinder are still in the U.S. military inventory and were used in Desert Storm. Current missile inventories rely heavily on non-imaging IR seekers. The trend of future systems is toward imagers. Most missile seekers employ infrared detection in some form. About half rely exclusively on IR. An additional 20 percent are dual mode, relying on the IR and something else [Lennox and Rees, 2]. Several nations already possess some level of indigenous missile capability. These nations include Brazil, Canada, CIS, France, Italy, Israel, Pakistan, PRC, South Africa, Taiwan, UK, and the USA. This group is growing and will soon include several other nations.

The newest generation of smart standoff weapons relies heavily on real-time target identification, discrimination, tracking, navigation, and night vision provided by their imaging infrared electro-optics. Infrared is ideally suited for this standoff mission because it is passive (and therefore clandestine) and works at night. Moreover, the EO nature of smart weapons will tend to replace the radar emphasis as stealth platforms are increasingly used for low-intensity conflicts. It is more difficult to perform IR missile warning than radar guided missile warning, giving governments and police advantages over terrorists and guerrillas. Lastly, IR outperforms the visible part of the spectrum because of the latter's clutter problems during the day, the availability of visible camouflage, and lack of light at night.

These seeker/missiles are being designed for basing on aircraft, ships, ground, and space vehicles. Figure 9.1 is a photograph of the IR imaging Lightweight Exo-Atmospheric Projectile (LEAP). The optical telescope at the front of the unit focuses the target image on the miniaturized focal plane array in the nose of the seeker column, at the extreme left. The unit's electronics are housed within the metallic ring behind the optics. The remainder of the unit, the largest section, contains the propulsion system and fuel supply. The divert engines, visible on the top and at the side of the propulsion section, provide the thrust that maneuvers the unit to intercept the target [U.S. Army Strategic Defense Command, 3].

A representative, but notional, architecture of a "strap-down" staring seeker is shown in Figure 9.2. The seeker head often contains only the minimum hardware needed to sense the scene. This keeps its volume, weight, and power requirements low. The seeker head must contain all the optics, the focal plane, and its control electronics, as well as basic signal processing and FPA cooling systems.

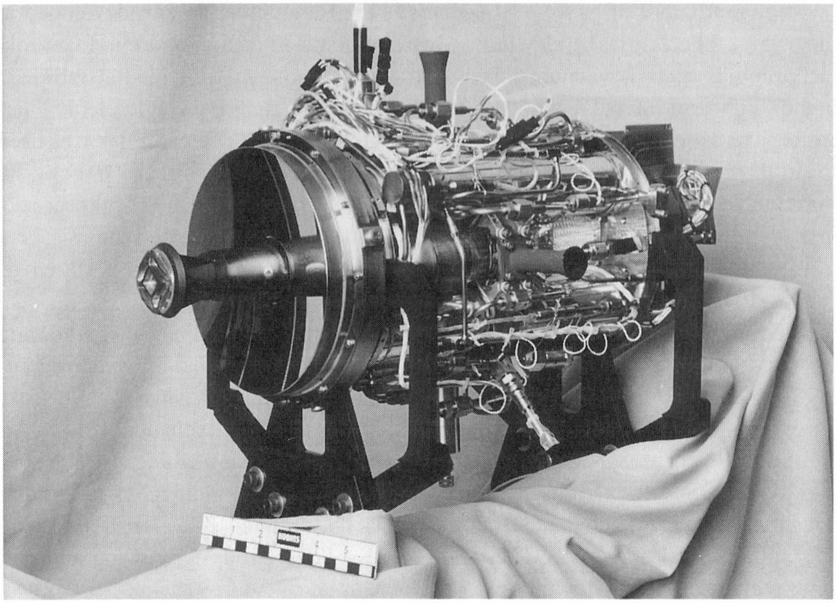

FIGURE 9.1 LEAP, A State-of-the-Art Infrared Seeker (photo courtesy of U.S. Army Strategic Defense Command)

The architecture illustrated in Figure 9.2 has the seeker's output going to a missile-based processor to perform tracking and aimpoint selection. There is a general trend to place more and more of this processing directly behind the focal plane in the seeker.

Figure 9.2 depicts a conceptual/generic representation of the actual architecture of an individual seeker. Most systems depart from this architecture. For example, several production systems employ gimbals and/or scanning mechanisms.

An electro-optical seeker's function is usually to provide information for missile guidance to the general area of the target, to recognize or "type" the target, to discriminate the target from other objects, and to perform aimpoint selection. Some systems include the function of supplying information for warhead detonation timing. A typical concept of operations for an infrared missile includes a standby turn-on, followed by a commit, which cools the focal plane. The seeker is first locked onto its target by an external, wide-FOV sensor or a human. This implies a given range to track a point source that is usually determined by radiometric considerations. Then, the missile is launched and flies out locked onto its target, matching any target movement. This phase is usually called the "main chase." Finally, when the target is close and imaged, the missile either chooses an aimpoint and conducts final maneuvers to get to the target or selects a point and

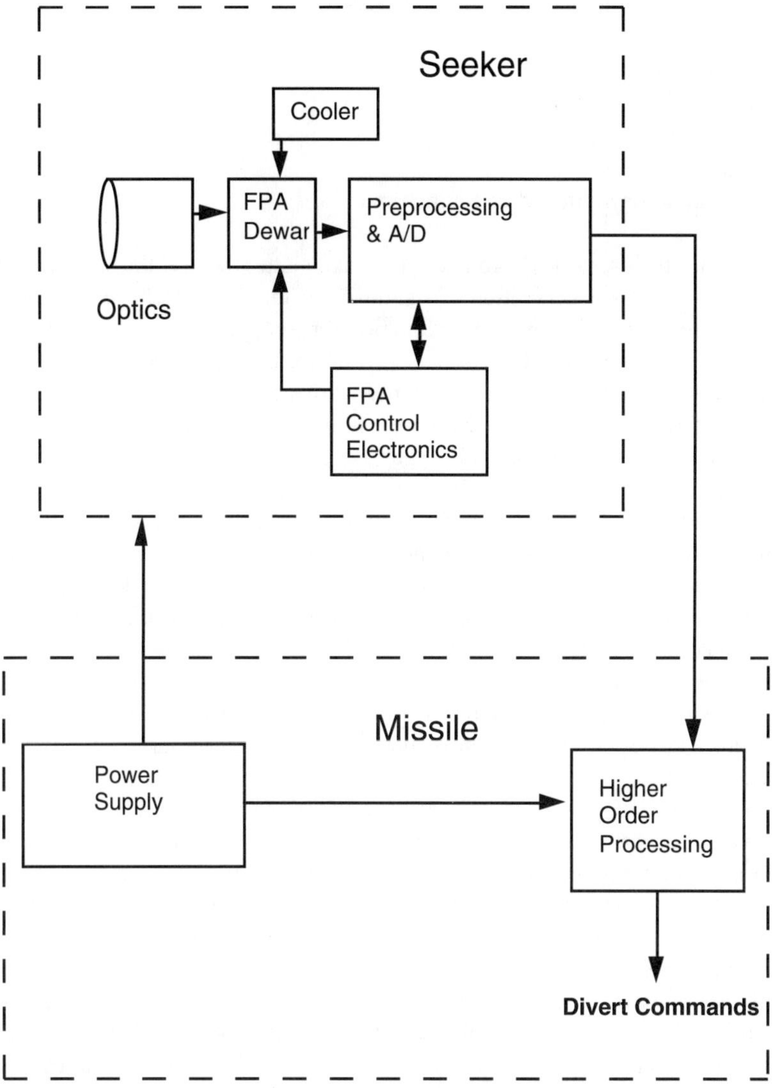

FIGURE 9.2 Representative Seeker Architecture

time to fuse and explode. Aimpoint selection occurs after the target is imaged by at least a few pixels. That is, the target extends over several pixels in each direction, vertically and horizontally in the sensor field of view. This implies a critical range based on the target size and seeker resolution, not radiometry. This last functional interval is called the "end-game."

There are five salient features in seeker design. Seekers must be

1. lightweight
2. low in power consumption
3. producible at high rates
4. rugged
5. allowed to have limited operating times

The first four features usually conflict. To make matters worse, the last one is not always true. The next generation of standoff weapons and electro-optical cruise missiles may require that the seeker function for long periods, as the main chase may last hours during which the seeker is used for navigation. Other developing concepts include having a vehicle dispense warheads and return to a home base, again using the seeker for navigation. This mission requires the seeker to be designed for multiple uses, like a FLIR or camera.

Usually, seeker requirements flow from the weapon's probability of kill (Pk), the probability that the missile will destroy its target. The calculation of a Pk must include effects of warheads, countermeasures, missile dynamics, and operators, and all are out of the scope of this text on electro-optics. The seeker parameters enter this calculation through miss distance. Although analytical calculation and simulation of miss distance is complicated, it is usually dominated by the seeker's measurement and aimpoint accuracies [Bass et al., 4]. Table 9.1 lists some typical seeker specifications and their implications for miss distance. Seeker requirements are a close tradeoff with missile dynamics, warhead, target, and mission operations.

There are sundry points that call for keen attention from engineers and managers. Missile seekers must have a well balanced mass distribution. This is because they are sometimes spin stabilized and always operate in high-G and highly maneuverable environments. Although this is not necessarily difficult with on-axis optical systems, it is something the engineer must be aware of from the beginning. This can be difficult with off-axis systems or systems with extreme volume constraints. Missiles and missile seekers must be designed for upgrades and long shelf life. Missile weapon systems can be used after more than 40 years ["TOW: An Example of Continuous Upgrade," 5], and the missions and targets will change greatly in such a time span. They are often adapted for missions and targets for which they were not designed. This is likely to become more common in an age of reduced budgets. Passmore published a "best guess" for future multipurpose seekers as having the following parameters: 256×256 InSb FPA, approximately 200 μrad IFOV, diamond turned Cassegrain optics of about a 15 cm aperture with a $\pm$ 7.5 degree search FOV, an NEDT of about 0.07 K, and a rate-stabilized gimbal [Passmore, 6]. Engineers designing seekers have one of the most difficult electro-optical jobs. The basic requirements are mutually opposing, except for the short life. Seekers must be low in cost (tens of thousands of dollars), producible at high rates

TABLE 9.1 Seeker Specifications and Miss Distance

Seeker Specification	*Effect on Miss Distance*	*Comments*
Data latency	Usually a strong driver if more than two frame times	Affects guidance.
Field of view	Usually not a strong driver on miss distance, but it can be critical for acquisition and tracking.	Depends on targets and missions.
Frame rate	The faster the frame rate, the smaller the miss distance or more accurate the fusing.	This is especially true for techniques that require several images for aimpoint selection, discrimination, or typing.
Number of data samples available for processing when target is an extended source	The more the better. Miss distance is usually a strong function of useful frames if there are less than approx. five, and a weak function if there are more than ten.	This is a product of the frame time, target size, resolution, and relative closing velocity.
Processing rate	The quicker, the smaller the miss distance or more accurate the fusing.	It should be able to process a full frame in less than a frame time.
Range of the last useful data	The smaller the better.	The missile must fly blind and predict ahead.
Resolution	The better the resolution, the smaller the miss distance or more accurate the fusing.	This is usually a strong, nonlinear function.
Stiffness	The stiffer the seeker, the smaller the miss distance or more accurate the fusing.	It increases accuracy.

(tens to hundreds per month), highly reliable (0.99 seeker mission success), low in mass (a kilogram or less), and rugged.

9.2 ENVIRONMENTS

The environments for seekers are quite severe. Often, they are fired out of a gun or ride on a rapidly accelerating and maneuvering missile. This leads to requirements for seekers to survive and operate under several thousand Gs in acceleration and hundreds to thousands of Gs in shock in any axis, with linear and angular accelerations of thousands of radians per second squared. They must also be able to operate in likely temperature ranges, encountered from stockpile to target, that vary from –46 to +63° C [Pleikies, Wittmer and Lindner, 7]. Seekers must also be immune to severe microphonics, since they experience tremendous vibration during the missile's thrust.

Weapons are often carried outside a fast moving airplane, on a ship, on a bouncing vehicle, or lugged around on a person's back. The storage and transportation environments can be severe. This trend of abusive basing may be changing somewhat. The U.S. is developing the AIM-9X, which will be "heat seeking with improved maneuverability designed for compressed or internal carriage on future Navy planes as well as the Air Force's F-22" ["Air Force Quits Optically Guided Sidewinder Effort," 8]. An internal carriage eliminates the aerodynamic effects on the aircraft and allows the aircraft greater maneuverability. An internal carriage also provides for a more benign environment, with the exception of heat. In the hot desert sun, internal carriages can get to 60 or 70° C. This poses problems for detectors, since these temperatures approach typical HgCdTe's dewar temperatures. The seekers are usually capped or sheathed to provide a measure of protection.

9.3 TARGETS AND BACKGROUNDS

Targets and background vary because missiles and their seekers are normally used for specific applications; for example, as an "air-to-air" missile. However, budget constraints are requiring multi-use. Thus, targets and backgrounds are becoming more numerous for any given seeker. Ideally, an imaging seeker with a programmable image processor can be programmed for a variety of targets and backgrounds.

Operating within the atmosphere, seekers usually have a well defined bandpass in an atmospheric transmission band. However, the optical bandpass also depends on the targets. The principle combustion products are water and CO_2. Therefore, it is easy to see that IR emission lines of water (around 2.7 μm) and CO_2 (around 4.4 μm) are typically selected [Hudson, 9]. Bandpass tuning is crucial to seeker performance. The bandpass should be selected to maximize target-to-background signals. Although many seekers operate in the standard atmospheric transmission bands (e.g., 3 to 5 μm), they frequently have a much narrower bandpass within the atmospheric window that allows the target signature through while minimizing the background signature. Improved system performance can be realized by carefully tuning the spectral band to maximize target-to-clutter or target-to-background signal.

9.4 FIGURES OF MERIT

The appropriate figure of merit for a seeker is closely tied to its targets, missions, and backgrounds. As seekers become more and more "front-end thinking" information processors, their figures of merit will reflect more systems and less radiometric consideration. Figures of merit such as the probability of recognizing a target, aimpoint selection accuracy, probability of detection (P_d) versus range, and probability of false alarm (P_{fa}) versus range will gain usage over the more typical radiometric figures.

A common figure of merit used for seekers is the noise equivalent target (NET), although one may find any EO figure of merit applied to a given seeker and situation. No figure of merit is correct for every combination of hardware and use.

NET is an excellent figure of merit when all noise sources are considered and properly applied, and when the sensor is not background limited. However, it has a key parameter that is independent of sensor hardware: range. NET must be defined at a given range. Usually, this range is the maximum range at which that the seeker must detect a given target. This requirement is derived from system considerations flowing from missions and scenarios.

For low-background systems, NET can provide a quick "back-of the-envelope" estimate of sensitivity. It is useful for design trade-offs about a rigid mission scenario. One should keep in mind that it is idealized, and that other noise sources will creep in. Thus, sensitivity will never be as good as the calculation indicates, unless one grossly underestimates an efficiency term. Seekers are rarely background limited when viewing the sky; however, they are almost always background or clutter limited when viewing the ground.

NET can be defined (see Equation 9.1) as the intensity (in watts per steradian) a target must have at a given range for the resulting signal coming out of the detector to equal the average noise coming out of the detector. NET can be defined in terms of watts or photons. When internal noise dominates (not background limited), NET can be defined in terms of watts as follows:

$$NET_r = \frac{R^2 (A_d \Delta f)^{1/2}}{A_o D^* K_e K_o K_b} \tag{9.1}$$

where

NET_r = required power (in watts) of the target to give SNR = 1 at the given range

R^2 = range from target to sensor in centimeters

A_d = area of the detector in square centimeters

Δf = noise bandwidth (see Chapter 8, Section 8.4)

A_o = clear aperture area (in square centimeters). Remember to subtract the central obscuration for on-axis reflective systems

D^* = detector D^*, a parameter typically measured by the vendor. It should include as many noise sources as possible. NEI can take into account the dc background if the quoted D^* includes this.

K_e = electronic efficiency, usually 0.5 to 0.9

K_o = In-band average optical train transmission, including all elements and filter, usually 0.5 to 0.7

K_b = optical blur efficiency, as defined in Chapter 8, Section 8.4

As mentioned in previous chapters, the user of this (or any) equation is cautioned to double check the units to verify that they are consistent. It is easy and common to mix microns with centimeters or meters. The K terms are efficiency terms but also can be used as fudge factors to modify simple NEI calculations to gain increased fidelity in the representation of the system. The key message of this equation is that NET depends strongly on range, a mission parameter. Table 9.2 lists some common system degradation effects and their relationship to NET. It should be noted that, for any given system, only some (or none) may apply.

The resulting signal-to-noise ratio can be computed by comparing the target's radiant intensity with the NET. Usually, a signal-to-noise of five to ten is required for detection. Once detected, a smart tracker can track with a signal-to-noise of about two, while a dumb tracker needs ten or more.

A fundamental drawback of this (and most) figures of merit is that it does not include the more system-wide consequence of using the image processor. It is a classic electro-optic figure of merit preferred by engineers who like to cut their career knowledge at the point where the electrons leave the FPA dewar.

A useful variation of the NET is the clutter equivalent target (CET). This is an analogous figure of merit that includes clutter. As with any infrared sensor, the BLIP versions and clutter-limited versions rely greatly on resolution, and this is often the sensitivity driver. Janssens gives a NET or CET expression for a clutter-dominated imager using background subtraction techniques that allow clutter leaking based on LOS drift during integration [Janssens, 10].

$$CET_{df} = KL_{\sigma}Fp\left(\frac{V_d}{V_t}\right)^2 \tag{9.2}$$

where

CET_{df} = CET as a function of the clutter drift in W/sr

K = a unitless, localized background constant that, unfortunately, may vary across the scene (If uncertain, use 0.2 [Janssens, 11].)

L_s = standard deviation of the background radiance in W/cm^2/sr

Fp = area of the pixel footprint in both dimensions in square centimeters

V_d = background velocity with respect to the LOS in meters per second

V_t = target velocity with respect to the background in meters per second

It is interesting to note that the only sensor parameters in this version of a CET are the ability to track the target and background, and the resolution (the latter is inherent in the footprint). Equation (9.2) does not account for other methods of clutter rejection by the processor. Also, the equation does not include FPA nonuniformity, but this tends to increase the clutter. This indicates that seeker performance in high-clutter conditions is enhanced by larger and more uniform FPAs (not necessarily more sensitive ones), a viable niche for the Schottky barriers.

There are two forms of this CET definition. The first is raw CET. That is the signal a target must present at the aperture for the focal plane to output a signal that is equal to the average clutter in the scene. Clutter can be greatly reduced via signal processing, so a processed signal-to-clutter ratio is also sometimes quoted. This is more useful from a system perspective. Simple spatial and temporal filtering can usually reduce the clutter by factors of three to ten, while complicated processing can do better if a lot of *a priori* knowledge about the target or clutter exists.

Another figure of merit sometimes used is the detectable equivalent irradiance (DEI). Usually, it is the same as the NEI discussed in Chapter 11. However, some users include a factor that brings the DEI up to that of the desired SNR. In other words, if an SNR of ten is required, they will multiply the NEI by ten to achieve DEI. As with any figure of merit, it is best to be absolutely sure and clear when discussing it by including all assumptions and accurately defining the figure of merit.

For proper use, a figure of merit should include all deleterious effects. Seekers tend to have special effects that diminish the useful photons they capture. They frequently suffer from target smearing over more than one pixel, especially for long integration times, and excessive scatter and transmission loss due to window degradation. Background noise generated by hot optics must be included, as seekers rarely have cold telescopes. Seekers are often stored in high-temperature environments. At elevated temperatures of 60° C, even low-emissivity mirrors (0.02) can dump many photons in the LWIR to the focal plane. In addition, aeroheating of the dome or window during the missile flight can drive the window temperature higher than 100° C, which contributes far more noise. This high background can

TABLE 9.2 Concerns with NET

Concern	*Options*	*Comments*
1/f noise	Include in D*.	Although it is reasonable to compensate for 1/f noise by degrading D* for a given readout rate, the effects from 1/f noise are not constant (or easily scaled) if readout rate changes.
Aging effects	They degrade everything.	
Atmospheric absorption/ scattering/emission	Use a fudge factor to reduce NET based on the range and expected atmospheric effects.	
Clutter effects	These can only be included as a fudge factor in this figure of merit.	NET should be augmented with a CET for applications where clutter is a driver.
DC background effects	Use a BLIP version.	
Fill factor	Should be included in K_b.	This is especially critical for monolithic FPAs.
Filter corner effects	These are best considered in determining target's in-band radiant intensity for SNR or SCR.	This is especially critical for very narrow bandwidths.
Not square detectors	This concern is addressed in A_d and K_b.	
Nuclear radiation noise	Include in D*.	
Optics noise effects	These may be included in D* or fudge factor.	They are not properly addressed by this FOM in Eq. (9.1).
Other than circular aperture	Include as a fudge factor or in K_b.	This is not properly addressed by this FOM in Eq. (9.1).
Spectrally sensitive detectors	One must integrate in D* by sensitivity function.	This FOM does not properly represent Schottky barriers or quantum wells with bandpasses of more than about 0.5 microns.
Stray light	Extra noise can be included in an "effective D*" or as a fudge factor.	This is not properly addressed by this FOM in Eq. (9.1).
Target contrast	The target radiant intensity can be reduced to consider contrast in the SNR calculations.	This must be considered for high-background conditions.
Target crossing pixel in less than dwell time	Can be included as a fudge factor.	
Target spectral effects	Determine weighted average over bandpass.	

lead to a cluttered image out of the FPA resulting from focal plane nonuniformity. A high-sensitivity, low-uniformity focal plane viewing through an excessively hot dome may indeed be a poor system choice.

9.5 COMPONENT CONSIDERATIONS

9.5.1 Optics

Modern seeker optics must produce a stable image on the focal plane, and they are usually limited in FOV to a few degrees on a side. Seeker optics are usually simple, stiff, rugged, fast (F# < 3.5), and lightweight. Stabilization mirrors may be included to steady the image. Refractive and reflective on- and off-axis designs are common. Multi-bandpass and even multisensor seekers are also a wave of the future, requiring reflecting afocal assemblies. Dual-mode seeker telescopes have been developed to gather both millimeter wave RF and IR.

A seeker's field of view is often limited to allow high resolution. As a consequence, they are normally tasked and initially pointed by a pilot, FLIR, or wide field of view surveillance sensor. Often, a seeker's FOV is just a few degrees on a side with a resolution of 500 μrad or less. However, as FPA sizes increase and processing penalties decrease, there is a trend toward utilizing fields of view spanning tens of degrees.

Infrared scanners have been used in seekers. The Maverick is such a seeker. The array is scanned in two dimensions by rapidly rotating faceted wheels. This provides a reliable and producible way of scanning. Scanners suffer from the drawbacks of building a "scene" in which every scene pixel does not represent the same instant in time. Each scene pixel was scanned at a different time and represents a slightly different temporal position. This can lead to confusion with flares and rapidly maneuvering targets. The typical brute-force mitigation of this is to scan so rapidly there is little time difference across the scene. This can lead to large signal/image processors to balance the data latency across a scanned scene. The current trend is to exclusively employ starers for smart weapon seekers. The advantages include rapid frame rates, fewer moving parts, and the most suitable data for advanced target recognition, decoy discrimination, and clutter rejection.

Cooling seeker optics is not standard practice but is being explored to increase sensitivity. Next-generation IR seekers require minimum optical degradation during rapid cooling of the optics to cryogenic temperatures. In the LATS program, Lockheed's research center built a mirror to be cooled from 300 K to less than 120 K in less than five minutes. The mirror is made from I-70 HIPped beryllium with channels internal to the mirror for the flow of the J-T supplied cryogen. Figure is maintained to less than 0.027λ at 10 μm by assuring that stress is not applied to the surface during the rapid cool-down [Reiss, 12, 13]. At times during the flyout (main chase), a seeker cannot gather useful information even when the target is in range. This may be due to the target being out of the field of view, excessive jitter because of divert or axial thrusting, or environmental considerations such as flying through smoke, jammers, or nuclear noise. As the seeker approaches the target, the signal can increase to a level of saturation. Also, the target becomes grossly out of focus.

The latter usually happens just before the seeker hits the target. It arises from the fact that seekers are usually focused at infinity with no focus adjustment. Yet, unlike other IR sensors, the seeker must function not at a single range but from near zero to the acquisition range requirement. Although it is possible to design in a moving element to focus on the target, the processing and reduced reliability (remember that seekers are riding a controlled explosion) usually make this impractical. Lloyd describes a procedure to estimate this close-in defocus [Lloyd, 14].

$$R_{df} = \frac{D_o}{\delta} \tag{9.3}$$

where

R_{df} = range where defocus equals the tolerable amount. Anything under this range (for a system focused at infinity) will have intolerable focus. The units are the same as input for D_o.

D_o = diameter of the aperture

d = allowable defocus in radians. Usually, the image processor can extract useful information until the defocused blur is close to two pixel IFOVs, but SNR concerns may limit the blur to only one pixel.

Sometimes, for image processing and calculating kill probabilities, it is more useful to express the range in the time domain. This can be done simply if the resultant closing velocity is known.

$$T_{df} = \frac{D_o}{\delta V_c} \tag{9.4}$$

where

T_{df} = time before impact, where defocus equals the tolerable amount. All images during this time (for a system focused at infinity) will have intolerable focus.

D_o = diameter of the aperture

d = allowable defocus in radians

V_c = resultant closing velocity in units per second, where the units are the same as input for the diameter of the aperture. This velocity is the result of the components of both the missile and target velocity along the line of sight.

Equation (9.4) is appropriately applied to stationary targets or moving targets where the closing velocity is constant. The important message from this equation is that the time the seeker becomes inoperable is dependent on the aperture size, the allowable blur, and the closing velocity.

To some it may be surprising that the defocus range is only a function of the aperture diameter and allowable defocus. Lloyd derives these equations from first principles, and one can see how other terms cancel out. Those interested should refer to his book.

These simple equations underlie a disturbing trend for the systems engineer. Because of this defocusing effect, the seeker must be designed to ensure adequate resolution at close ranges, whether the resolution is limited by unit cell size (detector) or optical blur. At the last range to go before defocusing blur occurs, this blurred footprint must be acceptable for guidance. This is because no increase in resolution will occur as the seeker approaches the target. The footprint at the target remains a constant because the optical blur increases in proportion to the decreasing range.

For example, a 10 cm diameter aperture would have a diffraction limit of 244 μm at 10 microns. If the tolerable blur is 300 microradians, the range where this is reached would be a third of a kilometer! At this range, the footprint on a target (driven by the blur) would be 10 cm. As the seeker approached, the blur would get worse; it would be a milliradian at 100 m. However, at 100 m, the footprint on the target is still 10 cm, but it is blurred over more FPA unit cells. As the wish to type targets and select specific aimpoints increases, the resolution required to do this increases. Since one is fighting the diffraction limit in the infrared, the diameter of the optics increases, and the allowable blur decreases. The contradiction facing modern designers is that by increasing resolution, the miss distance does not necessarily reduce accordingly because the blind range also increases.

9.5.2 FPAs

As time goes by, missile customers demand higher sensitivity and effectiveness for less weight and size. This leads designers to move from the classic single-element or quad cell to large format staring arrays. The trend to larger and larger staring arrays has been occurring throughout the eighties and nineties. As FPAs increase, so do the integration times, while at the same time the IFOV decreases, all making image stabilization increasingly important. However, seekers must also be produced in large quantity and, at the present time, the feasibility of large-scale production of staring FPAs is somewhat questionable. The producibility impacts of staring FPAs may be the biggest challenge facing the new generation of infrared seekers. This is underscored by the difficulty in obtaining in-spec FPAs for the Javelin seeker. Seekers are often cued off another sensor or sight. These wide-FOV sights are evolving to GEN II scanning FPAs and, eventually, to large staring arrays. Javelin, slated to replace the aging TOW, is expected to use a GEN II FLIR for an IR sight [Smith, 15].

FPAs for seekers are often required to perform rapid integration and readout. "Fast framing" the FPA in this manner has several advantages. It can reduce smear and data latency, provide more data, and provide more images of a target before the aimpoint needs to be selected. Data latency and temporal consistency across the FPA are often key issues. The focal planes may use a double layered readout, in which the collected charge can be dumped (within a millisecond) to a buffer CCD, then read out while the array is busy converting new photons to electrical signals. This affords almost no dead time between frames and a minimal data latency. Other FPAs read out every other line on every readout, which is fine for an interlacing display but can add several extra circuit boards to the image processor. Integration of electronics processing directly to the FPA to increase speed and reduce power and mass is being pursued.

A key concern, resulting from the high background caused by dome heating, is FPA response uniformity. It is desirable to have uncorrected uniformity near ±3 percent. This is difficult to achieve with LWIR HgCdTe, borderline with InSb, and only readily achievable with Pt:Si. If processing can greatly increase uniformity, then less uniformity is acceptable.

The number of pixels, hence focal plane size, is usually less than required for a FLIR or camera display. Most seekers are being designed with arrays from 64 × 64 to 256 × 256. For instance, a 128 × 128 pixel, 3 to 5 μm staring HgCdTe is being considered for the Sidewinder missile, which has high dome heating and substantial required background suppression [Kurer, 16]. Large arrays have the disadvantage of requiring large processors. Their resolution sometimes cannot be fully used with seekers due to the diffraction limit of the required small optics. The only case that can be made for large arrays is to employ a large field of view, which usually isn't necessary with a seeker, unless it is performing additional functions (such as surveillance).

Most seeker FPAs are cooled by single-gas, blow-down J-T systems. This is appropriate for IR systems that must be used just once. The weight impact depends mostly on the weight of the gas carried to blow the FPA down. Key to obtaining rapid cool-down is a reduction in cold end thermal mass. Metal dewar components can decrease this mass substantially, for instance, as was done with the Javelin seeker. Replacing the glass components with metal ones (e.g., a nickel/copper/nickel structure that replaces the glass coldshield) reduced the thermal mass from 298 to 97 Joules [Whicker, 17]. All components, no matter how arcane, should be examined in detail to reduce thermal mass. A case in point is the thin film polyamide coating over FPA wires providing thermal isolation and electrical conduction that can reduce heat loss to 0.5 mW per lead [Whicker,18].

9.5.3 Image Processors

As smart weapon seekers attempt to mimic the eye-brain combination, the image processor has increasingly higher-order functions levied on it. With a half million (or so) lines of code operating on data from a 128 × 128 (or larger) array being read out 100 to 1,000 times per second, the required throughput per gram and/or watt becomes staggering. Moreover, the image processor's performance is critical to system performance, and useful figures of merit to the customer must include processing performance. Z planes have been considered for several seeker programs, and an MWIR 128 × 128 HgCdTe coupled with a VHSIC is being developed by NRL [Struck, 19].

Again, the level of the technology must be balanced with the producibility for each individual application. The processor must have a high throughput per gram and watt, but also must be producible on a high-rate basis. The signal processors are often custom made for the seeker application. It is common to use parallel processors, three-dimensional computers, application specific integrated circuits (ASICs), and other forms of exotic technology. The benefits of these lie with their high performance for their weight and power—exactly what the doctor ordered for seeker applications.

The number of lines of code for a seeker processor is large and growing. The DARPA smart weapons program showed, with 200,000 lines of code, the "ability of systems to not only recognize targets but to comprehend them in an environmental context, recognizing fields, forests, and roads, as well as the ability to refer to digital maps" [Haystead, 20]. Specially coded algorithms can greatly reduce cost and weight by providing efficient use of processing hardware. The Thirsty Saber smart weapons program has heavy reliance on such efficient algorithms and compact processing. It is expected that the Thirsty Saber will fuse both radar and IR imagery to autonomously navigate to its target [Adams, 21]. This implies sensor fusion for seekers.

9.6 SYSTEM CONSIDERATIONS

9.6.1 Producibility

The wonderful perplexity about IR seekers is the rate at which they can be produced. More than any other IR system, seekers are made at rates high enough to be called "production." Table 9.3 gives some production histories and forecasts. Seeker production gives the manufacturing operations manager a real challenge.

TABLE 9.3 Recent Missile Cost and Production Projections

Missile	*Total Forecast Production*	*Approximate Expected Cost per Missile*	*Source*
ASRAMM	15,000	Not available	(A)
GBI/ERIS	750-1,000	Not available	(C)
IR Maverick	25,514	$122,000	(J)
Javelin	79,620	$73,000	(J, F)
Maverick	Not available	$74,000	(H)
NLOS	16,500 (down from 90,000)	Not available	(E)
SADARM	300,000	Not available	(G)
SLAM RMP	Not available	$1,000,000	(H)
Stinger	Not available	$26,000	(I)
Tomahawk	Not available	$1,300,000	(H)
TOW	>500,000	$20,000 TO 26,000 (9)	(A, B, D)

Sources: (A) C. Miller. September 30, 1991. "BAe, Hughes to Team on ASRAAM Program." *Defense News,* 41. (B) B. Smith. June 10, 1991. "Army to Award Development Contract For Improved TOW Missile Infrared Sight." *Aviation Week and Space Technology,* 60. (C) J. Asker. February 4, 1991. "Army ERIS Interceptor Destroys Dummy Warhead in SDI Test." *Aviation Week and Space Technology,* 22. (D) February 1991. "TOW An Example of Continuous Upgrade." *National Defense,* 30–33. (E) March 27, 1989. "Non-Line of Sight Missile Will Use Platinum Silicide Infrared Detectors." *Aviation Week and Space Technology,* 67–70. (F) C. Baker. March 4, 1991. "Army May Delay Antitank Weapon." *Defense News,* 4, 29. (G) January 18, 1988. "Aerojet Technology is Key to Development of High Rate Smart Munitions." *Aviation Week and Space Technology,* 68–69. (H) C. Adams. April, 1991. "The Impact of Desert Storm: Morale and Profits Rise." *Military & Aerospace Electronics,* 1. (I) September 16, 1991. "U.S. Army Taps GD for Stinger Missiles." *Defense News,* 27. (J). August 24-30, 1992. "Program Acquisition Cost Summary," *Defense News.*

Scheduling supplier delivery of components such as focal planes and optics is often a rate limiter. Testing is a concern and opportunity. In-process testing can determine where a problem is occurring before much value is added to the seeker. However, testing can be exceptionally costly and must be done in a manner that does not affect the life of the seeker or drive production rates.

9.6.2 Cost

Although the electro-optical seeker represents a small percentage of the weight of a missile system, it represents a large percentage of the cost—up to 50 percent or more. Therefore, a missile seeker should be produced for tens of thousands of dollars, which is cheap by IR standards. However, seeker development is still expensive and can run from tens to hundreds of millions of dollars. The critical factor in seeker cost is the number of units to be made.

More than any other IR system, seekers are subject to a learning curve during production and a continuous per unit reduction in cost as the quantity produced is increased. High learning curves of 95 or 98 percent are typical for IR systems in general, meaning not much gain is realized. However, seekers frequently experience learning curves in the 80 to 90 percentage realm. One can expect the curves to be higher (i.e., less reduction in price) for highly complex seekers. Some highly complex seekers can still experience very desirable learning curves if their complexity is in algorithms, software, or easily produced items. Development programs can be very expensive, costing tens to hundreds of millions of dollars, depending largely on the number of algorithms and heuristics that require development. A caveat on low learning curves (remember, low is better) is that, in the future, fewer systems are likely to be procured. This is the result of declining military budgets, and the outcome will be higher per unit costs.

As a rule, it is difficult to package a gimbaled telescope in a missile body where the ratio between seeker aperture and missile outside diameter exceeds about 0.6. Although unglamorous and not considered in the realm of high technology, packaging can make development costs rocket exponentially and the production costs increase greatly, but with a smaller slope. Packaging is frequently costly for developmental units because of the extensive non-recurring engineering effort. Production programs are affected by packaging because of the complex procedures and tooling required to cram everything into a small volume. As such, packaging is often a cost driver. The desired aperture diameter to body diameter ratios for seekers is high. Diameter ratios of 0.9 for on-axis and 0.6 for off-axis or gimbaled seekers are common. Packaging a strap-down seeker is far easier if it can be mounted to the center of the missile. In this case, aperture diameter to missile diameter ratios of 0.8 to 0.9 are frequently achieved.

Gimbaled and off-axis seekers usually cost more than on-axis or body-fixed seekers. However, for specific applications, the extra cost may be offset by reduced fuel consumption, improved performance, or less expensive optics. Likewise, an off-axis seeker design may save weight and cost for sunshades, aperture size, or packaging. Although ASICs are costly on a one-time basis, there is little cost differential between them and military-grade electronics for quantities of a few thousand or more, which are typical seeker quantities. When in doubt, one should assume ASICs for electronics and image processing. It is important to consider the costs of software and how software affects the hardware.

9.6.3 Weight

The weight of a seeker is usually reduced to a minimum, with the total weight usually being a kilogram or less for a simple strap-down, on-axis system and a few kilograms for complex systems. Increased seeker weight has a damaging effect on missile performance because it reduces the available divert capability that a vehicle of a given size and weight may have. It is not uncommon for the weight of a missile to increase by a factor of ten for a comparable increase in the weight of the seeker. A weight increase also ripples into the holding fixtures and launch mechanisms and, finally, the platform carrying the missile. As a method to optimize missile-wide weight, cost, and performance, it is almost always wise to spend the extra money on making the seeker lightweight and efficient.

The best way to estimate weight is to scale from a system that is as similar as possible, using the component relationships in Section 9.2. That is, to scale the optics by relationships presented in Chapter 3. The signal processors can be estimated from the throughput and processing from scaling relationships in Chapter 6. Another way is to generate a parts list, weigh or estimate the weight of each part based on other seekers, add them together, and add a few percent for overhead. As a third alternative, develop a concept and add up the absolute weight of the optics via the algorithms in Chapter 4 (or scale from them). Then add the weight of the scanners and gimbals via the algorithms in Chapter 7 (or scale from them). Next add the weight of the signal processors via the algorithms in Chapter 6 (or scale from them) and the open-cycle cryocooler via the algorithms in Chapter 5 (or scale from them). With each estimate, one should give the seeker the benefit of the doubt for low weight.

9.6.4 Power

The signal processor consumes most of power in an IR seeker. This is because seekers do not require some of the major power consumers of other IR systems (e.g., blow-down cryostat cooling may be used instead of mechanical). It is also because the image processor must accomplish more and more of the higher-order functions that one normally attributes to a person. The power can be scaled from other seekers linearly by the number of operations and the number of bits, excluding from the scaling a watt or two for overhead. Some seekers have integrated laser rangers that do require an enormous amount of power, since lasers usually run at a few percent in power efficiency. Unfortunately, this tends to be true of laser diodes as well, because of the thermal stabilization required.

9.6.5 Scheduling

Scheduling must be addressed from the beginning to ensure that component specifications are producible and achievable. Component delivery schedules are principle causes of system-wide schedule and cost overruns. Component costs and delivery should be monitored early in the program by high levels of management. For critical and long lead time components, having multiple qualified sources is always a wise idea. Purchasing from more than one vendor is more costly but is usually financially workable with large production runs (1,000 or more). Some key components in which one should always pursue multiple vendors include FPAs and image processing hardware. Typical rates for seeker production are tens to thousands per month. Full-rate production almost always entails massive facilitization.

9.6.6 Reliability

Another seeker attribute is its high reliability for one-time use. Traditionally, this is achieved by keeping the moving parts to a minimum, employing a simple design, using proven component technologies, and through rugged construction. The cooling is often provided via a Joule-Thompson system, since the seeker only needs to be used once. The single use makes the design of high reliability easier. However, the seeker must function while getting hit with divert thrusters, axial boost, and a plethora of other nasty system functions. Therefore, scan mechanisms are extra strong, and the telescopes are very stiff. Often, the electronics are "potted" in a compound that holds the parts in place and provides a good thermal path. Seekers are thermally designed for high reliability for the duration of the mission. Since the duration is so short, thermal engineers will often play a cost and weight reduction trick by designing the seeker to be in a constant state of thermal runaway when operating. It will continue to heat up and eventually break. The catch is that the time it takes until it overheats is longer than its mission. As anyone who has had a radiator leak knows, one can safely drive a car a few minutes from a cold start until it starts to overheat. Seekers usually last less than a few minutes.

9.6.7 Critical Trade-Offs and Analysis

The key trade-offs in the design of a seeker revolve around a balance of performance for specific missions, producibility, and cost. Specific critical trade-offs to consider and analyses that must be performed in detail include:

- Active, semi-active or passive
- Strap down, gimbaled or pointed by a mirror
- Staring versus scanning
- Stabilization methodology
- End game algorithms and image processing hardware
- Optics size versus FPA performance versus mission constraints
- Bandpass and bandpass tuning

References

1. D. Hughes. October 21, 1991. "GE Missile Warning Systems Rely on Staring Arrays." *Aviation Week and Space Technology* 42–49.
2. Based on missiles described in D. Lennox and A. Rees, eds. 1990. *Jane's Air Launched Weapons.* Surrey, England: Jane's Information Group.

3. 1993. Information and photograph courtesy of the U.S. Army, Strategic Defense Command.
4. R. Bass, J. Carson, J. Landry, and J. Ratovic. 1990. "Establishing Requirements For Homing Applications." *Proc. SPIE* 1339: 53–76.
5. February 1991. "TOW an Example of Continuous Upgrade." *National Defense,* 30–33.
6. R. Passmore. 1992. "Infrared Research and Development Activities at the U.S. Army Missile Command." *Proc. of TMSA Electro Optics and Infrared Systems Conference,* Orlando, Florida.
7. R. Pleikies, D. Wittmer and F. Lindner. 1989. "IR-Detector Types for Short Distance Seeker Ammunition Application." *Proc. SPIE* 1157: 267–274.
8. September 16, 1991. "Air Force Quits Optically Guided Sidewinder Effort." *Defense News,* 1
9. R.D. Hudson. 1969. *Infrared Systems Engineering.* New York: John Wiley & Sons, 87.
10. T. Janssens. January/February 1980. "Step-Stare Detection of Moving Targets." *Optical Engineering* 19: 145–150.
11. Ibid.
12. R. Reiss. January 1993. "Rapid Mirror Cooling." *OE Reports,* 12.
13. R. Reiss. November 1991. "Rapid Lens Cell Cooling." *OE Reports,* 11.
14. M. Lloyd. 1973. *Thermal imaging Systems.* New York: Plenum, 275.
15. B. Smith. June 10, 1991. "Army to Award Development Contract For Improved TOW Missile Infrared Sight." *Aviation Week and Space Technology,* 60.
16. M. Kurer, D. Scribner, and J. Killiany. March 1987. "Infrared Focal Plane Array Technology Development For Navy Applications." *Optical Engineering* 26, 182–190.
17. S. Whicker. 1992. "New Technologies For FPA Dewars." *Proc. SPIE* 1683: 102–112.
18. Ibid.
19. M. Struck. March 1991. "Naval Research Lab, Developing Future Technology Today." *Defense Electronics,* 27–41.
20. J. Haystead. April 1991. "Thermal Imaging Technology Has Versatile And Bright Future." *Defense Electronics,* 48-52.
21. J. Adams. April 1991. "DARPA Puts Squeeze on Sensor Packages." *Military And Aerospace Electronics,* 14.

Appendix 9A

Data Sheets for Representative Seeker Systems

High Endoatmospheric Defense Interceptor (HEDI) Seeker

Manufacturer: Hughes, El Segundo, California

Intended Application: SDIO

Availability: Developmental

Description:

HEDI is an SDI project to develop kinetic kill vehicles for terminal defense in the high atmosphere. Prototype designs have been built and tested.

Characteristics:

Bandpass: MWIR

Cooling: 77 K

FPA Format: Scanning 18 rows of 120 elements with the seeker using only two rows to enhance yield

Image: 240 × 240

Signal processing: A readout chip and three commercial-grade 80386 microprocessors

Information from W. William Scott. November 28, 1988. "Hughes Prepares Model of SDI Interceptor's Seeker for Flight Tests." Aviation Week and Space Technology, *75–76.*

Javelin Seeker

Manufacturer: Texas Instruments, Dallas, Texas, and Martin Marietta, Orlando, Florida

Intended Application: Anti-tank missile, replacement for Dragon

Availability: In development phase

Description:

Javelin is the new name for the A/AWS-M system. The system is a fire-and-forget missile with a separate command launch unit (CLU). The missile seeker employs an imaging LWIR FPA. The focal planes are made by SBRC as a primary source and Loral as a secondary. The command launch unit contains a thermal sight and a day E-O sight. The missile contains a high-performance image tracker, digital autopilot, and solid state arming system.

Characteristics:

Seeker Bandpass: LWIR

Seeker FPA Format: 64 × 64 staring

Seeker FPA Material: HgCdTe

CLU FPA: 240 × 1

Cool-Down Time: To 77 K in less than 9 seconds

Dewar Lifetime: 10 years

Projected Average FPA Cost: Approximately $13,000

FPA Manufacturers: Texas Instruments (25 percent) and SBRC (75 percent)

Development Costs: $350 million

Total Program Costs: > $2 billion

Information from C. Baker. June 24, 1991. "Army Relaxes Sensor Requirements on AAWS-M." Defense News, *39; July 15, 1991. "AAWS-M Scores Bullseye."* Space News, *2; May 11, 1992. "TI, Martin Select Loral as Second Source of Javelin Focal Plane Array."* Inside The Army, *11; and S. Whicker. 1992. "New Technologies For FPA Dewars,"* Proc. SPIE *1683, 102–112.*

LEAP Seekers

Manufacturer: Boeing, Seattle, Washington, and Hughes, El Segundo, California

Intended Application: SDIO, U.S. Army

Qualification: Space

Availability: Developmental

Description:

LEAP is an advanced technology integration program that is managed by the U.S. Army Strategic Defense Command for the Strategic Defense Initiative Organization. The program aims at developing, integrating, and demonstrating, through flight tests, the accuracy and high performance of lightweight kill vehicle subsystem technologies. The kill vehicle is designed for launch by ground- or space-based rockets and destroys its target by kinetic energy. The projectile consists of an advanced IR imaging sensor; a compact electronics unit, and inertial sensor assembly; a miniaturized high-performance divert propulsion system; thermal batteries; and a command link. All of these components are lightweight and state of the art. Sub-orbital flight tests are to be conducted out of White Sands Missile Range. A version of a Leap kill vehicle is pictured in Figure 10.1.

Characteristics:

Acceleration: 3.5 g

Dimensions: 40.6 cm in length and 15.2 cm in diameter

Seeker Aperture: 12.7 cm

Total Electronics Weight: 171 grams

Weight: About 6 kg with a 170 gram optical system and a 28 gram guidance computer

FPA: 128 × 128 imaging HgCdTe

Information courtesy of the U.S. Army Strategic Defense Command

---

Maverick Seeker

Manufacturer: Hughes, El Segundo, California

Intended Application: USAF, NATO, SDIO

Qualification: Full flight qualification

Availability: Off the line. Production line is open and operating.

Description:

The Maverick seeker contains a linear array of HgCdTe that can be scanned to produce an image. Sensitivity is about twice that of the TV-guided Maverick.

Characteristics:

Bandpass: LWIR

FPA: 16 element scanned with mirrors spinning at 3,600 rpm

Weight: A few kilograms

Information from D. Morrison. April 1991. "The Maverick and the Mark 1 Eyeball." Lasers & Optronics.

N-LOS (Non-Line Of Sight) Seeker

Manufacturer: TBD

Intended Application: U.S. Army weapon missile

Qualification: Flight qualification

Availability: Highly developmental

Description:

The Non-Line of Sight missile is an anti-helicopter and anti-armor missile designed to hit targets tens of kilometers away that are hidden by terrain.

Characteristics:

Missile Weight: 52 kg

FPA: Pt:Si of 256 by 256 or larger

Production: Forecast of 90,000 units

Diameter: About 15.3 cm

Length: About 31 cm

Information from March 27, 1989. "Non-line Of Sight Missile Will Use Platinum Silicide Infrared Detectors." Aviation Week, *67–70.*

10

FLIRs and IRSTs

"A tool is an extension of a man's hand, and a machine is but a complex tool, and he that invents a machine augments the power of a man and the well-being of mankind."

Henry Ward Beecher

10.1 TECHNOLOGY BASICS

10.1.1 Introduction

Like pornography, FLIRs are sometimes hard to define precisely, but a seasoned infrared engineer usually knows one when he sees one. The term "FLIR" probably should be eliminated from IR techno-speak, but so many people in the industry use it that it is likely to remain in the jargon for awhile. "FLIR" is archaic sixties jargon for forward-looking infrared. The term originated to distinguish these systems from IR line scanners, which look down rather than forward. There is nothing to stop a FLIR from looking backward. Conversely, most sensors that do look forward are not considered to be FLIRs (e.g., cameras and astronomical instruments). An IR sensor that views perpendicular to the motion, sometimes using the motion for one of the dimensions of the scan, is sometimes (though, fortunately, rarely) referred to as a side-looking infrared (SLIR). Modern FLIRs are characterized by real-time imaging with real-time ergonomics, high backgrounds, self-scanning, imaged targets, and image stabilization.

So how can you tell a FLIR when you see one? Well, typically, the salient features include the following:

- FLIRs display their data in real time on a cathode ray tube (CRT) or LED array. The display is updated at TV-like rates for a real-time human interface.
- FLIRs are produced in large quantities (tens to thousands).

- FLIRs can be used for rescue, search, track, navigation, target identification, and to lock a weapon onto a target.
- FLIRs typically cover large fields of view, and have larger fields of regard, in which they are manually steered.
- FLIRs frequently are gimbaled.
- FLIRs are normally mounted on moving military platforms such as aircraft, ships, or tanks, or on special mobile weapon systems.
- FLIRs generally operate in a wide atmospheric transmission band—usually 3–5 μm, 8–12 μm, or both.
- FLIRs ordinarily are characterized but not calibrated and, hence, are usually of little use for scientific endeavors.
- FLIRs are meant to be turned on and off frequently (several thousand times in their lifetime).
- FLIRs have mean times between failure of 200 to 1,000 hours.
- FLIRs frequently have zoom lenses or multiple fields of view.
- FLIRs usually have tactical military (rather than strategic, commercial, or scientific) uses.
- FLIRs provide target detection by sensing a difference in temperature between the target and background.
- FLIRs rely on real-time human interaction, interpretation, and control.
- FLIRs are usually designed to operate against multiple targets and backgrounds.
- FLIRs are composed of several line replaceable "black boxes" and several shop replaceable assemblies.

A recent outgrowth of military FLIRs are the infrared search and track (IRST) systems. To the operator, these passive systems are used in a radar-like manner (usually with a radar-like display) to detect and track objects. Although not driven by the need to form a pleasing image, many of their design features, requirements, targets, and weight/cost/power drivers are similar to those of FLIRs. Moreover, there are several advanced-concept FLIRs that perform IRST functions, and vice versa. For these reasons, IRSTS are discussed in this chapter along with FLIRs. However, IRST systems are developing into a specialized class of their own.

The earliest FLIRs were serial scanners. Most currently operational FLIRs are referred to as "first generation" or "Gen 1" FLIRs. They employ first-generation common modules, including a dewar containing 60, 120, or 180 discrete elements of photoconductive HgCdTe. "Gen 2" (second-generation) FLIRs employ a dense linear array of photovoltaic HgCdTe, usually 480 or 960 elements long with 2 or more detectors in TDI for each element. Second-generation FLIRs are expected to leverage from technology development, providing a reduced mass and cost while increasing performance. One can expect that retrofitting Gen 2 systems into packages to replace Gen 1 systems will be a big business opportunity throughout the nineties. Additionally, there are full-framing FLIRs (sometimes called Gen 3) that employ a staring Pt:Si, InSb, QWIP, or HgCdTe to deliver a full CRT frame every time they are read out.

Worldwide, there are over 100 different FLIR systems in operation. Tens are in development, and there are about 15 nations with indigenous FLIR manufacturing capability. These include Australia, Canada, Denmark, France, Germany, India, Israel, Italy, Norway, PRC, Russia, Sweden, and, of course, the United Kingdom and the United States.

FLIRs usually come in several discrete packages referred to as line replaceable units (LRUs) or weapon replaceable units (WRUs). Such packages include the scanner head, power supply, image processor, recorder, display, and controls. They can take the form of black boxes spread around the host platform. The controls and display must be mounted in the cockpit with the humans, while the other "black boxes" can be placed just about anywhere. Most currently deployed displays are conventional black and white CRTs, but flat panel displays are becoming the dominant technology.

A representative, but notional, FLIR architecture is shown in Figure 10.1. It has five distinct hardware pieces (LRUs). These consist of (1) a gimbaled IR receiver/sensor head assembly, (2) an auxiliary electronics box, (3) an image processing box, (4) a power supply box, and (5) the control and display station. The receiver/sensor head must contain all the optics, including collecting, imaging, zoom, focusing, and spectral filtering assemblies, plus electronics and motors to control any moving parts. Some form of thermal reference is often included for self-characterization and uniformity correction. The focal plane dewar assembly and its control electronics are in this LRU. The control electronics usually consist of communication circuits, bias generators, and clocks. If the focal plane needs cooling, some form of cooler must be included. A focal plane's output is of low voltage and amperage, making it delicate signal. Therefore, before it can be sent across a long line, it usually requires analog preprocessing that includes amplification, control, and correction. The basic analog electronics must be physically near the FPA and included in the camera head. Often, the A/D is also included here so the data can be sent digitally.

The architecture of Figure 10.1 has the video signal output to a support LRU for image and higher-order processing. These functions are usually performed by large, power-consuming electronics. Therefore, it is convenient to locate them off of the gimbal assembly and in their own LRU. This support box performs control, processing, power conversion, and video formatting. Processing is becoming increasingly sophisticated, requiring more volume and controls than are convenient for location in the camera head.

The auxiliary support electronics LRU performs basic system control and serves as the interface between the control and display LRU and the rest of the system. Again, the control electronics are usually heavy and power hungry, so it is convenient to locate them away from both the sensor head and the control and display LRUs.

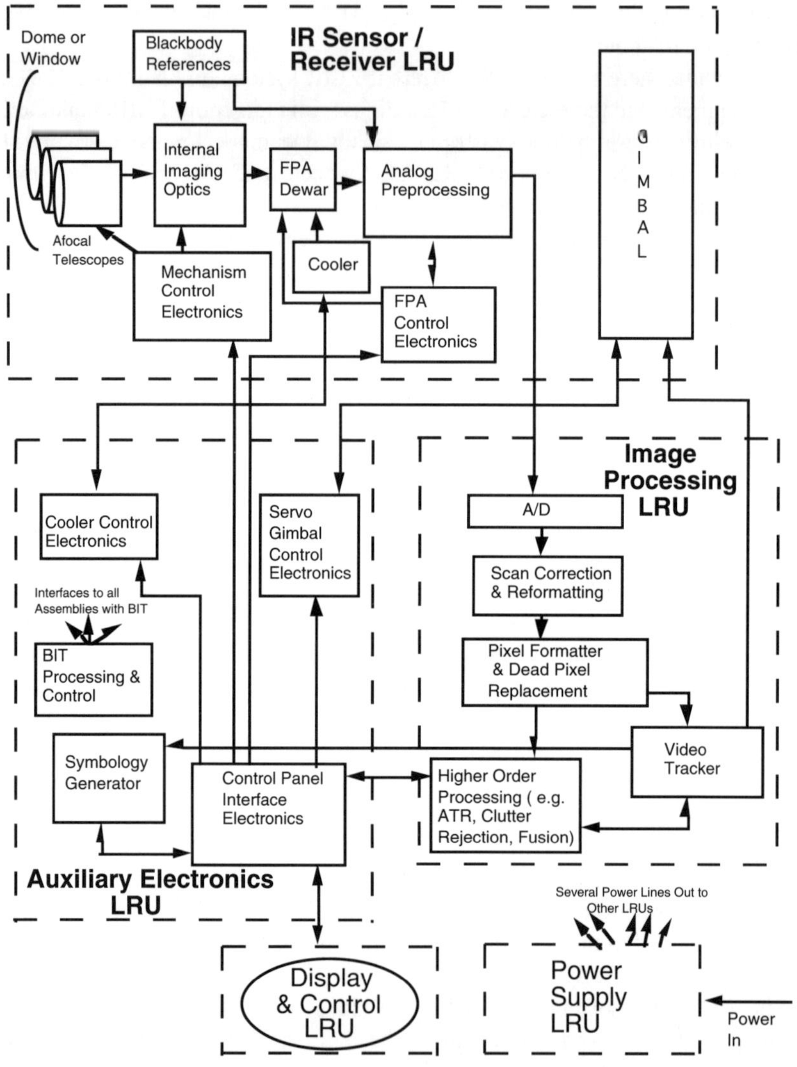

FIGURE 10.1 Representative FLIR System Architecture

Most systems depart significantly from the architecture of Figure 10.1. There is no intent in Figure 10.1 to depict the actual architecture of any individual camera; in fact, no FLIR is likely to have that exact architecture.

The number of LRUs and the divisions between them are often determined by the packaging and basing requirements. The sensor head must have a clear view from the platform, free of obstructions. This sensor head is often packaged in turrets or pods. Figure 10.2 (courtesy of Texas Instruments) is a photo of the Texas Instruments' AN/AAQ-17 turret style sensor head and the associated LRUs that compose this entire FLIR system.

Another popular package for aircraft is the pod-mounted FLIR such as the TIALD pod of Figure 10.3 (courtesy of GEC Ferranti Defense Systems). The pod allows easier installation and removal and provides easier retrofit for older airframes. Aircraft manufactures and users always fret over the aerodynamics of add-ons to their planes; therefore, conformal packaging is likely to be dominant in the near future. Pods often limit aircraft maneuverability and acceleration, and these limitations may be severe. However, pods tend to be more aerodynamic than turrets and even are suitable for supersonic flight. For example, U.S. Navy AN/AAS-38 has been designed for supersonic applications [Burke, 1].

The best and most expensive basing mode is to imbed the sensor into the airframe. In this concept, the FLIR is housed entirely within the structure and looks out through a window which is conformal to the structure or has a slightly protruding dome. The protruding dome of an otherwise conformal FLIR is the case with the Falcon Eye, pictured in Figure 10.4 (photograph courtesy of Texas Instruments) and some CIS FLIRs. Figure 10.4 also shows the Falcon Eye LRUs.

Newer IRST systems such as the EOSS will probably employ embedded sensor packaging with a conformal fuselage aperture [Goodman, 2]. This type of packaging provides the least disturbance to the airframe and missions. More and more systems will require this. There are two drawbacks to conformal packaging. First is that the packaging becomes "application specific." If a FLIR is designed to be shoe-horned into a given platform, it may not be easily adapted to another platform in the future. The second drawback is that it may be difficult to ensure a large field of regard.

Several FLIRs integrate a laser ranger or target designator. An example is the target acquisition designation sight and the pilot night vision sensor (TADS/PNVS) FLIR system found on the Apache helicopter. The laser is used to illuminate a target for the laser-guided Hellfire missile. Like many other FLIRs, the TADS may be equipped with day television vision systems. These titivations add weight, cost, power, and capability while enhancing user effectiveness in specific missions.

Most FLIRs have multiple fields of view that are user-switchable during operation (e.g., the SAFIRE). This gives both a wide, general surveillance mode as well as a higher magnification and narrower field for targeting, designating, or detailed intelligence gathering. Often, a user activates the wide field of view to survey a large area and switches to a narrow FOV to check "hot spots" and eliminate false alarms. Other ergonomic considerations are the changes in magnification and FOV acceptable when switching. A community-wide rule of thumb indicates that the magnification change probably should not exceed five, and probably should be on the order of a factor of four, from one step to the next. If it is greater, then the human operator may have difficulty tracking from one field of view to

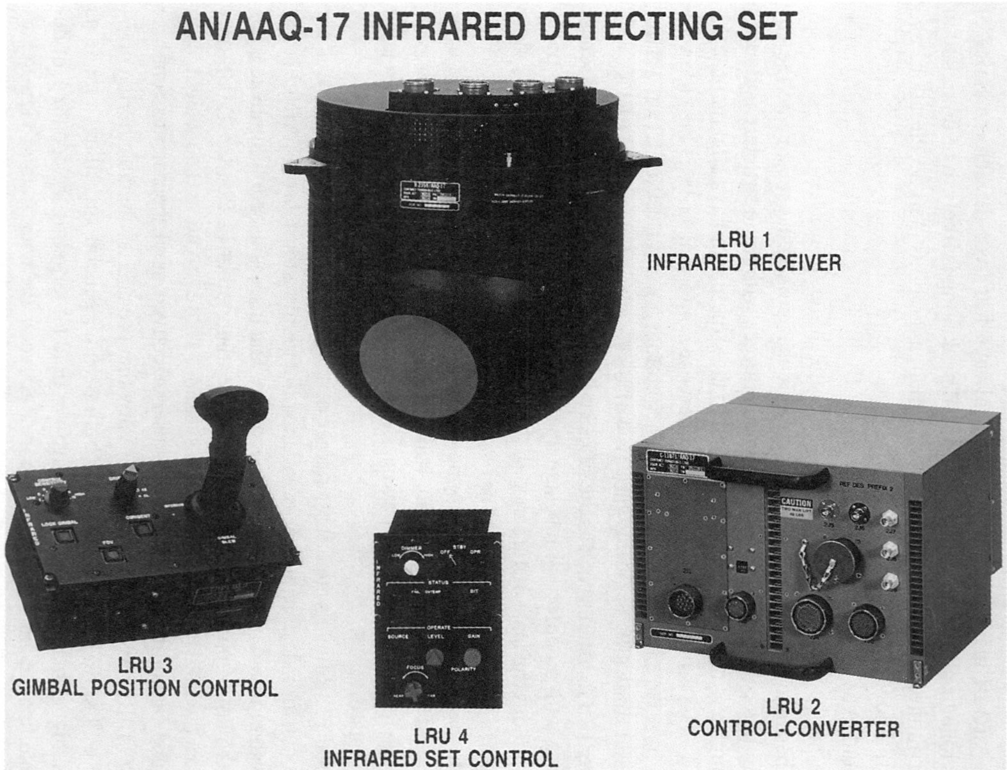

FIGURE 10.2 A Representative Turret-Style FLIR and Associated LRUs (photograph courtesy of Texas Instruments)

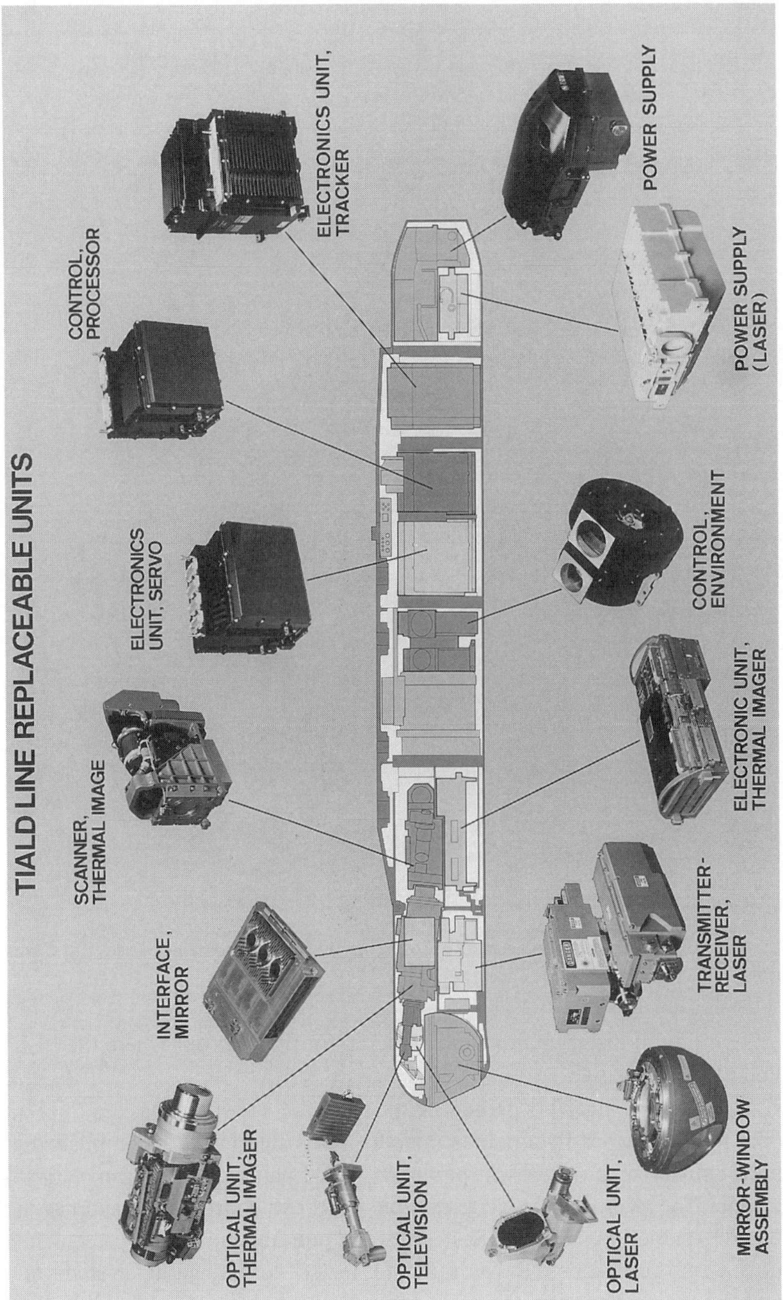

FIGURE 10.3 A Representative Pod-Style FLIR (photograph courtesy of GEC-Ferranti Defense Systems)

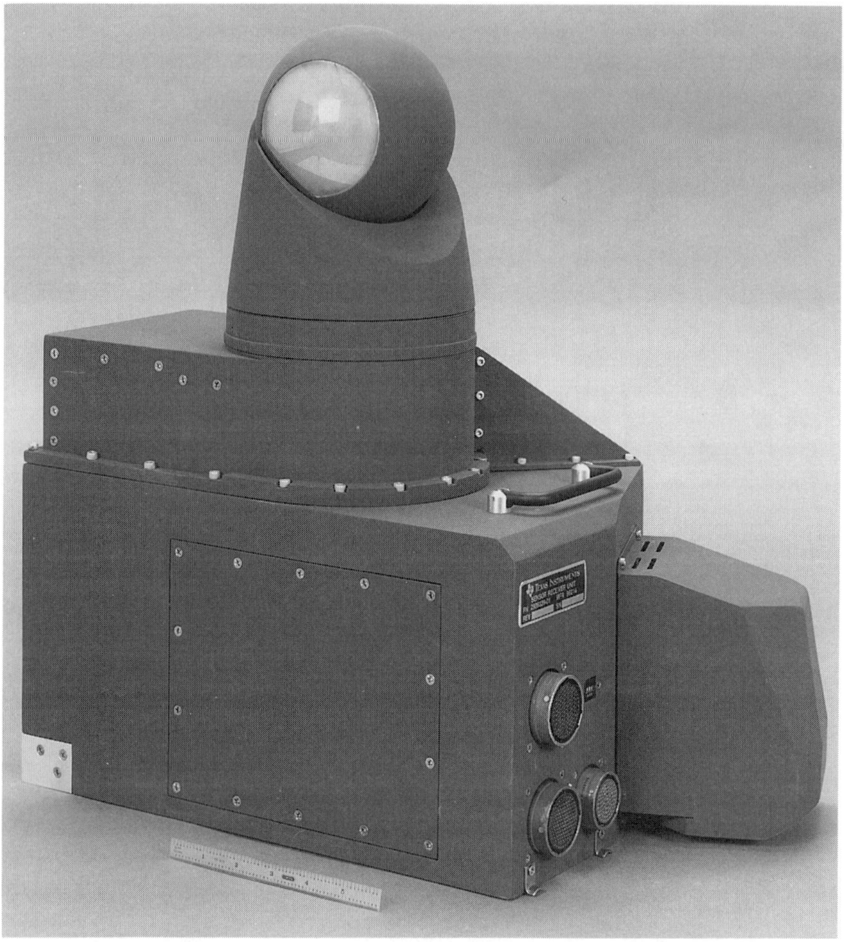

FIGURE 10.4 Falcon Eye, a Representative Conformal FLIR (photograph courtesy of Texas Instruments)

another, resulting in disorientation and difficulty in figuring out where the FLIR is pointed after the change.

A FLIR's field of regard is greater than its field of view; it is tasked to point within the field of regard by a human operator. This steering is accomplished in real time—usually with a joystick, preprogrammed pattern, or head movement. The "joystick" or "handgrip" requires simple hand-eye coordination and has the feel of a video game. As with PNVS, steering and pointing the FLIR by a human's head movement is considered to be the most effective for navigation systems and the most expensive steering method. Slaving a FLIR to a pilot's head and displaying the image in front of his eyes provides simulated night vision. It can be described as a "pseudo-virtual reality" of infrared vision. Humans tend to track with their eyes, but steering and tracking with the head seems to take only minor adjustments. It has been the Falcon Eye experience that head steering is an effective way of controlling and tracking with a narrow field of view, and it only takes two or three flights for the pilots to become proficient [Lydick, 3]. Conversely, It can take as many as ten times the number of flights for a pilot to become proficient in steering a field of view with aircraft maneuvers or joysticks. Other emerging steering methods include use of keyboards, trackballs, and computer-style mice.

In the seventies, the U.S. DoD initiated the common module concept to reduce production costs. The British followed the American lead, implementing common modules in the 1980s. The French also have adopted a common module approach to tactical military FLIRs with their systeme modulaire thermique (SMT). These FLIR subassemblies (or modules) are completely interchangeable with units of the same FLIR, and potentially interchangeable with subassemblies of a different FLIR design. However, a common module from one country is not readily interchangeable with one from another, as each nation has pursued its "pet" technologies. The common modules include detector/dewar units, optical units, scan converters, cryocooling units, and signal processing units. Although this reduces engineering options and system performance, it has enhanced producibility and lowered cost, thus allowing the IR industry to mature more rapidly than it otherwise would have. The common module philosophy has been a great success and should be considered an architect for reducing costs across the electro-optics industry.

In the United States, the common module program began in 1968, sponsored by the Night Vision and Electro-Optics Directorate (NVEOD). The production of FLIRs has steadily increased while their cost has decreased, owing to the beneficial impact of the common module approach. For example, TI has produced over 15,000 common module FLIR systems [Texas Instruments, 4], and total national production has reached many hundreds per month. Often, the common modules are supplied as government-furnished equipment (GFE) to the FLIR manufacturer, who has no say in whether they use SBRC or TI detectors.

U.S. common modules for detectors are 8–12 micron PC HgCdTe arrays with optional lengths of 60, 120, and 180 discrete elements [Haystead, 5]. The U.K. Thermal Imaging Common Module Detector is an 8×6 SPRITE array of HgCdTe.

Second-generation common module focal planes are being developed in various sizes, including 480×4 and 960×4 PV devices with on-chip TDI, multiplexing, and conditioning. This reduces the number of leads out of the dewar [Haystead, 6]. Like the common module, they will be packaged in a common dewar assembly called the Standard Advanced Dewar Assembly (SADA). (See Chapter 5 for details.) Detectors of the second generation also have on-FPA multiplexing and automatic gain and level controls, as opposed to the common modules with their individual leads per detector element and manual gain and level controls.

10.1.2 Applications

FLIRs have various uses, depending on the platform and user. They can be used anywhere a minute temperature differential provides detection utility. Non-military uses include generic search and track, snow rescue, mountain rescue, illegal border crossing detection, and pilot assistance at night or in bad weather. Since radar detectors are so cheap and effective, FLIRs have grown into a position where they can replace radar for police and border patrol missions. Law enforcement missions are subject to specific requirements that the evidence gathered be of admissible or prosecutable quality. This means that it should have high resolution, good image quality, and time and position tags. Being able to record the culprit's voice is another nice feature in many instances. Additionally, FLIRs can be used for forest fire detection, fire fighting, inspection and discreet surveillance, and evidence gathering. An example would be an infrared video of a drug deal for court evidence.

One of the primary military uses is navigation, which allows attack helicopters to fly "nap of the earth" and fixed-wing aircraft to perform low-level penetration in adverse weather and at night. With proper bandpass selection, FLIRs can be used to navigate through rain, dust, fog, and smoke. FLIRs offer long-range of target detection. "Obtaining long-distance visual contact with an object is a key element in the success of close-in air support missions" [Scott, 7]. FLIRs offer pilots such a high level of "situational awareness and vision" at night that they are even used during the day. Moreover, their passive nature is an advantage in real combat situations as they are less vulnerable to countermeasures. With a FLIR, there is no emitted signal. This compares favorably to radar, which screams out its position with an RF pulse. Because of this, FLIRs such as the LANTIRN are "tailored for high precision strikes and deep interdiction missions" [Scott, 8].

FLIRs are increasingly important in aircraft. For instance, the F-117 Stealth fighter's cockpit is centered about the FLIR display, which has a downward-looking and forward-looking IR imager ["F-117A Cockpit Systems Are Focused on Attack Role," 9]. In the military, FLIRs are now considered a necessary augmentation for radar. In situations where weather conditions permit, they can replace radar. Newer military jets have low radar signatures, and IR detection provides crew awareness and functionality with no radar emissions. It doesn't make sense to spend millions or billions on a stealth aircraft and then install radar that screams out its position.

A derivative of FLIR systems that is becoming a type of sensor in its own right is the infrared search-and-track system (IRSTS or IRST). Airborne, shipborne, and ground-based IRSTs are desirable to replace radar because of the stealthy nature of all passive IR. Although they share many similar features, IRST systems are not true FLIRs in the strictest sense. Often, there is no imaging display at all. They are really retrofits or augmentations for radars. They must track tens to hundreds of targets at once in large fields of view, which places emphasis on their processing to reject clutter and to declare and track targets. They have a larger field of view, coupled with slower update rates, than FLIRs. Additionally, IRST performance is usually increased by better vertical resolution, while increased horizontal resolution usually makes a better image for a FLIR.

The bandpass depends on the targets and background. IRSTs have been successfully tested in the MWIR and an 8–10 μm bandpass, but it is the "long term (U.S.) Navy approach to develop 8–12 μm FPA technology for the IRST" [Kurer et al., 10]. The large field of view and slow frame times are illustrated by the development of a dual-band (3–5 and 8–12 μm) advanced air defense electro-optical system (AADEOS) with a 360° per second scan rate [Haystead, 11]. Several other countries have been developing infrared search-and-track systems for aircraft and ships to detect hostile planes and missiles. The (former) Soviets have equipped their late-model fighters with a conformal IRST system. It is a dome-shaped device located right in front of the pilot canopy (like the Falcon Eye) on the MIG-29 [Goodman, 12].

Automatic target handoff systems have been integrated with FLIRs by using data links to provide target locations directly to the aircraft and FLIR [Scott, 13]. Missile warning may also see integration into FLIRs. It has been the U.S. experience in Vietnam and Iraq that pilots require early warning of missile launches to effectively maneuver and deploy countermeasures. Several companies have been making missile warning sensors using staring MWIR and LWIR focal planes ["Hughes, GE Missile Warning Systems Rely on Staring Arrays," 14].

There are several shipboard military uses for FLIRs. The French navy's nuclear submarines have miniature IR thermal imagers mounted in their PIVAIR periscopes to provide a day/night surveillance and identification capability. In addition, there are several fielded systems that perform infrared search and track for sea skimming missiles and enemy aircraft. The French have had such a search-and-track system operational for some time, called VAMPIR. It is based on older component technology but boasts complicated multispectral detection and signal processing. The Israelis have the DS-35 SPIRTAS ship IRST. It is designed to detect sea skimming missiles and can provide a 1.25 second scan in an anti-missile mode and an 8.0 second scan of 360° × 22° in the search mode [Blake,15]. Additionally, Thorn EMI has developed a 3–5 and 8–13 μm ship IRST for the United Kingdom ["Thorn EMI Unit Develops Missile Tracking System," 16]. The Canadians have a shipboard IRST (AN/SAR-8) featuring large pixels, but it is compatible with a future upgrade such as a 480 × 12 that could scan 360° × 20° [Struck,17].

FLIRs are being employed on tanks and jeeps for night driving and target location. What is camouflage in the visible is not necessarily camouflage in the infrared. This is part of the principle that you "can fool all of the spectrum some of the time, and some of the spectrum all of the time, but you can't fool all of the spectrum all of the time."

Following in the path of the B-52, several aircraft platforms (e.g., the B-2) are being considered for lifetimes of 50 to 100 years [Scott, 18]. Obviously, no FLIR designer is sitting in his cubicle thinking about how his FLIR will work against threats and targets in the twenty-second century. Therefore, it is assumed that expensive, modern air platforms will have several avionics and FLIR upgrades during their lifetimes. This means that a major aircraft acquisition such as the F-22 or B-2 may present opportunities for FLIR vendors for nearly 100 years!

10.2 ENVIRONMENTS

A sensor's intended environment is crucial to its design, cost, weight, and performance. FLIRs operate on jittering platforms, in the atmosphere, and in conditions ranging from jungles to arctic deserts. They must be "jacks of all environments" (unlike space sensors, which must be masters of one).

FLIRs are meant to operate on vibrating and unstable platforms such as a helicopter, jet, moving tank or small plane. As a result, image stabilization is required for high resolution. Platform-induced line of sight vibrations have several deleterious effects on a FLIR system, including a reduction in SNR, image smearing, increase in minimum detectable temperature (MDT), and reduced display comprehensibility. Plus, the vibrational effects on a FLIR can manifest themselves as a peculiar stroboscopic effect in which resonance causes the modulation transfer function to actually increase with increasing frequency, then fall off again after the resonance is passed. These problems increase with the integration time, so serial scanners are often not affected, but parallel scanners are, and starers are affected the most. The performance of the actual scan mechanisms can be degraded as vibration is coupled into the scanning.

The stabilization and pointing assembly can be contained in the FLIR's dome as with the head-steered Falcon Eye FLIR [Marr et al., 19]. The Falcon Eye has a two-axis image stabilization mechanism that also compensates for thermal focus shifts [Marr et al., 20]. The stabilization can also be contained outside the sensor head using the initial object space pointing, steering mirror or even the gimbals. Electronic motion compensation is also used. This has a large impact on required memory, electronics weight, and power. The drawbacks include a more limited response than opto-mechanical systems. The reader is referred to Chapter 7 for a more detailed and technical discussion of opto-mechanical image stabilization.

FLIR operating temperatures vary greatly due to the different environments encountered by a mobile platform. Temperature specifications are also extreme (–40° to 70° C) because the military usually buys a FLIR with the intent of using it in African jungles and on the arctic ice caps. This deployment diversity will likely expand as FLIRs are built to last longer, with the dissolving of the Soviet Union making regional conflicts more of a threat, and with fewer military assets being pushed to do more. A military commander can no longer be confident that a helicopter will be used in only one part of the world. The temperature range in which a FLIR must operate is often –30 to 80° C and may vary within that range during operation. The temperature range requires detailed thermal design of the telescope. Temperature variations may cause focus shifts, tilting of elements, LOS shifts, and increased aberrations. Unlike space systems, most FLIRs do not possess completely athermalized telescopes. This leads manufacturers to include moving thermal compensation optics, as with the Falcon Eye to ensure that the effects are not significant. If approached early in the design, incorporating thermal compensation has little impact on schedules, but if forgotten or relegated to the end of the design process, this requirement will blow the entire agenda. Advanced composite metering structures work well to provide a lightweight, low production cost athermalized design.

FLIRs often must function after extensive use in jungles, deserts, cities, and oceans. As anyone who has scraped bugs off of a car knows, this poses problems for the outer surfaces—especially optics. FLIRs and IRSTs usually employ a window in front of the telescope for protection. However, this window must be resistant to bugs, dust, salt, leaves, jet fuel, and so forth, and it may be cleaned with a cloth rubbed over its surface. Just surviving high-speed rain impact is often difficult. "A typical requirement for an IR sensor system is to survive a barrage of 1.8 mm water droplets falling at a rate of an inch per hour and impacting the exterior surface at 400+ miles per hour at a 90° impact angle for 20 minutes" [Fink, 21]. Diamond coatings have greatly increased the life of the windows and make FLIRs more usable in the situations where they are needed most.

FLIRs operate within the earth's atmosphere, and the atmospheric effects must always be considered. Operating an IR sensor in the atmosphere requires consideration of several atmospheric effects. Degradation of image quality and MTF results from atmospheric aerosol scattering, and the atmosphere gives rise to infrared "foreground" radiation and foreclutter. All of these effects are highly affected by weather as well as fundamental differences between geographic features such as oceans, lakes, and ground; deserts versus jungles, and so forth. For example, scatter exhibits a bimodal distribution with wavelength over the ground and trimodal over oceans. Even wind speed in excess of 5 m/s can affect detection range [Milne et al., 22].

Classic radiometry indicates that the range of a FLIR is proportional to the fourth root of the number of detectors. However, FLIRs operate in the atmosphere. There is empirical evidence that when atmospheric attenuation is taken into consideration, the range increases to the logarithm of the total number of detectors times a proportionality constant as in Equation (10.1) [Kurer et al., 23]

$$R \approx k \log_{10} N_d \tag{10.1}$$

where

R = range of the FLIR

k = proportionality constant

N_d = number of detector elements on focal plane

This departure from classic radiometry illustrates the importance of the atmosphere's contribution to FLIR performance. Ranges should not be quoted without proper consideration of the atmosphere and a statement on conditions assumed. The relative humidity, or mass of water, along the line of sight is critical, and its effects are dependent on the spectral bandpass. Atmospheric attenuation in LWIR usually varies depending on visibility (weather) from approximately 0.6 to 0.9 extinction per kilometer. FLIRs tend to loose image quality in poor weather such as rain, fog, or high relative humidity. This is because of reduced atmospheric transmission, increased scatter, and the effective scene temperature becoming more homogeneous as surface temperature differences blend. On a side note, this is not necessarily true for smoke. In military, police, and fire department applications, there is an increasing desire to see through smoke. Selected infrared bandpasses offer some capability in this arena.

The key concept to remember when considering a FLIR's range is that the atmosphere will always degrade its performance. Increasing sensor capability may offset this degradation but, in the real world, it will never work as well as classic vacuum radiometry would indicate.

10.3 TARGETS AND BACKGROUNDS

A given FLIR may be designed for several targets. However, there is one underlying phenomenological difference, and that is that the target's in-band radiance is different from that of its localized background. This difference in radiance is usually a result of a temperature/emissivity difference, but it can also be a chemical excitation that causes the target to be a spectral emitter in the bandpass. Several likely FLIR targets are spectral emitters that depend on the specialized chemical phenomenology of engine plumes. Also, supersonic targets react with the atmosphere by creating a bow shock wave that excites specific atmospheric molecular emissions in narrow, discrete bands. Even an object with a small IR cross-section should produce supersonic bow shock if its velocity exceeds the speed of sound.

Any IR display tends to look different from what the common observer is accustomed to seeing on television. This is because we usually see visible band television and visible band eye-brain images. In the visible, the perceived scene is based on reflected light, with no emission from most objects. IR imagery is the opposite; most of the energy is emitted by the object, with just a little reflection. This results in fascinating features of FLIR imagery such as latent heat detection. A temperature differential can persist after the object that caused it is removed. FLIR imagery can detect a handprint on a desk seconds after the hand is removed, the outline of airplanes on an airfield after the planes have taken off, and residual heat on a car body long after the engine has been turned off. This also results in scenes that confuse inexperienced operators.

Another implementation is for diagnoses based on minute temperature differentials across a single object. For example, the Hughes AN/AAS-33 TRAM FLIR is said to be sensitive enough to "detect the quantities of oil in storage tanks on the ground, solely through temperature differentials caused by varying tank levels" [Burke, 24].

Table 10.1 contains representative targets and backgrounds for FLIRs. Most FLIRs are resolution limited rather than minimum resolvable temperature (MRT) limited. Table 10.1 assumes that the target is "imaged" or extends overfills at least one pixel. One should assume that the MRT for the case when people are targets will be a factor of two to five lower if the person is properly clothed. This is due to the insulating effect of clothing.

The relative difference between the background and the target is crucial to detection. "In contrast to most visible imagery which has low background and high contrast, infrared imagery features very small contrast variation on large pedestals or background" [Silverman et al., 25]. These properties lead to dynamic range problems and require careful settings and grey scale adjustments.

Normal video processing electronics for a FLIR will normalize the scene display based on some mean or average of the scene. This is done to provide a usable display (i.e., one that is not completely black or completely white). For a target to show up, it must be different from the "average" scene (which is defined by the processing algorithms and might not be average at all). Fundamental to detection and contrast is target-to-background contrast, which is a function of both temperature and emissivity. With FLIRs, it is important to consider emissivity for ambient temperature targets and backgrounds with small (say, < 5 K) temperature differences. For example, Farmer points out that the thermal contrast difference at 10 μm between a 300 K target and 302 K target goes from about 0 to 500 percent, with less than a 10 percent change in relative emissivity [Farmer, 26]! Moreover,

TABLE 10.1 Representative Targets and Background Combinations for FLIRs

Function	*Representative Targets*	*Backgrounds*
Air surveillance	Aircraft	Sky, ground
Border patrol, drug interdiction, and law enforcement	People	Vegetation
	Cars/vehicles	Vegetation, dirt roads
	Boats, small planes	Water, sky
Buried object detection	Ground	Ground
Military ground targeting	Power plants, radar dishes, buildings, hangars	Hard earth
Search and rescue	People	Ice, snow, water, vegetation, deserts

he points out, “Brightness temperature calibrations obtained for a particular target location and temperature can change significantly as the target changes spatial location or becomes covered with environmental dirt and grime residues, which can accumulate on the target during normal operation” [Farmer, 27].

Even though FLIRs operate in an atmospheric transmission band, there are still small but real losses in transmission of the target signal. Likewise, the unwanted background also experiences only small losses, and most of it is transmitted to the FLIR. This results in cluttered situations. Modern FLIRs are often limited by the background noise and clutter. Clutter in FLIR systems tends to increase false alarms, fatigue the human operator, and reduce the system’s effectiveness. However, the offset is that FLIRs typically utilize the greatest clutter processor known—a human. However, newer FLIRs and IRSTs attempt to automate the target-in-clutter identification function to ease operator load.

Sky backgrounds from high-altitude aircraft are low in clutter and dc background, but not totally free of them. The Argus is reported to view cold sky background with a dc level of 160 K [Witte, 28].

There are two common bandpasses for FLIRs: the MWIR 3–5 μm atmospheric window and the LWIR 8–12 μm window. Much recent discussion revolves around which is best. The 8–12 μm is almost always better from a target and background consideration but almost always worse from a technology standpoint. However, recently published data indicates about a 10 percent increase in transmission for the MWIR in high relative humidity conditions such as sea basing. Oddly enough, other data indicated that the 8–12 μm band does better in water fog, depending on droplet size.

Multiple-band FLIRs offer discrimination capability against decoys, flares, and backgrounds. If the bands are completely separate (no overlapping), a target temperature can be determined. The farther apart and the more narrow the bands, the higher the accuracy (if all bands have sufficient signal-to-noise ratios). One problem encountered with using both the 3–5 μm and 8–12 μm bands is that the SNR for typical targets is much greater in the 8–12 μm, so the MWIR band becomes a range limitation.

10.4 FIGURES OF MERIT

Sensitivity is usually driven by the field of view, update rates, NEDT, and range. As explained in Chapter 8, noise equivalent delta temperature (NEDT) is a commonly cited figure of merit. NEDT is also called the *noise equivalent temperature difference.* This is equal to the temperature difference in Kelvins that must be present in a scene to produce a change in signal equal to the noise of the FLIR.

NEDT does not show all noise sources (e.g., spatial and noise generated when switching from one row to the next) on a display. MRT usually implies people and is therefore more subjective than other figures of merit. Recent advances in automated testing remove direct involvement of people in the test loop. The test fixtures accomplish this by "calibrating" and including the effects of the human eye-brain in the final test results. The MRT is really a measure of how good a sensor/display/human is at detecting a standard four-bar pattern. This, in turn, is an indicator of field performance, albeit frequently a poor indicator.

One important parameter in the NEDT is the $\partial P/\partial T$. It's value is determined by bandpass and target temperature, and it is target dependent in that the target's emissivity and temperature affect the result. It is beneficial to maximize this derivative to reduce the NEDT. The only way the engineer can maximize this is with bandpass selection. The 8–12 micron bandpass usually has greater target detection capability than the MWIR 3–5 μm. This is because the derivative of inband radiance with respect to temperature maximizes in the 8–12 μm band for room temperature objects. Thus, a small differential in temperature gives a larger change in radiance. Also, the 8–12 bandpass is less effected by colder winter temperatures of man-made objects than the 3–5 μm MWIR [Haystead, 29].

As in most figures of merit, there are several subtleties. First is that the NEDT depends on the optics, and FLIRs often have different fields of view and F#s. Thus, a FLIR will have a different NEDT for different fields of view. Second, the perceived performance could be background limited and depend totally on the scene and not the sensor. Third, the clutter may produce a signal passed by the signal processor to the screen, which may result in false alarms and missed detections that are not predictable by the NEDT. Fourth, for point sources, the NEDT is a function of the range and background. NEDT assumes the target is larger than a pixel. Fifth, the NEDT must be quoted with a temperature reference to have meaning. Standard temperatures are 300 and 500 Kelvins. Sixth, the NEDT is very sensitive to bandpass, so do not expect to have the same NEDT after even a slight bandpass modification.

Since FLIRs operate in real time with human interaction, the interaction with the display and the human eye-brain must be considered in design. The human eye-brain interaction with a FLIR is critical to performance in real situations. A human response is a function of brightness, contrast, and spatial frequency. "The response of the eye lens is a combination of diffraction and aberration and can be approximated by a decaying exponential function for spatial frequencies exceeding 0.4 cycles per mrad" [Vortman, 30]. For lower spatial frequencies, there is a decrease in the response that is attributed to signal processing in the brain—probably akin to an edge detection [Holst, 31]. The eye-brain peaks at about three cycles per degree and, in tests, people have a tendency to move their heads closer or farther from the display to increase target detection. The temporal effects must also be considered. It is easier for people to detect and track with low-frequency jitter (less than 5 Hz or so) than it is for high-frequency jitter. These effects and person-to-person variations are part of the reason why analysis sometimes predicts a better response than is realized in the field.

Temporal effects of the human-machine interface also effects the update rate. Updates of less than 30 Hz are usually irritating to watch, with the worst frequency being about 10 Hz. Update rates lead to some subtle imaging results. For some conditions, such as rapid target movements, a 240-element focal plane can produce a more pleasing image than a 480-element one. This is because it is interlaced and updates a portion of the screen at 60 Hz, while a 480 will only update at 30 Hz. So how does one account for all of these human interaction factors? Well, the community consensus would probably be "not very effectively."

However, the MRT or the minimum resolvable temperature difference (MRTD) is an attempt to incorporate the human and display into the figure of merit. MRT is a measure of the FLIR-display-human system that includes empirical human testing and the resulting spatial and temporal contributions. This include the effects of the display and the human eye-brain. This is a better system of merit for comparing the practical effectiveness of one FLIR system with another.

MRT is usually measured by using the Johnson technique of human subjects recognizing four bars of a pattern. This is a pattern that has four (usually vertical) bars cut into a sheet. The pattern allows blackbody radiation of a slightly different temperature to pass through. A human then attempts to recognize the bars on a display. The four-bar test should be done at various spatial frequencies or resolutions. At resolutions approaching that of the FLIR systems, the MRT will degrade. It is best to be provided with a curve of MRT verses spatial frequency. Unfortunately, MRT is usually defined at a given spatial frequency. It is also usually defined, as in Eq. (10.2), at a spatial frequency f_0 of a system equal to

$$f_0 = \frac{1}{2\ \text{IFOV}} \tag{10.2}$$

where

f_0 = zeroth-order spatial frequency

IFOV = pixel IFOV (should be ≈ resolution of the system)

In effect, the higher the f_0 and lower the MRT, the better the system/ human is at detecting a target.

It is difficult to calculate the MRT analytically, as all component MTFs must be known, including the display and eyeball. Based on empirical data, Holst gives an equation Eq. (10.3) to predict MRT from NEDT if head movement is allowed (for the user to choose the spatial frequency on his retina) [Holst, 32]

$$\text{MRT}(f) = \frac{k\,(\text{NEDT})}{\text{MTF}_{\text{sys}(f)}} \tag{10.3}$$

where

MRT(f) = MRT at the spatial frequency f

k = empirical proportionality constant (Common knowledge indicates this should be between 0.5 and 2.)

NEDT = noise equivalent temperature difference of FLIR

$MTF_{sys(f)}$ = modulation transfer function of entire system at spatial frequency f

A general rule of thumb is that MRTD varies inversely to the system MTF in a FLIR. The vibration reduces the MTF in a system-specific manner, and the MRTD increases proportionally. During testing, if the MRT or NEDT is not being achieved, the culprit may be a reduction in MTF from improperly assembled optics, image forming and interlace problems, or improperly functioning stabilization rather than a purely electronic sensitivity/noise problem. LOS stability is very critical in determining the range at which an operator can identify a target. Table 10.2 shows various MRT concerns and suggested solutions.

The minimum detectable temperature (MDT) is another figure of merit sometimes applied. It is similar to MRT except that a different test setup is used to derive it. MRT is defined as the smallest temperature difference between a target and the background that still allows a human observer or computer tracker to detect the target using the system. Because this figure of merit includes qualities of both the user and the specified target, it has great pragmatic advantages but terrible test consistency. MDT also has the disadvantage of being dependent on target size and resolution, background temperature, and the way the system is set up. It is appropriate for IRSTs and FLIRs when considering specific targets.

FLIRs are the most ergonomic of IR sensors. In the future, they will become almost bionic. Therefore, useful figures of merit must include the hard-to-quantify human being. MRT and MDT do this, and they do have useful bearing on field performance, although not as universally as engineers would like. More factors than those driven by radiometric concerns determine a FLIR's usefulness and user friendliness. The controls must be easily actuated, with the symbology clear, understandable, and user selectable. The image must be interpreted by a human, so it should look pretty. A high-quality video picture is not trivial; it is essential to maximize user effectiveness and reduce fatigue. The picture should be free of lines, streaks, blooming, aliasing, and wobbling. Noise is an overriding concern, as the human will prefer a 200-line picture with a 40 dB SNR to a 600-line picture with a 15 dB SNR. The picture should be as good as today's cable television. None of these factors directly relate to any figure of merit discussed in this text, but they must be considered for overall system utility. Currently, the best staring and second-generation scanning FLIRs produce video, that, under the best circumstances, looks like a black and white TV. Color FLIR displays are in their infancy but developing. By using multiple-bandpass FLIRs, a false-color coded image can be presented.

Users frequently demand to know the range at which they can do a particular job with a particular system under a specified set of conditions. Often, this is the

TABLE 10.2 Concerns with MRT

Concern	*How to Handle It*	*Comments*
1/f noise	Include in D*.	Effects are normally included; for variable read out/integration times, the measurements should be taken at several different rates.
Aging effects	Degrade MRT.	MRT measurements are usually made when the system is new.
Atmospheric absorption/scattering/emission	Use a fudge factor to reduce MRT, based on the range and expected effects from the atmosphere, or perform computer simulation.	The standard four-bar pattern method of determining MRT does not account for atmospheric effects due to the short range to the four-bar pattern generator.
Clutter effects	Use a different clutter figure of merit.	MRT does not account for clutter.
DC background effects	Use a different figure of merit.	MRT measurements can be set such that dc backgrounds are somewhat simulated.
Fill factor	Normally compensated for by the MRT measurement methodology	
Filter corner effects	Normally included in MRT measurement methodology for wideband thermal targets	This is especially critical for narrow bandwidths.
Not square detectors	Normally compensated for by the MRT measurement methodology	
Optics' noise effects	Normally compensated for by the MRT measurement methodology	
Other than circular aperture	Normally compensated for by the MRT measurement methodology	
Spectrally sensitive detectors	Normally compensated for by the MRT measurement methodology for wideband thermal targets	
Stray light	Normally compensated for by the MRT measurement methodology for low-intensity scatter	Unfortunately, high-intensity scatter (e.g. from the sun) is usually the problem and not usually accounted for by MRT measurements.
Target crossing pixel in less than dwell time	Use a different figure of merit for such conditions.	
Target spectral effects	Use different figure of merit if target is not a normal thermal emitter.	

figure of merit they care about. When a human operator is involved, the range becomes more empirical. In certain circumstances, a human can easily pick up an anticipated target with an SNR of 1 or 2, while completely missing an unexpected target with an SNR of 5 or 10. FLIR operators often talk about three distinct ranges: detection, recognition, and identification. The detection stage is usually when the target is unresolved or only has a few pixels across it and is at the longest detectable distance. Here, the operator simply determines that something is there and that it is almost certainly a target. If there are multiple fields of view, the operator is likely to center the target and switch to a higher magnification for more detail. The recognition range is when there are several pixels across the target, and the operator can identify its generic nature, such as an ox cart, car, or plane. Identification, the final level, is when the operator can identify the type of car, such as a 1978 Corvette. The resolution/spatial frequency, image clarity and stability, and MRT all add up to allow the human operator to recognize and identify a target. The ranges for these operations depend on the FLIR, operator, immediate weather conditions, and target/background characteristics. However, the detection range for a given target is always the longest, and the identification range is always the shortest. The situation must be analyzed to determine these ranges—preferably with a well controlled hardware test or a computer model such as the dual-dimension FLIR 90 developed at NVEOD. Traditionally, it has been thought that six to seven cycles (pixels) were needed across a target in the visible for target identification, but recent data indicates a lower number of cycles may be required in the infrared [Sanders, 33].

Since IRSTs usually do not have a human component in their detection hierarchy, the minimum detectable target irradiance (MDTI) is sometimes used. The MDTI is the minimum target irradiance required to produce a desired probability of detection and false alarm. MDTI is similar to NEI except that it includes the effects of clutter and processing gains.

10.5 COMPONENT CONSIDERATIONS

The classic system driver for a FLIR is to meet performance with a manufacturable design whose costs are in line with program objectives. Given the platforms that the FLIRs ride on, power and weight usually are not stress-inducing requirements. Packaging can be a driver in the case of trying to retrofit a FLIR into an existing envelope originally used to house an older FLIR, visible camera, or whatever. Resolution is key to FLIR clutter rejection, target recognition (be it automatic or human), and night navigation.

10.5.1 Optics

As a class, FLIRs have the most complicated and highly folded optics of all infrared sensors. This is because FLIRs have large fields of view, severe packaging constraints, wide bandpasses, and the need to produce images of excellent quality. To achieve the above, it is not uncommon for a single system to approach 50 optical elements with several separate optical paths. As such, the optics are frequently a weight, cost, and producibility driver. FLIRs typically have a broad bandpass several microns in width. Such a wide bandpass frequently requires correction for chromatic distortions. Diffractive optics have been used to accomplish chromatic correction with only one element, as opposed to two or more elements using refractive techniques. Owing to the large field of view, the optics are usually refractive or some combination of refractive, reflective, and diffractive. The platform-imposed packaging constraints often limit the aperture to 15 cm or less. This places the sensitivity demands on the focal plane.

Narcissus effect can be a problem in FLIRs and cameras. Being unique to IR systems, it frequently arises when optical designers do not give proper consideration to this effect. It is a back reflection of the cold detector; i.e., the detector seeing itself. It usually manifest itself as a dark shadow in the center of the display. It can be effective mitigated by titling flat components, minimizing the effective cold area, constraining the curvature of refractive elements, applying anti-reflection coatings, and adding appropriate signal processing [Howard and Abel, 34]. Other techniques to counter Narcissus are to use the most effective anti-reflection coatings possible and to ensure that the afocal does not return possible in-focus back reflections to the FPA.

FLIRs frequently view through a window or dome. This protects the other optics, provides a windshield, and furnishes an aerodynamic surface. In most cases, the window does not contribute anything to the power of the optical train because it is a flat. As such, FLIR system performance may be degraded only slightly when the window is damaged, and even improved if it is removed. "Even when the protective lens was shattered, apparently by striking a bird, tapes revealed no discernible difference in the LANTIRN's infrared picture" ["LANTIRN Equipped F-15Es Pose Strong Deterrence to Iraqi Threat," 35]. The ideal qualifications for the window are for it to be nonexistent for the optical, cost, manufacturing, and radiometric engineer but like a sheet of armor for the reliability engineer. This means it should be

- stiff, strong, hard, and perfectly transparent
- resistant to bugs, rain, and dust
- of a uniform index of refraction of 1.00000000000

It should *not* be

- fungus forming
- hygroscopic
- prone to inband reflection
- subject to bending, tilting, diffraction, or refraction

In addition, the material should perform to total perfection with no cost or weight impact. Obviously, this perfect material does not exist, but some come close. Typically, germanium, zinc sulfide, or zinc selenide is used. The front window must also be designed for easy replacement. Bugs, particles, and dust will erode its surface, making it one of the most frequently replaced FLIR components. Being exposed to the outside atmosphere, windows and domes tend to form condensation in certain environments. As little as 200 nm of ice on a window coating can cause noticeable decrease in window transmission below 11 microns, and a 2,000 nm film can absorb 40 percent of the transmission in a 10.5–12.5 bandpass [Jacyszyn and Pellicori, 36]. Amazingly enough, one effective solution is to heat the dome or window slightly. Many IR transparent materials are semiconductors, so this can be accomplished easily by passing a current though the window.

Finally, it is wise to store the FLIR in a position that protects the window when not in use or during landings and takeoffs.

Nature has blessed the IR engineer by giving germanium an index of refraction twice that of diamond. Ergo, diamond-like carbon is a near-perfect anti-reflection coating for germanium. Plus, it is low in cost and highly durable. Using RF plasma deposition, diamond-like carbon coatings are being applied to the TADS/PNVS germanium windows, M-1 tank, and Leopard tank FLIR. It is expected that the coatings and windows should last 20 years, yielding a $30 million savings for the army for the TADS/PNVS system ["U.S. Army Likes the Look of Hard Carbon Coatings," 37]. Other concerns with FLIR widows and domes are the increase of the radar cross section of the platform and the reduction in its aerodynamics.

If scanning is required, there are serious implications for the optical train. Most are scanned after a large afocal telescope condenses the beam to a reasonable size and before an imaging optical assembly. Image space scanners suffer from defocus as a function of scan angle and increased aberrations as the image is moved to an off-axis position. To allow large field angles, the optical designs are usually complex, with multiple refractive element assemblies to provide flat fields of several tens of degrees.

10.5.2 Pointing Mechanisms

Since FLIR field of regards are usually large, most FLIRs have an object-space multiple-axis gimbal and may also have some form of a pointing mirror. The field of regard and aperture size determine the implementation. Usually, FLIRs are mechanically gimbaled in object space. These gimbals must be reliable, accurate, and compact. The reader is referred to Chapter 7 for a more detailed discussion of gimbals.

FLIRs almost always employ two-axis stabilization, and sometimes three or four axes of stabilization are inherent to the gimbals. More than two axes is advantageous to prevent gimbal lock and to provide stabilization at all gimbal angles. Nesting a fine gimbal inside a coarse one provides a large possible area for the FLIR to be pointed, and rapid and fine stabilization once the FLIR is pointed. LOS stabilization usually needs to be on the order of 1/10 to 1/2 of a pixel per integration time. This means that focal planes with higher density and greater resolution will put a greater burden on the stabilization features of FLIRs.

Head-steered FLIRs involve special considerations and challenges. The servo mechanism is a challenge but not impossible, with accelerations of only a few radians per second squared resulting. Yes, a person can whip his head around from one side to the other faster than that but, unless suffering from a PCP fit, people usually don't. The trick is to establish head movement with enough fidelity to control the gimbal.

Scanning large field angles from small packages often gives rise to convoluted optical trains that produce image rotation as the pointing is changed. Although much work is being done for electronic compensation, production systems now employ a de-rotation mechanism, usually consisting of a moving prism or tilted mirror. These can be located in image space in a place where the beam is small. They tend to add a kilogram or more to the FLIR weight, owing to the drive electronics, mountings, and prism. The cost impact can be large, as de-rotation mechanisms use prisms that may cost several tens of thousands of dollars, and they must be accurately rotated and controlled. It is difficult to compensate for image rotation to less than a milliradian, and the most advanced systems typically can't do it to better than a couple hundred microradians. Image rotation is especially troublesome with TDI focal planes, as the TDI effectiveness decreases rapidly as registration becomes less accurate. Image rotation and skew also tend to increase the effective pixel footprint, making the background and clutter signal increase while having no effect on the target signal. However, this reduces the signal to background or signal to clutter.

10.5.3 Image Scanning Mechanisms

Image space scanners have the lowest weight but suffer from the lowest scan angles and increased aberrations at the edges of the field. FLIRs use several scan patterns to cover the scene at which they are pointed. Table 7.3 lists the most common, and the reader is referred to Chapter 7 for a more detailed discussion.

FLIRs often view an internal thermal reference source (blackbody) during the dead portions of their scan. To allow advantageous use of a tiny (usually only a few cm) moving mirror that reduces power, weight, and vibration, scanning is usually done after the afocal telescope. Serial scanning FLIRs usually employ spinning mirrors that can rotate faster than 45,000 rpm. This causes several negative system effects, including reduced reliability, gyroscopic effects, and higher cost. With the recent reduction in staring focal planes (InSb, HgCdTe, and Pt:Si), a large-format staring array can be implemented to eliminate the need for the rapid scanner. This can translate to system cost savings. The designer should be careful not to be penny-wise and focal plane foolish when he can trade the cost, weight, and power of a scanning system for the denser focal plane.

Scanners can be either closed loop or open loop. Usually, the scanner position is determined at the edges and the electronics interpolate the scanner's position between the edges by assuming that the scanner is moving linearly. One of the Gen 1 U.S. common modules is such a scanner assembly. The linearity of these scans is often insufficient to meet Gen 2 system performance, and some FLIRs employ special scan encoders to provide a more closed-loop approach. Second-generation scanning FPAs require a much more linear scan for their TDI than do the common modules.

10.5.4 FPAs

As time goes by, customers have demanded higher sensitivity for about the same weight and size. This led designers to implement more highly populated focal planes. The 180-element common module dewar FLIR is now being replaced by systems with larger TDI line scanners. A television screen can be updated with a single sweep of a 480 linear array (or a double sweep, if the elements are staggered) allowing smaller, slower scan mechanisms and longer integration times. As FPA size increases, so do integration times, while the IFOV decreases. All of these factors increase the importance of image stabilization.

Search-and-track FLIRs and IRSTs have large fields to cover and will probably employ TDI arrays until very large-format, high fill factor and high-sensitivity staring arrays become cost effective in the next century. Search systems commonly have staggered focal plane arrays. This is because it is imperative that there be no dead zones where a target might miss detection. Additionally, being scanners, they can easily accommodate the larger FPA because the optics need only produce image quality in one axis over a large field. For example, an U.S. Navy shipboard search-and-track system has been suggested with a 360 × 20 degree field of regard covered by scanning two 480 × 12 arrays through an f/1, 10 inch aperture [Kurer et al., 38]. Moreover, the scanning FLIR on the B-52 is being replaced with a full-framing, staring 480 × 640 Pt:Si MWIR FPA with a 4,000 hour MTBF resonant cryocooler [Haystead, 39].

FLIR users and engineers have grown accustomed to scanning arrays and FLIR systems that use unscanned two-dimensional arrays. These "exhibit several curious effects, including the obscuring of high-frequency scene detail by image aliasing," causing this information to appear at lower spatial frequencies [Dann et al., 40]. They also display uniformity variations that cause a confusing temporal display (i.e., a target's display increases and decreases in brightness as it cross pixels, while the range and radiant emittance have not changed) and more confusing clutter from the increased fixed pattern noise. The aforementioned aliasing, and to a lesser extent the nonuniformity effect, can be reduced by using a dither mirror to, in effect, scan the pixel across its dead space during integration to yield a high fill factor.

Obviously, another system effect of the focal plane choice is the sensitivity of the FLIR. Several figures of merit for FLIRs are presented in Table 10.5, and the FPA sensitivity is key to all of them. The reader is referred to several excellent texts and papers on range calculations for standard FLIR systems. However, the wide bandpass of several microns leads to special considerations for range/sensitivity calculations when the detector used has a strong spectral sensitivity dependance. Unfortunately, there are few published references to this issue. This is especially true for Schottky barriers and quantum well detectors, which both lend themselves to full framing densities. As mentioned in Chapter 4, it is necessary to consider the spectral content of the target signature as well as the spectral response of the detector. Doing so implies an integration by wavelength. Pellegrini has published the following generalized "detection" range equation for FLIRs using Schottky barrier detectors and point source targets [Pellegrini, 41]:

$$R^2 = \frac{A_t A_l t_s \tau_o}{N_{f(SNR)}} \times \frac{1.24\ C_1 C_{R1}}{\pi} \int_{\lambda 1}^{\lambda 2} \left(1 - \frac{\lambda}{\lambda_c}\right)^2 \left[\frac{1}{\lambda^5} \exp\left(\frac{c_1}{\lambda T} - 1\right)\right] d\lambda \qquad (10.4)$$

where

R = range in centimeters

A_t = emitting surface area of the target in square centimeters

A_l = clear area of aperture in square centimeters

t_s = stare time in seconds

τ_o = telescope transmittance (usually ≈ 0.5)

C_1 = Schottky emission constant (usually 0.28 to 0.35 for Pt:Si)

CR_1 = radiometric C_1 constant for blackbody radiation of 3.7×10^{-12} W-cm^2 (This is usually abbreviated only as C_1; however, this causes confusion when performing calculations with Schottky barriers due to the Fowler constant C_1.)

λ_1 = spectral bandpass cut-on

λ_2 = spectral bandpass cutoff

λ = infinitesimal wavelength for the calculation in the integral in microns

λ_c = FPA cutoff (1.24/work function—usually about 5.6 μm for Pt:Si)

C_2 = second blackbody radiation constant of 14,388 μm K or 1.44 cm K

T = temperature of target in Kelvins

N_f = noise floor of detector in electrons, including all possible sources

SNR = desired SNR (usually 5 to 10)

The author suggests that the technically oriented reader perform numerical integration with a spreadsheet to solve the above equation. The nontechnical reader merely must remember that the spectral content is important in any range/sensitivity calculations. Equation (10.3) has several assumptions built in, including

- unity in atmospheric transmission
- nonuniformity of noise and clutter not being a driver
- spectrally flat optical response
- a grey/blackbody target

Should the target be a spectral emitter, a spreadsheet can be easily set up that calculates the IRST/FLIR's response as a function of wavelength. Again, the important point is that some newer focal planes—especially the Schottky and quantum wells—have a strong sensitivity dependence on wavelength. This means that classic radiometric equations that assume a flat response over the bandpass cannot be used. Unfortunately, one must dust off the calculus books and integrate [as in Equation (10.4)]; fortunately, personal computers and spreadsheets can perform the integral easily via numeric methods.

10.5.5 Signal and Image Processors

Signal processing can make or break a modern FLIR system. Older FLIRs merely corrected the focal plane output and conditioned it for proper display on a television screen. Newer FLIRs will function as expert systems, aiding the user in decision making and reducing the load. In general, enhancing mission effectiveness while reducing the human operator load will be the key to successful FLIRs of the future. Accomplishing this is largely the task of the signal and image processor.

The U.S. Army Comanche helicopter FLIR will have sophisticated signal processing and computer targeting to reduce the crew workload. Its targets will be prioritized, stored, and tracked [Baker, 42]. According to *Defence News*, "The sensor would allow the two Comanche pilots to safely fly at low altitudes at night or in poor weather while enabling them to detect targets" [Baker, 43].

Another key feature of future FLIRs will be the ability to identify a target with a single frame of data, or at most within a few frames of data. Generally, IRSTs (and future FLIRs) will identify and track targets without human involvement. Much effort is going into temporal and spectral processing of several frames of data in the signal processor to aid the human user in automatic target detection and classification. Several target features can be extracted with single-bandpass FLIRs, including those based on shape, intensity, and contrast. In addition, multiband sensors can determine temperature and multiband intensity, and they can infer absolute size and emissivities. This is accomplished through extensive signal processing with advanced computer hardware running custom code. Parallel processors and neural nets are being employed in developmental units to accomplish these tasks.

Most FLIR systems include alphanumeric data directly on the display overlaying the image. In the FLIR world, these display graphics are called "symbology." The symbology is usually recorded directly with the image. The type of symbology is driven by the missions and the operator desires. For law enforcement, the time and date are important, while military users tend to like to see the azimuth, elevation track point coordinates, and range. Highlighting and outlining potential targets can be useful for any user. Sometime this can be done with minimal added electronics of only one or two boards.

10.5.6 Cooling

Like all IR sensors, the typical FLIR has a dewar and focal plane that are cooled to cryogenic temperatures. The cooled assembly usually includes some baffles and filters but rarely includes much of the optics. This is usually because FLIRs must operate within the atmosphere, and cooling must take place in a vacuum to eliminate condensation.

FLIRs are one of the largest users of the cryocoolers discussed in Chapter 5. The cryocoolers used in FLIRs usually supply 0.5 to 2 W of cooling between 70 and 80 Kelvins, with a MTTF of 1,000 to 3,000 hours.

Cooling of the FLIRs extends beyond the focal plane. Because things are packed so tightly and missions can last so long, forced air or fluid cooling usually is included. Components that become hot require a thermal path to the heat exchanger. These usually include the cryocooler compressor, scanner motor, and densely packaged electronics.

10.6 SYSTEM CONSIDERATIONS

A FLIR's cost, weight, power, and producibility are highly dependent on the specific customer requirements. For example, by averaging parameters from 63 current operational/production, passive, first-generation FLIRs, an average weight of 70 kg (14 kg one sigma) is obtained. These same FLIRs required about 200 W (51 W one sigma) and an average unit production cost of $1.1 million ($492,000 one sigma) dollars. The large standard deviation in weight, power, and cost is clearly due to different requirements, capabilities, technologies, manufacturers, and platforms. Nevertheless, one can see that a FLIR is between a seeker and a space sensor in cost, between a camera and seeker in electrical power consumption, and the most massive of the three.

Whether one designs or requires a serial scanner, parallel scanner, or starer is a tradeoff that is implemented at the system level. Although this tradeoff encompasses FPA (cost, producibility, etc.), cryocooler, gimbal, and telescope, final determination of the lowest cost, highest producibility, lowest weight system must be made at the system level. Weight, power, and cost are sensitive to clutter rejection trade-offs. Clutter changes as a function of resolution by the second or third power. Processing is usually linear to the number of detectors and frame time. This means that a clutter-limited system will usually retain lower weight and power requirements by adding more detectors—if the resolution can be added without a significant impact to the optics. If the optics must be increased, the system will almost always cost and weigh more because the dome and entire path must be increased. Unfortunately, detectors are expensive, so the system will usually have a higher cost.

When Gen 2 systems are in production, it is likely that the signal processor will grow in portion to weight, power, and cost consumed by a FLIR system. The FPA unit cell count increases, and the FLIRs are pushed to act like expert systems. This is especially true for smaller production runs because code tends to run about $100 per line. One can also expect Gen 2 focal planes to be a larger cost and weight driver, especially in the nineties, when learning curves on dense arrays have not had a change to significantly reduce the price.

Some FLIR systems have a laser integrated with the FLIR for ranging and/or target designation. Traditionally, system weight and power increase greatly when a laser is added. However, with recent advances in solid state lasers, the weight and power increases can be minimal or, in some cases, barely noticeable. Cost, weight, and power are highly interwoven and can effectively be traded against each other in the definition phase of a FLIR program.

10.6.1 Cost

It costs more to develop and produce a FLIR than a hand-held camera, but usually less than a space system. Typical recurring unit prices range from $200,000 to a few million dollars per system when in production. These cost levels are usually reached after a development program of a year or more, costing several to hundreds of millions of dollars. Obviously, the actual cost depends on performance specifications and the level of image processing required. Some key system cost drivers include automatic target recognition, amount of clutter processing, sensitivity, Mil-Spec level, MTBF, number to be produced, and logistics support needed.

Development costs can range from a few hundred thousand dollars for a commercial FLIR based on existing technology, to several hundred of million for pushing the state of the art for a military system. A notable example of advanced FLIR development cost is the F-14D IRST, which cost GE and Martin Marietta approximately $20 million [Baker, 44]. Martin Marietta is building five prototypes of the Comanche helicopter FLIR for $300 million. The infrared search-and-track system for the F-22 is being developed by Martin Marietta under a $50 million, five-year contract that follows previous development ["Martin Marietta Will Develop," 45]. The LANTIRN development process cost $94 million, and the Comanche FLIR development ran $300 million [Baker, 46]. Production cost is more of a key issue for this FLIR since the entire Comanche helicopter is advertised not to exceed $7.5 million in production, and the Falcon Eye for fixed-wing aircraft is expected to cost about $1 million each [Scott, 47]. As indicated by the data sheets in Appendix 10A, several FLIRs cost less than $1 million each in production.

Since FLIRs are designed and meant for production of tens to hundreds of units, benefits can be realized the from learning curve. Using data from congressional budgets for purchases of FLIRs, an approximate learning curve can be calculated to be in the neighborhood of 80 percent.

As with any IR system, the best way to estimate the cost of a FLIR is to cost a completed unit, apply a learning curve, and estimate methods to decrease or increase the curve weighted if requirements, costs, or schedules change. Another way is to generate a parts list, cost or estimate the cost of each part, add them together, and add 15 percent for unforeseen but expected increases. Yet another method is to

1. Develop a concept and add up the cost of the optics via the information in Chapter 3.
2. From information given in Chapter 7, estimate the cost of the scanners and gimbals.
3. Using Chapter 4 data, estimate the FPA cost.
4. Estimate the cost of the signal processors via the information in Chapter 6.
5. From Chapter 5, estimate the cryocooler cost.
6. Add these costs together.

An additional 10 percent should be added for a turret-mounted FLIR, 15 percent for a pod-mounted unit, and 25 percent for a package-constrained FLIR. Additionally, FLIR systems tend to require 200 to 300 drawings, and one can usually estimate nonrecurring mechanical engineering at 40 to 60 engineering hours per drawing at about $100 per hour.

10.6.2 Weight

FLIR systems tend to weigh tens of kilograms to hundreds of kilograms, depending on their requirements, performance emphasis, and whether a laser is included. Being frequently air platform mounted, weight is more of a concern than for ship-mounted FLIRs or ground-based cameras, but less of a concern than for space-based systems. If weight is a key FLIR requirement, more money and effort must be spent to reduce these items. Packaging can add significantly to the weight.

The best way to estimate the weight of a FLIR is to weigh a completed unit and estimate changes to this weight by scaling relationships, if modifications are required. Another way is to generate a parts list, weigh or estimate the weight of each part based on other FLIRs, add them together, and 10 percent for all the forgotten stuff. Barring this method, develop a concept and add up the weight of the optics using the algorithms in Chapter 3, or scale from them. The weight of the scanners and gimbals can be derived from the algorithms in Chapter 7 (or scaled from them), the weight of the signal processors via the algorithms of Chapter 6 (or scaled from them), and the cryocooler via the algorithms of Chapter 5 (or scaled from them). An additional 5 percent should be added for a turret-mounted FLIR, 10 percent for a pod-mounted unit, and 20 percent for a package-constrained one. Additionally, the power supply should be estimated at 5 to 20 kg. The focal plane itself is usually insignificant in the weight estimates, although it may have a huge impact because of its effects on other components.

Current systems tend to favor a 1.06 μm laser over a CO_2, given the lower weight and power consumption of the 1.06 μm unit and the large inventory of 1.06 μm laser-guided weapons. Usually, a 1.06 laser adds 10 to 30 kg and 100 W, while a CO_2 adds 100 kg or more and up to 1,000 W. If a diode laser is used the impact is far less—usually only tens of watts and a few kilograms.

10.6.3 Power

FLIR systems without integrated lasers usually require from a few tens of watts to a thousand or so. If power is a key FLIR requirement, more money and effort must be spent to reduce the classical power consumers such as the general-purpose electronics, cryocoolers, and image processors.

The most accurate way to estimate the power requirements of a FLIR is to scale from a completed unit. Another way is to generate a parts list, determine or estimate the power of each part, add the figures together, and 25 percent for overlooked components, margin, and power conversion inefficiencies. If this method is inapplicable, develop a concept and add up the power consumed by the components. Power can be scaled for the scanners and gimbals via the methods in Chapter 7, the FPA via methods in Chapter 4, the power of the signal processors via the methods of Chapter 6, and the cryocooler via the methods of Chapter 5. An additional 25 percent should be added for power conversion and miscellaneous items.

10.6.4 Producibility

A typical FLIR production house acquires the individual optical elements, focal plane, and cryocooler from other vendors or another division. The FLIR house conducts the initial and manufacturing design and development procedures, performs system integration, (usually) assembles the telescope, mates the focal plane, builds the gimbal, and performs integration and test.

The alignment of tens of optical elements is tricky and critical to system performance. A slight change in spacing or tilt may decrease the MTF enough to have a noticeable effect on the MRT. It is always wise to design such a system with production in mind, which means specifying placement tolerances that are easily achievable in a manufacturing environment and using the lowest possible number of elements. Frequently, these are competing requirements that must be balanced as early in the development cycle as feasible. Binary optics can form a bridge between these competing desires by providing chromatic correction and aberration cancelation.

Automated testing is crucial to high-rate production, because testing FLIRs and space sensors often creates a serious bottleneck. Installing multiple automated FLIR test stations costs a great deal, but this is the only way to exceed a rate of 10 or more per month. It will reduce costs on runs of 100 or more FLIRs.

Ensuring timely component delivery is also critical to meeting production goals. Special attention should be paid early in the program to ensure that focal planes, cooling units, ASICS, and large or unusual optics can be supplied by at least two vendors. Additionally, one should understand subtle effects that the use of specialized capital equipment, beryllium optics, or exotic diamond turning machines can have on scheduling. For example, diamond turning machines will tend to increase production rates and shorten the schedule; however, they must be available and in operating condition. Otherwise, machining of a part may be delayed for weeks.

Large-scale production of electro-optics is a success story with the LANTIRN pods. LANTIRN is made in a paperless factory in about 70 days (with half of the time being allotted for testing), from a crew of about 100 ["No Paper in Pod Plant," 48]. A touch-screen computer dispenses worker assembly instructions and cartoons. The computer also orders parts, which are delivered by a robot ["No Paper in Pod Plant," 49].

10.6.5 Schedule Drivers

The schedule must be addressed from the beginning to ensure that components can be produced to specifications and on time. Interface control is the key to schedule and cost overruns, and it should be monitored early in the program by high levels of management. Although not glamorous, merely achieving an understanding of how parts and pieces mate is a major scheduling milestone. Having to remake parts or reorder from a supplier is almost always a schedule breaker.

In contrast to basic cameras, integration and test are frequently cost and schedule drivers for FLIRs. This is especially critical if the test facility is an obvious bottleneck in the production flow. A FLIR house should have at least two fully operating altitude-temperature chambers and radiometric characterization stations. Hopefully, only one will be needed to meet the desired production rate, with other serving as a backup. These stations are notorious for breaking down or exhibiting unexplained behavior at the most critical time of a production run.

Using multiple qualified sources for critical components is always a wise idea, and it usually makes financial sense with large program runs of 100 or more. Some key components for which one should always pursue multiple vendors include FPAs, cryocoolers, large or unusual optics, filters, and large or unusual motors or bearings. If the program is small, funding multiple vendors may not be an option. In this case, the FLIR primary source should at least make sure that multiple potential suppliers are available, even if they can't be placed under contract.

Typical FLIR production rates are 1 to 10 per month. Attempting to make more usually entails massive facilitization and second sources for cryocoolers and focal planes. Entire FLIR systems have been integrated in six months [Burke, 50] and a LANTIRN takes only 70 days to produce. However, for the Canadian National Defence operations in the Persian Gulf, FLIR Systems, Inc., was able to equip five Sea King helicopters with night vision within 10 days of the initial request [FSI, 51].

10.6.6 Reliability and Maintainability

Given that FLIRs are used in a military or paramilitary operations where mission success and crew survival may depend on FLIR operation, reliability is at a higher premium than with held-held cameras. The U.S. Air Force, along with paramilitary users, requires increased maintainability and reliability out of FLIR systems. The current MTBF is on the order of a few hundred hours, with the goal for second-generation systems of a thousand hours or so. For example, the first-generation AN/AAQ 9 FLIR made by Texas Instruments reports a MTBF of 730 hours [Burke, 52]. Determining the MTBF requires several thousand hours of reliability test and several hundred hours of flight test. This means that it can only be estimated in the design and prototype stage and really can't be determined until several units are in the field.

One of the drivers of the MTBF is cryocooler reliability. Although several are produced with a specified MTBF of several thousand hours, rarely is that achieved in the field. MTBF of between 1,000 and 2,000 hours is more typical. Vast improvement is expected the nineties in cryocooler reliability, which will translate into improved FLIR reliability.

FLIR customers are concerned about maintaining the system for several years after purchase. Ideally, a FLIR system would have several line replaceable units (LRUs, usually called *weapon replaceable units* by the U.S. Navy) that can be changed in the field, by the user, with a few man-hours of work. LRUs typically include the window, turret or pod, display, power supplies, controls, and sometimes the signal processor and digital scan converter. The minimum time to replace (MTTR) is what the customer can expect to expend on LRU exchange. The shorter the time, the better, and users almost always want it to be less than one day.

Shop replaceable units must be replaced at the manufacturer's facility, but they can greatly extend the FLIR's life span. Ideally, there should be few of these, and replacement times should be no more than a few days. Typically, these are smaller components with high rates of failure. They can include the cryocooler, FPA, dewar, stabilization mirror assembly, and motors.

Sometimes the designers will be given a specification on availability instead of, or in addition to, MTBF. Availability is usually defined as

$$\text{Availability} = \frac{\text{MTBF}}{\text{MTBF} + M_{et}} \tag{10.5}$$

where

Availability = expectation to failure in hours

MTBF = mean time between failure in hours

M_{et} = mean elapsed time in hours to replace LRU

FLIRs usually have specific built-in test (BIT) features. These are preprogrammed tests that the FLIR system conducts on itself when started up, shut down, and even during operation. These are typically performed in the operational background and are transparent to the user unless a failure is detected. When the FLIR detects a failure, it flags the operator that something is malfunctioning and that a test is required.

10.6.7 Critical Trade-Offs and Analysis

The key trade-offs in the design of a FLIR revolve around a balance of performance for specific missions, producibility, user friendliness, reliability, and cost. Specific critical trades and analysis that must be analyzed in detail include

- pointing architecture (gimbal, mirror, body fixed)
- bandpass
- staring versus scanning FPA
- stabilization methodology and performance
- level of image processing
- display quality, symbology, and enhancements
- optics size versus (FPA performance versus mission constraints)
- logistics support

References

1. D. Burke, ed. 1990. *Jane's Avionics, 8th Edition, 1989–1990.* Coulsdon, Surrey, England: Jane's Defence Data, 102.
2. G. Goodman. March 1989. "Infrared Search and Track System Heat Up The Air Battle." *Armed Forces Journal,* 87–88.
3. L. Lydick. January 1989. "Building the System." *Code One,* 19–21.
4. 1992. Information courtesy of Texas Instruments.
5. J. Haystead. April 1991. "Thermal Imaging Technology Has Versatile and Bright Future." *Defense Electronics* 23, 4: 48–52.
6. Ibid.
7. W. Scott. April 17, 1989. "Falcon Eye FLIR, GEC Helmet Aid F-16 Mission Flexibility." *Aviation Week and Space Technology.*
8. Ibid.
9. June 8, 1992. "F117A Cockpit Systems Are Focused On Attack Role." *Aviation Week and Space Technology,* 64.
10. M. Kurer, D. Scribner, and J. Killiany. March 1987. "Infrared Focal Plane Array Technology Development For Navy Applications." *Optical Engineering,* 182–90.
11. J. Haystead. April 1991. "Thermal Imaging Technology Has Versatile and Bright Future." *Defense Electronics* 23, 4: 48–52.
12. G. Goodman. March 1989. "Infrared Search and Track Systems Heat Up The Air Battle." *Armed Forces Journal,* 87–88.
13. W. Scott. April 17, 1989. "Falcon Eye FLIR, GEC Helmet Aid F-16 Mission Flexibility." *Aviation Week and Space Technology.*
14. October 21, 1991. "Hughes, GE Missile Warning Systems Rely on Staring Arrays." *Aviation Week and Space Technology,* 42–49.
15. B. Blake, ed. 1991. *Jane's Radar and Electronic Warfare Systems 1991–92,* 3rd ed. Surrey, England: Jane's Information Group, 301.
16. September 16, 1991. "Thorn EMI Unit Develops Missile Tracking System." *Defence News,* 27.
17. M. Struck. March 1991. "Naval Research Lab Developing Future Technology Today." *Defense Electronics,* 27–41.
18. W. Scott. August 3, 1992. "Whitman AFB Upgrade Tailored to Efficient B-2 Operations." *Aviation Week and Space Technology,* 64–65.
19. L. Marr, S. Sagan, and C. Fuchs. 1987. "Internal Stabilization Mirror With Novel Gimbal Set For Falcon Eye FLIR." *Proc. SPIE 779,* 88–94.
20. Ibid.
21. R. Fink. July 1991. "For the Military, Optics Also Play a Protective Role." *Photonics Spectra,* 97–101.

22. E. Milne et al. 1991. "Sensitivity Analysis of Navy Tactical Decision Aid FLIR Performance Codes." *Proc. SPIE* 1486: 151–160.
23. M. Kurer, D. Scribner, and J. Killiany. March 1987. "Infrared Focal Plane Array Technology Development for Navy Applications." *Optical Engineering* 26, 3: 182–190.
24. D. Burke, ed. 1990. *Jane's Avionics, 8th Edition, 1989–1990.* Coulsdon, Surrey, England: Jane's Defence Data, 102.
25. J. Silverman, J. Mooney, and V. Vickers. February 1990. "Display of Wide Dynamic Range Infrared Images From Pt:Si Schottky Barrier Cameras." *Optical Engineering* 29: 97–106.
26. W. Farmer. November 1991. "Analysis of Emissivity Effects on Target Detection Through Smokes/Obscurants." *Optical Engineering* 30: 1701–1708.
27. Ibid.
28. D. Witte. "An Assessment of The CD-FLIR as an Argus Sensor" ERIM Report No. 223500-1-F, AD-A220 841, 1.
29. J. Haystead. April 1991. "Thermal Imaging Technology Has Versatile and Bright Future." *Defense Electronics* 23, 4: 48–52.
30. J. Vortman and A. Bar-Lev. June 1987. "Improved Minimum Resolvable Temperature Difference Model For Infrared Imaging Systems." *Optical Engineering* 26: 492–498.
31. G. Holst. 1989. "Minimum Resolvable Temperature Predictions, Test Methodology and Data Analysis." *Proc. SPIE* 1157: 208–216.
32. Ibid.
33. J. Sanders, C. Halford, and K. Krapels. 1990. "Human Recognition of Infrared Images." *Proc. SPIE* 1309: 106–113.
34. J. Howard and I. Abel. 1982. "Narcissus: Reflections on Retroreflections in Thermal Imaging Systems." *Applied Optics* 21: 3393–3397.
35. November 12, 1990. "LANTIRN Equipped F-15Es Pose Strong Deterrence to Iraqi Threat." *Aviation Week and Space Technology,* 63.
36. F. Jacyszyn and S. Pellicori. 1992. "Influence of Water Ice Absorption on Transmission For a 10.5 to 12.5 Filter/Window." *Optical Engineering* 31: 1952–1954.
37. February 1990. "U.S. Army Likes The Look of Hard Carbon Coatings." *Laser Focus World,* 9, 10.
38. M. Kurer, D. Scribner, and J. Killiany. March 1987. "Infrared Focal Plane Array Technology Development for Navy Applications." *Optical Engineering* 26, 3: 182–190.
39. J. Haystead. April 1991. "Thermal Imaging Technology Has Versatile and Bright Future." *Defense Electronics* 23, 4: 48–52.
40. R. Dann et al. 1986. "Sampling Effects in CdHgTe Focal Plane Arrays—Practical Results." *Proc. SPIE* 685: 123–117.
41. P. Pellegrini. December 1989. "Range Calculations For Staring Schottky Barrier Sensors." *Optical Engineering* 28: 1288–1293.
42. C. Baker. September 1991. "Comanche Team Picks Martin For EOSS." *Defence News,* 20.
43. Ibid.
44. Ibid.
45. November 16, 1992. "Martin Marietta Will Develop." *Aviation Week and Space Technology,* 53.

46. C. Baker. September 1991. “Comanche Team Picks Martin For EOSS.” *Defence News,* 20.
47. W. Scott. August 3, 1992. “Whitman AFB Upgrade Tailored to Efficient B-2 Operations.” *Aviation Week and Space Technology,* 64–65.
48. May 30, 1990. “No Paper in Pod Plant.” *Flight International,* 26.
49. Ibid.
50. D. Burke, ed. 1990. *Jane's Avionics, 8th Edition, 1989–1990.* Coulsdon, Surrey, England: Jane's Defence Data, 97.
51. 1993. Information courtesy of FSI.
52. D. Burke, ed. 1990. *Jane's Avionics, 8th Edition, 1989–1990.* Coulsdon, Surrey, England: Jane's Defence Data, 97.

Appendix 10A

Data Sheets for Representative FLIRs and IRSTs

AADEOS (Advanced Air Defense Electro Optical Sensor) IRST

Manufacturer: Martin Marietta, Orlando, Florida; sponsored by U.S. Army CE-COM NVEOD

Intended Application: Generic surface-based search and track with various platform and vehicle options

Qualification: Proof-of-principle

Availability: 1995 production

Description:

AADEOS is a scanning infrared search-and-track system providing targeting and detection information with several potential applications. It consists of two separate units, a turret based sensor head and a small electronics/processing box. Detection and tracking information on simultaneously prioritized multiple targets provides essential data to the operator and fire control system. Full scan imagery (360° × 20°) or smaller subsets can be simultaneously displayed with overlayed target track information.

Characteristics:

Bandpass: Both 3–5 and 8–12 μm

Field of Regard: 20° × 360°

Scan Rate: 360° per second

Track Establishment: 3 to 5 seconds

False Alarm Rate Goal: One per hour

Weight: 110 kg

Processor Volume: < 0.04 cubic meters

Sensor Volume: < 0.14 cubic meters

Information supplied by General Electric

Advanced IRST

Manufacturer: Spar Aerospace Limited, Brampton, Ontario, Canada

Intended Application: Shipborne IRST

Qualification: U.S. Mil-stds

Description:

The Advanced IRST is optimized for detection of anti-ship missiles. It is a lightweight version of the AN/SAR-8, retaining the large aperture and high sensitivity along with the basic hardware and software architecture.

Characteristics:

Bandpass: 3–5 μm and 8–12 μm

FOR: 360° in azimuth, –1° to 24° (optional) in elevation

Scanner Rotation Rate: 60 rpm

Detection Range: 30.5 km against a supersonic sea-skimmer under typical weather conditions

Declaration Range: 28 km against a supersonic sea-skimmer under typical weather conditions

Angular Accuracy: Better than 0.1°

Range Estimation Accuracy: Better than 20%

Track Capacity: >200 tracks

Scanner Dimensions: 81 × 140 × 81cm

Scanner Weight: 182 kg

Control and Processing Rack Volume: 56 × 183 × 64 cm

Control and Processing Rack Weight: 443 kg

Power Condition Unit Dimensions: 56 × 91 × 61 cm

Power Condition Unit Weight: 204 kg

Power required: 10 kW

Information supplied by Spar Aerospace Limited

AN/AAQ-17 FLIR

Manufacturer: Texas Instruments, Dallas, Texas

Intended Application: USAF Special Operation Forces

Qualification: Mil-E-5400T, class 1A

Availability: In production

Description:

The AN/AAQ-17 is a DoD common module system. It was developed in 1986 for fire control, search and navigation applications. It is composed of four major assemblies: The turret style IR receiver, the control converter, the gimbal position control, and the infrared set control. The system features an automatic video tracker, symbology generator, 1553B data bus, and an integrated helmet and display sighting system interface.

Characteristics:

WFOV: 13.7° × 18.2°

NFOV: 3° × 4°

Field of Regard: +15° to –105° el, ±200° az

MTBF: > 350 hours

Power: 100 W max @ 28 Vdc and 1,900 W max @ 115 Vac

Turret Weight: 43 kg

Control Converter Weight: 22 kg

Gimbal Position Control Weight: 2.3 kg

Infrared Set Control Weight: 1.8 kg

Information courtesy of Texas Instruments

AN/AAQ-18 FLIR

Manufacturer: Texas Instruments, Dallas, Texas

Intended Application: USAF MH-53J helicopter, adaptable to other missions and platforms

Qualification: Mil-E-5400T, class 1A

Availability: In production

Description:

The AN/AAQ-18 is a DOD common module based FLIR. It was developed in 1987 for special operations and search-and-rescue applications. It is composed of five major assemblies. It has self-contained stabilization.

Characteristics:

WFOV: 13.8° × 18.2°

NFOV: 3° × 4°

Field of Regard: +15° to –105° el, ±190° az

MTBF: > 280 hours

Power: 100 W max @ 28 Vdc and 2.5 KVA, three phase, 115 Vac (400 Hz)

Weight: 98 kg

Information courtesy of Texas Instruments

AN/AAS-33 A TRAM Target Recognition and Attack Multi-Sensor Detecting and Ranging Set (DRS) FLIR

Manufacturer: Hughes Aircraft Company, El Segundo, California

Intended Application: U.S. Navy

Qualification: Flight qualification

Availability: Custom made per contract

Description:

The DRS is a scanning FLIR providing targeting and detection information from a 51 cm diameter spherical turret. The turret also contains an active laser ranger/ designator. The telescope is a continuous zoom with a magnification of ×2.6 to ×13.

Characteristics:

Magnification: ×2.6 to 13

Diameter: 51 cm

Video Output: Composite 525 or 87

Information courtesy of Hughes Aircraft Company

B-52 FLIR

Manufacturer: Loral

Intended Application: USAF B-52 bombers

Qualification: U.S. military flight worthy

Availability: Custom made per contract

Description:

This is a replacement/ retrofit FLIR for the AN/AAQ-6 FLIR. The FLIR is cylindrical is shape with fold/pointing mirrors suspended below the structure. It employs a staring Pt:Si FPA similar to one developed under a DARPA manufacturing program to reduce costs. The FLIR has two fields of view and a high reliability cryocooler.

Characteristics:

FPA: 640 × 480 Pt:Si

FPA Operating Temp: 74 K

Bandpass: MWIR

Prototype Cost: $1 million each

Military Production Cost: \$310,000

Commercial High Volume Production Cost: Projected at \$20,000 to \$30,000

MTBF: 2,700 hours with possible improvements to 5,700 hours

Information from B. Henderson. May 4, 1992. "Loral to refit B-52 FLIR System with Platinum Silicide Sensor." Aviation Week & Space Technology, *81.*

Chaparral Air Defense System FLIR

Manufacturer: Texas Instruments, Dallas, Texas

Intended Application: Control for military anti-aircraft system

Qualification: Military

Availability: Custom made per contract; delivery ≈ 11 months ARO

Description:

The FLIR is part of the Chaparral air defence system. The FLIR is retrofitted to provide increased range while maintaining a passive nature to the system. The entire FLIR system is highly modular in design to facilitate repairs and maintenance. It is able to track multiple targets simultaneously.

Characteristics:

FLIR FOV: Selectable, 2° × 2.65° and 15° × 20°

FPA: HgCdTe Common module

Cooling: Cryostat demand flow, <3 minute cool-down time

System Weight: 66 kg

Power Required: 420 W at 28 Vdc

Information courtesy of Texas Instruments

Comanche Helicopter EOSS-Electro Optical Sensor System FLIR

Manufacturer: Martin Marietta, Orlando, Florida

Intended Application: Navigation, targeting and detection for the Army's Comanche Light Helicopter

Qualification: Military flight

Availability: Developmental

Description:

The Comanche EOSS has a second generation advanced focal plane technology FLIR which dramatically improves capabilities over current systems. The EOSS is aerodynamic, lightweight and, like the TADS/PNVS, is on the front of the helicopter. It will have sophisticated signal processing, improved maintainability, and computer targeting to reduce the load on the crew. Its targets will be prioritized, stored, and tracked.

Characteristics:

FLIR FOV: Selectable, 1.5° × 2.0°, 6° × 8°, 30° × 40°, and 30° × 52°.

FPA: HgCdTe 480 by 4

Bandpass: LWIR

Cooling: Closed-cycle Stirling

Development Cost: $244 million for three FSD systems

Information courtesy of Martin Marietta Corporation

Ericsson Thermal Imager FLIR

Manufacturer: Ericsson, Moindal, Sweden

Intended Application: Military

Qualification: Army and navy requirements

Availability: Custom made per contract

Description:

This is a self-contained unit based on a SPRITE HgCdTe detector. Several options are available including additional signal processing and lens wiper.

Characteristics:

LWIR Bandpass: 8–12 μm

FOV: 3° × 2° and 12° × 8°, other FOV available

Thermal Sensitivity: <0.1 K

Cooling: Closed-cycle split Stirling

Camera Head Dimensions: Length, 60 cm; width, 19 cm; height: 32 cm

Output: 50 Hz 1:2 interlaced CCIR std

Weight: <25 kg

Alert Time: < 45 seconds

Power Consumption: 280 W @ 28 Vdc

Information supplied by Ericsson

Falcon Eye FLIR

Manufacturer: Texas Instruments, Dallas, Texas

Intended Application: USAF, F-16

Qualification: Flight qualification with Mil-Std 210A, 785A

Availability: Proof of principle demonstration hardware

Description:

Falcon Eye is a FLIR for a fixed-wing fighter aircraft to provide for air-to-ground, air-to-air and air-to-ship roles. The gimbaled sensor head and all electronics are made to be housed inside the F-16 as a distributed system. A protruding 13 cm dome, housing the optics, is only exterior change to the aircraft and provides minimal impact to aircraft operations. The line of sight is steered by the pilot's head, and the image is displayed on his helmet in front of his eyes. This gives the pilot a more natural way of seeing in the dark compared to the "fixed knothole" phenomena achieved by fixed forward navigation FLIRs that are displayed on the aircraft's heads-up display. Falcon eye is pictured in Figure 10.4.

Characteristics:

Dome Diameter: 13 cm

WFOV: 22.5 × 30 ° (1× magnification)

NFOV: 4 × 5.4 ° (5.63X magnification)

Total Weight: 65 kg

Power Requirement: <50 W @ 28 Vdc; 1 kV, 3 phase, 400 Hz, 115 Vac

System Cooling Method: Aircraft supplied cooling air

Slew Rate: 173 degrees/second

Slew Acceleration: 35 rad/s^2 (az) and 15 rad/s^2 (el)

Image Stabilization: < 50 μrad (NFOV) and <200 μrad (WFOV)

Information courtesy of Texas Instruments

FSI 2000 Series (AN/AAQ-21) FLIR

Manufacturer: FLIR Systems Inc., (FSI) Portland Oregon

Intended Application: Para-military and military

Qualification: Flight, Mil-STD-810C

Availability: Custom made per order; delivery ≈ 4 months ARO (or less)

Description:

FSI is a leading manufacturer of simple basic generic FLIRs that are priced for the common man. They have increasing sales to police, border patrols, and U.S. and international military organizations. FSI is willing to modify as a customer requires; active stabilization, embedded CCD TV cameras, laser illuminators, and staring arrays are potential options. The 2000 series has been mounted on various planes, helicopters, land vehicles, and ships. The 2000G has a HgCdTe TDI FPA made in house. They use a closed-cycle split Stirling cryocooler. Unlike many FLIRs costing much more, there is no scheduled servicing. The warranty is for 1,000 hours or one year, with many users achieving 3,000 hours before the first need for repair.

Characteristics:

Bandpass: 8–12 μm

FPA: HgCdTe 2 × 4 TDI

FOV: 5° × 2.7° or 7° × 3.8° to 28° × 15°

FOR: +120° to –120° (az) and +30° to –180° (el)

Turret Diameter: 36 cm

Weight: 23 kg for gimbal and sensor head, 35 kg for entire system

Resolution: 250 μrad narrow field to 1.4 mrad in the wide field

FOV Change Time: 0.12 seconds

Footprint: 350 × 343 active pixels in frame

MRTD: 0.2° C @ 2 cy/mrad in narrow and 0.16°C @ 0.36 cy/mrad in wide

Video output: RS 170 or CCIR

Gimbal Slew rate: 60°/ second

Information courtesy of by FLIR Systems Inc.

FSI SAFIRE (Shipborne/Airborne Forward-Looking InfraRed Equipment) (AN/AAQ-22) FLIR

Manufacturer: FLIR Systems Inc., (FSI) Portland, Oregon

Intended Application: Military ship and air

Qualification: Mil-Std-810D and Mil-Std 461

Availability: Custom made from continuing line; delivery ≈ 6 months ARO

Description:

SAFIRE does for the military user what the 2000 series does for police and paramilitary users—makes thermal imaging affordable. It is a fully qualified military system with all the features and options that are expected, such as active stabilization, digital image processing, fully interchangeable LRUs, interfaces, BIT/FIT, wash/wipe and de-ice. SAFIRE is suitable for long-range target detection, classification, and tracking utilizing advanced digital processors. The digital architecture permits built-in standard and optional features such as adaptive filtering, freeze frame, digital zoom, panoramic scan and display, auto scan, and advanced tracking with thermal cuing.

Characteristics:

Bandpass: 8–12 μm

FPA: HgCdTe, 4 by 4 TDI

Active Stabilization: Active 3 or 4 axis

FOV: 5° × 3° and 28° × 16.8°

FOR: 360° (az) and +30° to –120° (el)

Resolution: 500 true pixels with a 4:3 aspect ratio, providing 166 μrad in the narrow FOV and 930 μrad in the wide FOV

Turret Weight: 40.5 kg

Information courtesy of FLIR Systems Inc.

HNVS (Hughes Night Vision System) FLIR

Manufacturer: Hughes, El Segundo, California

Intended Application: Various basic FLIR applications

Availability: Custom made per contract; delivery ≈12 months ARO

Description:

The An/AAQ-16 HNVS is a generic-purpose FLIR adaptable to many applications. It is a stabilized turret-style sensor head with a system electronic unit for power, control, and data formatting. It has been effectively employed on helicopters and airplanes. The AN/AAQB model of HNVS, which added performance and reliability improvements to the original system, was first delivered in 1989. A new version of that system, called Hi-Mag, has also been developed and offers a third, high-magnification capability for long-range target identification. It also provides an enhanced autotracker, improved stabilization, and growth capability for a laser.

Characteristics:

Magnification: 1×, 6×, 16× (Hi-Mag option)

Turret Diameter: 32.9 cm

Video Output: Composite 525, 875, 625 optional @ 50 or 60 Hz

Turret Weight: 24.5 kg

Total Subsystem Weight: 45.5 kg

Information Supplied by Hughes Aircraft Company

IRSCAN IRST

Manufacturer: Hollandse Signaalapparaten, Hengelo, The Netherlands

Intended Application: Ship-based IRST

Qualification: International military standards For shipborne equipment

Availability: Custom made

Description:

The IRSCAN is a ship-based IRSTS. However, it does offer an imaging mode akin to a standard FLIR. The system consists of a stabilized, post-mounted sensor head, a a cooling cabinet, and a electronics cabinet. The IRSCAN is a fully passive infrared surveillance system capable of automatically detecting and tracking air targets. It detects the IR radiation form the surface skin aero-heating of missiles and aircraft. Advanced signal processing techniques result in reliable and accurate detection and tracking performance. The system provides automatic alerts on up to 32 tracks.

Characteristics:

Bandpass: 8–12 μm

Scan Speed: 78 rpm

Elevation Coverage: 14°

FPA: 1,024-element staggered HgCdTe

Cooling: Via closed-cycle Stirling

Detection Range: Typically 15 km for aircraft and 12 km for missiles

Spatial Accuracy: < 1 mrad

Reaction Time: Typically 2.5 seconds from first detection until alert is presented

Sensor Head Dimension and Weight: 48 × 104 × 37 cm and 95 kg

Electronics Cabinet Dimensions and Weight: 75 × 189 × 45 cm and 200 kg

Supply Cabinet Dimensions and Weight: 75 × 79 × 44 cm and 75 kg

Chiller Cabinet Dimensions and Weight: 74 × 113 × 61 cm and 190 kg

Control Panel Dimensions and Weight: 26 × 21 × 53 cm and 10 kg

Power Requirements: 4.4 kW max

Information courtesy of Hollandse Signaalapparaten

IRTV-445G FLIR

Manufacturer: Inframetrics, North Billerica, Massachusetts

Intended Application: General purpose LWIR FLIR

Qualification: Flight qualification

Availability: Custom made per contract; delivery ≈ 1 to 2 months ARO

Description:

The IRTV-445 G is a small turret-mounted FLIR system. It is easy to install and use and is TV compatible. It employs serial scanning of a four-element LWIR PC HgCdTe. The 23 cm gimbal provides rate stabilization for dc to 30 Hz. A 2:1 electro-optical zoom is offered as an option. The basic package contains three LRUs: the turret, a pistol grip for control, and an electronics control unit.

Characteristics:

WFOV: 21° × 28°

NFOV: 5.25° × 7.0°

IFOV: 2 milliradians in wide, 500 μrad in narrow

MRT: 0.3° @ 0.25 cy/milliradian

FOR: Continuous 360° in azimuth and elevation

Operating Temperature: –20°C to +50°C

Gimbal Assembly Weight: 13.6 kg

Support Module Weight: 5 kg

Control Grip Weight: 0.9 kg

Price: ≈$125,000

Information courtesy of Inframetrics

LANTIRN (Low Altitude Navigation and Targeting Infrared System for Night) FLIR

Manufacturer: Martin Marietta, Orlando, Florida

Intended Application: USAF, UKAF

Qualification: Flight qualification

Availability: Custom made per contract

Description:

LANTIRN is an advanced dual POD-mounted system. The pods are completely self-contained—one for navigation and the other for targeting. The navigation contains a wide FOV FLIR and a radar. The targeting pod allows aircraft to navigate, acquire, track, and provide fire control against ground targets and provides an automatic handoff to the IR Maverick missile. The entire LANTIRN production is accomplished in a paperless computerized factory.[A, B]

Characteristics:

FPA: HgCdTe common module[A]

Cost: ≈$4 million each[B]

Number of Pieces in Nav Pod: 9500, ≈twice for the targeting pod[A]

Number of Working Days Needed to Complete Nav Pod: 70[A]

Dev Cost: $94 million, 700 pods purchased under $3.16 billion authorization [A, B]

Information from (A) May 30, 1990. "LANTIRN Lights the Night." Flight International, *22–26; and (B) W. Scott. April 17, 1989. "Falcon Eye FLIR, GEC Helmet Aid F-16 Mission Flexibility."* Aviation Week and Space Technology.

Mast Mounted Sight FLIR

Manufacturer: McDonnell Douglas, Santa Ana, California

Intended Application: Generic search and track with various basing options

Qualification: U.S. military

Availability: Custom made per contract

Description:

McDonnell Douglas adapted a helicopter "above-rotor" turret-style system to ship and vehicle mounted systems. The Mast Mounted Sight is a day/night reconnaissance and targeting electro-optical search, acquisition, verification and tracking system. The sensor package includes a television sensor for day missions and a thermal imaging providing infrared images for detection and recognition at night and in adverse weather. Additionally, a laser rangefinder designator has been integrated for targeting missions with "on-gimbal" autoboresight capability for unsurpassed targeting accuracy. The sensor are integrated with user-friendly displays and controls that can be handled by a single operator. The system employs a patented superior stabilization, which extends recognition and hands-off targeting ranges.

Characteristics:

None Available

Information supplied by McDonnell Douglas

MOSP (Multimission Optronic Stabilized Payload) FLIR

Manufacturer: Israel Aircraft Industries LTD, Tamam Plant, Israel

Intended Application: Generic EO and IR payloads and applications

Qualification: Flight Worthy

Availability: Currently In production

Description:

MOSP is a turret-style stabilized FLIR/TV and/or laser sensor system. The MOSP modular design and payload structure allow interchanging of sensors, enabling various configurations. It has been successfully used on Helicopters, UAVs, and Aircraft. The basic configuration of the day/night payload was the winner of the short-range program of the DoD UAV in the USA. It features a look-down capability through nadir, a four-axis gimbal system, and an automatic TV/FLIR tracker.

Characteristics:

Stabilization: Better than 25 μrad

Elevation: +15° to –105°

Azimuth: Unlimited (n × 360°)

Two Axis Slew Rate: 60°/second

Two Axis Tracking Rate: 6°/second

Two Axis Acceleration: 60°/second

Turret Diameter: 36 cm

Turret Height: 54 cm

FLIR Bandpass: 8 μm to 12 μm or 2 μm to 5 μm

Dimensions: 36 cm in diameter and 54 cm in height

Weight: 32–40 kg

Power Required: 200 W avg, 500 W max

Information supplied by IAI

Sea Owl FLIR

Manufacturer: GEC Sensors Ltd., Essex, United Kingdom

Intended Application: Helicopter-borne target detection, identification and tracking of surface targets

Qualification: U.K. Ministry of defence standards for naval helicopters

Availability: Custom made per contract; delivery ≈6 months ARO

Description:

The Sea Owl is a helicopter-based thermal imager consisting of a turret-based sensor head and two remotely located electronic boxes. The system employs dual field of view optics and advanced signal processing to provide wide-area surveillance, detection, and identification.

Characteristics:

Bandpass: 8–3 μm

FOR: +20° × –30° in el and +120° × –120° az

Max Slew Rate: 60°/second

Stability: 50 μrad rms typical on both axes

MRTD: 0.1°C or better

FOV at 625 Lines, 50 Hz: 8° × 12° and 1.33° × 2°

FPA: 8 bar SPRITE / TED

Turret Weight: 64 kg

Signal Processor: 17 kg

Tracking Unit Weight: 13 kg

Compressor Weight: 10 kg

Power Required: 900 W

Information supplied by GEC Sensors

Second Generation Tank Sight (SGTS) FLIR

Manufacturer: Hughes, El Segundo, California

Intended Application: Passive long-range target acquisition for U.S. military tanks

Qualification: U.S. military

Availability: Developmental

Description:

The SGTS is applying second-generation FLIR technology to a tank sight. The system consists of a sensor head with a FLIR, TV sensor, and eye-safe laser ranger in a periscope-like configuration. This is a modern, high-performance upgrade to the Electro-Optical Tracking System (EOTS). EOTS is equipped with a dual-mode tracker that can maintain track on aircraft flying through clutter such as hills or trees.

Characteristics:

Stabilization: Two-axis, digital

Elevation Travel: –12° to +22°

Azimuth Travel: n × 360°

LOS Stabilization: Less than or equal to 50 μrad

FPA: Second-generation 480 × 4, PV HgCdTe in a standard advanced dewar assembly

Bandpass: 8–12 μm

NFOV: 2.5° × 3.3°

WFOV: 7.5° × 10°

Magnification: 8.65 X/2.9 X

Information courtesy of Hughes Aircraft

Simrad TC10 Thermal Imager FLIR

Manufacturer: Simrad Optronics A/S, Norway

Intended Application: Long-range naval and coastal artillery applications

Availability: Custom made per order, series production since 1986

Description:

The TC10 is designed for high resolution and accurately distinguishing thermal differences. The output is displayed on a TV monitor, and it is normally mounted on a stabilized platform. The detector is cooled by a closed-cycle cooler.

Characteristics:

Bandpass: 8-12 μm

Focal Plane: HgCdTe

FOV: 1 × 3°

Spatial Resolution: 0.25 millirad

Thermal Resolution (MRTD): 0.08°C @ 1 line pair/mrad

Focusing: Infinity

Weight: 14 kg for camera, 10.5 kg for electronics unit, and 16 kg for control and display unit

Information courtesy of Simrad Optronics A/S

TADS/PNVS (Target Acquisition Designation Sight/Pilot Night Vision Sensor) FLIR

Manufacturer: Martin Marietta, Orlando, Florida

Intended Application: Military helicopter

Qualification: Flight qualification

Availability: Custom made per contract

Description:

TADS/PNVS is a helicopter-based navigation and fire control system. It is used to provide vision to the pilots during night operations or in adverse weather conditions as well as to lock onto and track targets. The sensors are a visible, a FLIR, and a laser rangefinder mounted on a frontal turret. The PNVS is head steered. The TADS/PNVS development and production has cost over $1.5 billion.

Characteristics:

TADS:

FLIR FOV: 3.1° narrow, 10.1° medium, 50° wide

FOR: ±120° (az), +30 to –60° (el)

PNVS:

FOV: 30 × 40°

FOR: ±90° (az), +20 to –45 (el)

Information courtesy of Martin Marietta

Thermovision® 1000 FLIR

Manufacturer: AGEMA Infrared Systems, Sweden; marketed through regional sales representative and direct sales office in Secaucus, New Jersey

Intended Application: Surveillance, search and seizure, and rescue

Qualification: Industrial

Availability: Regular production

Description:

The Thermovision 1000 is a high-definition FLIR effective in long range, nighttime surveillance operations and under hazy daytime conditions. All imager components, including its 12-bit signal processor, microprocessor, and integrated dual field of view lens assembly, are housed in a single unit. In its specially configured, gyro-stabilized, gimbaled pod, the Thermovision 1000 is approved for on-board and external installation on a variety of helicopters, fixed wing, and remotely piloted aircraft. It can meet the exacting requirements of full military missions and covert and civil operations.

Characteristics:

Cooling Method: Stirling microcooler

FOV: Dual, 5° × 3.3° and 20 ° × 13.2°

Spectral Bandwidth: 8–12 μm

Price: $90,000 to $100,000

Information courtesy of AGEMA

TIALD (Thermal and TV Imaging Airborne Laser Designator) FLIR

Manufacturer: GEC Ferranti, United Kingdom

Intended Application: Military aircraft

Qualification: Military

Availability: Custom made per contract; delivery negotiable

Description:

Like the U.S. LANTIRN, TIALD is a pod system which hangs below the aircraft. A Television, FLIR and Laser are packaged in the front and share a common optics and forward window. The electronics for the servo control and image processing are in the middle of the pod with the power supplied at the end. Active de-rotation of the image is included to provide the user with a horizontal flat field. TIALD is pictured in Figure 10.3.

Characteristics:

FLIR/TV FOV: Selectable, 3.6° × 2.4° and 10° × 6.7° with ×2, ×4 zoom

Field of Regard: + 30° to –150° in el and continuous roll

Pod Dimensions: 2.9 m length, 0.305 m diameter

Pod Weight: 210 kg

Power Required: 3.3 kW peak, 1 kW average, and 0.2 kW transit

Information supplied by GEC Ferranti

Tiger FLIR

Manufacturer: SATEL Group, led by SAT, Paris France

Intended Application: Trigat anti-tank/medium range weapon system

Qualification: French military

Availability: Custom made per contract from line, delivery ≈ 12 months ARO

Description:

Tiger is a thermal viewer/sight designed for integration into the fire control system. It provides a thermal picture reconstructed on a flat-panel LED screen. Cooling can be provided wither by a Joule-Thompson blow-down or a split Stirling.

Characteristics:

Fields of View: 4 × 8° and 1.5 × 3°

Focusing: 50 m to infinity

MRT: 0.1 K

Detector: 48 × 4 HgCdTe IRCCD

Spectral bandpass: 8–12 μm

Power Consumption: 15 W for J-T version

Cool-Down Time: < 1.5 minutes with J-T, less than 5 minutes for Stirling

Dimensions: 46.5 × 32 × 18 cm

Weight: 6.5 kg without cooler

Information courtesy of SAT

VAMPIR (Veille Air Mer Panoramic Infrarouge) ML Thermal Imager IRST

Manufacturer: SAT and CSEE, Paris, France

Intended Application: Various military

Qualification: Fully applicable to French military specs

Availability: Development by SAT and SCEE, nearing completion

Description:

The VAMPIR ML family of lightweight infrared systems is designated for panoramic search applications aboard surface vessels. This equipment is derived from the VAMPIR currently in service with the French Navy. It is based on a modular design concept, enabling it to meet the requirements of any type of warship. It can be configured in a monospectral or dual bandwidth version (3–5 or 8–12 μm). The unit is fully passive, thus solving problems posed by electronic warfare (radar silence, jamming, antiradiation missiles).

Characteristics:

Normal Detection Range for Sea Skimming Missile: 10 km

Number of Threats Processed Simultaneously: 40 max

Reaction Time between Initial Detection and TD: 2 seconds

Elevation Coverage: –10° to +60°

Azimuth Coverage: 360°

Elevation Field of View: 6°

Target Designation Accuracy: 3 mrad

Information courtesy of SAT

WF-360TL FLIR

Manufacturer: Westinghouse, Orlando, Florida

Intended Application: Fixed-wing air platforms

Qualification: U.S. military flight

Availability: In inventory; delivery ≈1 to 2 months ARO

Description:

The WF-360 is a generic, gen one, turret-designed FLIR for general surveillance, based on U.S.A. common modules. Options include a laser rangefinder, various televisions, a digital tracker, zoom lens, and a reduced-weight, single-FOV FLIR. Versions of the FLIR have been flying drug interdiction and coastal surveillance missions for several years. Over 25,000 hours of successful operation have been conducted. It can be equipped with a laser and/or television camera.

Characteristics:

LWIR Bandpass: 8 to 10 μm

NFOV: 2.7° × 3.6° with 11X

WFOV: 11.1° × 14.8° with 2.7X

NEDT: <0.11 K, MRT @ 1/2 Fc0 < 0.33 K

Cooling: Split Stirling

Turret Size: 41 cm dia., 56 cm H

Azimuth Freedom: 360°

LOS Stability: 20 μrad

Approximate Cost: $550,000; day TV and eye-safe laser rangefinder plug-in modules available at extra cost

Power Consumption: 120 to 1,400 W

MTBF: 335 hours

Weight: 91 kg excluding display

Information courtesy of Westinghouse

11

Space-Based Sensors

"Space isn't remote at all. It's only an hour's drive away, if your car could go straight upwards."

Sir Fred Hoyle

11.1 TECHNOLOGY BASICS

11.1.1 Introduction

In 1967, the Hughes Far Infrared Search and Track Sensor became the first long-wavelength sensor to be successfully flown in space ["Eye In The Sky," 1]. Since then, several nations have orbited hundreds of electro-optical sensors. The several advantages to space basing infrared sensors include

- the ability to specifically tune the orbit to cover a ground swath in the most optimal spatial or temporal way
- a lack of atmospheric effects on observation
- global coverage
- the ability to engage in legal clandestine operations

Because international rules of aviation do not extend into space, there are presently no legal constraints concerning overflight of satellites. Satellites are relatively safe from attack, because no anti-satellite weapons are currently fielded by any country. Among the disadvantages are protracted schedules and excessive costs to manufacture, launch, and maintain satellites. Space basing facilitates covert operations, but maintenance, repair, and upgrade is difficult, if not impossible, and very expensive.

A representative (but notional) space sensor architecture is shown in Figure 11.1. Although frequently completely self-contained, the architecture of 11.1 does indicate that some sensor control processing and image processing can be done by separate units located within the spacecraft, but separate from the sensor. To reduce mass and power consumption, control of many different components are often done by a single box. There is no intent in Figure 11.1 to depict the actual architecture of any individual space sensor; in fact, no sensor is likely to have this exact architecture.

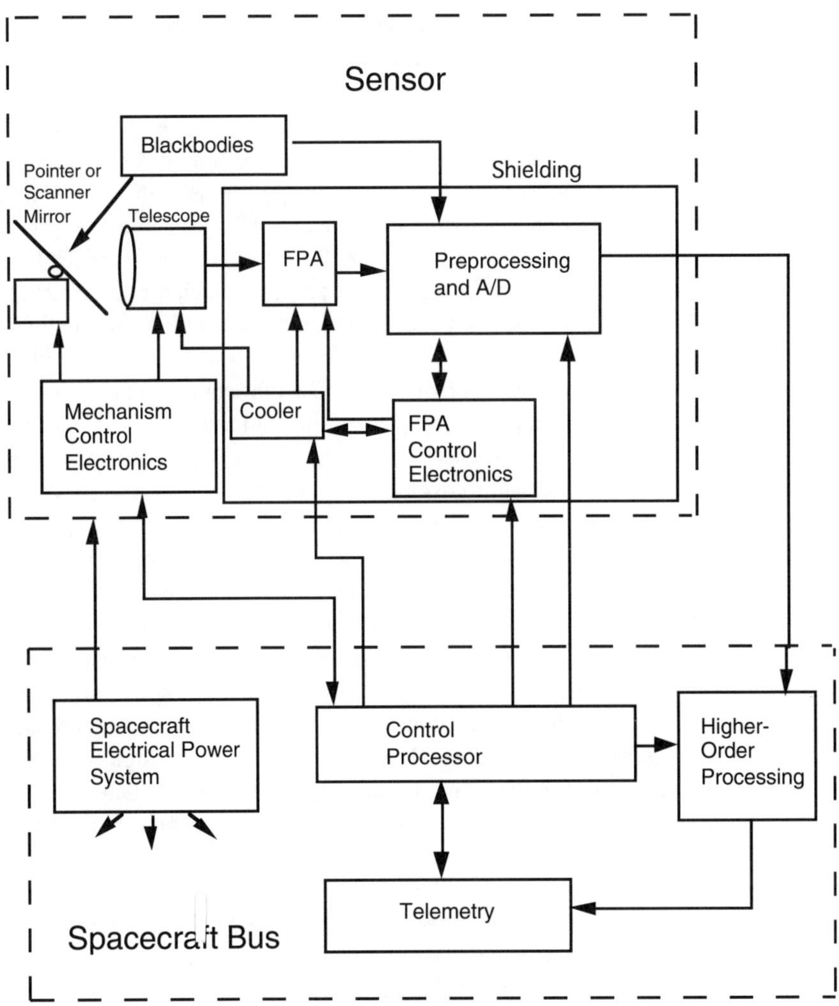

FIGURE 11.1 Representative Space Sensor Architecture

11.1.2 Design and Fabrication Process

Perhaps the biggest limiter in the widespread use of space sensors is the inaccessibility of space itself. Launch costs are steep, plus there are limited launch opportunities and few spacecraft manufacturing and test facilities. A space vehicle is, unfortunately, still a rather rare creature. The culture and situation feeds on itself in a circular pattern, analogous to the situation described in Chapter 1. The fewer the existing opportunities for launch, the greater the desire to test and qualify the vehicle. This, in turn, leads to shortages in S-level parts, test installations, and fabrication facilities. The more a space mission costs, the fewer space missions humankind can afford to conduct. The fewer space missions to be conducted, the more they cost. On and on, until commercialization, SDI, or something else breaks the circle.

A new design for a traditional Earth-orbiting space sensor now takes about a decade to develop from funded concept to operation. This time is approximately doubled for an interplanetary probe. The sensor design begins with excruciating detail to materials, interaction with other subsystems, and a gargantuan labyrinth of studies and trade-offs leading to a final design. Mechanisms and devices tend to grow (and grow) in complexity throughout the program. Each time a design review is held, another back-up part or procedure seems to get added. Simple lightweight concepts turn into massive Rube Goldberg monsters. Complicated or critical electrical boards and mechanisms are verified through computer simulation and breadboarding. During this entire time, the engineer's work is checked and rechecked by internal reviews, customer reviews, and independent companies specifically hired by the customer to oversee technical work.

Space sensor hardware is fabricated by engineers and highly skilled technicians, in clean rooms, with largely custom-made tooling. Fabrication is highly controlled and documented with paperwork, which literally can weigh more than the spacecraft. Every precaution imaginable is taken and documented.

Continuous testing of piece parts, subassemblies, integrated sensors, and spacecraft are the hallmark of a space project. Even for a single spacecraft, there are often one or more identical "qual" units which are extensively tested and used for engineering exercises. This results in low risk, high reliability, high costs, and excellent exhibits at the Smithsonian Air and Space Museum. There are two schools of thought on this. First is the traditional excessive testing philosophy, bolstered by the Hubble fiasco, and second is the new commercialization philosophy. The traditionalists state that the testing is done to ensure functionality and operability and that this is the best course for future systems. They point out that integrated testing would have revealed the spherical aberration of the Hubble. The commercialists believe that many of the excessive tests do not add value—just cost. The resulting paperwork does not add reliability; it only documents the reliability that was already inherent in the hardware. Time will tell.

Successful alternatives to the traditional processes have been used by amateur satellite builders and by the SDIO. These schools of thought evolved in the eighties in reaction to the high costs and long schedules imposed by the detailed and deliberate traditional process. SDIO, universities and amateur scientists have been successful by merely doing things right without the detailed paper trail and verification testing process. Several scientific sensors have been built in less than a year and have successfully operated in space with limited lifetimes. The key to success with this method is to base these systems on modified designs and to use experienced space engineers, managers, and technicians who know the pitfalls to avoid and how to select the minimum of critical tests to ensure proper operation.

Another opposing duality in space systems is the use of advanced technology. Americans traditionally push the state of the art for space systems, while the defunct Soviet Union (and currently the CIS) adapts existing technology to allow lower cost and faster deployment, usually at a measured performance sacrifice. The Soviet SS-6 launch vehicle is an example. More than 1,000 have been produced over the last quarter century, modified as needed from existing hardware [Johnson, 2]. For the past 30 years in the Soviet Union, and now in the CIS, concepts of generic spacecraft buses have been put into practice. The generic Soyuz space vehicle can be configured for various manned or unmanned missions. Development of a generalized bus provides for large schedule compression, risk reduction, and cost reduction. The basic subsystems of a space vehicle can be built and assembled from previous designs in months instead of years and be highly adaptable to various payloads. This is a wonderful commercial idea; it is analogous to a car company placing a given engine in several models. Actually, this is how most modern space launch vehicles are made—relatively independent of payloads, then custom configured with special interface modules, fairings, and additional strap-on boosters if needed. Although movements have been afoot in the USA and Europe since the seventies for a "standard" or pseudo-standard spacecraft bus, few practical applications have resulted. In contrast, the former Soviets concentrated on economy and stockpiling spacecraft subassemblies—and even whole spacecraft—for rapid turnaround times. For instance, in 1981, a navigation satellite was replaced in less than two months, and it has been said that more valuable military assets could have been replaced in a manner of days [Johnson, 3].

The commercialists feel that launch costs will decrease through innovative, low-cost launch systems such as Pegasus. When launch costs are low, it may become cost-effective in the long run to test less and to require less stringent controls on payloads. The result will be a low-cost payload and launch vehicle with a relatively high failure rate. An increase in failures will result in replacement of the spacecraft. However, it can be cost-effective to discard and replace space assets if the assets and the launch costs are low enough. Generally, to provide a life-cycle cost savings, each replaceable satellite must be cheaper than its predecessor by means of a learning curve, and the launch costs should be around $1,000 per kilogram or less. Trends and developments indicate that it is quite possible that these conditions will be ripe for exploitation in the late nineties for small, low Earth-orbiting satellites.

11.1.3 Launching and Orbiting

There are several operating launch sites throughout the world. Plesetsk, in the CIS, is still the busiest spaceport in the world. Other major Earth spaceports include KSC (Florida), Baikonur (CIS), Vandenberg AFB (California), ESA's Guiana Space Center (Kuroor, French Guiana), Tanegashima (Japan), Wallops Island (Virginia), White Sands (New Mexico), Poker Flats (Alaska), Kapustin Yar (CIS), San Marco (Kenya), Palmachim (Israel), Shar Center (India), Jiuquan (PRC), Tiayuan (PRC), Xichang (PRC), Kwajalian Atoll, and others in located in Norway, South Africa, and Australia. Most launch facilities are located near the equator to gain a kick from the Earth's rotational velocity. This tends to place them in the tropics or subtropics, where it is usually hot and humid, and always subject to nasty weather.

Regardless of the manufacturing process and test philosophy, after final testing, the spacecraft is integrated with the launch vehicle. This is usually done at the launch range. The space vehicle is encapsulated in an aerodynamic shroud for protection. The shroud is jettisoned on the way to orbit.

After integration and check-out, the rocket is deployed on its individual launch pad. This deployment causes special concerns for infrared sensors. The perfect launch range would be clean, dry, and cold. Only a few are cold, and none is especially clean or dry. For example, KSC and the French Guiana sites are both on coastal swamps. Even some new sensors (e.g., the Infrared Space Observatory shown in Figure 11.2) employ perishable cryogens, and sitting in the hot sun causes even more rapid loss of stored cryogens. Water vapor can condense on various surfaces, causing excessive outgassing in orbit and potentially resulting in optical, electronic, or mechanical failure. Additionally, particulates such as sand have frequently blown inside shuttle bays and unmanned rocket shrouds. Even when bad weather does not damage the spacecraft, attempts to protect it from the environment can inadvertently cause damage or failure. This occurred when hurricane Andrew potentially threatened the Mars observer deployed at the Kennedy Space Center. The hurricane veered south and never touched the cape, but the space probe was damaged by hurried workers trying to protect it from the storm.

During launch, sensors must survive extreme vibration and acceleration (e.g., 10 G acceleration and 10 G rms vibration). Solid rockets usually produce a rougher ride than do liquid propelled systems. Sub-orbital flights follow a ballistic parabola path and fall back to Earth, usually without obtaining an altitude of more

FIGURE 11.2 The Infrared Space Observatory (ISO) Spacecraft (figure courtesy of ESA)

than 200 km. Interplanetary vehicles usually are placed into an Earth obit then essentially launched again toward their destination. Earth orbiters are usually placed into an orbit that closely approximates the desired final trajectory and eventually tweaked to their exact orbit by firing engines and thrusters.

A spacecraft's speed in Earth orbit depends on the orbit altitude by a simple equation of

$$V \approx \left(\frac{398200}{6370 + A} \right)^{1/2} \tag{11.1}$$

where

V = orbital velocity in kilometers per second

A = altitude above Earth's surface in kilometers

Thus, a spacecraft's orbital velocity is determined only by the altitude. This is true because any falling object's acceleration is determined only by gravity and not by any "object-controlled" feature. Also note that the constants in Equation (11.1), which are in units of kilometers for convenience, have been combined for ease of use. Therefore, this equation only applies to the Earth. More generalized equations can be found in astronomy and physics texts which can be used for any body orbiting another.

One can see that the lower the orbit, the faster a satellite must move. Equation (11.1), then, shows that an IR sensor will cover more of the Earth in a given time, in a lower Earth orbit because it must move faster. Since it is at a low altitude, the range to the ground is less than it is for a high altitude. This results in a smaller footprint, meaning a higher ground resolution, than the same sensor would at an higher altitude. The satellite will also complete an orbit of the Earth faster, which may allow more rapid revisit rates. However, a low orbit also means that the flight over a specific area (like a target) is rapid, and the total field of view is small. Other downsides to lower orbits are the increased interaction with atomic oxygen, induced drag from the rarefied (but existent) upper atmosphere and glow effects (see Section 11.2 for a discussion of the environments). The lowest orbits used today are usually 200 km or so, and the highest orbits are usually those at the geosynchronous altitude of ≈39,000 km.

If one plugs in 38,500 km as the altitude into Equation (11.1), the velocity becomes a little less than 3 km/s. At this circumference, it would take just about 24 hours to complete an orbit—approximately the same time it takes the Earth to rotate, thus matching the Earth's east-west rotation. The net effect of this orbit configuration is that the spacecraft hovers continuously at one point above the earth. About one-third of the Earth can be usefully viewed from a geosynchronous orbit. At this high altitude, there is no effect from the atmosphere. Also, radiation effects are lower than in orbits that cross through the Van Allen belts, but there is reduced protection from solar radiation. The main disadvantages of geosynchronous orbits are high launch costs owing to the extra fuel that must be carried to lift the satellite that high and poorer ground resolution (larger footprint) due to the great distances. Also, the polar regions of the Earth are not visible.

To confuse matters a bit, orbits really are not circular, as frequently shown in drawings and news shows, and as implied by Equation (11.1). They are elliptic, with the object they orbit at one foci. This means their altitude varies, and so does their velocity! Intelligence sensor platforms from the defunct Soviet Union, and presumably now CIS, use an ingenious orbital solution to the competing requirements for low-drag, low-altitude observation. They use a Molniya orbit, which is extremely elliptical, taking the vehicle very low then high again. The orbit is configured so the low points occur over the area they wish to observe, giving high resolution. The orbit quickly takes the satellite to a higher altitude. The satellite is exposed to effects from the rarefied atmosphere for only a short time on each orbit. One drawback is that these satellites frequently cross the radiation belts and, during long missions, could accumulate a large dose. The ex-Soviet satellites of these types never had long missions, so this was not an issue.

Regardless of orbits, satellites usually die from the effects of drag depleting fuel, ground controllers using up fuel, catastrophic failure, or slow death by accumulated radiation effects. Deep space probes suffer the same effects except for that of drag; however, they have the added problem of drawing energy from a dimmer and dimmer power source (if solar powered) as they fly away from the sun.

For Earth orbiting satellites, orbital maintenance can be performed to extend their lifetime and improve performance. The high cost of orbital maintenance can be justified by performance improvement and the increased lifetime of a satellite. Recently, this became a reality for low Earth-orbit systems with the American Space Shuttle. The Solar-Max and Hubble satellites were serviced by astronauts. In fact, the Hubble is the first electro-optical space system designed for instrument replacement and minor repairs by astronauts. For orbital maintenance to be justifiable and practical, it must be considered from the beginning. Plans, provisions, and designs must be made early in the concept phase of a program to accommodate in-orbit serviceability by ensuring the key attributes of modularity and accessibility. Orbital maintenance is largely a ground operation, with the activities taking place in orbit being "just the apex of a pyramid that envelops flight spares, tools, designs, logistics systems, space and ground support equipment, and skilled manpower" ["Designing an Observatory for Maintenance in Orbit," 4]. These efforts must complement and overlap each other for an effective operation.

11.1.4 Applications

Infrared sensors on space platforms usually perform one of four basic functions: military/intelligence gathering, astronomy, Earth environmental/resources sensing, or weather monitoring. All except astronomy can be classified as forms of Earth remote sensing.

Intelligence and military services from wealthy nations have long employed space-based sensors to acquire information. It is believed that the U.S. spends about $5 billion per year on space reconnaissance, with a total of about $100 billion since 1960 [Richelson, 5]. This traditionally has been linked to long-term strategic monitoring to assess military and economic might. Although the cold war is over, these are still important and are being recognized as such by less wealthy nations. There is an ongoing proliferation in space-based reconnaissance, with entering countries including France, Germany, India, Israel, Italy, Japan, PRC, Saudi Arabia, South Africa, Spain, U.A.E., U.K., and even a European consortium [Richelson, 6]. A proliferation of space surveillance makes it much harder for a nation to prepare for war in a clandestine manner or to cheat on a treaty. Gathering basic intelligence data from space now transcends the traditional use by the military and intelligence agencies, and the USA's National Reconnaissance Office (NRO) is now relying on new technologies to meet increasing demands in a decreasing budget environment [Scott, 7]. Intelligence gathering of crop data and weather trends from space has also been used by hunger relief organizations to more effectively forecast droughts and famines. Corporations are even buying Sattelite Probatorie d'Obervation de la Terre (SPOT) data to access the development of their competitors' facilities.

The military has used space-based surveillance for missile launches with the Defense Support Program (DSP) spacecraft pictured in Figure 11.3 (photograph courtesy of TRW). The military has also considered space sensors for weapon fire control. Since Desert Storm, this has also evolved to include timely tactical military information. Additionally, space basing provides excellent viewing geometries for global events such as nuclear explosions and environmental changes that the military is concerned about. Despite the collapse of the Soviet empire, missile warning is still important. General Kutyna, commander of the U.S. Space Command, said, "Today the missile warning function is perhaps the most important of those performed by military space systems" [Kutyna, 8]. He also indicated that knowledge of weather provided by defense weather satellites greatly optimized the application of aircraft and their weapon systems in Desert Storm [Kutyna, 9].

There are several unique reasons for conducting astronomy in space. The most important is to get above the effects of the Earth's atmosphere, which is an absorber, emitter, and scatterer of infrared radiation. Space-based infrared sensors allow viewing in colors that the atmosphere naturally absorbs. IR astronomy is needed to answer basic cosmological and astronomical questions concerning the formation of stars, protoplanetary disks, extra-solar planets, brown dwarfs, dust and interstellar media, protogalaxies, the cosmic distance scale, and ultra-luminous galaxies. The newest astronomical space sensor is ESA's ISO shown in Figure 11.2. ISO has an infrared telescope servicing several instruments in an architecture similar to the Hubble. It employs a dewar containing a cryogen to cool infrared instruments to near liquid helium temperatures.

Perhaps the most critical use of IR sensors in the nineties is scientific observation of the Earth's environment from space. Uncorrupted, continuous data gathering from the vantage of earth orbit greatly facilitates detecting the subtle changes

FIGURE 11.3 The DSP Spacecraft (photograph courtesy of TRW)

indicating environmental stresses and trends. Additional information concerning environmental and weather sensors is covered in Chapter 12. Oddly enough, space can actually be more economical for these long-term global measurements. A single platform can provide information of greater spatial resolution than could hundreds of airplanes or thousands of ground stations.

One interesting future application may be space vehicle collision avoidance. The geosynchronous and low-Earth orbits are becoming increasingly populated with debris. The debris consists of remnants of satellites, garbage from manned flights, booster separation parts, dust, loose pieces, strips of metal layer insulation (MLI), coatings, and paint. The larger pieces are tracked and their orbits documented. The smaller pieces are too small to track via ground-based radar yet are still large enough to cause damage. This poses a serious threat to anything operating in these orbits. A potential future use for IR sensors is to view the forward path of a satellite or space station for junk and debris and provide information to steer the spacecraft clear.

11.2 ENVIRONMENTS

The space environment is unique and contradictory. Dynamically, it is the most benign, but the ride to orbit is the most severe. There is no atmosphere, yet atomic oxygen is a serious concern for low Earth orbits. Although space is excessively clean from an IR perspective, spacecraft tend to be dirty, foul, belching assortments of nasty contaminants. And although space is the best vacuum known to man, it contains high-velocity molecules and dust. Finally, radiation is strong and persistent.

11.2.1 Thermal-Mechanical Environments

The high vibration of launch must be survived, but the IR sensor doesn't have to function during launch. This allows moving assemblies to be held firmly in place. This one-time securing is called "pinning," "caging," or "locking." Any secured device generally will be required to have backup release to prevent failure. It is of great embarrassment and cost for a safety mechanism designed to be used only during launch to subsequently prevent the sensor's proper function. Although these devices are needed, the engineer and manager should expect to use pyrotechnics and high-reliability, multiple removal mechanisms. Requirements for redundant, reliable mechanisms have become mandatory in the IR industry since CIRRIS's failed cover on its first launch on the Shuttle. The sensor worked fine, but no light could enter the telescope through the stuck cover. Everyone remotely involved in the mission was concerned and upset that such a trivial item could jeopardize the entire sensor's mission.

In addition to caging devices, if space and mass are available, shock absorbers may also be implemented on the launch vehicle or sensor. Generic launch environments include steady-state accelerations of 5 to 20 g, shocks of 500 G for <1 millisecond during staging, thermal variations of +50 to –50 °C within a few minutes, vibration of 0.5 g^2/Hz from 50 to 1,000 Hz, and a pressure drop from sea level to vacuum in a few minutes.

Once deployed, the vibration on a spacecraft is usually very low, especially when a sensor needs to be operating. Image stabilization on a space sensor usually requires less amplitude and bandwidth than that of a terrestrial FLIR. The only vibration that occurs is the result of spacecraft mechanisms or attitude control thrusters. Oddly enough, one large contributor to spacecraft vibration is often the sensor itself—the cryocoolers, scan mirrors, gimbals, and so on. Often, observation can be scheduled around the relatively short-lived vibration transients from these mechanisms. Spacecraft are vibrationally quiet enough to support sub-microradian optical pointing, as has been openly demonstrated [Scott, 10]. A properly designed spacecraft will dampen out any serious jitters within a few minutes after a vibration event. Major sources and durations of these "jitter events" are tabulated in Table 11.1.

The temperature of space is surprisingly warm to many people. A spacecraft in low Earth orbit usually has a temperature unexpectedly close to a comfortable room temperature. Interplanetary spacecraft traveling away from the sun will get much colder, and those traveling toward the sun may get much hotter. Thermal control engineers of an Earth-orbiting spacecraft bus are usually more concerned with getting the heat out, instead of generating heat. Continuously running electronic boxes buried in 50 to 100 cm of spacecraft stuff tend to get very hot, which requires excellent thermal conductivity to an exterior radiator. IR sensors tend to hang on the outside of a spacecraft bus, so they are, thankfully, a little cooler. The sensor's ambient temperature depends on the spacecraft geometry and the sensor's orientation. If a hot radiator is near the sensor, or the sensor is exposed to the

TABLE 11.1 Non-instrument Causes of Jitter in Space

Vibration	*Duration*
Antenna pointings	Transient or constant
Attitude control thruster firings	Transient
Drag make-up firings	Transient
Mid-course burns for interplanetary flights	Transient
Orbital changes	Transient
Reaction wheels	Constant with peaks and valleys
Solar array pointings	Constant or transient
Spun or de-spun components	Constant
Terminator crossings	Transient (depends on orbit altitude; can be as frequent as every half hour or so)

sun, it will be much hotter than if it were pointed at cold space. Solar array placement can be another bugaboo. The arrays must not reflect the sun into or onto the sensor or be allowed to radiate excessively onto the sensor. Occasionally, motors or other mechanisms may require heaters.

It is no surprise to anyone that space is a good vacuum. However, this fact must not be overlooked in design. Engineers must realize there will be no thermal transport via convection, requiring conductive or radiative heat sinking of all heat generating parts, including electronics to radiative elements. Second, the sensor must be designed to allow rapid outgassing after assembly. This dictates unrestricted paths for the flow of gas. It is unwise to include extremely high surface area materials (e.g. carbon) that will require days or weeks to outgas in a test vacuum chamber. If items like carbon desiccants are included, they should be kept cold when a vacuum is applied. Additionally, MLI, which is used frequently, can take long periods to properly outgas. Although most spurious failures encountered in testing are temperature related, there do exist gremlins dedicated to inducing vacuum failures. Such failures usually occur in connectors that work in the lab but fail by separation under vacuum, separation of thermal straps, and misaligned optics. Vacuum failures are especially irritating because they can't be fixed under a vacuum. One has to repressurize the sensor, fix the likely problem, evacuate the chamber, and retest.

11.2.2 Atomic Oxygen

Managers and engineers new to spacecraft design are frequently astounded to find their more seasoned counterparts so concerned about what appears to be an atmospheric effect. Effects resulting from the minuscule, rarefied atmosphere experienced in orbit are of great concern. Although minuscule and rarefied, something is there. In fact, at the shuttle orbit of 250 to 300 km, there are about a billion particles per cubic centimeter. As Hunton puts it, "Objects in low Earth orbit travel through a tenuous and yet highly reactive atmosphere. Some materials barely survive the trip; others glow along the way" [Hunton, 11]. Any surface exposed to atomic oxygen in low Earth orbit must be designed to be tolerant.

One effect germane to IR sensors is surface erosion, which can make a mirror or baffle coating disappear. When atomic oxygen combines chemically with other atoms or molecules (especially metals), it changes their optical and mechanical properties. Another specific effect on IR sensors from atomic oxygen is its attack on baffle coatings, which results in surface coating property changes. It can attack mirror coatings, changing a highly reflective aluminum surface to absorptive aluminum oxide. There have also been reports of highly kinetic oxygen migrating into materials and causing bulk effect problems. But perhaps the most notable effects are erosion of coatings and bulk material, whereby "most reactive organic materials lost about 12 microns of thickness during a 40-hour exposure" [Hunton,

12]. Both NASA and the ESA have much data on the effects of atomic oxygen and should be consulted both frequently and early in the design phase. Additionally, several high-energy atomic oxygen bombardment facilities exist around the country in which limited testing can be done on new coatings.

Silver is a poor choice for outer coatings in low Earth orbit due to its high reaction rate with atomic oxygen. MgF_2 and diamond react 1,000 times less [Musikant and Malloy, 13] and should be considered as coating materials. The effects of atomic oxygen vary greatly, depending on whether the molecules impact directly from the side or migrate down a labyrinthine path. Surfaces exposed in the "ram" direction or along the velocity vector are far more susceptible to damage. The effects are orders of magnitude less for surfaces not directly exposed in the ram direction. This is analogous to motorcycle riders getting more bugs in their teeth than in their ears or on the back of their necks. Optimally, a sensor would be mounted on the back of a satellite and look backward.

The shuttle and other spacecraft in low Earth orbit tend to glow because of reaction with atomic oxygen. This can be a limiting noise source for sensitive IR observations. It has been proposed that cold surfaces promote glow more than hot surfaces [Hunton, 14]. If true, this could develop into a serious concern for cool scan mirrors in low Earth orbit.

11.2.3 Radiation

Another harsh attribute of the space environment is nuclear radiation. In space, the sensor is bombarded with gamma rays, x-rays, UV, neutrons, electrons, protons, alpha particles, and higher mass ions—everything but the kitchen sink (and that might even become a problem in the future). The two types of effects are spurious transients and slow death from accumulated total dose.

Transient effects usually are limited to a single piece part experiencing a brief malfunction or "upset." When properly designed, this will not cause a "latch-up" of the electronic controls, nor will it make major spacecraft controls go awry. Total dose effects cause a slow degradation in performance until total failure of a piece part, subassembly, or the whole spacecraft occurs. With properly designed electronics, total dose is usually the worse of the two problems. These integrated effects depend on the dose rate, which is highly dependent on orbits. The dose rate, along with the length of time spent at all the dose levels, determines the total dose. For example, over a four-year mission, the MSX electro-optical satellite will receive exposure to between 2,300 and 77,000 rads (Si) [Anderson et al., 15]. The cumulative radiation dose can be far worse for solar orbiters and vehicles visiting near (or in orbit around) other planets such as Jupiter and Saturn. For instance, the Galileo spacecraft is expected to receive a dose of a megarad in only one hour when it is closest to Jupiter [Rasmussen, 16].

Total dose damage in electronics usually occurs via a cumulative buildup of charges and/or lattice displacement damage. Lattice defects can occur when the radiation knocks an atom from its normal lattice position to an interstitial position. This degrades the lattice and therefore the semiconductor properties that the manufacturer finely tuned. Atomic displacements in the semiconductor can be annealed out with an elevation in temperature. Radiation can also cause excess charge to build up in reaction to oxides. In extreme cases, the extra charge can lead to insulators becoming conductors and inducing short circuits. The less dramatic and more common effect is that threshold voltages will dramatically increase. This will result in simple semiconductor transistor failure. Most radiation-tolerant CMOS can withstand 15 kilorads (Si). Silicon can be hardened to about a megarad, and GaAs at least an order of magnitude harder [Fredricks, 17].

Several sources and kinds of electromagnetic and energized particle radiation must be considered. These include the solar wind, cosmic rays, trapped particles in the Van Allen belts, high-energy electromagnetic radiation, and UV.

The solar "wind" is a rarefied plasma of charged particles and ions flowing out from the sun. It is largely shielded by the Van Allen belts (actually, it pumps the belts). The effect for lower-altitude orbiters is increased Van Allen belt radiation levels. Solar flares emit protons and some alpha particles, heavy ions, and electrons. Solar flares are associated with the solar sunspot cycle and can be statistically predicted. Solar flares can douse the orbital environment with 100 rad/hr [Rasmussen, 18]. The magnetic field change induced by solar activity affects the Van Allen regions as well as geosynchronous and very high altitudes. The radiation fields take on a dynamic nature, with changes measured in minutes. Sporadic magnetic storms at geosynchronous orbit can produce a modulation of the electron flux above 50 keV by an order of magnitude within 10 minutes and are a common feature of the midnight-to-dawn portion of the orbit [Stassinopoulos and Raymond, 19].

Cosmic rays actually impart very little total dose but can cause many single-event upsets. This irritating radiation originates from outside our solar system and is largely attenuated on Earth by the atmosphere. Galactic cosmic rays are composed of about 80 to 85 percent protons and 10 to 14 percent alpha particles, with the remainder being heavier ions (e.g., carbon, aluminum and iron). All travel at relativistic speeds. All semiconductors (including detectors, charge coupled devices, microprocessors, and memory devices) are susceptible to cosmic ray-induced effects. High-atmosphere instruments such as used in astronomical observatories or airborne platforms may also suffer an increase in spurious transient upsets from cosmic rays.

The Van Allen belts that encompass the Earth are bands of high-energy trapped particles. Horseshoe shaped in cross-section, they consist of an inner zone (lower altitude) of trapped protons and an outer zone of trapped electrons. The inner belt starts as low as a couple hundred kilometers and extends a few thousand kilometers. The outer belt extends 6,000 km, starting at an altitude of roughly 16,000 km. Observed dose rates in the belts are as high as 1,000 rad/hr, while the average dose rate in the geosynchronous orbit is about 10^5 rad/year [Sasmussen, 20].

There is a special low-altitude region off the coast of Brazil that has an abnormally high level of radiation. It has plagued spacecraft designers for years and is called the South Atlantic Anomaly (SAA). It occurs because the Earth's magnetic field is inclined 11° to the spin axis and offset by some 500 km, causing part of the inner Van Allen belt to descend into the upper atmosphere [Sherrill, 21].

Without the protection of an atmosphere, the flux from the high-energy portion of the electromagnetic spectrum is high. X-rays, gamma-rays, and especially UV photons abound. They are generated by our sun as well as galactic and even extragalactic sources. A sensor and its associated electronics are exposed to levels much higher than the engineer is used to working with on the Earth.

High-energy particles from solar wind, the Van Allen belts, and cosmic rays cause transient effects in electronics, such as a bit-flop that can erase memory or cause a switch to turn on or off. Cosmic rays tend to cause singe event upsets when an ion at high speed passes through the p-n junction of a semiconductor. When the ion streaks through the semiconductor material, it creates an ionization path and generates free electron-hole pairs. Enough energy may be deposited to neutralize the charge and cause the bit to change from a zero to one, or vice versa. High-energy particles can cause scintillation in refractive optics resulting from bremsstrahlung radiation as the particles are slowed down. Bremsstrahlung emissions can also be a concern for high-Z materials near the FPA, such as In FPA bumps; gold connectors; and InSb, PT:Si, and HgCdTe focal planes. Spurious hits on the focal plane by natural gamma rays will occasionally cause a noisy pixel, even to the point of saturation.

Radiation effects tend to have significant degradation effects on focal plane readouts and general electronics. MOS and MIS technologies, from which most CCDs are made, are vulnerable to charge buildup effects that decrease charge transfer efficiency [Srour and McGarrity, 22]. The drop in transfer efficiency is often a few percent per year, depending on total dose. After one year, an image can have severe shadowing at the edges where the charge was repeatedly transferred. Also, radiation can wreak havoc on the passivation layer of infrared detectors, causing an inversion that shorts unit cells together [Srour and McGarrity, 23]. FPAs can be permanently damaged by total ionizing dose and displacement damage. Displacement damage cause an increase in dark current or noise [Srour and McGarrity, 24]. In MOS semiconductors, the threshold voltage shifts in response to the cumulative effects of gamma rays. A shift of about 0.5 V occurs at about 10^4 rads (Si). The radiation sensitivity depends on the impurities in the gate oxide, and hardened devices can survive a megarad [Glasstone and Dolan, 25]. Other cumulative dose effects from spaceflight include microprocessor binary "bit-flops" where a "0" changes to a "1" or vise versa.

Radiation effects can happen to all the electronics in a sensor package. Digital technologies are vulnerable to single-event upset, and anything using an oxide insulator contacting an active device is susceptible to total dose effects. Even on general, less glamorous electronics, these effects should not be overlooked. When it flies through the SAA, the Hubble telescope's low-power bipolar random access memory chips suffer from single-event upsets that cause crucial focusing elements to lose data [Burgess, 26]. Solar cell arrays degrade, and optics form color centers that result in darkening or transmission loss in the material. Radiation can also damage optics by producing an increase in absorption, luminescence, and scattering, as well as changes in the refractive index [Srour and McGarrity, 27]. Lubricants loose their viscosity and metals become brittle. Table 11.2 summarizes some common adverse effects of the space radiation environment.

We have all noticed color fade on objects kept in the car for several hot, sunny days. Because there is no ozone in space to provide absorption, the solar radiance level is higher and far more damaging, as its UV content is much stronger. Specifically, the UV can bleach coatings and paints and thus upset the thermal equilibrium. It can also damage coatings. On the positive side, it can anneal color centers from some transmissive optics.

In general, radiation effects are all bad and should be avoided like the plague or a "red team" of safety engineers. Either can cause massive increases in sensor program cost, risk, performance uncertainty, and schedules.

TABLE 11.2 Radiation-Induced Failures

Item	*Radiation Level of Significant Damage*	*Primary Cause*	*Failure Mode*
Commercial ICs	Tens of kilorads	Total dose, prompt gamma, x-rays, and particles	Latch-ups, bit flops, electrical failure
GaAs ICs[A]	10 to 100 Mrad	Prompt gamma, x-rays, particles, and total dose	Single-event upsets and electronic effects
Lubricants[B]	Megarads	Total dose	Increased viscosity, acidity, and volatility
Metals[B]	Many megarads	Total dose	Embrittlement
Petroleum[B]	Kilorads	Total dose	Increased viscosity, acidity, and volatility
Rad-hard electronics[C]	Megarads to tens of megarads	Total dose, prompt gamma, x-rays, and particles	Electronics failure
Teflon®[A]	Tens of kilorads	Total dose	Embrittlement and increased conductivity
Transmissive optics	Approximately hundreds of kilorads (material dependent)	Total dose	Color center formation reduces transparency

Sources: (A) J. Seur and J. McGarrity. November 1988. "Radiation Effects on Microelectronics in Space." *Proc IEEE* 76, 11:1,442–1,467; (B) Carpenter et al. 1972. *Environmental Space Sciences.* Northbrook, IL: Whitehall Co., 23, 229–230; (C) J. Raymond and E. Peterson. 1987. "Comparison of Neutron, Proton, and Gamma Ray Effects in Semiconductor Devices. *IEEE Trans.* NS-34, 6: 1,622–1,628.

There are five ways to offset radiation effects:

1. Design orbits to reduce the total dose.
2. Shield susceptible components.
3. Use radiation-hardened materials and circuits.
4. Greatly overdesign the product.
5. Employ annealing procedures.

You will probably use all of the above if your sensor must survive in a high-radiation environment. Let's examine each in a little more detail.

Spacecraft flight paths should avoid regions of high radiation, whether they be the Van Allen belts of the Earth, radiation zones of other planets, or flight paths close to the sun. Sometimes this is not possible. When Galileo's flight path was changed to orbit the sun and Earth for gravity assist, it was exposed to significantly more radiation than it would have encountered on a direct flight to Jupiter.

When in Earth orbit, the time spent in the Van Allen belts should be minimized by ensuring that the sensor's altitude avoids high-radiation regions. The path of an Earth orbit can greatly affect the dose rates, as it may pass through the SAA. If not required by the satellite's mission, it is best to design an orbit that avoids this region. Flying through the SAA causes the integrated total dose to build up rapidly. Additionally, the SAA can be a source of bremsstrahlung radiation.

Simple shielding can prevent many unwanted effects by reducing the integrated effect on the enclosed electronics. Thin shields of aluminum can do much, depending on the type of radiation and the energy spectrum. High-energy electrons are heavily attenuated by an aluminum shield with an areal density of 2 g/cm^2. However, the bremsstrahlung flux levels are not significantly affected by any reasonably sized aluminium shield [Stassinopoulos and Raymond, 28]. One should be careful that the shield does not overly induce bremsstrahlung radiation or other forms of secondary emissions. Shielding is a brute-force method that works well in space but can add significantly to the mass of a spacecraft. Spacecraft of old were often made of aluminum or titanium housings that provided a good deal of shielding for the electronics nestled inside. However, current trends for lightweight space structures call for composites, epoxies, and plastics. Lightweight materials of graphite polymers and ceramic fibers are replacing aluminum as the structures of satellites, with resultant weight savings of 40 to 50 percent [Marcus, 29]. Some concepts have the electronics mounted right on the skin to eliminate the weight of printed circuit boards. Obviously, there is far less "natural" shielding with direct skin mounting, and any proposed weight reductions in material or architecture must be carefully examined in terms of higher radiation effects.

Some optical materials are "harder" than others, and some experience annealing of the color centers at a lower temperature. These types should be employed if possible. Reflective optics are much more resistant than refractive ones, and they do not have transmission variance problems. Coatings usually do not suffer noticeably from total dose but are susceptible to damage from extreme transient levels of pulsed radiation.

Integrated circuits can be made hard by increasing the pre-radiation threshold voltage. Unfortunately, this results in performance degradation [Marcus, 30]. Additional hardening techniques include circuit design changes and process changes such as using an epitaxial silicon layer on top of a heavily doped silicon substrate or on an insulating layer [Srour and McGarrity, 31]. The first process is called *silicon on silicon (SOS)*, and the second is *silicon on insulator (SOI)*. Sapphire can also be used as a substrates for increased hardness. Unfortunately, the use of sapphire is also sometimes called *silicon on sapphire* or (you guessed it) *SOS*.

There are numerous fault tolerant techniques. Two ICs can be employed for any function, and the function only happens if both agree, or the IC may need to repeat the command several times before it is executed. Other fault tolerance techniques include the use of current limiters and radiation detectors to sense the radiation and rapidly shut off power for a moment to prevent a permanent "latch-up" failure [Rasmussen, 32]. Mil-Handbooks-279 and -280 give guidelines for radiation-hardened circuit design.

Another brute-force way of dealing with radiation is simple sensor overdesign. As with most brute-force approaches, this increases cost and weight. Unlike most other overdesign implementations, care should be taken to expose subtle effects that otherwise might be overlooked. Merely oversizing the aperture may offset degradation in transmission and charge transfer efficiency, but it cannot accommodate digital electronic failures.

Finally, some of these effects can be annealed out by warming the electronics to above room temperature for several hours or days. It was found that the PC IR detectors of IRAS suffered a tenfold increase in noise after each SAA passage. This effect was erased by passing a large current through the detectors [Sherrill, 33]. The darkening of refractive optics can sometimes be annealed out by heating or "bleaching" with UV radiation by pointing at the sun.

Another issue is spacecraft charging. The space environment, coupled with the isolation of spacecraft, can cause static charges to build up. No part can be allowed to build up a significant charge, as arcing may occur. This can be mitigated by ensuring that all parts are grounded, and special conductive IR black paints have been developed for this purpose [Ximenez-de-Ferran, 34].

11.2.4 Particles and Contamination

Space vehicles tend to be dirty. Particles and dust that stay put in a gravity environment tend to float to the most critical areas when in space. When a spacecraft arrives in orbit, it undergoes a period during which it flakes particles and outgasses water and other obnoxious substances. Particles can cause increased thermal emission and scatter on the optics, and the gases can condense on cold surfaces. Even though these surfaces are often inside and hidden from direct access, some molecules can still migrate down. Over time, this may develop into a problem. If mission constraints allow, it is wise to provide a time period in space for the spacecraft to clean up its act before opening or cooling down the sensor. Although the escaping particles and outgassing are reduced after a few days or weeks, it still persists at some lower level and could be a problem for an extended period of time. It has been noted that with some space sensors (notably the Thematic Mapper) there tends to be a 5 to 10 percent decrease in sensitivity in the first few months after it reaches orbit. It is believed that this is owed to contaminants outgassing from the spacecraft and coating the optics.

Extreme care must be taken in material selection and preparation. Outgassing can come from odd sources such as the wiring harness material and wire insulation. Any plastic, epoxy, glue, or lubricant should be considered a potential culprit and analyzed thoroughly before actual use. IR black paints tend to have large surface areas and can harbor substantial amounts of gas that can take months or years to escape. Under no circumstances should such black paints get wet or be touched by human hands while on the ground.

Even for atheists, cleanliness is next to godliness for space sensor fabrication. One of the most foul smelling and visually repulsive sights the author can remember occurred as a result of concentrated body oils in a test chamber. We were getting a slight contamination of cryogenic optics in a very large chamber and suspected residual human oils as the source of the problem. Before becoming a clean and dedicated facility, the chamber had been used as a general thermal-vacuum test facility with people crawling in and out all day. To investigate the problem, the author attached a one-inch germanium flat to a liquid nitrogen line. The chamber was heated and the sample cooled. After several days and a slow cycle to bring everything back to ambient, the sample was removed. It had a noticeable smell. Other engineers avoided the author for some time, fearing he would whip out his malodorous testament to the human condition. A spectroscopic analysis revealed that the problem was in fact human oils. The entire chamber was scrubbed clean, and strict policies of full clean room bunny-suits was instituted. After the corrective measures, the contamination problem never returned.

Fasteners can generate particles during assembly that can come back to haunt when zero G is achieved. As they are frequently not lubricated, minute particles can rub off when connecting and disconnecting them. Care must be taken to get the particles out. An effective way is to hold a vacuum hose near the fasteners as they are being tightened [Ray, 35]. Porous ceramics can be another subtle generator of particle contamination. Porous ceramics can have ceramic dust left over from machining and imbedded deep in the material [Ray, 36].

To keep the environment in the close region of the spacecraft as clean as possible, every piece part should be cleaned with a solvent, handled only with gloves, and "baked-out" at temperatures of 70 to 100° C. The cleanliness policy should then be applied to the subassembly and entire sensor and spacecraft, and even to

MLI and packaging cartons. True, this costs a lot, but it is less costly than replacing a failed satellite.

Plume constituents from attitude control system (ACS) firings can coat optics and surfaces. Hydrazine thrusters kick out up to 10 percent unburned hydrazine. Additionally, NH, NH_3. N_2, H_2, CO, CO_2, and H_2O can be emitted by some thrusters. Many of these will freeze on cryogenic surfaces such as a cold filter or FPA. This contamination results in "condensation of exhaust species on baffles and optics, pitting of optical surfaces, film absorption from the condensation and surface etching from acid deposits" [Miller et al., 37].

Although a great vacuum, space is not empty; it is full of tiny micrometeorites just waiting to wreak damage on optical surfaces. These particle have the very low mass (10^{-12} to 10^{-10} g) but travel at high velocities and may impact at 20 km/s [Mirtich et al., 38]. The accumulated effects of micrometeorites can cause substantial performance degradation. Their effect on optics is usually pitting of the coatings that they hit. There is some good news about micrometeorites. Instrumented satellites have indicated that a mirror in low Earth orbit may be hit by 350 micrometeorites/cm^2/year, which is several orders of magnitude less than some model predictions of 3.5×10^5 hits/cm^2/year. Loss in total reflection may be lower than 1 percent in 11 years [Mirtich et al., 39].

11.3 TARGETS AND BACKGROUNDS

Targets and background are highly dependent on the application, as shown in Table 11.3. Background is usually driven by the spectral bandpass and the viewing geometry. Viewing the hard Earth presents the most clutter and highest background and is usually called viewing below the horizon (BTH). Conversely, a geometry pointing the sensor into deep space is called above the horizon (ATH) and has the lowest level of background and clutter. Limb viewing (i.e., viewing a portion of the atmosphere) is ambiguous with clutter and background, but it is usually worse than expected.

There are subtleties with space backgrounds. First of all, an ATH space background is not a zero radiance background as is often assumed. The background radiance from the big bang is near a 4 Kelvin blackbody, but there exists significant clutter plus other sources of dc background in our corner of the universe. There are many astronomical sources of IR radiation—IRAS identified over 250,000 point sources. Additionally, the ATH view into deep space is becoming cluttered with false targets caused by space debris, other satellites, and spent booster parts. Sensitive sensors will be able to detect paint flecks and bolts hundreds to thousands of meters away, which could lead to elevated false alarm rates or corrupted scientific data.

There also exist extended, diffuse regions of the sky that contain elevated radiance levels generated by zodiacal dust, dense star fields, and intra- and extragalac-

TABLE 11.3 Targets and Backgrounds

Function	*Representative Targets*	*Background*
Astronomy	Hot: stars, galaxies Warm: planets, asteroids Cool: dust Cold: dust, gas	Space, infrared cirrus, zodiacal light
Cloud monitoring	H_2O emission lines	Hard earth, limb
Environmental	Ozone, nitro-oxides, water, carbon monoxide, carbon dioxide	Hard earth, limb
Intelligence	Ground-based targets Space assets	Hard earth Deep space
Military	Ground-air-based targets	Hard earth, limb, high-altitude atmosphere
Military	Satellites	Sky, zodiacal
Weather sounders	Carbon dioxide or water emission bands	Hard earth, limb

tic sources. Figure 11.4 plots some of these sources. As demonstrated by IRAS, the sky is simply not black at IR wavelengths. There are several sources of background radiation. IRAS detected diffuse emission from dust in the ecliptic plane, from comet trails, and from interstellar dust [Davis et al., 40]. There is a distribution of particles around the sun that reflect and emit in the infrared. They are roughly distributed throughout the ecliptic plane, which is close to the zodiacal plane. Their emission/reflection is often referred to as *zodiacal light.* AFGL has done numerous measurements on this, and it seems to peak between 10 and 20 μm, with a peak radiance of 2×10^{-9} to 5×10^{-9} W-cm^{2}sr^{-1} [Spiro and Schlessinger, 41]. Zodiacal light can be the dominant noise source for high-sensitivity LWIR space sensors. Finally, infrared cirrus are extended fuzzy patches with filamentary structures ranging in size from tens of degrees to a few arcminutes [Kessler et al., 42]. These extended backgrounds pose a fundamental background limit on useful sensitivity. None of these issues can be ignored for a sensor with an NEFD of 1×10^{-15} W/cm^2 or less. Note that in some wavelength regions, a sensor may well be background limited even when viewing away from the Earth.

One odd attribute about Earth observation from space is that in the infrared, one can choose the desired penetration altitude, as listed in Table 11.4. This phenomenon results from the atmospheric gases being somewhat layered in concentration and highly absorbent in certain IR bandpasses.

Table 11.4 is only representative. Actual clutter, dc temperature, and altitude depend on precise bandpass edges.

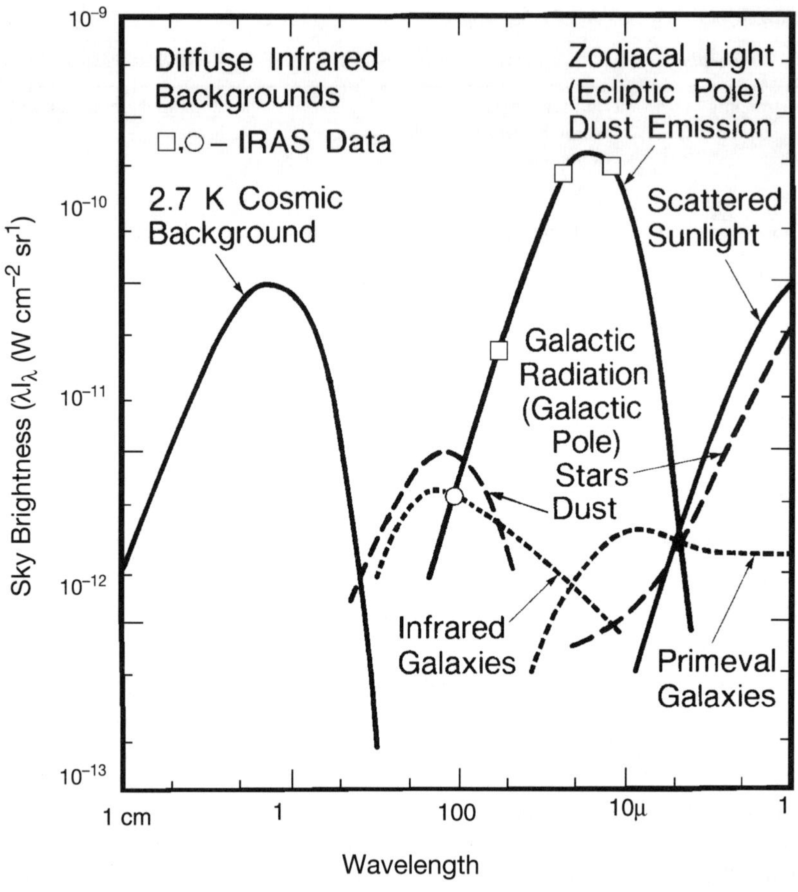

FIGURE 11.4 ATH Space Background Radiation (figure courtesy of Prof. P. Richards, University of California, Berkeley; adapted from data prepared by the COBE team)

11.4 FIGURES OF MERIT

As with any sensor, the industry uses several figures of merit to describe performance. A common figure of merit used for a space sensors is the noise equivalent irradiance (NEI), also called the noise equivalent flux density (NEFD). However, one may find any EO figure of merit applied to a given sensor and situation. As mentioned in previous chapters, no figure of merit is correct for every combination of hardware and use.

NEI is an excellent figure of merit when all noise sources are considered and properly applied and the sensor is not background limited. For low-background systems, this can provide a quick "back-of the-envelope" estimate of sensitivity

TABLE 11.4 IR Bandpass, Clutter, and Altitude from Space

Bandpass (μm)	*Approx. Altitude (km)*	*Dominant Molecule*	*Approx. DC Background Temperature*	*Clutter*
2.7	5	Water	250	Moderate
3–5	0	Narrow CO_2	300	Extreme
4.3	20	CO_2	200	Low
6.3	5	Water	250	Moderate
8–12	0	Some H_2O	300	Extreme
14.7	20	CO_2	200	Low

and is useful for design trade-off studies. One should keep in mind that it is idealized, and other noise sources will creep in. Sensor sensitivity will never be as good as NEI indicates, unless one incorrectly estimates some term. Space sensors are rarely background limited when viewing deep space, as is the case for astronomical sensors. However, they are almost always background and/or clutter limited when viewing the Earth or the atmosphere, as are many military sensors. NEI can be defined by Equation (11.2) as the power per square centimeter (in watts per square centimeter) that the target must deliver at the aperture for the signal coming out of the focal plane to equal to the average noise out of the focal plane (or, in other words, the signal delivered by the target at the sensor aperture to produce an SNR of one). NEI can be defined in terms of watts or photons. In terms of watts

$$\mathrm{NEI} = \frac{(A_d \Delta f)^{1/2}}{A_o D^* K_e K_o K_b} \tag{11.2}$$

where

A_d = detector area in square centimeters

Δf = noise bandwidth as defined in Chapter 8

A_o = clear aperture area in square centimeters (Remember to subtract and optical obscuration, if applicable.)

D^* = detector D^* evaluated with as many noise sources as possible (NEI can take into account the dc background if the quoted D^* includes this.)

K_e = electronic efficiency, usually 0.5 to 0.9

K_o = in-band average optical train transmission, including all elements and filter, usually 0.5 to 0.7

K_b = optical blur efficiency

As mentioned in previous chapters, the user of this or any other equation is cautioned to double check the units to verify that they are consistent. It is easy and common to mix microns with centimeters or meters.

The K terms represent efficiency but also can be used as fudge factors to reduce simple NEI calculations to what is more representative of the system.

Equation (11.2) can have the same substitutions as described in Chapter 8 to give an expression with F#, if so desired. Additionally, it can be defined in terms of the background, which is appropriate if the background is the noise driver. NEI for BLIP conditions can be written as

$$NEI_{BLIP} = \left(\frac{L_b FPhc}{R_b^2 \lambda A_o K_o K_b K_e}\right)^{1/2} \tag{11.3}$$

where

NEI_{BLIP} = noise equivalent irradiance in W/cm^2 for severely dc background limited conditions

L_b = radiance in W/cm^2/sr from the background

FP = pixel footprint on the background in square centimeters at the target distance

h = Planck's constant of 6.63×10^{-34} W/s^2

c = velocity of light of 3×10^{10} cm/s

R_b = range to the background in centimeters

λ = center or average wavelength (the wider the band, the less accurate this becomes, and eventually one will have to integrated)

A_o = clear area of the optics in square centimeters

K_o, K_b, K_e = efficiency fudge factors as defined after Equation (11.2)

Note that NEI_{BLIP} depends primarily on the pixel footprint, which is a function that depends solely on the range and sensor resolution.

Table 11.5 gives a checklist of special considerations when using NEI. Not all of these considerations apply to every application.

TABLE 11.5 Issues and Concerns with NEI

Concern	*Solution*	*Comments*
1/f noise	Include in D*.	Degrading D* is fine for a given readout rate, but it is not constant if readout rate varies.
Aging effects	Degrade everything.	
Atmospheric absorption/scattering/emission	Fudge factor.	Not sufficiently handled by this figure of merit.
Clutter effects	Can be included only as a fudge factor in this FOM.	Really should not be considered in this FOM, and NEI should not be used to evaluate/describe a sensor where clutter is a driver.
DC background effects	Use BLIP version.	BLIP and non-BLIP versions may be added via RSS.
Fill factor	Should be included in K_b.	Especially critical for monolithic FPAs.
Filter corner effects	Best considered in determining a target's in-band radiant intensity for SNR or SCR.	Especially critical for narrow bandwidths.
Flicker or display effects	One could attempt to degrade the result by a fudge factor.	Not handled by NEI.
Not square detectors	Addressed in A_d and K_b.	
Nuclear radiation noise	Include as degraded D*.	
Nuclear darkening of optics	Include as degraded K_o.	
Optics noise effects	May be included in D* or fudge factor.	Not properly addressed by this FOM.
Other than circular aperture	Include as a fudge factor or in K_b or adjust A_o.	Not properly addressed by this FOM.
Spectrally sensitive detectors	Must integrate D* by sensitivity function.	NEI based on D* does not properly represent Schottky barriers and quantum wells with bandpasses of more than approx. 0.5 microns.
Stray light	Extra noise can be included in effective D* or as a fudge factor.	Not properly addressed by this FOM.
Target contrast	The target radiant intensity can be reduced to consider contrast in SNR calculations.	Must be considered for high-background conditions.
Target crossing pixel in less than the dwell time	Can be included as a fudge factor.	
Target spectral effects	Use weighted average over bandpass.	
TDI	One should decrease the NEI by the square root of the number of TDI stages (or elements).	

The resulting SNR can be computed by comparing the target's irradiance at the aperture of the sensor with the NEI as

$$\mathrm{SNR} = \frac{(I_t/R^2)}{\mathrm{NEI}} \tag{11.4}$$

where

I_t = in-band radiant intensity of the target in watts per steradian

R = range (must be in the same units as NEFD)

NEI = the most representative noise equivalent irradiance (or NEFD) that you can devise

Usually, an SNR of 5 to 10 is required for detection, and 10 to 100 for clean imaging. Once detected, a smart target tracker can track with SNRs of about 3 or more, while a dumb tracker needs 10 or more.

11.5 COMPONENT CONCERNS

Maybe no man is an island to himself, but every spacecraft is. It must be understood by all involved in a space project that every subsystem affects every other subsystem and that they all must be optimized in unison. Designing an IR sensor for space is like conducting a symphony. It is the responsibility of the project office to ensure that all components function harmoniously, complementing and enhancing each other. Concerns about subsystem interaction are crucial to space systems. For example, if not properly positioned, the solar array may dump heat through radiation or reflection to a radiator, or a sensor may block the sun from a solar array.

Space platforms and buses have many subsystems supporting the sensor payloads. These include data systems, attitude control systems, propulsion systems to maintain orbit, communication systems, thermal systems, power generator systems, and support structures. Attitude control systems use star or horizon sensors for attitude information, and they must be given a clear view. The attitude and navigation system invokes control by periodic firing of small thrusters to orient the spacecraft, thus compensating for atmospheric drag losses or to alter the orbit. Although the thrusters are small, the fuels used can have highly reactive by-products that can bond, erode, and change surface properties. Cold gas thrusters are the best from an IR sensor standpoint, because they usually just expel nitrogen. However, they produce the weakest impulse, and it is usually mass/life effective to use monopropellant or bipropellant fueled thrusters. The thermal control system is integral to the spacecraft and may use radiators, heat pipes, thermal straps, coatings, multilayer insulation, and louvers. A typical satellite structure may employ honeycomb or corrugated aluminum bounded on both ends by an aluminum face sheet. This technique results in extreme stiffness and rigidity at a very low weight. Aluminium, the quintessential spacecraft structure material, is being replaced by advanced material composites and epoxies because they offer higher stiffness and lighter weight. Outgassing is always a concern with IR sensors and is especially critical when using epoxies or other organic compounds.

The optics must be capable of surviving the ride to orbit and the space environment. This implies an athermalized design with composite or beryllium reflective optics preferred. The optics tend to be a cost and weight driver for the space sensor. As a result, great emphasis is usually placed on lightweight telescope assemblies.

The focal plane must be relatively hard to radiation effects. This often entails a specially made FPA with extra masking steps and layers to isolate the detectors from the bulk material. Because everything is so costly on a spacecraft, FPAs are not a strong cost driver. It is often wise to use custom-made FPAs to optimize system-level performance. This may even reduce total cost by allowing smaller, lighter optics.

Potential cryogenic requirements are always a strong driver for any space bus that supports an IR instrument. Most contemporary science instruments require a 1 W cryocooler at 55 to 75 K and a 5 to 10 year life-span with a reliability of greater than 95 percent [Ross et al., 43]. Using a cryocooler affects the spacecraft's stiffness and vibration isolation, electrical power system, and thermal system requirements. The vacuum of space eases telescope cooling. Many systems require either actively or passively cooled optics. Since the optics do not need to be as cool as the FPA, the spent FPA cryogen can sometimes be used to cool the optics. Radiative cooling of telescopes can provide a significant drop in temperature and is often adequate. Isolated telescope optics and structures will naturally cool to 200 to 250 K, and it has even been proposed that telescopes could be radiative cooled to the 40 K region [Davis et al., 44].

Mechanisms are usually required to be fail-safe and redundant. All bearings are usually required to be ABEC class 7 or better and sometimes are required to be in a laminated phenolic vacuum-rated cage. The lubricant must be vacuum certified and is frequently a dry lubricant like MoS_2. Lubricants must be of low vapor pressure, not outgas, and not splatter about. Often, the lubricants are infused into the metal of gears. As the gears wear, new lubricant is exposed. In this manner, they are self-lubricating.

Electronic effects such as spacecraft charging, emission, magnetics, cold flows, and effects of dissimilar metals make material selection for housings and structures very critical. Often, there are specific exclusions from using high electromotive force elements like cadmium, zinc, and magnesium.

Much effort has been expended to generate approved material lists for spacecraft design. These material lists should be adhered to, whenever possible, as they tend to mitigate many of the concerns discussed above.

11.6 SYSTEM CONSIDERATIONS

To guess at a space sensor cost, weight, power, and so forth, use rough rules of thumb to estimate the weight, power, or cost of the components. Then add them together, with overhead, for a first cut. For a second cut, rely on analogy to similar systems, with scaling factors. Use estimates from vendors, detailed parts and functions lists and analyses for third cut. Proposals often rely on the first or second cuts, and sometimes the third, if it is a long proposal process. (But governments and companies tend to be moving away from such lengthy processes.) Funded feasibility studies usually rely on the second or third cut approach, which can be highly accurate if not much is neglected.

11.6.1 Cost

Safety is not the only factor that affects cost and schedule. The results of a survey conducted for this book are included in Table 11.6. Leaders in IR space sensor production were asked to approximate the impact on cost and mass of various subsystem attributes. Twenty-seven replies were received and averaged. The results are presented in the table.

TABLE 11.6 Effects of Subsystems on Total Cost, Weight, and Power

Subassembly or Service	*Approx. Avg. Percentage of System Cost*	*Approx. Avg. Percentage of System Weight*
Focal plane assembly	20	4
Integration, assembly, and test	10	N/A
Cryocooling	9	19
Management/administration	9	N/A
Optical train	8	13
Signal/image processor	8	10
Gimbal/scanner/stabilization mechanism	7	11
Software	6	N/A
General electronics, power supplies, etc.	5	11
Housing and support	4	18
Power supply	3	9
Miscellaneous	2	7

One notes that the general impression of people working on space systems is that the FPA and integration, assembly, and test (IA&T) are the largest cost components. General management, optics, gimbals, signal/image processors, and cryocooling were all significant and, on average, about equal in impact. It is to be expected that focal planes are expensive components. Typically, space sensors push the state of the art in FPA technology to allow for reduced size of the optics and thereby reduce the size of the gimbal. Since large-scale production has not occurred on an IR space system, it can be cost-effective to screen FPAs to select the very best in performance. IA&T ranks high in cost due to the extensive testing and qualification discussed earlier.

Space-based sensors simply cost a lot. There is often ample logical justification for the cost. When system factors such as mass, power, launch costs, reliability, and performance are considered, it usually makes sense to spend more on the sensor payload and to optimize these factors to the detriment of cost. This is owed to several factors. First is that almost any space sensor is still engineered from the ground up, optimized, and packaged for the given mission and space vehicle. This makes sense given the cost of launches and lost opportunities as discussed earlier. Second is the level of standards often placed on space components. S-level parts and systems can cost a factor of two to ten above their military-rated counterparts. A factor of ten increase is not uncommon between a commercial part and a space qualified part. Third is the level of environmental testing required to ensure and test for reliability. Excessive testing and screening forces low yields and high costs.

Launch costs are high. Typical 1991 costs were $400 million for a Shuttle and $200 million for a Titan [Covault, 45]. Launch costs to low Earth orbit are around $10,000 to $20,000 per kilogram. The higher the orbit, the higher the cost, as more fuel and larger launch vehicles that must be used. Given the per-kilogram launch costs to low Earth orbit, the weight of a space sensor is a large cost driver.

The complete development of a space sensor usually costs between $10 million for a simple instrument to a few hundred million dollars for a large multispectral imager, excluding platform, launch, and operation costs. Merely adapting a camera, seeker, or FLIR for space basing can cost ten times the "commercial" price of the sensor. When a constellation of sensors is considered with launch and operational costs, the total price is usually in the billions to hundreds of billions of dollars. For example, it is expected that the demonstration/validation phase of Brilliant Eyes will cost $885 million, with the full constellation costing $4.7 billion in 1991 dollars [Smith, 46].

Learning curves are largely non-existent, as every sensor made usually features some "improvement" or tweak to the design. Also, space sensor programs are measured in decades of time, and even a few percentage points of inflation can wipe out any cost savings apparent in "then-year" dollars. Still, the price of additional sensors can be reduced in terms of constant dollars, and learning curves of 80 to 95 percent have been estimated.

Jamieson defines a cost index for space sensors of simply [Jamieson, 47]

$$J_\delta = (A_o N_d)^{1/2} \tag{11.5}$$

where

J_δ = cost index

A_o = aperture area

N_d = number of detectors

He also points out that the cost of a space sensor is dominated by launch, electronics testing, and data processing. The aperture size is a good indicator from which to scale the weight (and, hence, launch costs), and the number of detectors is a reasonable quantity by which to scale testing. Jamison does point out that the state of the technology will change this. Currently, very lightweight optics do not scale directly to the sensor's weight; however, the more lightweight the optics, the more they cost. Additionally, the number of detectors on an current FPA does not necessarily correspond to increased weight or cost as clearly as it does with old-style discrete detectors. However, some FPAs are costly, as is all FPA testing. This simple back-of-the-envelope rule of thumb for cost scaling still has some applicability. Obviously, the reader is cautioned not to use this to scale vastly different systems or for large changes in optics or detectors.

The life cycle cost must include the overall operational costs, which can be significant. Currently, several operators are required around the clock for most spacecraft. Although many plans have been made to reduce this number, the operational ground crews will always be a considerable cost contributor.

Costs can even be extremely high during the initial deployment and check-out period. However, recent advances reflect a general trend to distributed mission operations and improved data link capabilities for cost reduction. Such measures were implemented on the U.S. Defense Support Program (DSP) infrared sensor-based early warning satellite. During a recent deployment of the TRW-built DSP spacecraft, 80 percent of the orbital test team remained in Redondo Beach, California, viewing the launch and monitoring satellite health on large screens using the TRW Orbital Test Station (TOTS). At their fingertips were controls that can call up system configurations, data management drawings, and specifications, enabling the team to resolve anomalies instantly. As part of a cooperative total quality management effort with the Air Force, the spacecraft data is brought to the engineers, rather than the engineers to the data. TOTS saved hundreds of thousands of dollars in travel costs and time. TOTS managers estimate that the cost of operating TOTS for a year is more than offset by the savings realized by bringing data to engineers for just a single launch. The manufacturers remain deeply involved with DSP during on-orbit testing that usually occurs in the first 30 days after launch. During this process, TOTS receives and processes telemetry from the DSP satellites through special Air Force-supplied data links, just as they do during the launch and deployment sequence. If a system parameter wanders off limits, it is immediately called to the TOTS operator's attention. The problem can be analyzed and the corrective action recommended to the Air Force through TOTS data links. Using U.S. Air Force-supplied communication hardware with the systems integration test equipment on the launch pad, data from pre-launch tests was conveyed directly to engineers in California for real-time analysis. Known as "factory-controlled launch support," it normally would occur when weather or other issues delay testing or launches. Since the engineers are still at the manufacturing facility, they can work on other aspects of the program instead of simply waiting at an off-site location to resume work [TRW, 48].

11.6.2 Weight and Mass

To estimate the weight of a space sensor, use the rough relationships provided in Part II of this book, and be as optimistic as possible. Multiply the sum by 1.2 to 1.5 for structure, margin, overhead, and things that you forgot. Usually, a weight estimate for any given part is not far off; it is just that things are forgotten. Typical forgotten items are those unglamorous things that hold everything together such as subassembly structures, motor mounts, thermal and mechanical washers, wires, glue, paint, and metal layer insulation.

The survey tabulated in Table 11.6 indicates that the community perceives (1) cryocooling and (2) housing and support to be the major weight drivers. The general electronics, optical train, signal/image processor, and gimbal were also perceived as large contributors to total weight. The perception for the cryocooling may be unusually high due to respondents considering cryogen filled dewars. Nevertheless, this is interesting because, in the early stages of a program, both cryocooling and housing and support frequently are overlooked.

Modern space sensor architecture includes a cryocooler to replace the dewars of systems like ISO and IRAS. Cryocooler weight can be estimated by scaling the relationships in Chapter 5. If a dewar is involved, simply multiply the density of the cryogen by its volume and add 35 to 50 percent for the dewar container. Focal planes are often customized to squeeze out every last bit of performance, so the telescope can be smaller and lighter. Telescopes are usually made of the lightest materials.

Space structure mass ratios range from 15 to 30 percent of the mass that they support. Microsats using stiff composites rate near the lower number, and large spacecraft using more traditional aluminum structures approach the higher number. However, a spacecraft's weight usually will be three to seven times the science or payload weight when the necessary subsystems to support the payload are considered. The total dry weight of the supporting spacecraft bus is usually a function of the payloads. Figure 11.5 (figure courtesy of NASA JPL) plots several spacecraft weights as a function of the instrument weight. The wet weight (including propellants) is a function of the payload, the expected life of the satellite, and the expected maneuvers.

It is common for spacecraft parts to have multiple functions to save on volume and weight. Weight savings can occur by using the natural structure or housings for additional purposes such as for mounting electronics or even as printed circuit boards. For example, ISO's sunshade functions both as a sunshade and support for the solar cells.

Space sensor are usually power misers. For low Earth-orbiting satellites, at the margin it usually takes an addition 100 to 200 g to produce an extra watt, 35 to 70 g to dissipate the residual heat, and added structure for support. This adds up to 150 to 300 g/W [Miller, 49]. The added weight penalty for an additional watt can be more than an order of magnitude greater for interplanetary flights.

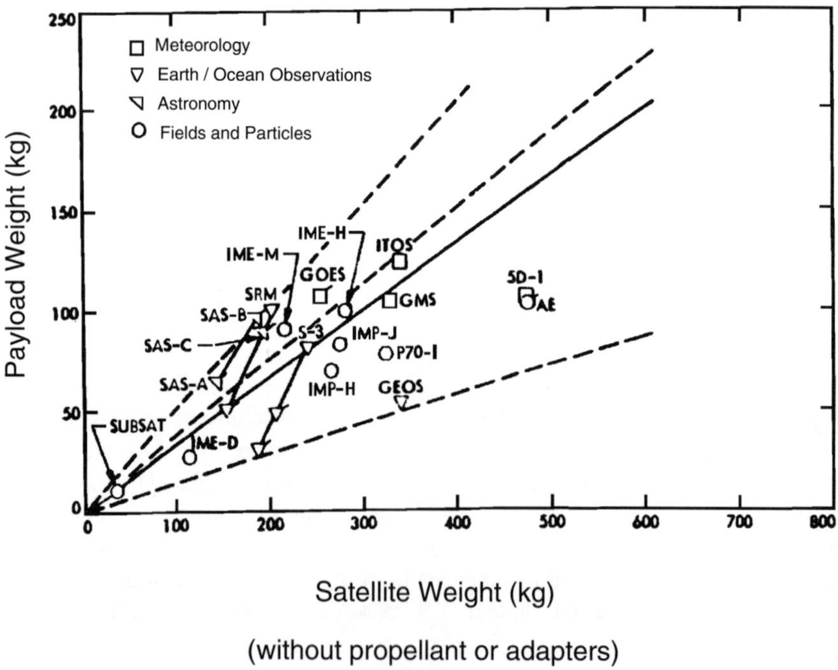

FIGURE 11.5 Spacecraft Mass as a Function of Instrument Weight (figure courtesy of NASA JPL)

11.6.3 Electrical Power

As noted, power consumption is also a strong driver. Any spacecraft is like an island unto itself. It must generate all its electrical power, then reject the resulting thermal increase via Planckian radiation to space.

Electromagnetic compatibility is an important issue in the electrical design of a space-based sensor. Sensors must not suffer damage nor affect other payloads or subsystems because of poor EMI/EMC design. The entire sensor and all components must have low and constant ripple. There are usually specific and definite requirements regarding the permissible electromagnetic emissions, often directed by Mil-Std-461.

11.6.4 Producibility

Space sensor producibility is still an oxymoron. There are few IR sensor spacecraft designs of which more than one or two have been made. Those few, like DSP, have incorporated so many upgrades from one spacecraft (or block of spacecraft) to the next that learning curves and producibility estimates seem meaningless without a great understanding of the modifications. There are tens of organizations around the world that produce a sensor or two at a time. This means that the planet's combined output is only a few per year. Facilities, funds, and skilled workers are the limiters. It would be entirely possible for the USA, CIS, France, and England to produce several space sensors per month each, if the will and the money were there.

11.6.5 Schedule

Scheduling space sensor design is a black art that relies greatly on luck, experience, and circumstance. Traditionally, a typical space sensor takes a year or more to reach a preliminary design review; another year or more for the critical design review; several years each for fabrication, test, and integration into the spacecraft; and a year or two awaiting launch. Lead time on space-qualified electronic parts can easily span 12 to 18 months, even for connectors.

Measuring a space project's life in decades has frustrated military and scientific users alike while discouraging commercial users from entering the arena. When flight time is considered, a planetary astronomer can spend his entire career on a single mission and have only one or two chances at developing and flying a space sensor. An entire military career can be spent on a single system, without the chance to promote an upgrade. This sad circumstance is being attacked in the United States by NASA and DoD via efforts to reduce the time from an idea until space based data is received. SDIO has had fantastic success with their Delta programs by cutting through the red tape and qualification procedures.

11.6.6 Reliability

Reliability is another system driver. Because launch and lost opportunity costs remain high, reliability is still essential. After a space vehicle spends a few years waiting for a launch and five years going to Saturn, if a sensor fails, the investigating team members are likely to have retired before they get another chance. As discussed previously, recent advances in launch vehicles may lead to relatively low-cost launches. This may relax the requirements for extreme reliability in Earth orbiting vehicles in the next century.

11.6.7 Resolution

Resolution is often another driver for space systems. If the space system is for Earth viewing, it is probably clutter limited, needing better and better resolution to increase performance. If the space system is for space viewing, one of the reasons it is there may be to exclude the resolution-limiting characteristics of the atmosphere. Again, better resolution is demanded. With infrared systems, this often translates into large optics and large focal plane, both of which affect cost and weight.

11.6.8 Safety

Safety can be a system driver, and proving that a system is safe is a universal irritant. There are two types of safety: range and spacecraft. Range safety is concerned with safety requirements imposed on the sensors and other systems while on the ground and during launch. Usually, the range officials will insist that the sensor impose no particularly odd safety problems or that such problems be identified and neutralized. Misunderstandings leading to disagreement with range safety officers have, on occasion, led several IR engineers to question the cranial capacity and genealogy of the range officers and safety engineers. It is wise for the engineering team to contact range safety early in the design process. Failure to develop a close working relationship early enough may lead to unexpected delays in the launch schedule. Although irritating, it is always easier to work with someone from the beginning rather than fight with him at the end of a program.

A recurring problem with space sensors is the use and handling of potentially hazardous materials. Some IR materials may be hazardous, but not in the form in which they are used on the sensor. Space sensors often employ nasty and poisonous materials such as beryllium. Additionally, focal planes contain outright poisons like arsenic and many heavy metals. When the safety engineer spots these things on a material sheet, he will flip into a mode that requires massive paperwork and procedures. Problems arise when safety engineers simply do not understanding the use of these materials. The small amounts of a hazardous elements chemically combined into stable materials in the sensor is not a concern, but it is an issue until everyone understands the situation and application. Safety engineers are not IR engineers and don't realize that the crews won't be touching or eating the HgCdTe focal plane. This must be explained to them with respect. After all, it is their job to ensure safety, even though it often seems like their job is to drive IR engineers crazy.

The only real range safety issues normally imposed by IR sensors are pressurized containers, cryogenic dewars, and J-T bottles. These require special precautions. Any high-pressure container might explode. Additionally, the IR engineer should be prepared for lengthy debates and procedures if the sensor uses any explosive or flammable cryogen such as hydrogen, oxygen, or methane.

Spacecraft safety is another subset of general safety. If the sensor is one payload among many others, it must not be allowed to damage the spacecraft in any failure mode. The project team should be prepared with adequate budget and schedule to conduct demonstrations of electronic, mechanical, and cryogenic failure modes. Again, the best advice is to work with the safety engineers early in the design to ensure that catastrophic failures are precluded in the design stage. There is an order of magnitude increase in safety concerns for any sensor riding on the Shuttle or any manned spacecraft. In general, the sensor will be considered guilty until proven innocent of any safety issues. Rarely will paper analysis sufficiently answer a safety concern.

11.6.9 Critical Trade-Offs and Analysis

The key trade-offs in the design of a space sensor revolve around the exact application and mission requirements. Generally, the specific critical trade-offs and analyses that must be considered in detail include

- bandpass edge definitions
- calibration
- contamination control
- cryocoolers versus dewars
- level of on-board image processing versus telemetry of raw data
- level of on-orbit maintenance
- material trade-offs of all kinds
- one (or few) large sensors versus several small sensors to meet high-level coverage requirements
- orbit parameters
- pointing architecture (gimbal, mirror, body fixed)
- radiation analysis and circumvention trade-offs
- reliability
- stabilization methodology and performance

- staring versus scanning FPA
- warm versus cold optics
- weight versus performance (considering launch costs)

References

1. November 3, 1987. "Eye In The Sky." *Vectors* 24: 2–9.
2. N. Johnson. April 1981. "The Baikonur SS-6 Space Launch Facilities." *Spaceflight,* 109–116.
3. N. Johnson. 1987. *Soviet Military Strategy in Space.* London: Jane's Publishing, 51.
4. 1987. *Designing an Observatory for Maintenance in Orbit.* NASA Marshall Spaceflight Center Publication.
5. J. Richelson. January 1991. "The Future of Space Reconnaissance." *Scientific American,* 38–44.
6. Ibid.
7. W. Scott. February 1, 1993. "High Demand Stretches NRO Intelligence Assets." *Aviation Week And Space Technology,* 49–52.
8. D. Kutyna. September-October 1991. "The Military Space Program and Desert Storm." *Space Times* 30: 3–7.
9. Ibid.
10. W. Scott. June 17, 1991. "RME Spacecraft Demonstrates Precision Tracking and Pointing." *Aviation Week and Space Technology,* 204.
11. D. Hunton. November 1989. "Shuttle Glow." *Scientific American,* 92–98.
12. Ibid.
13. S. Musikant and W. Malloy. 1990. "Environments Stressful To Optical Materials In Low Earth Orbit." *Proc. SPIE* 1330: 119–130.
14. D. Hunton. November 1989. "Shuttle Glow." *Scientific American,* 92–98.
15. J. Anderson, G. Downs, and P. Trepahnier. 1991. "A Signal Processor for Space Based Visible Sensing." *Proc. SPIE* 1479: 78–100.
16. R. Rasmussen. November 1988. "Spacecraft Electronics Design for Radiation Tolerance." *Proc. of the IEEE* 76, 11: 1,527–1,537.
17. W. Fredricks. 1987. "The Infrared Focal Plane Array Producibility Initiative And Related Technologies." Presented at the UCLA short course 823.29.
18. R. Rasmussen. November 1988. "Spacecraft Electronics Design for Radiation Tolerance." *Proc. of the IEEE* 76, 11: 1,527–1,537.
19. E. Stassinopoulos and J. Raymond. November 1988. "The Space Radiation Environment for Electronics." *Proc. of the IEEE* 76, 11: 1,423–1,424.
20. R. Rasmussen. November 1988. "Spacecraft Electronics Design for Radiation Tolerance." *Proc. of the IEEE* 76, 11: 1,527–1,537.
21. T. Sherrill. February 1991. "Orbital Science's Bermuda Triangle." *Sky and Telescope,* 134–139.
22. J. Srour and J. McGarrity. November 1988. "Radiation Effects on Microelectronics in Space." *Proc. of the IEEE* 76, 11: 1,442–1,467.
23. Ibid.
24. Ibid.
25. S. Glasstone and P. Dolan. 1977. *The Effects of Nuclear Weapons.* Washington, DC: U.S. Government Printing Office (prepared and published by The U.S. DoD and DOE), 351.

26. L. Burgess. January 1991. "Cosmic Radiation Zapping Troubled Hubble Memory Chips." *Military and Aerospace Electronics,* 1, 16.
27. J. Srour and J. McGarrity. November 1988. "Radiation Effects on Microelectronics in Space." *Proc. of the IEEE* 76, 11: 1,442–1,467.
28. E. Stassinopoulos and J. Raymond. November 1988. "The Space Radiation Environment for Electronics." *Proc. of the IEEE* 76, 11: 1,423–1,424.
29. D. Marcus. October 14–20, 1991. "Composites Increasing Payload Capacities." *Space News,* 8.
30. J. Srour and J. McGarrity. November 1988. "Radiation Effects on Microelectronics in Space." *Proc. of the IEEE* 76, 11: 1,442–1,467.
31. Ibid.
32. R. Rasmussen. November 1988. "Spacecraft Electronics Design for Radiation Tolerance." *Proc. of the IEEE* 76, 11: 1,527–1,537.
33. T. Sherrill. February 1991. "Orbital Science's Bermuda Triangle." *Sky and Telescope,* 134–139
34. S. Ximenez-de-Ferran. August 1991. "The ISO Spacecraft." *ESA Bulletin* 67: 17–26.
35. D. Ray. 1989. "Contamination Control Program for The Extreme Ultraviolet Explorer Instruments." *Proc. SPIE* 1118: 136–143.
36. Ibid.
37. J. Miller, G. Steiner, and G. Dryden. 1988. "Front Surface Optic Contamination from Small Rocket Plumes." *Proc. SPIE* 967: 320–331.
38. M. Mirtich, H. Mark, and W. Kerlake. 1990. "Effects of Eleven Years in Earth Orbit on a Mirror Surface." *J. Spacecraft and Rockets* 27, 3: 258–266.
39. Ibid.
40. J. Davis, T. Hawarden, and C. Mountain. 1991. "Radiatively Cooled Telescopes: A New Direction for Infrared Space Astronomy." *Acta Astronautica* 25: 223–228.
41. I.J. Spiro and M. Schlessinger. 1989. *Infrared Technology Fundamentals.* New York: Marcel Dekker, 87.
42. M. Kessler, A. Heske, L. Metcalfe, and A. Salama. August 1991. "The ISO Mission—A Scientific Overview." *ESA Bulletin* 67: 8–16.
43. R. Ross, D. Johnson, and R. Sugimura. 1991. "Characterization Of Miniature Stirling Cycle Cryocoolers For Space Application." *Proc. Sixth International Cryocoolers Conference* (DTRC-91/002), 27–38
44. J. Davis, T. Hawarden, and C. Mountain. 1991. "Radiatively Cooled Telescopes: A New Direction for Infrared Space Astronomy." *Acta Astronautica* 25: 223–228.
45. C. Covault. December 2, 1991. "Shuttle Deploys DSP Satellite, Crew Performs Reconnaissance Tests." *Aviation Week and Space Technology,* 23–24.
46. B. Smith. January 11, 1993. "TRW Rockwell to Conduct On Orbit SDI Tests by 1997." *Aviation Week And Space Technology,* 56–57.
47. J. Jamieson. 1976. "Passive Infrared Sensor: Limitations on Performance." *Applied Optics* 15, 4: 891–909.
48. 1992. Information courtesy of TRW.
49. J. Miller. 1990. "Key Attributes and Generic Requirements for Cryocooler Application on Microsats." *Proc. of the Sixth International Cryocooler Conference,* 285-93.

Appendix 11A

Data Sheets for Representative Systems

CIRRIS 1A (Cryogenic Infrared Radiometer In Shuttle)

Manufacturer: Utah State, Logan, Utah; Space Systems Engineering, Logan, Utah; and SSG, Waltham, Massachusetts

Intended Application: Space, Space Shuttle based

Qualification: Full space qualification

Availability: Completed mission

Description:

CIRRIS-1A is an atmospheric measurement experiment designed to obtain simultaneous spectral and spatial infrared measurements of a variety of upper atmospheric phenomena. The CIRRIS 1A payload includes several sensors and subsystems. The primary sensors are the high-resolution Michelson interferometer and a dual-channel radiometer that share the common collecting optics of a high off-axis rejection, 0.3 m telescope.

Characteristics:

Aperture: 30.48 cm "D" shaped for a 182 cm^2 clear area

Image Quality: 0.5 millirad on axis

Optical Throughput: 60 percent

Optics, Detectors, and Preamp Temperature: 12 K

Out of Field Rejection: 1×10^{10} @ 2.5°

Radiometer Dynamic Range: 1×10^7

Interferometer FPA: Five elements of Si:As

Gimbal Capacity: 715 kg-m

Positional Accuracy: ± 0.167 degree

Power Requirements: 852 W in joystick mode

Gimbal Weight: 930 kg

Information from: Ahmadjain, Nadile, Wise, and Bartschi. November-December 1990. "CIRRIS 1-A Space Shuttle Experiment." Journal of Spacecraft, *666–674.*

DSP (Defense Support Program)

Manufacturer: TRW, El Segundo, California; and Aerojet, Azusa, California

Intended Application: Space-based USAF early warning

Qualification: Full space qualification

Description:

TRW Space & Technology Group builds the DSP spacecraft under contract to the U.S. Air Force Space Systems Division and integrates the sensor payload provided by Aerojet ElectroSystems Company of Azusa. The DSP satellites use the sensor to sense IR radiation from missile plumes against the Earth's background. DSP satellites have been the spaceborne segment of NORAD's Tactical Warning and Attack Assessment System since 1970. DSP has repeatedly proven its reliability and potential for growth. DSP satellites have exceeded their specified design lives by some 30 percent through five upgrade programs. These upgrades have allowed DSP to provide accurate, reliable data in the face of changing requirements (greater number, smaller targets, advanced countermeasures) with no interruption in service. Evolutionary growth has improved satellite capability, survivability and life expectancy without major redesign. The original DSP weighed 954 kg, required 400 W of power, had 2,000 detectors and a design life of three years. In the 1970s, as mission requirements changed, the satellite was upgraded to meet the increasingly complex national needs. As part of this upgrade, the weight grew to 1680 kg, the power to 680 W, the number of detectors to 6,000 and the design life aimed at a goal of five years. The current DSP has a weight of 2,360 kg, uses 1274 W and incorporates significant improvements on survivability. [A] DSP is pictured in Figure 11.3.

Characteristics:

Bandpass: Infrared[B]

Orbit: Geosynchronous[B]

Scan: The satellite spins like a top to provide scanning[C]

Cost: $300 million for one, $180 million for multi-procurements,[D] total program costs over $8 billion[E]

Telescope Assembly Length: 3.6 m[F]

MWIR Focal Plane: HgCdTe[G]

SWIR Focal Plane: Lead sulfide[G]

Information sources: (A) Information courtesy of TRW; (B) C. Covault. November 18, 1991. "Astronauts to Launch Warning Satellite, Assess Manned Reconnaissance From Space." Aviation Week and Space Technology, *66–69; (C) December 9–15, 1991. "A Defence Support Program Satellite."* Defense News; (*D) C. Covault. December 2, 1991. "Shuttle Deploys DSP Satellite, Crew Performs Reconnaissance Tests."* Aviation Week and Space Technology, *23–24; (E) A. Velocci. December 2, 1991. "TRW, Aerojet Joint New Teams in Competition for FEWS Pact."* Aviation Week and Space Technology, *46–47; (F) C. Covault. February 4, 1989. "New Missile -Warning Satellite to be Launched on First Titan 4."* Aviation Week and Space Technology, *34–40; (G) V. Kiernan. October 5, 1992. "Bush Reports Reveals Payload Aboard Titan."* Space News, *17*

EDO/Barnes Thermal Imaging System

Intended Application: Space, on the Shuttle

Qualification: Full space qualification

Availability: Custom made; delivery ≈ 12 months ARO

Description:

Barnes, in cooperation with Inframetrics, offers an infrared camera for application in orbital experiments aboard manned or unmanned spacecraft. The infrared imaging system provides non-contact measurement and analysis of static or dynamic thermal patterns in either fully automated or user-interactive experiments This ruggedized scanning camera evolved directly from the Inframetrics model 600 IR Radiometer. The system is composed of an electronics processing unit and an optical sensor head enclosing a two-dimensional scanning mechanism, self-contained Stirling cooler, single-element detector, and interface electronics. A temperature reference in the sensor head allows repeatable, quantitative radiometric measurements.

Characteristics:

Spectral Response: 8 to 12 μm with options for 3 to 5 μm and 3 to 12 μm

NETD: 0.15° C

MDTD: 0.01° C

Frame Rate: 30 frames per second

IFOV: 1.8 millirad square

Target Temperature Range: 0 to 400° C

Dynamic Range: 48 dB

Focus Range: 30 cm to infinity

Storage Temperature: –10°C to 55° C

Operating Temperature: 10° to 40° C for the optical head and 10° to 50°C for the electronics

Power Requirement: 30 W @ 12 Vdc

Weight: 4.1 kg for the optical head and 9.6 kg for the electronics

Volume: 20.3 × 15.2 × 22.9 cm for the optical head and 30.5 × 17.8 × 33 cm for the electronics

Information courtesy of EDO/ Barnes

FEWS (Follow On Early Warning Satellite)

Manufacturer: TRW/Grumman and Lockheed/Hughes/Honeywell/IBM[D]

Intended Application: USAF space based early warning[A]

Qualification: Expected to be full space qualification

Availability: Developmental, made per contract

Description:

This is a replacement for the DSP satellite based on near-term BSTS technology. FEW uses the hot emission from a rocket plume to provide accurate information on the trajectory and launch point[A] with an initial operational capability in 2003.[B] Initial concepts for conceptual design, demonstration and validation are $240 million apiece.[D] The Lockheed concept retains a scanning FPA while the TRW concept has a "toggle starer" in which short quick movements of a mirror change the field of view of the million pixels.[E] The two teams were awarded $240 million to conduct ground demonstrations and begin the design. One team will be selected to actually build the sensor.[E]

Characteristics:

Expected Program cost: 10 billion dollars[C]

Constellation: 12 to 18 satellites[B]

Orbit: Geosynchronous[D]

Information sources: (A) V. Kiernan. July 27–August 2, 1992. "AF Picks TRW, Lockheed Teams to Design Satellite." Defence News, *8; (B) A. Velocci. December 2, 1991. "TRW, Aerojet Joint New Teams in Competition for FEWS Pact."* Aviation Week and Space Technology, *46–47; (C) C. Covualt. November 18, 1991. "Astronauts to Launch Warning Satellite, Assess Manned Reconnaissance From Space."* Aviation Week and Space Technology, *66–69; (D) L. Coleman. August 1992. "Air Force Awards Dual FEWS Contracts."* Military & Aerospace Electronics, *4, 13; (E) B. Smith. August 31, 1992. "TRW, Lockheed to Test FEWS Sensors in Simulator."* Aviation Week and Space Technology, *69–71.*

IRAS (Infrared Astronomical Satellite)

Manufacturer: JPL, California Institute of Technology, Ball and Rockwell

Intended Application: Space based astronomy

Qualification: Space qualification

Availability: One made, mission completed

Description:

IRAS was a joint mission of the US, Netherlands and the United Kingdom. It was launched into a 900 km polar orbit on Jan 25, 1983 from a Delta launch vehicle. The IRAS identified over a quarter million IR sources, discovered a comet, detected objects as cold as 15 Kelvins and provided evidence of solid material around the stars Vega and Fomalhaut.

Characteristics:

Aperture: 57 cm

Cryogen: 513 liters of superfluid helium

FPA: An array of 62 detectors cooled to ≈ 2 K

Bandpasses: Four bands between 8 and 119 μm

Weight: 748 kg

U.S. Cost: ≈ $120 million

Information from NASA (JPL Fact sheet: IRAS/2/84 MBM)

ISO (Infrared Space Observatory)

Intended Application: Space

Qualification: Full space qualification

Availability: Highly developmental

Description:

The ISO spacecraft is an European-funded, large-aperture follow-on to IRAS and is pictured in Figure 11.2. It contains several instruments, including a general-purpose infrared camera, polarimeter, spectrometers, and a far IR camera providing data from 2.5 to 200 μm. The gas from evaporating cryogen is routed through the optical support structure to cool it, and the optics before it are vented to space. The telescope mirrors are made of lightweighted fused silica coated with gold. The mirrors are mounted to the metering structure via invar pads and cooled via copper heat straps. The solar aspect angle must be kept in the plane of symmetry of the sun shield and between 60° and 120°. The directions of the Earth, moon, and Jupiter must be outside cones of 77°, 24°, and 5°. First call for observatory proposals were issued 18 months before launch.

Characteristics:

Provided Detector Cooling Interface: 1.7 to 1.9 K ± 0.05 in 1,000 seconds

Primary Mirror Cooling: < 3.2 K +/ = 0.1 in 1,000 seconds

Secondary Mirror: < 4 K

Lower Baffle Cooling: < 5 K

Upper Baffle Cooling: < 7.5 K

Spectrometer Bandpass: From 2.5 to 180 μm

Far IR Camera Bandpass: 30 to 200 μm

Spectrophotometer Bandpass: 2.5 to 12 μm

Pointing Accuracy: A few arcsec for 10 hours

Cryogen: 2,000 L of liquid helium

Orbit: Elliptical about Earth with 1,000 km perigee and 70,500 km apogee

Diffraction Limited Performance: At 5 microns

Lifetime: 18 months, limited by cryogen depletion

Telescope: Richey-Chretien with 9 m focal length and 0.6 m aperture

Launch Weight: 2,400 kg

Dimensions: 5.3 m long, 3.5 m diameter

Aperture: 60 cm

Telemetry Rate: 32 kb/s

Information from: August, 1991. S. Ximenez-de-Ferran. "The ISO Spacecraft." ESA Bulletin, 67; M. Kessler, A. Heske, L. Metcalfe, and A. Salama. August 1991. "The ISO Mission—A Scientific Overview." ESA Bulletin 67; and ESA brochure on ISO.

ISOcam

Intended Application: Space, on ISO

Qualification: Full space qualification

Availability: Highly developmental

Description:

The ISOcam camera is a primary instrument of the ESA ISO spacecraft. The 32 by 32 FPA is made from InSb and Si:Ga. The readout transistor cannot operate at 2 K, so it is heated to about 60 K by dissipating 150 μW of electrical power through a resister thermally coupled to the read out JFET. A 3 pixel wide guard ring of unused pixels is implemented around the entire Ga:Si array. Each FPA views through a filter wheel mechanism which supports discrete filters and CVFs, as well as a lens wheel supporting 4 lens (like eyepieces) that give different resolutions. The camera employes a space qualified superconductor stepper motor to operate at low temperatures.

Characteristics:

Array Size: 32 × 32

Array Material: InSb for MWIR, Si:Ga for LWIR, two channels of 32 × 32 array

Pixel IFOV: 1, 5, 3, 6, and 12 arcseconds

Si:Ga Quantum Efficiency: 25%

Si:Ga RoA: 10^{16} ohms cm^2

Information sources: A. Davidson, A. Seidel, and M. Warner. February 1989. "The Cryogenic System of The ISO Satellite: Achieving Very Low Temperatures for a Large Payload." ESA Bulletin, *52–58; C. Cesarsky, F. Sibille, and L. Vigroux. 1989. "ISOcam, a Camera for the Infrared Space Observatory."* Proc. SPIE *1130: 202–213; and the ESA brochure on ISO.*

ISOPHOT

Intended Application: Infrared astronomy, using the ISO cryogenic space infrared telescope

Manufacturer: Dornier, Germany (Subcontractors: C. Zeiss, Battelle), for the Max-Planck-Institut fur Astronomie, Heidelberg, Germany

Qualification: Full space qualification

Availability: Dedicated development effort as an ISO instrument

Description:

ISOPHOT is an instrument of ESA's Infrared Space Observatory (ISO) spacecraft. It consists of two units, a sensor unit operating at 3 K in the focal plane area, and an electronics unit at ambient. A focal plane chopper provides linear scanning within the telescope FOV to within 1 arcsecond accuracy. The instrument contains several focal planes.

Characteristics:

Wavelength Coverage: 2 to 240 μm

Dynamic Range of Detectors: 10^{-12} to 10^{-18} W

NEP of Detector: ≈ 10^{-17} W/Hz$^{-1/2}$

P1 Detector (Photo-polarimeter): Single Si:Ga

P2 Detector (Photo-polarimeter): Single Si:B

P3 Detector (Photo-polarimeter): Single Ge:Ga

C200 Detector (Camera): 2 × 2 stressed Ge:Ga pixels, operates out to 240 μm @ 1.8 K

PHT-S Detectors (Spectrophotometer): 2 linear arrays of 64 Si:Ga pixels

Focal Plane Power Required: 5 mW

Instrument Weight: 21.8 kg (9.4 kg sensor unit + 12.4 kg electronics unit)

Instrument Power: 18 W

Instrument Cost: Approx. 90 million Deutsch Marks, funded by the German Space Agency (DARA) for one engineering qualification model and two flight models

Information courtesy of MPIA/ Dornier

SIRTF (Space Infrared Telescope Facility)

Manufacturer: TBD, managed by NASA JPL

Intended Application: Space-based astronomy

Qualification: Space qualified

Availability: Highly developmental, in concept stage since the late seventies

Description:

SIRTF is a large follow-on complement to IRAS and ISO. It is expected to be the fourth in a series of astrophysics observatories to be launched in the 1990s by NASA. Current thoughts on SIRTF include a 1 meter aperture cooled to 40 K with a chopping secondary. NASA plans to complement the Hubble with the SIRTF and eventually a 10 to 20 meter deployable reflector for far IR studies. The altitude is exceptionally high to ease thermal load on the spacecraft and preserve cryogen.

Characteristics:

Back Focal Distance: 100 cm[A]

Aperture: ≈ 90 cm[D]

Diffraction Limited Wavelength: 3 μm[D]

Angular Resolution: 0.7 arcseconds @ 3 μm[D]

Spectral Capability: 1.8 to 700 μm[A], capable of BLIP operations below 100 μm[A]

Life Expectancy: 5 years, limited by cryogen depletion[B]

Expected Cost: ≈ $1.3 billion[B]

First Light: 2001 (if approved by 1994)[B]

Orbit: 100,000 km high Earth orbit[B]

FPA: Mostly doped silicon and germanium bolometer arrays[C]

Dimensions: 5.5 meters long and 2.8 meters in diameter[C]

Weight: ≈ 3450 kg without cryogens and 4020 kg with cryogens[C]

Cryogen Volume: 4,000 liters[C]

Sensitivity: ≈ 10,000 times better than IRAS[C]

Pointing Accuracy: 0.25 arcseconds[D]

Pointing Stability: 0.15 arcseconds[C,D]

Information sources: (A) A. Dinger. 1988. "Straylight Analysis of an Aperture Shade Off-Set From The SIRTF Optical Axis." Proc. SPIE *967: 64–71; (B) G. Gillett, I. Gatley, and D. Hollenvach. August 1991. "Infrared Astronomy Takes Center Stage."* Sky & Telescope, *148–153; (C) NASA JPL Fact Sheet on SIRTF; (D) D. Cruikshank, M. Werner, and D. Backman. 1992. "SIRTF: Capabilities For Planetary Science."* Adv. Space Res. *12,11: 187–193*

12

Weather and Environmental Monitoring Sensors

"Now there is one outstanding important fact regarding spaceship earth, and this is that no instruction book came with it."

R. Buckminster Fuller

12.1 INTRODUCTION

In the late nineties, the orbiting espionage sentries, defense satellites, and older weather satellites will be joined by orbiting environmental and weather watchdogs. Additionally, more and more specially equipped aircraft and ground vehicles will be used to sense pollution and weather phenomena. These instruments will have the ability to monitor what cities, countries, or even factories are doing to the Earth.

Resulting from a hundred years of industrial pollution, the atmosphere is in a stressed condition. It is increasingly important for mankind to understand the environment and the weather to live in a more symbiotic relationship with the planet. The problem is that there exist gaps in mankind's knowledge and understanding, as well as insufficient early warning of problems. IR instrumentation can provide detailed measurements of specific gas species important to modeling and predicting atmospheric changes. Current atmospheric weather models have reached a level of maturity where their spatial and radiometric accuracy is comparable to (or better than) that obtained by existing satellite instruments. Increased spatial resolution and measurement accuracy are required to further refine environmental and weather forecast models.

Local and global economies are sensitive to both severe weather and long-term environmental changes. Economic reactions to weather and the environment are becoming more severe as international trade expands and capital equipment, which may be damaged or delayed by storms, becomes more expensive. Also, the geometric increase in population requires efficient food production and distribution, both dependent on environmental changes and weather.

Moreover, understanding weather and the environment is key to understanding mankind's role and its effects on our planet. Natural perturbations in the environment, such as the effects of increased solar wind or volcanoes, tend to be episodic and transient in nature. Investigators are concerned that the effects of human activities are not so transient and are slowly causing long-term changes that may take centuries to correct—if they can be corrected at all. Environmental research can yield wonderful knowledge, through which we can build an industrial society more in harmony with the Earth. Some may view the goal not so much as an attempt to protect the Earth as an attempt to protect our own hides, as long-term effects of abuse can be devastating to the economy or even to the human race.

The weather in the United States is particularly bad. The U.S. tends to have a few orders of magnitude more tornadoes and severe thunderstorms than any other country. There are also factors of two to five more hurricanes, blizzards, hailstorms, and even earthquakes than most countries. Such violent phenomena, coupled with the value of U.S. real estate and the emphasis placed on life, means that a significant focus on early warning of bad weather is well justified.

Key to completing the above is a thorough understanding of long-term weather and environmental patterns, such as the El Niño effect. Unfortunately, this is where meteorologists have the least skill and knowledge. They have become very proficient at predicting local weather for a few days but are still poor at telling a farmer whether the next year or two will be wetter or dryer than the past year.

As part of the answer to these problems, the United States has embarked on a major space program through NASA called the Earth Observing System (EOS). It will result in several satellites coordinated with ground and airborne systems spanning decades of observation of the Earth's environment. NASA expects to spend about $30 billion, half in the nineties, on this endeavor. However, EOS is just part of a major U.S. government effort to learn more about the environment. Other major agencies with similar tasks include the DoD, the Department of Agriculture, the Department of Commerce, NOAA, the National Science Foundation, and even the Smithsonian Institution. Additionally, the U.S. effort is only part of a larger international effort involving hardware from all the leaders in weather and environmental satellites, including the ESA, Japan, Canada, and the CIS.

The EOS goal is to understand the Earth as an integrated environmental system: "To advance understanding of the entire Earth system on the global scale, by developing a deeper comprehension of the components of that system, the interactions among them, and how the Earth system is changing. To quantify changes in the Earth system, EOS will provide systematic, continuing observations from low Earth orbit for a minimum of 15 years" [NASA, 1].

Several objectives supporting this goal rely in part on infrared observation. These include the efforts to create an integral scientific observing system and to acquire and assemble a global database for remote sensing measurements including [NASA, 2]

- global distribution of energy input and output from the Earth
- structure, state variables, composition, and dynamics of the atmosphere
- circulation, surface temperature, wind stress and status of the oceans and seas
- extent, type, state, elevation, roughness, and dynamics of glaciers, ice sheets, snow, and sea ice

To satisfy these aims and goals, a series of multispectral instruments are being developed. Multispectral imaging is a powerful tool allowing comparison from UV to IR representing different chemical species, air movements at different altitudes, or other phenomena. Sensor fusion (see Chapter 6), radiometric calibration, pointing accuracy, and registration are typical key issues. Table 12.1 lists the major IR instruments currently being developed for the space segment of EOS.

In addition to the specific missions mentioned above, general environmental surveillance in the infrared is beneficial for several reasons. For example, sensors on both GOES and polar orbiters have found remote forest fires and provided the initial direction to fire fighting crews. Gross global IR monitoring of ocean temperature helps meteorologists predict the severity of hurricane seasons and of the development of El Niño effects and other phenomena.

Satellite imagery is being used to locate areas where rare plants grow. Even some common plant species lack current, detailed geographic mappings of their locations. For example, no one paid much attention to the locations of the Pacific Yew tree until a cancer drug, Taxol, was derived from its bark. Satellite imagery effectively pinpointed growths of the trees [Polsky, 3]. No one can tell what plants we may need to identify next from orbit or from aircraft.

IR sensors play pivotal roles in the instruments used to understand and monitor the environment. IR observation can aid in solving these issues by supplying early estimates of crop failures, droughts, and global and regional food production, as well as changes in deforestation rates, desertification, and the greenhouse effect. Such phenomena must be quantified before far-reaching remedies can be implemented. IR sensors are generally used in one of four ways: (1) general thermal imagers to measure temperature, (2) spectrophotometers measuring the energy output (in terms of radiance) of specific atomic or molecular species' generated lines, (3) sounders providing temperature profiles as functions of altitude, and (4) as parts of multispectral collection efforts.

The basing of environmental sensors is not confined to space. Rather, it covers the gamut. Current plans call for IR environmental monitoring instruments based on planes, ships, balloons, on the ground, and in space. The reader is referred to the chapters concerning ground and space-based sensors for more information.

TABLE 12.1 Major EOS IR Instruments

Name	*Abbrev.*	*EOS Flights*	*Type*	*Accuracy*	*Bandpass (μm)*	*Spatial Res.*	*Cryocooling*	*Stability/Jitter*
Atmospheric IR Sounder	AIRS	Selected for EOS-A1, tentative for A2 and A3	Spectral sounder	0.05 emissivity, 1 K temp.	0.4 to 15.4	1 km @ nadir	Redundant 60 K Stirlings to a two-stage radiator	TBD
Advanced Spaceborne Thermal Emission and Reflection Radiometer	ASTR	Selected for EOS-A1, tentative for A2 and A3	Multispectral imaging radiometer	4%	0.5 to 12	18 to 128 μrad	80 K Stirling	2 pixels over 60 s with 1–2 pixel/s jitter
High Resolution Dynamic Limb Sounder	HIRDLS	Selected for EOS	Scanning IR limb sounder		6 to 18	1 by 10 km	Paired 80 K Stirlings	180 arcsec stability with 1–5 arcsec jitter
High Resolution Imaging Spectrometer	HIRIS	Tentative for EOS A2 and A3	Imaging spectrometer		192 channels from 0.4 to 2.45	30 m @ nadir	Radiators	1.08 arcsec stability with 7.2 arcsec jitter
Measurements of Pollution in the Troposphere	MOPITT	EOS A1	Gas cell correlation spectrometer	CO_2 concentration accuracy 10%, CH_4 concentration 1%	Narrow bands around 2.3, 2.4, and 4.7	22 by 22 km	80 K Stirling	TBD
Spectroscopy of the Atmosphere Using Far Infrared Emission	SAFIRE	Candidate for EOS B	Limb spectrometer		80 to 160, 310 to 390, and 630 to 1,560	1.5 to 3 km	Five 80 K Stirlings	1 arcsec/s
Stratospheric Wind Infrared Limb Sounder	SWIRLS	Candidate for EOS B	Limb viewing radiometer	1–2 K for temp. and 10% for species	Six narrow bands of about 7.8, 8.6, 8.9, 9.7, 15.4, and 16.5	3 km vertical, 200 by 350 km horizontal	Stirlings	1 arcsec/s
Tropospheric Emission Spectrometer	TES	Candidate for EOS B	Fourier transform spectrometer		2.3 to 16.7	7.5 by 0.75 mrad	Stirling	15 arcsec/32 s

Source: August, 1991. *1991 EOS Reference Handbook,* NASA GSFC Publication NP-144 (May 1991).

Much of this chapter concentrates on space basing, because this is the most glamorous and advanced technology, and because the most information is available on these instruments.

A ground-based IR camera with proper filters can do an excellent job of remote effluent detection, detecting heat pollution from specific sites, and estimating the bulk of effluent emissions. These instruments usually have well defined bandpasses, and probably contain interchangeable filter wheels, circular variable filters, or acousto-optical tunable filters. The cameras are easily serviced and modified for specific missions. One application allows environmental police to remotely monitor emissions of smokestacks from over a kilometer away. They can also be used to remotely monitor pollutants in tailpipe emissions.

Air-based sensors offer higher ground resolution than space-based sensors, and allow close-range, direct overflights of suspected polluters or environmentally stressed areas. The instruments are relatively easily modified for specific missions. Most instruments are line scanners requiring specific flight patterns.

Overflights must be approved by cognizant officials of governments and large area coverage is costly. Because of limited numbers of dedicated air platforms, inspection and data collection activities may be hindered. However, "airborne multispectral imagery can reveal changes in conditions that may not otherwise be visible to field investigators. Imagery can discriminate conditions such as differences in vegetation, soils, drainage, underground features, hydrocarbon features, and material characteristics" [Olson, 4]. Additionally, multispectral airborne observations have been effective in pinpointing chemical polluters [Olson, 5].

For weather monitoring, ER-2 high flying aircraft can provide in-situ measurements of thunderheads that extend up to 70,000 feet. Aircraft may also be used to observe severe weather: C-130s are routinely flown into the eyes of hurricanes for close-up measurements. Airborne sensors can also be very effective at identifying vegetation stress and quite useful for forest health check-ups.

The orbits for space basing run the gamut from low altitude polar to geosynchronous orbiters. Polar orbiting satellites circle the globe at an altitude low enough to permit several orbits each day. They have the advantages of overflying the environmentally interesting polar regions and of being low enough to have high spatial resolution, resulting in a small ground footprint. For instance, the latest generation of NOAA polar orbiting satellites operate in near sun-synchronous orbits with an inclination of 98.9 degrees and a period of 102 minutes. The geosynchronous orbits provide continual monitoring of about one-third of the Earth. Several satellites spaced around the globe provide complete coverage, except for the polar regions. However, because a sensor placed in geosynchronous orbit is so far away from the Earth, spatial resolution is poor.

Satellites of this class are general weather platforms or multi-use environmental platforms, such as the TIROS-N polar orbiting satellite. Old weather satellites were spin stabilized, while the more modern GOES Next and the EOS platforms are three-axis stabilized. Old weather satellites often used a single instrument for many purposes. Separate instruments, optimized for separate functions, are now used, as opposed to the "one instrument does all" philosophy of the past. Weather satellite instruments include high-resolution imaging radiometers that sense clouds and sounders that sense temperature and moisture data from the Earth's surface through the atmosphere. Instrument suites are not confined to the IR. Radar, UV, visible, and millimeter spectra provide complementary and important data. In fact, the European ESA-sponsored ERS-1 contained a scanning radiometer but mostly concentrated on radar instruments.

Earth monitoring sensors represent one of the IR growth fields in the 1990s. Understanding the requirements is of paramount importance, and it is difficult to break into an incumbent's territory. Newcomers to the field frequently fall far short when it comes to dealing with the intricacies and problems associated with calibration, and they propose exotic technologies to perform common missions. Customers are usually excessively conservative, not wanting to stray into the wilds of the state of the art.

To capture the growing market, many companies are setting up special centers of excellence to effectively port intelligence and military technology to remote sensing. Lockheed has established a Center for Remote Environmental Sensing Technology that will concentrate on spacecraft, instruments, and data processing [Saunders, 6]. A key point to remember is that the entire remote and environmental sensing fields are growing, not just the platforms and instrumentation. There will be so much data generated in so many bands that post processing for image fusion and interpretation is likely to be a vast effort with new information flowing from data decades after it was acquired. There will be major growth in specialized "third party" image processing, data handling, and interpretation.

There is a move to commercialize much of the general remote sensing data acquired by orbiting, airplane, and ground instrumentation. Commercial sales of instruments and data processing is a potential growth area for the late nineties. Currently, one can order photographs and other data from NOAA/NESDID/NCDC. They even take Master Card and VISA. Additionally, SPOT is going completely commercial, expecting to be free of all government involvement by mid-1998 [Saunders, 7]. This commercialization has even extended to the former Soviet Union—the CIS now offers orbital images, presumably from one of their spy satellites.

12.2 ENVIRONMENTS

The effect of the environment upon instruments' long-term stability is of concern because these instruments frequently measure long-term effects (up to five years for the EOS program). The environments are dependent largely on the basing and use of the sensors. Ground-based instruments will have environments similar to those discussed in Chapter 8 on ground-based cameras. Space-based sensors will have to survive and function in the environments described in Chapter 11.

12.3 TARGETS AND BACKGROUNDS

Environmental and weather monitoring occurs against a backdrop of chemical activity. The targets, as represented in Table 12.2, are usually a specific emission line of water for the weather satellites and a line of some pollutant (or reaction by-product).

Water pollutants can be spotted in the infrared by a change in temperature and transmissivity, usually as a function of elevation. This requires high resolution and an NEDT of less than a degree. Pollution on the surface of water can be detected by a change in reflectance at specific wavelengths. Usually the wavelengths are confined to the UV, visible, and SWIR, as sufficient solar radiance is needed to supply the reflection source.

Effluents in the lower atmosphere usually rise and diffuse. They may be detected by molecular emission, molecular absorption, or a change in temperature. Lower atmosphere pollutants are usually detected by ground-based or low-flying aircraft-based sensors.

Much recent attention to the environment is placed on the upper atmosphere, where monitoring is the domain of the space sensors. The upper atmosphere is a chemically dynamic place, consisting of a stew of source molecules, short-lived free radicals, and stable molecules—all energized by UV radiation and cosmic rays—undergoing thermal changes on an hourly basis. Source molecules migrate up from the troposphere into the stratosphere, where UV radiation disassociates them into reactive free radicals. Major source molecules are methane, nitrogen compounds, and chlorofluorocarbons. Free radicals can react with other molecules. In fact, it is the reaction of free radicals that, when acting like a catalyst, depletes the ozone. Eventually, all free radicals are combined into more stable molecules such as NO, CO, OH, H_20, which migrate back into the troposphere and get rained out. Common molecular species in the upper atmosphere include NO, H_2O, O_3, OH, CO, CO_2, CH_4, H_2, N_2O, HNO_3, H_2O_2, HCl, N_2O_5, and CH_3Cl.

TABLE 12.2 Representative Targets and Backgrounds

Function	*Representative Targets*	*Background*
Cloud monitoring	Water lines, thermal imaging	Hard earth, limb, various altitudes of atmosphere
Environmental	Ozone, nitro-oxides, water, carbon monoxide, carbon dioxide	Hard earth, limb, various altitudes of atmosphere
Sounders	Radiance from thermal emission	Limb, various altitudes of atmosphere
Weather sounders	Carbon dioxide rotational bands	Earth, limb, various nadir altitudes of atmosphere

12.4 FIGURES OF MERIT

Environmental sensors are like most IR sensors in that their performance is usually driven by the size of the optics, field of view, update rates, background, detector sensitivity, and range. The typical system figure of merit for environmental monitoring sensors may be the NEDT discussed in Section 9.4 or the NEI discussed in Section 13.4. Another frequently used figure of merit is the noise equivalent change in radiance, also called the noise equivalent delta radiance (NEΔN or NEDN) and sometimes the noise equivalent radiance difference (NEND). Remember that "N" is the old (but still frequently used) symbol for radiance. In this chapter, we will discuss NEDN because it seems to be a quoted radiometric figure of merit for many environmental and weather monitoring instruments.

NEDN is analogous to the NEDT of cameras, except the measurement is based on radiance, not temperature. Using radiance gives a better figure of merit with fewer assumptions and limiting conditions. It is also more appropriate for instruments that are measuring changes in radiance, in which temperature may be inferred if enough is known about the scene. NEDN is equal to the radiance difference that must be present in a scene to produce a change out of the focal plane equal to the resultant noise of the entire instrument. Again, it is similar to NEDT, except the minimal difference is measured in radiance, not temperature. One pixel can be distinguished from its neighbor when the average radiance of the scene composing the pixel differs by some amount times the NEDN. NEDN is an excellent figure of merit for an instrument that is built to detect changes in average scene radiance (N). It can be mathematically expressed as

$$\mathrm{NE\Delta N} = \frac{(A_d \Delta f)^{1/2}}{A_o \Omega D^* K_e K_o K_b K_s (\Delta \nu)} \tag{12.1}$$

where

NEΔN = increment in radiance that produces a signal equivalent to the system noise (SNR of 1)

A_d = Area of the detector (in centimeters)

Δf = noise bandwidth (in hertz) as approximated by Equation (8.5)

A_o = clear aperture area (in square centimeters)

Ω = solid field of view in steradians

D^* = detector D^* evaluated with as many noise sources as possible

K_e = electronic efficiency, usually 0.5 to 0.9

K_o = in-band average optical train transmission including all elements and filter

K_b = optical blur efficiency

K_s = scan and chopping efficiency

(Δv) = bandpass in microns (This may be omitted if the desired inputs are for just radiance and not spectral radiance.)

Ranges do not enter into the equation, because it is assumed that the object being observed overfills a pixel (is imaged). The quantity $A_0\Omega$ is called the *etendue invariant.* It is a useful quantity for spectroscopic instruments. Generally, the bigger, the better, and whatever the value, the user always wants more. Most interferometric devices have a small etendue driven by a small solid angle. This is especially true of modern spectroscopic devices for environmental and weather monitoring. This is one of the great benefits of crystal-based AOTFs (see Chapter 3) that permit large etendues with FOVs of up to tens of degrees.

Again, any figure of merit has issues and concerns when it is applied to a whole system for different sets of circumstances and conditions. Table 12.3 highlights some of the general issues associated with NEDN.

The signal-to-noise ratio can be derived using NEDN by simply comparing the differences in radiance between the required detected variations of the background with the NEDN. Typically, signal-to-noise rations of two to five are required for minimum usefulness.

12.5 SYSTEM EFFECTS OF COMPONENTS

Basically, most of the system effects mentioned in previous chapters apply to environmental instruments of the same basing.

Optics must be lightweight for space applications. Beryllium has been a traditional favorite of weather satellites. Silicon carbide is being considered for many EOS and second-generation weather instrumentation satellites. The instruments usually respond over a wide bandpass. This favors reflective systems for their lack of chromatic aberration and their flat transmission over large spectral bandpasses. The use of all-reflective optics also simplifies wideband spectral calibration. Usually, resolution requirements for airborne and ground systems are for a projected footprint of a few meters at a range of a few kilometers. This is not especially stressing. However, the typical requirement for resolution from space is a 1 km (or smaller) footprint, implying an approximately 28 μrad resolution for geosynchronous observation. This also necessitates large apertures and highly stabilized systems. The rapid orbiting of the Earth results in the instruments experiencing rapid thermal transitions as they go from sunlit conditions to darkness, implying that the optics must be athermalized or self-compensating for such space missions.

Environmental instruments rarely use large, two-dimensional arrays due to their dead pixels, low uniformity, high cost, and calibration problems. It is always easier to calibrate 10 or 1,000 pixels than it is to calibrate 100,000. The smaller the array, the easier it is to choose one that has high uniformity and calibration repeatability. Weather satellites use discrete PC HgCdTe detectors. They are oper-

TABLE 12.3 Concerns with NEΔN

Concern	*Recommendation*	*Comments*
1/f noise	Include in D* or as a separate noise term.	Including in the D* is fine for a given readout rate, but it is not constant if readout varies.
Aging effects	Degrade everything.	
Clutter effects	If it is an issue, a different FOM should be used.	A clutter-specific FOM should be used where clutter is a driver. NEΔN does not address clutter.
DC background effects	These are properly addressed by this FOM.	If saturation is not occurring.
Fill factor	Should be included in K_b.	This is especially critical for monolithic FPAs.
Filter corner effects	Degrade/adjust the value for the bandpass.	
Not square detectors	Addressed in A_d and K_b	
Optics noise effects	May be included in D* or fudge factor.	Must be considered for long-wavelength devices; not properly addressed by this version of NEΔN.
Spectral effects	Use weighted average over bandpass.	
Spectrally sensitive detectors	Must integrate D* by the detector's sensitivity function for a weighted average. This is easily done on a spreadsheet.	Equation (12.1), using a D*, does not properly represent Schottky barriers or quantum wells with bandpasses of more than approx. 0.5 microns.
Stray light	Extra noise can be included in effective D* or as a fudge factor.	This is not properly addressed by NEΔN.
TDI	One should decrease the NEΔN by slightly less than the square root of the number of TDI stages (or elements).	

ated at elevated temperatures of 105–115 K, with cutoffs as long as 14 μm. It is probable that this trend will begin to change in the nineties, and imaging arrays will enter the realm of environmental sensors, especially in ground-based units.

The cooling mechanism is largely determined by the sensor's basing. Ground instruments typically use pour-filled dewars of cryogen, although Stirlings are beginning to show up. Radiators have been the traditional coolers for weather satellites; long-life Stirlings are now being considered. The GOES Next weather satellite employs radiators to cool the IR detectors on the instruments. The radiators are concealed by doors before and during launch to prevent contamination. Even newer instruments are frequently required to have radiators, as weather bureau users feel comfortable with their reliability and lack of vibration. In contrast, many of the space environmental scientific IR instruments are using long-life, space-qualified Stirling coolers like the ones discussed in Chapter 5.

Space-based and airborne instruments usually do not have complex image processors—just signal conditioners, processors, and data compressors. The actual detailed analysis, image processing, and interpretation is done "off-line" on the ground, at a later time. There does not seem to be a great undertaking to enhance the real-time image processing capabilities of either weather or environmental satellites. This is probably because the scientists want uncorrupted, hard data to mull over (and over, and over...).

Since most environmental instruments have small FPAs, they must be scanned. The scanning, pointing, and stabilization mechanisms can become quite complicated and large. When space-based, they must be momentum compensated to exhibit no external torques. This is to prevent the spacecraft bus from moving in reaction to the instrument. The stability and registration requirements continue to increase as multispectral processing and high spatial resolution are required of more instruments.

Satellite data rates are normally limited by their telecommunication downlinks. This restricts the frequency of the obtained data. The GOES satellites nominally image every half hour or 15 minutes but have the capability of providing images every three minutes in times of severe weather.

12.6 SYSTEM DRIVERS

Infrared weather instrumentation and, to a lesser extent, environmental monitoring sensors are characterized by

- lack of risk in development
- narrow bandpasses centered on rotational lines
- high reliability
- high radiometric sensitivity
- high level of calibration

Usually, customers want a low-risk approach, using tried-and-true methods and proven components to solve specific requirements. Whether one designs or requires a serial scanner, line scanner, or starer is a trade-off decision that must be made at the system level. This decision encompasses FPA (cost, producibility, and so on), cryocooler, gimbal, and telescope. However, the final system exhibiting the lowest cost, highest producibility, and lowest mass can only be determined at the system level. Each sensor has application-specific requirements that can easily drive the decision one way or another.

Regardless of the emphasis on reliability and lack of developmental risk, EO instruments are not simple to make. There can still be schedule, cost, and performance problems. The long wavelength channel of the GOES sounder suffered from greatly reduced sensitivity because of higher than expected loss in the optical path and lower than expected detector performance. In the early nineties, it became obvious that the GOES Next instruments would not be ready for their proposed flight schedule. Many problems have been traced to underestimating the complexity of the spacecraft and instruments, coupled with insufficient up-front engineering. The first problem is common with EO instruments, as demonstrated by the Hubble, Teal Ruby, BSTS, SSTS, and the GOES Next instrumentation. The latter problem has only been recognized in the past few years.

Reliability is a key system driver. For scientific missions, it is important due to lost opportunity costs. For weather satellite missions, high reliability is needed to provide critical data to users. Reliability is usually quoted at a 0.95-0.99 confidence level for three to five years.

Absolute calibration is of key importance for scientific data gathering. Currently, neither government nor industrial facilities are capable of achieving the high accuracy of calibration required. Many instruments have absolute requirements of one to ten percent, which is extremely difficult to achieve in IR bandpasses. Calibration accuracies may also be quoted as within 1 K at 273 K or as an NEDN. Accurate absolute calibration calls for dedicated facilities and excellent, repeatable standards directly traceable to NIST. Experience counts when calibrating. It is practically impossible for the novice to achieve such calibration levels. In fact, this is such a large and important issue that NASA has established a special panel on EOS calibration.

Many newer instruments must be designed to fly on more than one spacecraft bus. The VIRSR instrument was required to be able to be placed on both the EUMETSAT and NOAA spacecraft. The Total Ozone Mapping Sensor (TOMS) has flown on American and CIS spacecraft. Many of the EOS instruments are scheduled for more than one spacecraft, and some are being considered for international missions. Designing for multiple platforms places increased emphasis on the need for the instrument to contain all functions in complete and small packages. This implies a "black box" approach to instrument design.

Cost can be dependent on the general basing and the number of sensors made. One should use space costs for space-based sensors, and ground costs for ground-based sensors. Environmental sensors will usually cost from 10 to 33 percent more than their counterparts, a product of the need for calibration and highly defined custom bandpasses. A rough guess would be that the development and prototyping of a ground-based camera will run between \$5 and \$10 million, between \$10 and \$20 million for an airborne instrument, and between \$20 and \$200 million for a space-based instrument.

Weight is always a concern but is somewhat less so for weather and environmental satellites than for many military satellites or interplanetary probes. Users will sometimes accommodate a larger weight to reduce risk and cost. It has been estimated that a spaceborne radiometer's weight could be reduced by 35 percent by using graphite epoxy for the structure, yet this idea is still waiting to be used in weather satellites [Hookman and Zurmehly, 8]. However, like any spacecraft, once the weight is budgeted, any increase will cause great concern, consternation, and potentially, the cancellation of the instrument or spacecraft. To scale the weights, use the weights of the examples in this chapter and scale each component by the scaling relations presented in Part 2.

Placement on the satellite or aircraft platform is a prime concern. For a satellite, the placement must provide the necessary clear field of regard, sufficiently small sources of stray light, proper heat sinks, and radiator clear view. For airborne instruments, the placement must provide the necessary clear field of regard, sufficiently small sources of stray light, proper heat sinks, and interfaces. The whole realm of platform integration should be addressed early in the program. Although unglamorous, matching the bolt patterns, thermal dynamics, vibration, and electrical interfaces is critical and time consuming. Intricate details are easily overlooked. The actual platform integration is usually a lengthy process for American satellites. However, American scientific instruments have been integrated onto the CIS' Meteor 3 spacecraft as little as two weeks prior to launch [NASA, 9]!

The key trade-offs and analyses in the design of an environmental sensor revolve around a proper flowdown of requirements from the scientific mission. Specific critical trades and analysis that must be analyzed in detail usually include

- verification that the science goals are satisfied
- calibration accuracy and repeatability
- instrument lifetime
- cooling methodology
- LOS stability and knowledge
- data compression and recording
- bandpass and fine tuning bandpass
- type of spectral filter (reliability, edge effects, calibration effects, etc.)

References

1. 1991. *The 1991 EOS Reference Handbook.* Greenbelt, MD: NASA Goddard Spaceflight Center, 9.

2. Ibid.
3. D. Polsky. October 28, 1991. "Satellite Images Assist Drug Makers in Cancer Quest." *Space News,* 7.
4. J. Olson. August 1992. "Airborne Multispectral Imaging Monitors Environment." *Photonics Spectra,* 71.
5. Ibid.
6. R. Saunders. September 16, 1991. "Lockheed Turns Attention to Environmental Research." *Space News,* 16.
7. R. Saunders. January 21, 1991. "SPOT Chooses Mid-1998 For Self Sufficiency." *Space News,* 33.
8. R. Hookman and G. Zurmehly. 1990. "The Applications of Composite Materials to Spaceborne Radiometer Instrument Design." *Proc. SPIE* 1303: 465–473.
9. Information from NASA GSFC Meteor-3/TOMS Press Kit. Greenbelt, MD: NASA Goddard Spaceflight Center Publication.

Appendix 12A

Data Sheets for Typical Instruments and Platforms

ABS (Airborne Bispectral Scanner)

Manufacturer: Daedalus Enterprises, Ann Arbor, Michigan

Intended Application: Various airborne infrared and ultraviolet line scanning

Qualification: Air flight

Availability: Custom made from continuing line; delivery ≈ 8 months ARO

Description:

The ABS is a complete airborne infrared line scanning system. The ABS consists of an line scanner sensor head and electronics. The rack mounted electronics consist of an operator console, digitizer, and a power distribution unit. The sensor scans in two spectral bands, a standard 8.5 to 12.5 μm and a band of the user's choice. The sensor head is remotely located from the other units and must view directly down for proper registration.

Characteristics:

Scan Rates: 100, 50, 25, 12.5, and 6.25 scans per second

IFOV: 2.5 mrad with options for either 1.25 & 5.0 mrad

Digitized FOV: 86°

Roll Correction: ±15°

Operating Temperature for Scan Head: –55° to +70°C

Power Required: 28 Vdc ± 3 V/40 A

Digitization: 8 or 12 bits, operator selectable

Scan Head Dimensions: 38 × 38 × 38 cm

Total System Weight: 81 kg

Price: ≈ $466,000

Information courtesy of Daedalus Enterprises

AIRS (Atmospheric Infrared Sounder)

Manufacturer: Loral, Lexington, Massachusetts

Intended Application: EOS and weather satellite platforms

Qualification: Full space qualification

Availability: Developmental

Basing: Space

Description:

AIRS is one of the main instruments and the chief atmospheric sounder on EOS. It is designed to meet the requirements of the NASA Earth science research programs and NOAA's operational plans. The design is a two pupil imaging, multi-aperture grating spectrometer providing contiguous spectral coverage throughout the range. The optical bench is passively cooled to 155 K.

Characteristics:

Scan pattern: Every 2.7 s, a ±49.5 degree swath across nadir in a direction perpendicular to the satellite's ground track.

Bandpass: 3,600 spectral measurements within 3.4–15.4 μm, plus six discrete visible and near-IR bands

NEDT: 0.2 K

FPA: Linear arrays of HgCdTe operating at 60 K

Data Rate: 1.85 Mbps

Volume: Approximately 95 × 80 × 117 cm (excluding sunshade).

Weight: 114 kg

Power required: About 300 W

Information from M. Cahine. 1991. "AIRS The Atmospheric Infrared Sounder." Optics and Photonics *2,10: 25–28.*

ATM (Airborne Thematic Mapper)

Manufacturer: Daedalus Enterprises, Ann Arbor, Michigan

Intended Application: Airborne simulation of Landsat's Thematic Mapper

Qualification: Air flight

Availability: Custom made from continuing line; delivery ≈ 9 months ARO

Description:

The ATM is a complete airborne line scanning system. The ATM consists of a line scanner sensor head and electronic boxes. The rack-mounted electronics consist of an operator console, digitizer, tape remote, and power distribution unit. The sensor head is remotely located from the other units and must view directly down for proper registration. Active temperature compensation is used to maintain band to band registration. The difference between the ATM and ATMX is larger collecting optics for low albedo targets and the exchange of other spectrometer assemblies.

Characteristics:

Scan Rates: 100, 50, 25, 12.5 scans per second

Spectral Bands: 11 from 0.42 to 13 μm

IFOV: 2.5 mrad with a 1.2 mrad option

Digitized FOV: 86°

Roll Correction: ±15°

Operating Temperature for Scan Head: –55° to +70°C

Power Required: 28 Vdc/45 A

Scan Head Dimensions: 45 × 40 × 40 cm

Scan Head Weight: 46 kg

Total System Weight: 193 kg

Price: ≈ $954,000

Information courtesy of Daedalus Enterprises

AVHRR/3 (Advanced Very High Resolution Radiometer)

Manufacturer: ITT, Ft. Wayne, Indiana

Intended Application: NOAA weather satellite

Qualification: Full space qualification

Availability: Custom made

Basing: Space

Description:

The instrument scans an annulus below the spacecraft. It is used to remotely determine cloud cover and surface temperature. Six discrete detectors simultaneously measure the scanned scene, providing registered multispectral data useful for oceanographic, hydrologic and meteorological studies.

Characteristics:

Spectral Range: 0.58–0.68 μm, 0.725–1.0 μm, 1.58–1.64 μm, 3.55–3.93 μm, 10.3–11.3 μm, and 11.5–12.5 μm

Detectors: Silicon, InGaAs, InSb and HgCdTe

Resolution: 1.3 milliradians

IR NEDT: < 0.12 K @ 300 K

Scan angle: ±55 degrees by a continuously rotating 360° cross track line scan

Optics: 20.3 diameter afocal Cassegrain

Scanner: 360 rpm hysteresis synchronous motor

Scan mirror material: Beryllium

Channel-to-Channel Registration: Better than 100 μrad

Cooler: Two-stage radiant cooler, IR detector controlled at 105 K

Data Output: 10 bit binary simultaneous sampling at 40 kHz rate

Design Life: 3 to 5 years

Power Consumed by Instrument: 28.5 W

Instrument Weight: 29.3 kg

Information courtesy of ITT

BAMM

Manufacturer: Talandic, Irwindale

Intended Application: Balloon background observations

Qualification: Flight qualification on balloon

Description:

BAMM IIA was a balloon measurement program of the early eighties. The sensor head contained a multispectral focal plane and z-plane electronics. A liquid nitrogen dewar supplied cooling.

Characteristics:

Bandpass: 2.6 to 5.1 μm with a 12-position filter wheel, 11 filters, and a blocked position FOV, selectable between three cooled F/2.3 objectives

Total FOV: 25.6° × 25.6°, 25.6° × 6.4°, and 6.4° × 1.6°

Largest Pixel IFOV: 1.7 millirads

Aperture: 2.5 cm

Data Rate: 0.5 Mbps

Cost: ≈ $3 to 6 million

Weight: ≈ 100 kg

Dewar Consumption: 2.8 liters in 16 hours

Information courtesy of Talandic Research

DSMP (Defense Meteorological Satellite Program)

Manufacturer: Team lead by GE/Astro (now part of Martin Marietta)

Intended Application: U.S. military weather satellite

Qualification: Full space qualification

Availability: Custom made

Basing: Space

Description:

The Air Force maintains at least two Block 52D spacecraft. The satellites orbit the globe about twice per day to provide detailed weather information to military forces. They are commanded by the Satellite Operations Group at Offutt Field, NB.

Characteristics:

Resolution: 2.9 km

Primary Imaging Instrument: Visible/infrared line scanner

Spacecraft Weight: 795 kg

Instrument Weight: 236 kg

Spacecraft Volume: 1.2 × 5.8 m

Solar Array: 12,500 cells providing 1,000 W

Information from C. Covault. November 26, 1990. "USAF Space Command Weather Satellite Images Persian Gulf Facilities at Night." Aviation Week and Space Technology, *28–29*

ER-2 High Altitude Earth Resources Survey Aircraft

Manufacturer: Lockheed

Intended Application: Remote sensing for NASA Ames Research Center

Qualification: Flight

Availability: Currently flying, can be used to support experiments

Description:

The ER-2 is an upgraded modernized version of the U-2 and is about 30 percent larger. NASA uses the ER-2 as a readily deployable high-altitude generic platform for sensors. It is especially useful in collecting remote sensing and in situ data on Earth resources, celestial phenomena, atmospheric dynamics, oceanic processes, and thunderstorm research. Flying at over 20 km in altitude, it is above 95 percent of the earth's atmosphere and has an effective horizon of 480 kilometers. More than 100 different sensors and experiments have been flown aboard the ER-2. NASA has an inventory of multispectral imagers including the AVIRIS and TMS.

Characteristics:

Crew: One pilot

Altitude: over 20 kilometers

Speed: 410 knots

Payload Capacity: 280 kg in nose, 340 kg in Q bay and 620 kg in wing pods

Viewing Capability: Zenith and nadir

Information courtesy of NASA

ERS1 IR Radiometer

Intended Application: Environmental sensing on ERS-1

Qualification: Full space qualification

Basing: Space

Description:

The primary objective is the measurement of sea surface temperature. Images assembled of the burning oil fields of Kuwait indicate an NEDT of 0.3 K.[B]

Characteristics:

Spectral channels: 1.6 μm, 3.7 μm, 11μm and 12 μm[A]

Spatial resolution: 1 km squared[A]

NEDT: <0.1 K[A]

Cooling: 80 K by a Stirling[A]

Optics: Rotating plane scan mirror feeding a f/2.3[A]

Calibration: Via two blackbodies on every scan[A]

Information sources: (A) F. Francis et al. February 1991. "The ERS-1 Spacecraft And Its Payload." ESA Bulletin *65: 27-48; (B) P. Jackson. November 1991. "Infrared Pictures From ERS1 Show Kuwait Oil Fires."* Lasers and Optronics, *15.*

GOES Next Platforms

Manufacturer: Loral Space Systems

Intended Application: NOAA, NASA

Qualification: Full space qualification

Availability: Custom made, based on previous work

Basing: Space

Description:

The GOES Next is an upgrade to the old geosynchronous GOES weather satellites. It is the first geostationary, three-axis stabilized operational weather satellite.

Characteristics:

Instrumentation: Separate imager and sounder

Weight: Approximately 1,000 kg

Volume: About 2.5 meter cube

IR detector cooling: Radiators

Basic Satellite Stabilization: About 200 μrad

Information courtesy of NOAA

GOES Next Imager Sensor Assembly

Manufacturer: ITT, Ft. Wayne, Indiana

Intended Application: NOAA weather satellites

Qualification: Full space qualification

Availability: Custom made

Basing: Space

Description:

The sounder sensor assembly contains a five-channel (bandpass) multispectral calibrated imager that is integrated into the spacecraft bus.

Characteristics:

Optical Aperture: 31.1 cm

Channel Alignment: Better than 28 microradians

Data Rate: 2.6208 Mbps

Detector Cooling: IR detectors radiant cooled and regulated to 102 K

Design Life: Five years orbital

Weight: 120 kg

Power: 119 W

Sensor Assembly Size: 79 × 76 × 114 cm

Channel One: 0.55–0.75 μm with a S/N of 150:1

Channel Two: 3.8–4.0 μm with NEDT of 1.4 K @ 300 K

Channel Three: 6.5–7.0 μm with NEDT of 1 K @ 230 K

Channel Four: 10.2–11.2 μm with NEDT of 0.35 K @ 300 K

Channel Five: 11.5–12.5 μm with NEDT of 0.35 K @ 300 K

Information courtesy of ITT

GOES Next Operational Environmental Satellite Sounding Subsystem

Manufacturer: ITT, Ft. Wayne, Indiana

Intended Application: NOAA weather satellites

Qualification: Full space qualification

Availability: Custom made

Basing: Space

Description:

The sounder sensor assembly contains a 19-channel (bandpass) multispectral calibrated radiometer that is integrated into the spacecraft bus.

Characteristics:

Optical Aperture: 31.1 cm

Spatial Resolution: 242 microradian (8 km nadir) diameter IFOV

Weight: 123 kg

Power: 106 W daily average

Design Life: Five years in orbit

Sensor Size: 79 × 76 × 137 cm

Electronics: 18 × 38 × 66 cm

Power Supply: 15 × 20 × 29 cm

Spectral Range: 0.7–14.7 μm

IFOV: ≤ 224 μrad

Information courtesy of ITT

HI-CAMP

Manufacturer: Lockheed, Palo Alto, California

Intended Application: Background and target observation from high altitude aircraft

Qualification: Flight

Description:

Hi Camp was an airplane-based target and background observation program of the eighties to complement space-based observations. HI CAMP used large-scale doped silicon mosaic focal plane arrays operating in the LWIR and SWIR. It had a series of spectral filters, both wide-band and narrow-band. It was operated on U-2, ER-2, and U.K. Canberra aircraft over a period of ten years. The United States' Defense Advanced Research Projects Agency (DARPA) sponsored this program.

Characteristics:

Aperture: 10 cm

FPA: Doped silicon mosaic FPAs

Sensitivity: 0.02 to 0.05 K

Program Cost: ≈ $2.5 million, development in 1980–1982

Weight: ≈ 50 kg*

Power Consumed: ≈ 100 W*

**IR Sensor only; does not include gimbals, data recording electronics and incidentals, but does include focal plane electronics.*

Information courtesy of Lockheed

HIRS/2 (High Resolution Infrared Radiation Sounder)

Manufacturer: ITT, Ft. Wayne, Indiana

Intended Application: NOAA weather satellite

Qualification: Full space qualification

Availability: Custom made

Basing: Space

Description:

The instrument detects energy emitted by the atmosphere in 19 infrared and visible bands. A vertical temperature profile can be constructed from the surface of the Earth to about 40 km.

Characteristics:

Detectors: Silicon, InSb (3.76–4.57 μm) and HgCdTe (6.72–14.95 μm)

Resolution at Nadir: 17.45 km

IFOV: 20.9 milliradians

NEDN: 0.03 to 0.96 for HgCdTe supported bands and 0.0002 to 0.003 for InSb supported bands

Optics: 15 cm Cassegrain

Scanner: 1.8° stepper with 56 scan steps

Cooler: Two-stage radiant cooler with IR detectors at 105 K

Data output: 13 bit binary

Scan Width from Nadir: ± 49.5 ° in 1.8° increments at 10 steps/second

Power Consumed by Instrument: 24 W

Size: 41 × 46 × 69 cm

Instrument Weight: 32 kg

Information courtesy of ITT

Meteor Class Weather Satellite Platform (Meteor 3 in particular)

Manufacturer: All-Union Research Institute for Electromechanics (VNIIEM), CIS

Intended Application: Weather satellite

Qualification: Full CIS space qualification

Availability: Custom made

Basing: Space

Description:

This is the latest version of the basic Meteor satellite bus from the early sixties. It is three-axis stabilized. Meteor 3 included the NASA and U.S.-made TOMS ozone mapping sensor.

Characteristics:

Contains a single-band (10.5–12.5 μm) high resolution radiometer and a 10-band spectrometer.

Launch Complex: Plesetsk

Booster: Cyclone

Weight: 2,150 kg

Payload weight: 750 kg

Information from the NASA Metero-3/TOMS Press Kit; and July 1990. "The Soviet Unmanned Spacefleet." Spaceflight *32: 236–239*

Meteosat Imaging Radiometer

Manufacturer: Team including Matra

Intended Application: ESA weather satellite

Qualification: Full space qualification

Availability: Custom made

Basing: Space

Description:

The instrument is the basic radiometer for weather data form the European Meteosat satellite.

Characteristics:

Bandpasses: 0.4–1.1 μm, 10.5 μm–12.5 μm, and narrow 5.7 μm and 7.1 μm

Image Segments: 32 × 32 corresponding to a 160 × 160 km footprint directly below the satellite

Digitization: 8 bits for IR, 6 bits for visible

Saturation Temperature: 330 K

NEDT: 0.25 K

Calibration Frequency: Daily

Information from J. Schmetz and M. Nuret. 1987. "Automatic Tracking of High Level Clouds in Meteosat Infrared Images With a Radiance Windowing Technique," ESA Journal *11, 3: 275–286.*

Multispectral Scanner

Manufacturer: Hughes SBRC, Santa Barbara, California

Intended Application: NASA and NOAA space-based environmental remote sensing and resource monitoring such as Landsat

Qualification: Full space qualification

Availability: Custom made based on previous work

Description:

The Multispectral scanner is a basic Landsat instrument with over 20 years of heritage. It has been modified and upgraded several times.[A]

Characteristics:

Scan Rate: scans fields 13 times a second[A]

Resolution: 30 to 80 m (depending upon the vintage of the instrument)[A]

Number Of Bands: Four to seven, depending upon the instrument vintage[A]

Basic Bandpasses: Narrow bands centered at 0.55, 0.65, 0.75 and 0.95 μm[B]

Distinguishable Levels of Brightness: 64[A]

Data Rate: 15 Mbps[A]

Information sources: (A) March 26, 1982. "Imaging the Earth: The Troubled First Decade of Landsat" Science *215: 1,600–1,603; (B) July 1982. "The Newest Landsat."* Sky & Telescope, *22-4.*

Post NOAA M Follow-On Imager

Intended Application: NOAA weather satellite

Qualification: Full space qualification

Availability: Custom made

Basing: Space

Description:

This is intended as an upgraded replacement to the AVHRR. The instrument will have the capability to form images and perform radiometry via a scanning mode. Calibration is performed by over-scanning into deep space and by an in-flight calibration unit. The imager will sample data between ±57 degrees each side of the nadir

Characteristics:

Spacecraft Altitude: 825 km

Resolution at sub-satellite point: 1 ± 0.1 km

Absolute Calibration Accuracy: 1 K for a scene of 173 K

Stability Within Orbit: 0.2 K

Interchannel Accuracy: 0.2 K

Average Orbital Power: 50 W or less

Uncompensated Momentum: Less than 0.005 newton meters squared

Reliability: 0.8 at the end of three years, 0.75 at the end of four years

Spectral Bands: 0.605–0.625 μm, 0.86–0.88 μm, 1.58–1.64 μm, 3.62–3.83 μm, 8.4–8.7 μm, 10.3–11.3 μm, and 11.5–12.5 μm

Information from requirements package for the NOAA Polar Environmental Satellites Instrument, "Requirements In The Time Frame 1997-2010," distributed by NOAA and NASA

SPOT (Satellite Probatorie d'Obervation de la Terre)

Manufacturer: Matra, France

Intended Application: CNES

Qualification: Full space qualification

Basing: Space

Description:

The French space agency has an ongoing Earth observing program called SPOT. The first launch of a SPOT was in 1986. The satellites carry infrared and visible instruments and image the Earth's surface for both commercial and government uses. SPOT images cover 60 × 60 km regions, in three spectral bands. SPOT can view in a nadir or oblique fashion. Oblique viewing, up to ±27 degrees from nadir, is provided via a rotatable strip selection mirror.

Characteristics:

Scan Pattern: Push-broom perpendicular to linear arrays, new line measured each 1.5 milliseconds[C]

Telescope Focal Length: 1,082 mm[C]

Current SPOT Resolution: 10 m in black and white, 20 meter in the multispectral mode[A]

SPOT 5 Resolution: 5 meter resolution in black and white and 10 meter in the multispectral mode[B]

Panchromatic Bandpass: 0.51–0.73 μm[A]

Multispectral Bandpasses: 0.5–0.59 μm, 0.61–0.68 μm and 0.79–0.89 μm[A]

Spot 4 Satellite Cost: Approximately $600 million each[D]

Development, Launch and Build Two SPOT 5s: 6-7 billion Francs[D]

Information from (A) C. Bernard. 1991. "The SPOT System And Defence Applications." Proc. SPIE *1521: 66–73; (B) R. Saunders. January 21, 1991. SPOT Chooses Mid-1998 for Self Sufficiency.* Space News, *33; (C) J. Schefter. February 1987. "SPOT...ing Earth From Space."* Popular Science, *78-81; (D) R. Riccitiello. January 25–31, 1993. "France, SPOT Plan New Satellite Design Before 2000."* Space News, *18.*

Thematic Mapper

Manufacturer: Hughes SBRC, Santa Barbara, California

Intended Application: NASA and NOAA space-based environmental remote sensing and resource monitoring such as Landsat

Qualification: Full space qualification

Availability: Custom made based on previous work

Description:

The Thematic Mapper (TM) is a multispectral line scanner based on the LandSat Earth observing platforms. SBRC has been making versions of the Thematic Mappers and Multi-Spectral Scanners for the LandSat program since its conception in the early seventies. As such, the TM line has had a legacy in space of over 15 years. The

Thematic Mapper 5 contains seven spectral bands in the infrared. An enhanced TM is being considered with increased spatial resolution.

Characteristics:

Relative Calibration Accuracies: 5 to 10 percent

Absolute Calibration Accuracies: About 10 to 20 percent

Footprint: 30 m

Image Size: 180 × 18 km

Bandpass: 0.45–0.52 μm, 0.52–0.60 μm, 0.63–0.69 μm, 0.76–0.90 μm, 1.55–1.75 μm, 2.08–2.35 μm, and 10.4–12.5 μm[A]

Power for Landsat 4 TM: 335 W[A]

Weight for Landsat 4 TM: 244 kg[A]

Dimensions for Landsat 4 TM: 210 × 66 × 109 cm[A]

Cost: Approximately $120 million 1982$[B]

Information courtesy of Hughes SBRC and from (A) 1983. J. Engel, "The Thematic Mapper-Instrument Overview and Preliminary On Orbit Results." Proc. SPIE *430: 75–84; (B) March 26, 1982. "Imaging the Earth: The troubled First Decade of Landsat"* Science *215: 1,600–1,603.*

TIMS (Thermal Infrared Multispectral Scanner)

Manufacturer: Daedalus Enterprises, Ann Arbor, Michigan

Intended Application: Geologic mapping

Qualification: Air flight

Availability: Custom made from continuing line; delivery ≈ 9 months ARO

Description:

TIMS is an optional spectrometer for AADS1278 System. The instrument collects line scanner data to form multispectral images. It can be used as an airborne geologic remote sensing tool by collecting mineral spectral signatures. The data has been used to discriminate silicate, carbonate, and hydrothermally altered rocks. The system contains six Infrared channels.

ATM Characteristics:

Scan Rates: 7.3, 8.7, 12 and 25 scans per second

IFOV: 2.5 mrad

Digitized FOV: 86°

Roll Correction: ±15°

Operating Temperature for Scan Head: –55° to +70° C

Power Required: 28 Vdc ± V/ 40 A

Digitization: 8 or 12 bits, operator selectable

Scan Head Dimensions: 48 × 40 × 56 cm

Total System Weight: 198 kg

Price: ≈ $1,100,000

Information courtesy of Daedalus Enterprises

Wildfire

Manufacturer: Daedalus Enterprises, Ann Arbor, Michigan

Intended Application: Provides capabilities for Infrared hyperspectral studies of geological and volcanic features, and environmental applications

Qualification: Air flight

Availability: Custom made from continuing line; delivery ≈ 9 months ARO

Description:

Wildfire is an optional spectrometer for the ATMX system. It was developed under a NASA SBIR program to monitor the gases carried into the upper atmosphere, to study nutrient transport phenomena, and to determine the intensity of wildfires. The system has 50 infrared spectral channels.

Characteristics:

Scan Rates: 12.5 and 25 scans per second

IFOV: 2.5 mrad (1.25 mrad optional)

Digitized FOV: 86°

Roll Correction: ±15°

Operating Temperature for Scan Head: –55° to +70° C

Power Required: 28 Vdc ±3 V/45 A

Digitization: 8 or 12 bits, operator selectable

Scan Head Dimensions: 48 × 40 × 56 cm

Total System Weight: 198 kg

Price: ≈ $1,400,000

Information courtesy of Daedalus Enterprises

Appendix A

Nomenclature

"If you're going to play the game properly, you'd better know every rule."

Barbara Jordan

The infrared industry suffers from inconsistent nomenclature, terminology, and specifications. This situation is likely to continue, because a large part of the industry is a discordant conglomeration of other disciplines. Unfortunately these various disciplines develop their own terminologies and nomenclatures, which are accepted by their own parochial journals. Additionally, there are those who still frequently use deprecated terms and units, such as inches and pounds. Admittedly, it is difficult not to do so in some cases, but use of English measures and archaic terminology should be discouraged. In general, the following units should never be used, either alone or as part of a figure or merit or radiometric quantity: inch, foot, mile, pound, degree Fahrenheit, or angstrom.

The SI units of measure do not specifically recognize the centimeter, preferring the meter or millimeter. This is unfortunate, because the centimeter (albeit a measurement non grata) it is often the most appropriate for real radiometric engineering. Most in the industry (including the author) believe this license is to be allowed to the IR artist. As such, centimeters are used frequently in this text, to the abhorrence of academic purists. Table A.1 gives the measurements in meters, but centimeters can be substituted. Photons are usually abbreviated as Ph.

Table A.1 lists common radiometric terms. Most of the terms in the table are recommended by the International Commission on Illumination (*Commission Internationale de'Eclairage,* or CIE) and adopted by major journals publishing IR material. Any of these terms can be expressed as a function of a narrow spectral bandpass by dividing by a bandpass unit, usually the micron. Radiometry may be calculated and expressed in power units (watts) or photon units. Both are correct.

TABLE A.1 Commonly Used Radiometric Quantities

Power-Based Term	*Symbol*	*Units*	*Photon-Based Term*	*Photon Symbol*	*Photon Units*	*Notes*
Emissivity	ε	None	Emissivity	ε	None	The ratio of the radiant existence of the surface to that of an ideal black body
Irradiance (or areance)	E_e or E (previously H)	W m^{-2}	Photon irradiance	E_e	Ph sec^{-1} m^{-2}	Arrived at area (at the aperture)
Radiance (or sterance)	L_e or L (Previously N)	W $m^{-2}sr^{-1}$	Photon radiance	L_p	Ph sec^{-1} m^{-2} sr^{-1}	
Radiant exitance	M_e or M	Wm^{-2}	Photon flux exitance	Mp	Ph sec^{-1} m^{-2}	Exited per unit area from a source (at the target)
Radiant flux	ϕ_e	W		ϕ_p or ϕ	Ph/second	
Radiant intensity (or pointance)	I_e or I (previously J)	W sr^{-1}		I_p	Ph sec^{-1} sr^{-2}	Leaving the target; what one really wants to know to describe a point source target.
Spectral flux	fl or ϕ_λ	W μm^{-1}		fl or ϕ_λ	Ph sec^{-1} μm^{-1}	

One is preferred over the other when the detector sensitivity or the phenomenology is better expressed in those terms. For example, if one is using a detector whose sensitivity is appropriately expressed by quantum efficiency, photon units should be used. Conversely, if one is using a detector whose sensitivity is appropriately expressed in power units of watts or D*, those should be used. Active systems employing pulsed lasers frequently express radiometric terms in joules. The reader must understand that radiometry expressions in watts, photons, or joules are all completely acceptable, depending on the situation.

People working in the field love to give names to spectral intervals. Unfortunately, many of these terms are subjective—what is short to one person is not necessarily short to another. Additionally, astronomers like to number atmospheric bandpasses using the alphabet. Table A.2 is offered as a general guide to spectral regions. Table A.2 is not blessed by any committee or national institute. Until such time, one should always ask another what he means by a given bandpass notation.

The nomenclature for sensitivity is confusing for several reasons. One is certainly that different sensitivity measurements have different applicability. Table A.3 lists some common figures of merit, their typical abbreviations, and proper units.

TABLE A.2 Generally Accepted Bandpass Notation

Notation	*Notation Abbreviation*	*Approximate Spectral Region (μm)*	*Notes*
Ultraviolet	UV	0.15 to 0.4	
Visible	Vis	0.4 to 0.75	
Near infrared	NIR	0.75 to 1.1	Limit of silicon
Shortwave infrared	SWIR	1.1 to 3	
J Band	J	Around 1.25	Used by astronomers
H Band	H	Around 1.65	Used by astronomers
K Band	K	Around 2.2	Used by astronomers
Midwave infrared	MWIR	3 to 6	
L Band	L	Around 3.6	Used by astronomers
Intermediate-wave infrared	IWIR	4-20	Rarely used
M Band	M	Around 4.7	Used by astronomers
Longwave infrared	LWIR	6 to 18	
Very longwave infrared	VLWIR	18 to 50	
Far infrared	FIR	25 to 100	
Submillimeter		100 to 1,000	

TABLE A.3 Commonly Used Sensitivity Measurements

Term	*Abbreviation*	*Typical Units*
Clutter equivalent irradiance	CEI	W/m^2 or $Ph/sec/cm^2$
Clutter equivalent power	CEP	W
Clutter equivalent target	CET	W/sr @ range or Ph/sec/sr @ range
Minimum detectable temperature	MDT	°C
Minimum resolvable temperature	MRT	°C
Minimum resolvable temperature difference	MRTD	°C
Noise equivalent spectral irradiance	NESI	$W/m^2/\mu m$ or $Ph/sec/\mu m$
Noise equivalent delta temperature	NEDT or NEΔT	°C
Noise equivalent flux density	NEFD	W/m^2 or $Ph/sec/cm^2$
Noise equivalent irradiance	NEI	W/m^2 or $Ph/sec/cm^2$
Noise equivalent power	NEP	W
Noise equivalent radiance	NER	$P/m^2\ sr^{-1}$
Noise equivalent spectral flux density	NESFD	$W/m^2/\mu m$
Noise equivalent spectral radiance	NESR	$W/m^2\ sr$
Noise equivalent spectral sterance	NES	$Ph/m^2/sr$
Noise equivalent target	NET	W/sr @ range or Ph/sec/sr @ range
Noise equivalent temperature difference	NETD	°C or K
Signal-to-clutter ratio	SCR	Unitless
Signal-to-noise ratio	SNR	Unitless

Appendix B

Glossary

"Listen, someone is screaming in agony—fortunately, I speak it fluently."

Spike Milligan

B.1 ABBREVIATIONS, ACRONYMS, AND SYMBOLS

AC, ac	Alternating Current
ACS	Attitude Control System
ACWP	Actual Cost of Work Performed
A/D	Analog to Digital
ADC	Analog-to-Digital Converter
AFGL	U.S. Air Force Geophysical Laboratory
AGC	Automatic Gain Control
AIAA	American Institute of Aeronautics and Astronautics
AOTF	Acousto-Optical Tunable Filter
APD	Avalanche Photo Diode
AR	Anti-reflective (usually in reference to an optical coating)
ARC	NASA's Ames Research Center
Arcsec	Arcsecond
ARO	After Receipt of Order
ASAT	Anti-satellite
AST	Airborne Surveillance Testbed, formerly referred to as AOA—the Airborne Optical Adjunct
ATH	Above the Horizon
AUPC	Average Unit Production Cost
BCWP	Budgeted Cost of Work Performed
BCWS	Budgeted Cost of Work Scheduled

BFL	Back Focal Length
BIB	Blocked Impurity Band
BIT	Built-In Test
BLIP	Background Limited In Performance
BMD	Ballistic missile defense
BMDO	Ballistic Missile Defense Organization (new name for SDIO)
BOM	Bill of Materials
BRDF	Bidirectional Reflectance Distribution Function
BRITE	Basic Research in Industrial Technology for Europe
BSTS	Boost phase Surveillance and Tracking Satellite
BTH	Below The Horizon
CAD	Computer Aided Design
CBD	Commerce Business Daily
CCD	Charge Coupled Device
CDR	Critical Design Review
CE	Cincinnati Electronics
CE	Concurrent Engineering
CET	Clutter Equivalent Target
CID	Charge Injection Device
CIE	Commission Internationale de l'Eclairage
CIP	Cold Isostatic Pressing
CIRRIS	Cryogenic Infrared Radiometer In Shuttle
CIS	Commonwealth of Independent States—the former USSR
CM	Concurrent Management or Common Module
cm	Centimeter
CMOS	Complementary Metal Oxide Semiconductor
CMT	HgCdTe in Eurospeak
CNVEO	Center for Night Vision and Electro-Optics (an old name for NVEOD)
CRADA	Cooperative Research and Development Agreement
CRT	Cathode Ray Tube
CSO	Closely Spaced Objects
CTE	Charge Transfer Efficiency
CTE	Coefficient of Thermal Expansion
Cups	Connections per second (in reference to neural networks)
CVD	Chemical Vapor Deposition
CVF	Circular Variable Filter
DC, dc	Direct Current
Δ	Delta, a general symbol for change in the quantity that follows
D*	Specific detectivity (pronounced "dee-star"). See Chapter 4 for detailed discussion.
DARPA	Defense Advanced Research Project Agency. Part of the U.S. DoD to promote R&D in defense technologies.

dB	Decibel
DEA	Drug Enforcement Agency
DEC	Digital Equipment Corporation
DEI	Detectable Equivalent Irradiance
ΔF	Symbol for electrical noise bandwidth of a detector
DLC	Diamond-Like Carbon
DMSP	Defense Meteorological Satellite Program
DoD	The U.S. government's Department Of Defense
DOE	U.S. Department Of Energy
DRA	Defense Research Agency
DRO	Direct Read Out (s)
DSP	Abbreviation for either Defense Support Program or Digital Signal Processing
e–	A symbol for the electron
E-O	Electro-Optics, Electro-Optical
El	Elevation
EMCD	Electro-Magnetic Carrier Depletion
EMD	Engineering and Manufacturing Development (like the old FSD)
EOS	NASA's Earth Observing System
EPA	Environmental Protection Agency
ERIM	Environmental Research Institute of Michigan
ESA	ElectroStrictive Actuator or European Space Agency
f/#	Symbol for F-number. The ratio of focal length to aperture
F#	Another symbol for F-number
Φ	Symbol for Photon Flux
FET	Field Effect Transistor
FEWS	Follow-on Early Warning Satellite
FFT	Fast Fourier Transform
FLIR	Forward Looking Infrared
FOM	Figure of Merit
FOR	Field of Regard
FOV	Field of View
FPA	Focal Plane Array or Focal Plane Assembly
FPU	Focal Plane Unit, used more frequently in Europe
FSD	Full Scale Development
FSI	FLIR Systems Inc.
FSM	Fast Steering Mirror
FWHM	Full Width Half Maximum
G	Gravity
g	(1) acceleration of gravity—9.8 meters per second, (2) gram
GFE	Government Furnished Equipment
GFSC	NASA's Goddard Space Flight Center in Greenbelt, Maryland

GN&C	Guidance, Navigation and Control
GOES	Geostationary Operational Environmental Satellite
GRIN	Gradient Index
h	General symbol for Plank's constant of 6.62×10^{-27} erg second
HEPA	High Efficiency Particulate Air
HgCdTe	Chemical abbreviation for Mercury Cadmium Telluride
HIDS	Hole Impeded Doping Superlattice
High Rel	High Reliability
HIP	(1) Heterojunction Internal Photoemission, used in discussions of FPAs, (2) Hot Isostatic Pressing, when discussing beryllium optics
HOE	(1) Holography Element, (2) Homing Overlay Experiment
HNVS	Hughes' Night Vision Sensor
HP	Hewlett Packard
HUD	Head Up Display
IAI	Israel Aircraft Industries
IBC	Impurity Blocked Conduction
IBM	International Business Machines
IC	Integrated Circuit
ICBM	Intercontinental Ballistic Missile
IDCA	Integrated Dewar Cooler Assembly
IEEE	Institute of Electrical and Electronic Engineers
IFF	Identification of Friend or Foe
IFOV	Instantaneous Field of View
InSb	Indium Antimonide detector material
IPS	The space shuttle's International Pointing System
IR&D	Internal Research & Development
IR	Infrared
Ir:Si	Iridium Silicide
IRAD	Internal Research and Development
IRAS	Infrared Astronomical Satellite
IRFPA	Infrared Focal Plane Array
IRST	Infrared Search and Track
IRSTS	Infrared Search and Track System
IRTF	Infrared Telescope Facility
ISC	Irvine Sensor Corporation
ISO	Infrared Space Observatory
J	Joule
JIT	Just In Time
JNLT	Japanese National Large Telescope
JPL	NASA's Jet Propulsion Laboratory
JSC	NASA's Johnson Space Center
J-T	Joule-Thompson effect or Joule-Thompson cooler

K	A Symbol for Kelvin
k	A generic symbol for one thousand of a unit. Also, a symbol for Boltzmann's Constant of 1.38×10^{-23} joules/K.
k	kilo (1,000)
kilorad, krad	A thousand rads
KAO	Kuiper Airborne Observatory
kg	Kilogram
KSC	NASA's Kennedy Space Center
λ	Lambda, general symbol for wavelength of light
LANTIRN	Low Altitude Navigation and Targeting Infrared at Night
LCC	Life Cycle Cost
LEA	Law Enforcement Agency
LEAP	Light Exo-atmospheric Projectile
LIRIS	Loral Infrared and Imaging Systems
LMS	Least Mean Squared
LMSC	Lockheed Missile and Space Company
LN2	Liquid nitrogen
LOS	Line of Sight
LPE	Liquid Phase Epitaxy
LRIP	Low Rate Initial Production
LRU	Line Replaceable Units
LSF	Line Spread Function
LSI	Large Scale Integration
m	Meter
Mb	Megabit, one million bits of information
MB	Megabyte, one million bytes of information
MBE	Molecular Beam Epitaxy
Mbps	Megabits Per Second, a measure of data transfer
MCT	Sometimes used abbreviation for Mercury Cadmium Telluride
MDT	Minimum Detectable Temperature
MDTI	Minimum Detectable Target Irradiance
Megarads	Millions of rads
Mfg	Manufacturing
MFLOPS	Millions of Floating Operations Per Second
MgF_2	Chemical abbreviation for Magnesium Fluoride, an optical protection coating
MHD	Magneto Hydro Dynamic
Microrad	Short for microradian
Microradian	One millionth of a radian
MicroWatt	One millionth of a watt
MIL-STD	Military Standard of the USA
Millikelvin	One thousandth of a Kelvin
Millirad	One thousandth of a radian

Milliwatt	One thousandth of a watt
MIMD	Multiple Instructions on Multiple Data
MIPS	Millions of Instructions Per Second
MIS	Metal Insulated Semiconductor
MIT	Massachusetts Institute of Technology
MLI	Metal Layer Insulation or Multiple Layered Insulation
mm	millimeter
MMT	Multi Mirror Telescope
MMW	Millimeter Wave
MOCVD	Metal Organic Chemical Vapor Deposition
MOPS	Millions of Operations Per Second
MOS	Metal Oxide Semiconductor
MRP	Materials and Resources Planning
MRT	Minimum Resolvable Temperature
MRTD	Minimum Resolvable Temperature Difference
MSFC	NASA's Marshall Space Flight Center in Huntsville Alabama
MTBF	Mean Time Between Failure
MTF	Modulation Transfer Function
MTTF	Mean Time To Failure
Mux	Multiplexer
mW	milliwatts
MWIR	Mid Wave Infrared
NASA	National Aeronautics and Space Administration
NASDA	National Space Development Agency of Japan
NATO	North Atlantic Treaty Organization
NCAR	National Center for Atmospheric Research
NEDN	Noise Equivalent Delta Radiance
NEDT	Noise Equivalent Delta Temperature
NEFD	Noise Equivalent Flux Density, synonymous to NEI
NEI	Noise Equivalent Irradiance
NEP	Noise Equivalent Power
NER	Noise Equivalent Radiance
NER	Noise Equivalent Radiance
NES	Noise Equivalent Sterance
NESA	Noise Equivalent Spectral Areance
NESFD	Noise Equivalent Spectral Flux Density
NESR	Noise Equivalent Spectral Radiance
NESS	Noise Equivalent Spectral Sterance
NET	Noise Equivalent Target
NETD	Noise Equivalent Temperature Difference
NIH	Not Invented Here
NIR	Near IR

NIST	The USA Government's National Institute of Standards and Technology, formerly the National Bureau of Standards
NLOS	Non-line of Sight
NOAA	National Oceanic and Atmospheric Administration
NRL	Naval Research Laboratory
NSF	National Science Foundation
NTVC	National Television Video Code
NVEOD	Night Vision and Electro-Optics Directorate (the new name for CNVEO or NVL)
NVG	Night Vision Goggles
NVL	Night Vision Lab, oldspeak for NVEOD
NVS	Night Vision System
OPV	Optionally Piloted Vehicle
OSA	Optical Society of America
OSC	Optical Science Center
OSHA	Occupational Safety and Health Administration
OTS	Off the Shelf
π	Pi, the ratio of a circle's circumference to its diameter, 3.14...
PbS	Lead Sulfide
PbSe	Lead Selenide
PC	(1) Photoconductor, (2) Personal Computer
PCM	(1) Phase Change Material, (2) Phase Conjugate Mirror
Pd	Probably of Detection
PDR	Preliminary Design Review
pF	picofarad
P_{fa}	Probability of false alarm
Ph	Photon
PLR	Percent Learning Rate
PLZT	Lead Lanthanum Zirconium Titanite
PM	Program Manager
Pod	A tubular packaging contained under an aircraft
PPM	Parts Per Million
PRC	People's Republic of China
PSD	Point Spread Distribution or Function
PSF	Point Spread Function
Pt:Si	Platinum Silicide
PV	Photo Voltaic
PZT	Piezoelectric Transducer
QA	Quality Assurance
QC	Quality Control
QWIPS	Quantum Well Infrared Photodetectors
R&E	Research & Experimentation, newspeak for IRAD
RADC	Rome Air Development Center

RAM	Random Access Memory
RF	Radio Frequency
RFI	Request for Information
RFP	Request for Proposal
RFQ	Request for Quote
RMS	Root Mean Squared
RPM	Revolution per Minute
RPS	Revolutions per Second
R_0A	Resistance Area product
ROI	Return On Investment
ROM	(1) Read Only Memory, (2) Rough Order of Magnitude
RPV	Remotely Piloted Vehicle
RSS	Root Sum Squared
S-Level	Space level
σ	(1) The symbol for the Stephan Boltzmann constant of 5.67×10^{-5} erg/cm^2-sec-K^4, (2) also a symbol for standard deviation
SAA	South Atlantic Anomaly, discussion in Chapter 11
SADA	Standard Advanced Dewar Assembly
SBIR	Small Business Innovate Research
SBRC	Santa Barbara Research Center
SCR	Signal-to-Clutter Ratio
SDC	Strategic Defense Center
SDI	United States' Strategic Defence Initiative to defend against ballistic missiles
SDIO	Strategic Defence Initiative Organization, the managing agency in charge of SDI programs (now called BMDO)
sec	Second or seconds
SETA	Scientific and Engineering Technical Assistance
SI	Society International system of units
Si	Silicon
Si:Ge	Germanium silicide pseudo-barrier detector
SiC	Silicon carbide, a composite applicable to mirror blanks
SIMD	Single Instruction Multiple Data, usually used in reference to parallel processors. Pronounced "sim-dee"
SIRTF	Space Infrared Telescope Facility
SLIR	Side Looking Infrared
SLM	Spatial Light Modulator
SMT	Systeme Modulaire Thermique
SNR	Signal-to-Noise Ratio
SOA	State of the Art
SOFIA	Stratospheric Observatory For Infrared Astronomy
SOFRADIR	Socete Francaise de Detecteurs Infrarouge
SOS	Silicon on Sapphire

SOW	Statement of Work
Specs	Specifications
SPIE	Society of Photoelectric Instrumentation Engineers
SPOT	Satellite Probatorie d'Obervation de la Terre
SPRITE	Signal Processing in the Element
sq	Square
SQUID	Superconducting Quantum Interference Device
sr	Steradian
SRU	Shop Replaceable Unit
SSE	Space Systems Engineering
SSG	Sensor Systems Group
SSTS	Space Surveillance and Tracking System. An old SDI space-based early-warning program.
STAMIDS	Standoff Minefield Detection System
Std	Standard
SWIR	Short Wave Infrared
TADS/PNVS	Target and Designation System/Pilot Night Vision System
TADS	Targeting and Designation Sight
TBD	To Be Determined
T_c	Critical temperature. This is the transition temperature where a material becomes a superconductor.
TCIM	Thermal Imaging Common Module—A British SPRITE common module
TE	Thermo-electric
TEC	Thermo-electric Coolers
TI	Texas Instruments
TIS	Total Integrated Scatter
TOTS	TRW Orbital Test Station
TQM	Total Quality Management
TTOF	Temperature Tunable Optical Filter
U of A	University of Arizona
UAV	Unmanned Areal Vehicle
UCB	University of California at Berkeley
UCF	University of Central Florida
UK	United Kingdom
USAF	United States Air Force
USC	University of Southern California
UV	Ultraviolet
V	Volt
Vac	Volts of Alternating Current
Vdc	Volts Direct Current
VHP	Vacuum Hot Pressed
VHSIC	Very High Speed Integrated Circuit

Vis	Visible
VLIWR	Very Long Wave Infrared
VLT	Very Large Telescope
VPE	Vapor Phase Epitaxy
W	Watts
WJSA	Wally J. Shafer Associates
WRDC	Wright Research and Development Center
ZnS	Zinc Sulfur
ZnSe	Zinc Selenide
Å	Angstrom
μm	Micron or micrometer unit of measure.
μrad	Microradian
≈	Approximately
≈ ≈	A crude approximation

B.2 DEFINITIONS

Aberration Imperfections of an image. The deviation from the ideal, because of physical properties such as diffraction, chromatic deviations or coma. It may also result from tolerances on lens prescriptions, misalignment in the optical system, or by limitations inherent in the optical design.

Absolute Calibration The task of determining the relationship of a sensor's outputs to a known input with traceable reference to known standards. This implies an exacting process of determining a sensor's output to a known input. A series (or matrix) of repeatable reference measurements traceable, with a known accuracy, to a primary-national (such as one from NIST) or international standard. The two keys to an absolute calibration are the traceability trail to an accepted standard and the repeatability accuracy for the entire test setup (including the instrument to be calibrated). Typically, radiometric calibration is only accurate to approximately 10 percent in the IR. Scientific instruments must have an absolute calibration to allow the investigator to properly interpret the sensor's output.

Absorption The assimilation of photons into the substance. The termination of a photon's journey by transforming the photon energy into another form. All photons are either absorbed, transmitted, or reflected when encountering a material.

AC, ac (1) Alternating current, (2) Sometimes sloppily used to refer to any modulation, regardless of whether it is current or not. For example the non-constant background component is sometimes called AC background.

Acousto-Optical Tunable Filter A solid state tunable spectral filter needing no moving parts. The principles are usually based on Bragg diffraction. An RF frequency transducer modulates pressure changes within a crystal. The resulting regions of rarifications and compressions act akin to a diffraction grating and separates light of a given color. The time required to change from one spectral region to another is typically in the tens of microseconds.

Achromat A lens (composed of different materials) that does not produce noticeable chromatic aberration.

Active Optics Optical systems or elements which compensate for wavefront errors. Usually used as the same as adaptive optics.

Actuals Aerospace industry slang for funds already irretrievably committed (money spent).

Adaptive Optics A technique of changing a light beam's wavefront by intentionally altering an optical surface (or a property such as the index of refraction) both spatially and temporally. Usually employs an element whose shape is adjusted to compensate for vibration, atmospheric modulation, or residual system aberrations. Usually implies a closed loop (feedback) servo system.

Adaptive Structures A structure whose geometric and/or structural characteristics can be rapidly changed to meet mission requirements in response to external stimulation.

Adaptive Threshold An image processing technique that adjusts the signal level based on scene statistics. It is performed in either the analog or digital domain, and is applicable to both starers and scanners. Several methods have been devised, including simple threshold estimation and standard deviation threshold estimation, to determine the appropriate local threshold levels.

Aeroheating The phenomenon of a rise in surface temperature resulting from friction with air molecules. This occurs in high speed flight and is a consideration for windows, domes, and targets.

Afocal Telescope A telescope with no focus. Both the object and image are at infinity. These are typically used as fore-optics for an imaging FLIR, space, or airborne instruments. Narcissus effects are particularly a concern with afocal telescopes.

Aimpoint Selection The process of choosing a final impact position on the target. The term is often used in conjunction with missile seekers.

Airmass A measure of the amount of linear atmosphere transmission induced radiation a target must pass through before reaching the instrument. In other words, the amount of atmosphere the LOS projects through until the target is reached. An airmass of one is defined as the total atmosphere from ground to space when pointed at the zenith. If a telescope is pointed directly overhead it would be viewing through an airmass of one. If pointed somewhat below the zenith, it would be looking through more atmosphere (because of its angle) and have an airmass greater than one.

Airy Disk The central spot on a focal plane produced by diffraction of an optical system when viewing a point source. For a diffraction limited, unobscured aperture system this includes about 82 percent of the energy.

Albedo The ratio of the sun's total energy on a surface to that reflected by the surface.

Alloy A mixture or solid solution of metal molecules not contained in a lattice structure.

Angstrom This is a tiny unit of linear measure equal to 10^{-10} m, 10^{-8} cm and 10^{-3} microns. A thousand angstroms equals one micron (or micrometer). It is frequently used to measure wavelength in the visible spectrum, and surface errors of optical surfaces. Recently this unit has been falling into disfavor, and some astronomical societies are discouraging its use.

Annealing A process of thermally stabilizing properties of bulk materials by heating and slow cooling. The process can reverse lattice damage caused by radiation or implants, thereby eliminating color centers in transmissive optics.

Antireflection Coating or Film A coating applied to the exterior surface of a lens that reduces reflection and thereby increases transmission. It should be 1/4 of a wave thick with a refractive index of a value that is the square root of the index of the material composing the lens.

Apogee The furthermost point from the earth in the orbit of a satellite.

Application Specific Integrated Circuit A custom-made single IC chip to perform specific user required functions.

Arcsecond A one-dimensional measurement of angle. A degree is divided into sixty arcminutes and an arcminute is divided into 60 arcseconds. Therefore, an arcsecond equals 1/3600 of a degree or 4.85 μrad

Areal Density The weight as a function of area. Usually this term is used for optics and expressed in kilograms of total weight per square meter of the optics' useful area.

Aspherics An optical element (lens or mirror) that does not have a spherical surface.

Athermalize The design process of insuring that system changes do not occur over a given temperature range. It is usually accomplished by balancing the expected environmental change with the thermal expansion and heat capacity properties of materials. The term is usually applied to optics and FPA mounting and mating apparatus.

Atomic Oxygen Molecular oxygen in a free state. Normally, oxygen will combine with itself to form O_2, but in the extreme upper atmosphere, conditions exist that allow a rarefied gas of single molecule oxygen.

Avalanche Photo Diode A solid state detector that produces a response by generating numerous hole-electron pairs for each absorbed photon.

Average Unit Production Cost This is usually defined as the total production price divided by the number of units. Non-recurring costs such as developmental costs, R&D, and maintenance costs are normally not included.

Axial Thrust The propulsion force (or thrust) along the long axis and flight vector of a missile.

Azimuth This term refers to the angular dimension measured from a local horizontal. Azimuth is perpendicular to elevation. The horizontal may be determined by the earth, the body of an aircraft, ship, or spacecraft. A 360 degree azimuth angle would represent a full circle along this horizontal, a 180 degree azimuth angle would equal a half circle, a 90 degree azimuth angle would equal a quarter circle, and so on.

Back Focal Length In general, the distance from the last optical element to the focus. Suitable back focal length is important in design to provide space for filters, mechanisms, and the detector package.

Background Limited in Performance This indicates that a sensor's or focal plane's sensitivity is limited by the scene that it is viewing. It usually means that the dominant FPA noise is a result of shot noise from random arrival times of photons. In this situation, added improvement in FPA sensitivity alone is useless.

Background Subtraction The image processing task of reducing the background so the targets stands out.

Baffle A structure that obstructs stray light from the desired image.

Bake Out A cleaning process where an item is taken to an elevated temperature during or after manufacturing to outgas contaminants. This often occurs in a vacuum or low pressure oven to facilitate the cleaning. An item should never be subjected to temperatures in excess of its bake-out temperature.

Bandpass An optical (or spectral) range of transmission. A 3 to 5μ bandpass will transmit IR radiation from 3 μm (inclusive) to 5 μm. Bandpass defines the transmitted color or wavelength. It is synonymous with passband.

Bang-Bang A servo control process that uses a square wave control drive. When a movement is required, the motors will be commanded to move in the desired direction with their maximum torque. The system will respond with maximum acceleration. When the position (or near it) is reached the motor will shut off. With a Bang-Bang system, the plant (or system) will frequently overshoot causing a maximum acceleration command to be issued for the opposite direction for a smaller amount of time. This is a "digital" control system since the motors have only two states, full on or off.

Bias Correction A signal processing technique to equalize the response from various nonuniform detector pixels. It is a simple matrix addition (one correction for every pixel). The corrections are stored from the last time the device viewed a uniform blackbody calibrator.

Bifrigent The property of refracting light in different amounts based on polarization.

Bill of Materials A list of every hardware part that makes up a system. In the strictest sense it would include every single piece, including screws, nuts, and washers.

Binary Optics Diffractive optics using a binary etched pattern. Chapter 3 contains a detailed discussion.

Bispectral The ability to detect in two optical bandpasses. The term usually implies simultaneous detection in both.

Black Box Technical slang for a component or subassembly that performs functions without requiring the user to have detailed knowledge of its internal design. This implies easily definable and accurate interface requirements and specifications. Frequently, these are self-contained modules that can easily be replaced with an identical one. Often (but not always) used in reference to electronics. Typical examples of black boxes are cryocooler control circuitry, FLIR symbology generator boxes, and power supply units. Typical examples of items that usually cannot be considered black boxes include telescopes, servo systems, and FPA assemblies, because of their complex interactions with other parts.

Black-Hot A mode of black & white display where the blacker the target the more inband IR radiation it is producing. In a black-hot display, the darker the object, the hotter it is. Most people feel this mode produces an IR picture that has a more natural feel and appearance.

Black (Material or Coating) A material or coating that has a high emissivity. Ideally the emissivity is > 0.98 for a wide range of angles through the spectral bandpass of interest.

Blackbody An ideal radiator and absorber with no reflection and a radiant exitance only determined by its temperature and described exactly by Plank's law. A blackbody has an effective emissivity of one. In the real world, nothing is a pure blackbody. Industrial blackbodies usually approach the ideal by using a cone shaped cavity with "black" coatings. These achieve an effective emissivity exceeding 0.99.

Blank The substrate used for a mirror after it has been made into the correct size and thickness, but without the optical figure ground. The grinding and lapidation then produces a mirror from a blank.

Blind Range The close-in distance where a seeker can either no longer properly function or where increased footprint resolution does not occur. It is usually caused by optical blurring of an image with a fixed focus telescope.

Blooming A phenomena of excessive bleeding and crosstalk of signal from one display pixel to others in a scene. Blooming can occur in both the time and spatial domains. It is common with CCDs when the unit cells are heavily saturated.

Blow Down Cooler A Joule-Thompson cryocooler.

Blur Circle The image of a point source at the focal plane.

Bolometer An infrared detector that uses a change in resistance, due to impinging infrared radiation, to provide a response.

Boost Phase Surveillance and Tracking Satellite An old SDI space-based early-warning program that evolved into the FEWS replacement of DSP.

Brassboard An engineering hardware mock up. It is used to verify a design and is implemented with specific components. It implies a state of near complete design. It is similar to a breadboard but has more fidelity to the end product. Brassboards are usually conducted near the CDR or in a post-CDR phase of a program as a proof-of-principle exercise.

Breadboard A functional, engineering, hardware mock up used to verify a design's feasibility. It implies a state of near complete design with some technical concerns or issues remaining. It usually demonstrates key performance of the end product with no regard to packaging, weight, or power. It is similar to a brassboard but has less fidelity to the end product. Breadboards are usually conducted near or prior to the CDR as a risk reduction determination.

Bremmstrahlung German for "braking radiation." It is radiation emitted by the change in acceleration of a particle. It is applicable to space based IR systems as particles are slowed down when they encounter the telescope or FPA materials.

Budgetary Estimate A crude (or ROM) estimate of the cost or price suitable for budgeting and getting a feel for the magnitude of the effort.

Bulk Effect An phenomena that is not limited to the surface, but occurs in the depth of the material. Usually the term is used to describe radiation effects. For example, the darkening of transmissive optics is a "bulk effect" because absorption occurs throughout the material, not just on the surface.

Bumping A process for providing an electrical connection between a detector unit cell and its associated readout circuitry in hybrid FPAs. Ordinarily, metal deposits are placed on the detector material unit cells and the associated locations on the readout circuitry. They are then physically pressed together at a slightly elevated temperature to form the connections. Indium is the most common metal used, although gold has also been employed.

Bumps The microscopic interconnects between detector unit cells and the ReadOut (or Mux) circuitry for an hybrid FPA. Chapter 3 includes a photograph and a more detailed discussion.

Buttability The potential to place a FPA immediately next to another FPA. Usually the separation between the two arrays equals no more than a few times the size of a unit cell. For an array to be "buttable" on one side, that side must be free of leads and overhanging substrate. A two-side buttable array has two sides of the array free from leads and overhanging substrate. Four arrays that are two-side-buttable arrays can be placed together in a square to make a focal plane four times the size of any single chip. A four-sided buttable array has all electrical connections coming out the back.

Calibration The task of determining the relationship of a sensor's outputs to a known input with reference to known standards. Space sensors and airborne instruments are usually calibrated, while cameras and FLIRs usually are not.

Cassegrain A reflecting telescope employing a paraboloidal primary and a hyperboloidal secondary. Usually, the secondary is suspended by a "spider" from the metering structure in front of the primary and the light is reflected through a hole in the center of the primary. The primary's pupil usually lies near the secondary.

Cathode Ray Tube A standard television picture tube. Typically used in the infrared industry for display of real time FLIR or scientific data.

Central Obscuration The blocked part of the aperture in an on-axis reflective system. It usually supports the secondary mirror.

Centroid An image processing technique of determining the "center" of an object. The center can be calculated various ways which provide different answers. For example, the center can be determined from the edges giving a geometric center to the projected image. Another "centroid" technique is to apply weights to the constituent pixels by their intensity, giving an intensity centroid.

Characterization A process of determining a sensor's output to a basic input. This is similar to a calibration but is less rigorous and not completely traceable. The process of measuring an output based on a blackbody with traceable temperature sensors but unknown radiant emittance is really a characterization. FLIRs, IRSTS, and cameras are usually characterized by comparison to a blackbody (assumed to have a uniform emissivity of one) rather than rigorously calibrated to reduce cost and shorten schedule.

Chop Frequency The frequency of the modulation provided by the chopper.

Chopping A technique to provide alternating detection modulation. The scene a detector views is alternated with the target scene and a blackbody source of known (or calibrated) properties. This allows constant (or slowing varying) noise characteristics to be easily subtracted. Astronomers frequently use this method by viewing deep, void space in place of the blackbody.

Chromatic Aberrations or Distortions A fundamental optical aberration resulting from optical materials having a varying index of refraction with respect to wavelength. A difference in magnification can also occur. These effects are particularly critical in IR systems, since bandpasses are typically several microns wide (many times that of UV or visible systems).

Closed Loop Pointing A pointing control system process that uses the sensor's output to adjust pointing via a feedback loop.

Clutter Leakage The amount of clutter that passes through the system regardless of all the employed techniques to mitigate it.

Clutter A structured background induced phenomena. Clutter is not spatially and/or temporally constant. Clutter is attributable to structured and time correlated events in the environment. Thus, it is not a stochastic process resulting in its signal levels combine linearly rather than in an RMS fashion. Its effect is to confuse operators, cause high false alarms, or require extremely high irradiance values from the target. It is often the dominant range limitation for FLIRs, IRSTs, and earth viewing space sensors.

Cold Finger Slang for a cryogenic cooled probe extending into the dewar to provide cooling. Usually, the expander of a Stirling or a thermal strap to a containment dewar.

Cold Shield A cold baffle that limits the solid angle viewed by the FPA. It is viewed by the FPA, so it is held at a low temperature. It may define the cold stop, or just shield unwanted stray radiation.

Cold Stop An aperture stop that is at a cold temperature. The detector will "see radiation" only from object space beyond the cold stop. It is usually cooled below ambient but somewhat above the detector temperature.

Commerce Business Daily A publicly-released periodical of the U.S. government's contract activity.

Common Module A concept developed in the U.S. and promoted by NVEOD of replaceable component subsystems for FLIRs. Typical common modules include detector dewar assemblies, scanners, and digital reformatters.

Configuration Management and Control The process of controlling the baseline of a design.

Constant Year Dollars Used with a price quoted without consideration for inflation.

Contrast The relative difference in brightness between adjacent areas of a display or image, or the relative defference in irradiance between a target and its background.

Cost of Money A fee normally charged to project-oriented customers to offset the cost of financing.

Cost Plus A type of contract in which the contractor gets reimbursed in full for all allowable costs including an additional percentage of these costs. The additional percentage represents the profit for the company. They usually come in two forms, a "cost plus fixed fee" or a "cost plus variable fee," where the variation is set by the customer as an incentive. These contracts are convenient and effective for R&D programs where the total cost is ambiguous, or projects with frequent requirement changes. They have recently been viewed in disfavor by many politicians and laymen because there is no definable incentive for the contractor to control costs. However, they provide a control on profit based on performance and independent of changes.

Costing Exercise A thorough effort to determine the cost of a project, piece of hardware, or system.

Critical Design Review A design review to evaluate the complete design of the program. Typically, 80 percent of the drawings are complete, breadboarding is complete, and the entire design is addressed in detail. The program should be ready to proceed to fabrication or a development phase.

Critical Path Typically a project will have several strings (or paths) of sequential activities that must occur to reach the desired end product. The critical path is the string of activities that takes the longest time to accomplish to reach the end goal. It is determined by flowcharts and schedules. The critical path is the one that will certainly delay the project if any schedule slips occur.

Crosstalk The phenomena of data or information that should be in one pixel going to one of its neighbors. This can occur by overfilling wells and having the electrons spill over. It can also occur by the optical blur spot directing photons to neighboring pixels. Additionally, there is always some electronic crosstalk from pixel to pixel.

Cryogenic An adjective to describe the science, process, components, or business relating to low temperature refrigeration and/or achieving, maintaining, and experimenting with low temperatures. In the electro-optics industry it usually is used when temperatures are lower than can be achieved with thermoelectric coolers (about 200 K).

Current Year Dollars Price quotes in dollars of the current year. Prices given in this book are current year dollars for 1992, 1993, and 1994.

Cut Off The longest wavelength a filter will pass or a detector or system will respond to. A 3–5μm filter's cutoff is 5 μm.

Cut On The shortest wavelength a filter will pass or a detector or system will respond to. A 3–5μm filter's cut on is 3 μm.

Dall-Kirkman Telescope A Cassegrain-like telescope with an ellipsoidal primary mirror and a spherical secondary.

Data Latency The phenomena in which data is gathered at a certain time is used by a processor at some later time. Data latency represents the elapsed time between the gathering and the using of the data. With IR systems, this time is a fraction of a second, but can be a driving system level concern for missiles and IRST applications. If sensitivity requires a 0.5 second integration time and 0.1 second to format the data for the computer, the data that shows a target is on a certain pixel is about 0.6 seconds old. It is similar to reading a newspaper in the afternoon that was printed in the wee hours of the morning. The information it contains is not exactly current, but a few hours old. If you cut out an article and send it to a friend the data latency becomes several days old. If he, in turn, sends it overseas to another friend, the data latency becomes several weeks. If the article was about a long term issue such as global warming the data latency is inconsequential. However if it was about rapid fluctuations of price in a volatile stock, the data latency makes the information useless.

DC Background A background with a constant level, or intensity. This is in contrast to "clutter" which is not spatial or temporally constant.

De-Bumping Slang for the separation of the detector material from the readout circuitry in a hybrid FPA. Sometimes referred to as delaminating.

Defense Support Program An early warning satellite made by TRW and Aerojet for the Air Force that scans the earth in the IR to warn of rocket launches. See Chapter 11.

Derivative A mathematical function describing the slope of a line. More accurately, it is the instantaneous rate of change of a function to a variable in that function. A larger magnitude derivative means that the function is changing more rapidly than a lower derivative.

Desert Storm The international effort to remove Iraq from Kuwait in 1991.

Dewar Generally, a thermos bottle for cryogenic materials. The IR industry usually calls the thermal isolation vessel assembly that hold the FPA, cold finger, and some cold electronics a dewar.

Diffraction Limit Occurs when the optical system is limited by diffraction. The best image that can be formed. The fundamental limitation of the sharpness of an image, determined by the wavelength and size of aperture, no matter how perfect the optics are.

Diffraction The phenomena of electromagnetic waves bending about an edge or surface toward the shadow of that surface. Diffraction causes an image of a point to be an airy pattern on the focal plane instead of an infinitesimally small point. Diffraction imposes a fundamental limitation of the sharpness of an image, determined by the aperture and wavelength.

Diffractive Optics Optical elements and/or systems using the theory of diffraction to control, focus and condition light. May be reflective or transmissive in nature.

Digital Scan Converter An electronics module that converts (or reformats) a parallel scan into a serial line in a standard (e.g., RS-170) video format.

Digital Signal Processor A chip (or chip set) that is specially made for image processing in the digital domain. Most of the effort in this area is tailored to the visible spectrum.

Dilate An image processing morphological filter that replaces the center pixel being operated on by the maximum of its two neighbors. Erode and dilate filters are used in conjunction to enhance or degrade particular scene characteristics as needed.

Direct Labor The labor that is charged to the project. Direct labor usually includes the program office, engineers, and technicians. It usually does not include corporate management, marketing, or general security. There is a trend to change classical indirect jobs such as secretaries, finance, and document control to direct labor.

Divert Thrust The propulsion of a missile orthogonal to the axial thrust used to move the missile toward the target.

Doppler Shift The phenomena of a change in frequency (and wavelength) caused by relative motion between the observer and emitter.

Dwell Time The total time a detector has to view a given target. Integration time is sometimes confused with dwell time. It may be the same, but not necessarily.

Edge Detection An imager processing technique for determining the edge of a target or scene.

Electro-Optical An adjective used to describe the science, hardware, and industry that is involved in electrically sensing light. In general usage, including this book, it is used to mean across the light spectrum from UV to far IR. Therefore, the infrared industry is a subset of the E-O industry. In a more strict scientific usage, the electro-optical effect is the specific phenomena of a change in index of refraction as a result of an applied electric field.

Electron Charge The electrical charge of a particle, ($= 1.6 \times 10^{-19}$ coulombs).

Elevation The angular dimension measured along a local vertical. Elevation is perpendicular to azimuth. The vertical may be determined by the Earth, the body of an aircraft, a ship, or a spacecraft. Usually 0 degrees in elevation represents the LOS being parallel to the azimuth plane. At +90 and –90 degrees, the LOS is perpendicular.

Emissivity A physical property of a material that (along with temperature) determines the magnitude of its radiant exitance. It is measured as a ratio of radiant output divided by the Planck function for the same temperature. In theory, every surface must have an emissivity between zero and one. Shiny objects have low emissivity and absorptive objects have high emissivity. In a sense, it can be though of as a fudge factor on the Planck equation for amplitude, but not the general shape.

End Game The final part of a missile's fight where the target is imaged and aimpoint selection occurs.

Epitaxy Growth of crystals in an overlaying fashion with definite orientation and lattice constants identical (or nearly so) to those of the substrate that they are being grown upon.

Erode A morphological filter that replaces the center pixel being operated on by the minimum of its two neighbors. This has the effect of making a peak more narrow in the erode direction, like a butte being eroded. A five pixel wide high intensity source would become a three pixel wide high intensity source.

Etendue The solid angle multiplied by the cross sectional area of the useful beam. Commonly used as a figure of merit for the useful FOV of a spectroscopic device. The larger the etendue, the larger the FOV that can be accommodated.

Eutectic Point or Temperature The temperature of an alloy where disassociation begins to occur. It is usually near the minimum boiling temperature of one of the constituent elements. In practice, no alloy employed in hardware should be subjected to a temperature near its eutectic point.

Extrinsic FPA A detector array that employs dopants into a materiel to produce a PC effect.

F-Number The ratio of focal length to aperture.

False Alarms The identification of noise or clutter artifacts as targets.

False Color (1) An image processing/display technique of assigning colors to intensity and displaying them. (2) Assigning colors to different IR bandpasses and displaying them as a color composite, in the visible.

False Targets Real items that are identified as targets because they have similar or identical characteristics. False targets are passed by the image processing as real targets owing to real attributes of the target. An example might be detecting forest fires from space for an early warning satellite or a missile seeker locking onto a passenger jet instead of a military jet. False targets do not necessarily indicate a failure of the image processor, but are more scenario and mission dependent.

Fault Tolerant Electronic design of nature that a single event upset or a single piece of hardware failing, does not significantly degrade system performance.

Field of Regard The maximum possible angular pointing ability extent of a sensor. Usually, is larger than the field of view because the field of regard includes gimbal limits.

Field of View The useful angular extent of the sensor.

Figure of Merit A mathematically calculated or measured indicator of performance. Usually, a measure of general applicability (or "goodness"). A figure of merit for a sensor's sensitivity may be the NEI, just as a figure of merit for a computer is its processing speed and a figure of merit for an automobile is it's gas mileage. Figures of merits are good for comparing one of a particular class to another of the same class. Unfortunately, with complex items like IR sensors, cars, and computers, a single figure of merit never represents the whole story.

Figure The general geometric shape of an optical surface Examples are parabolic mirrors or spherical lenses. The Hubble's primary mirror has an incorrect "figure."

Fill Factor The percentage of usable active area of a detector pixel to the entire area of a detector pixel. A 50 percent fill factor indicates that half of the area devoted to a detector pixel is capable of detecting IR radiation and the other half is not. Hybrids can approach 100 percent fill factor, while monolithic FPAs rarely achieve 60 percent.

Firm Fixed Price A type of contract in which the contractor gets paid a fixed price for the effort, which is agreed upon in advance. This is generally how consumers function in day-to-day life. These contracts require less paperwork and customer overseeing and are convenient and effective programs where the cost is well known. Traditionally, they have been only used in the IR industry for a purchase of a previously completed developed system (such as a seeker in production or a commercial camera).

First Light The first useful data taken by a telescope or instrument. Usually refers to the building of the telescope or instrument. In other words, the first useful day of operation.

Fixed Wing Refers to aircraft that develop lift by air moving across wings due to vehicle motion. This includes jets, airplanes, etc. Fixed wing aircraft are contrasted to rotary wing aircraft such as helicopters.

FLIR Symbology Notations placed on a FLIR display via screen graphics to aid the operator. Typical symbologies include the pointing direction, contrast settings, and boxes or circles around the tracked target.

Focal Length Generally, the distance from the first mirror or lens (objective) to the focus.

Fold Mirror (or Fold Optics) A mirror without power used to bend or tilt the light bundle. Usually, used to get the light-ray bundle in a desired packaging configuration.

Footprint The projected two-dimensional extent of a pixel on the scene. The area of the scene that contributes to a pixel. For example, if the aperture was parallel to the background scene, and at a range of 100 km, with a 10 μrad square pixel, the footprint would be 10 μrad × 100 Km = one meter by one meter square. If the view is tilted with respect to the optical system so the projected footprint intersects at an angle, the angular effects must be considered. If the same sensor viewed the same scene at a 45 degree angle, the footprint would be about 1.4 meters by 1.4 meters.

Forebaffle A baffle or light shield in object space, usually extending from the primary or objective, also called a sunshade.

Forward Looking Infrared Sensor Typically an airborne or shipboard scanning and imaging sensor with output displayed on a CRT in "real time" for navigation, night vision and targeting.

Frame Rate The rate at which a sensor can take a full frame of data. The number of frames per second.

Frame-To-Frame Differencing A signal processing technique to discriminate moving targets from constant background and background clutter. Typically, this applies to starers only because the subtraction must take place on the same footprint on the ground. Usually, for a scanner, this would require unreasonably large memory to store several complete scans of the field of regard. The order of differencing refers to how many frames are used and which is being compared with which. For instance, first order required two complete frames and simply compares one to the other. This is especially effective if one expects the target to cross a pixel (moving target indicator).

Free Radicals An atom or molecule with an unpaired electron which is "champing at the bit" to combine with another molecule.

Fudge Factor A multiplication quantity used to make an equation yield a more reasonable or correct result. Fudge factors usually have a bad name, which is unfortunate. Sometimes they are very appropriate, needed, and valuable if they are well understood and are not abused.

Fusion An image processing technique to constructively combine images from multiple sensors or different bandpasses to a single display.

Fuzz Button A thermally conductive washer or spacer between the FPA and the cold finger which isolates some vibration.

Gain Correction A signal processing technique to equalize the response from various nonuniform unit cells. It is a simple matrix multiplication (one correction for every pixel). The corrections are stored from the last time the device viewed the blackbody calibrator. Gain correction compensates for error in the slope, or "gain," of the response of a pixel. This can be done in either the analog or digital domain and must be performed for both starers and scanners.

Galilean Telescope A refracting telescope with a convex objective and a concave imaging lens

Gantt A bar chart.

GEN 1 Usually in reference to FLIRs, intensifiers, or focal planes. Includes the first generation common modules developed in the seventies with discrete detectors packaged in a dewar and a lead for every detector.

GEN 2 Second generation, usually in reference to scanning focal planes that are more dense and include FPA multiplexing and image processing

Geostationary Synonymous with geosynchronous.

Geosynchronous A orbit where the satellite takes the same amount of time to orbit the earth as the earth takes to rotate. The effect is that the satellite "hovers" above one spot on the earth's equator. This phenomena occurs at an altitude of 35,680 Km.

Getter A collection device that is held colder than most or all other surfaces. It is used to collect contaminants.

Grey Scale A range of available levels between black and white. often, used with eight-bit displays providing 256 levels of grey.

Greybody A radiator and absorber with a radiant exitance of the same distribution as a blackbody but shifted down by a constant. A graybody has a constant effective emissivity of less than one (blackbody).

GRIN Gradient Index. A lens with a varying index of refraction across its surface.

Groupthink The phenomena of synergistic self-reinforcement of ideas among a team. It has the undesirable effect of making a team closed to new ideas.

Heat Load The static input of heat or draining of cooling capability in an FPA dewar. This is caused by radiation of hotter surfaces and conduction across wires and mechanical interfaces.

HeNe A low cost common commercial laser using helium and neon gases. It lases at several wavelengths with 0.628 μm being the most common. It is indispensable in optics labs.

High Background A condition where the FPA views a scene of large radiance. It implies that the FPA must have large well capacity and will probably be BLIP.

Hubble The NASA visible space telescope launched in 1990 named after Mt. Wilson astronomer Edwin Hubble.

Hybrid Electronics A dense packaging method of arranging electrical components on a printed circuit board. Electronic packaging have multiple device technologies on the same substrates, including MOS and bipolar, analog and digital mixed on the same substrates, or silicon and GaAs. Hybrids often consist of a single IC chip with components crammed around it.

Hybrid FPA A focal plane architecture that consists of two physical pieces, made of different materials, electrically interconnected and bonded during manufacturing. The detector array is manufactured and optimized separately from the readout multiplexer. They are then bonded (usually bump-bonded) together. A monolithic FPA has the readout and detector made on a single piece of material.

Hydroscopic Dissolves or erodes with water.

Hyperspectral An adjective used to describe a sensor, scene, or technique that has many different spectral bands. Usually, hyperspectral implies tens to thousands of identifiable bandpasses generated by a grating, CVF, AOTF, or FFT spectrometer.

Hysteresis A memory or lag in a change of a characteristic. This results in a dependence of the output on the past use of the item. Common examples include the "memory effect" of a NiCad battery and magnetic hysteresis.

Image Processing Techniques and processes to condition a scene of data or extract information from the data. Image processing is a higher order of processing than signal processing, usually implying operations on portions of or the entire scene to extract information.

Image Processor The hardware that executes the algorithms and heuristics on the signals to perform image processing.

Image Space The part of the optical train that is forming images, where the rays are (or have been) bent. Roughly defined as the volume inside an optical system.

Imagers (1) A sensor system that forms a two-dimensional image. In this sense, all the sensors discussed in this book are imagers. There are IR radiometers and instruments that are not capable of forming images. (2) Sometimes used in the industry to mean a staring array (or system employing one) of 32 by 32 or larger. In this context, it is used to differentiate a starer from a scanner.

Inband Weighted Average A resulting quantity representative of the sensor's response across the spectral bandpass.

Inband Used as an adjective to indicate that the bandpass is considered. For example, inband radiance implies that the value is the radiance in the useful spectral bandpass of the sensor.

Inches An archaic English unit of linear measure equal to 2.54 cm. It is a deprecated unit whose use should be discouraged.

Indium Antimonide An infrared detector that can detect IR radiation from one to about 5.8 microns.

Inner Baffle Baffles close to the FPA, usually including the cold shield or cold stop.

Interlace The technique of raster scanning where alternative lines are scanned at different times. A scanning system will scan every other line and then display them. The next frame it will scan the other half of the lines and display them.

Intrinsic FPA A detector array that does not rely on dopants to produce a photoconductive effect.

Irradiance Radiance per unit area measured in W m^{-2} sr^{-1} or W cm^{-2} sr^{-1}.

Jet Propulsion Laboratory A NASA center that specializes in interplanetary flight and has nothing to do with jet propulsion anymore.

Jitter Spurious unpredictable movements in the LOS.

Johnson Space Center NASA center In Houston, Texas. JSC is mostly concerned with manned spaceflight.

Jones Accepted units for D* which are cm $Hz^{1/2}$/W.

Joule-Thompson Coolers Cryogenic coolers that employ the expansion of a non-ideal gas through an orifice to produce a cooling effect.

K Term Usually indicates an efficiency fudge factor.

Keck The 10 m telescope in Hawaii. For a period in the 1990s, it is the largest telescope in the world.

Kelvin A measurement of temperature. Zero Kelvin is absolute zero where no Brownian motion exists. Water freezes at about 273 K and room temperature is about 300 K. A change of one Kelvin is equal to a one degree centigrade change and a 1.8 degree F change. There are no degrees in the Kelvin notation.

Knots An English unit of speed. One knot is equal to about 1/2 meter per second, or 1.8 kilometers per hour. The use of this unit should be discouraged.

Laplacian Whitening Filter A signal processing technique to discriminate targets from constant background and background clutter. It is a three by three matrix spatial filter that squares the pixels instead of using a simple weighting. This is a form of spatial filter based on knowledge of the pixel to pixel correlation.

Lattice A regular spatial arrangement of atoms or molecules, such as in a crystal.

Learning Curve A technique to quantify the phenomena of reduced cost per unit as more units are made. The manufacturing process becomes more efficient as more and more units are built and the technicians become proficient with the process. There are several different types of learning curves, but the one most often used is based on the reduction in cost every time the amount of units is doubled. For example, an 80 percent learning curve means that a second "lot" of the same quantity will cost 80 percent of what it cost to produce the previous lot.

Least Mean Square Filter A signal processing technique to discriminate moving targets from constant background and background clutter. It is a high pass spatial frequency filter based on an LMS fit of the targets spatial characteristic. It has the drawback of "ringing" when an edge is encountered.

Life Cycle Cost The expected cost of an item or program including all costs throughout its life. Non-recurring costs such as developmental costs and R&D, as well as other costs like maintenance and replacement costs are normally included.

Limb A view (or portion of) the Earth's atmosphere between space and ground. A limb background has a "cross section" of atmosphere in it ranging from low altitude to high altitude atmosphere.

Line Scanner A sensor that provides scanning a line at a time and relies upon platform motion to provide movement in the other direction.

LN_2 Temperatures Ambient temperatures associated with liquid nitrogen. Liquid nitrogen is at 77 K, but it can be conveniently lowered to about 72 K by vacuum pumping on it. Conversely, important surfaces may never achieve the

77 K resulting from thermal heat loads, so a baffle or FPA may only reach 80 or 81 K. Thus, LN2 temperatures can be defined from 72 to 82 K.

Long-Lead Part A part, component or subassembly whose delivery time is long compared to the program or schedule. Generally, these are items which take six months or more for delivery. FPAs, optical filters, custom connectors, and hybrid circuit boards are common long-lead items.

Long Wave Infrared Typically a band from about 6 to about 18 microns. Sometimes used by the tactical folks to mean just the 8–12 atmospheric transmission band.

LOS Drift The undesirable movement of the line of sight caused by servo system inaccuracies.

Low Background A condition where the FPA views a scene of small radiance. It implies that the FPA must have low noise, the ability for long integration times, and usually does not need large well capacity.

Lyot Stop A specific type of optical stop in the telescope. It is important in IR telescopes because usually the optical train only has to be cooled between the Lyot stop and detector.

Main Chase The part of a missile mission where the missile seeker is locked onto the target and flying to an intercept. It is temporally between the search phase and the end-game.

ManScience A series of U.S. government programs to develop manufacturing sciences.

ManTech A series of U.S. government programs to develop manufacturing technologies.

MDTI The minimum detectable in-band target irradiance required to satisfy a desired probability of detection and false alarm. MDTI includes clutter and clutter processing effects, so it must be evaluated for a given clutter level.

Mercury Cadmium Telluride A PC or PV infrared detector that can be tuned at time of manufacture to detect between two and 20 microns.

Mesosphere A region of the atmosphere from about 50 km to 90 km where temperature decreases with increasing altitude.

Metallization The process of depositing metal on an integrated circuit.

Microlens A lens that mates with pixels on the FPA. These lens may or may not be binary.

Micrometer A unit of measure equal to one millionth of a meter. Synonymous with "micron" and "µm" and preferred over these terms in the International System of Units.

Micron A unit of length measurement equal to one millionth of a meter or 10^{-4} cm. Therefore, it takes 25.4 of them to equal a "mil." Usually used in the IR to express wavelength and unit cell (or pixel) spacing. Synonymous with "micrometer" and "µm."

Microradian A unit of angular measure equal to one millionth of a radian. A microradian is equal to 5.73×10^{-5} degrees or 0.21 arcsec.

Microyield The phenomena of a material being mailable on a microscopic level. Microyield usually does not affect its macroscopic properties. Microyield is a concern when grinding and polishing optical materials. It can limit the accuracy of their figure and induces flaws which cause scattering, especially with beryllium.

Mid Wave Infrared Typically from 3.0 to approximately six microns. Sometimes used by tactical folks to mean the three to five micron atmospheric transmission.

Mil Specs Short for USA Military Specifications. A series of wordy, exact, and lengthy specifications defining materials, processes, test procedures, and the like.

Millirad A unit of angular measure equal to one thousandth of a radian or 0.057 degrees or 3.4 arc minutes.

Molniya A type of highly elliptical orbit that takes the satellite to very high altitudes as well as very low altitudes.

Monolayer Used to denote a film of molecules that is only one molecule deep.

Monolithic FPA A FPA architecture where the readout and the detector are made on a single chip made from one material. Pt:Si and InSb are frequently made on a monolithic structure.

Monolithic Mirror A mirror made out of a single piece of substrate.

Multi-Color Array A focal plane made to detect more than one bandpass simultaneously. This is an infrared analogy to a color television camera.

Multiband An adjective used to describe a sensor with more than one spectral bandpass.

Multiple Instructions on Multiple Data A parallel processor architecture that allows several instructions to be carried out on multiple data line inputs simultaneously. In contrast, SIM-D employs only a single instruction at a time.

Multispectral Adjective to describe a sensor with more than one spectral bandpass. Usually it implies a capability to acquire data in four or more discrete bandpasses. Multispectral implies fewer bands than hyperspectral.

Nadir A point directly below the zenith, straight down.

Nanoradian A unit of angular measurement equal to one billionth of a radian, or 5.7×10^{-8} degrees or 2.1×10^{-4} arcseconds.

Narcissus Effect An undesirable shading effect in an IR system output. It is caused by the cryogenic detector "seeing" itself via imaged reflections from optical surfaces in its telescope. It usually appears as a dark circle in the middle of a display.

National Television Video Code A standardized code and format for a video or television display.

Near IR A generic bandpass usually from 0.7 to 1.1μm (the cut-off of silicon).

Near Net Shaping The process of making an optical blank with the approximately correct figure. Figuring and polishing are still required, but to a lesser degree.

Near Wave Infrared Generic bandpass, typically from 0.8 to 1.1 microns in wavelength.

Neural Net A nodal parallel computer processor architecture, where the nodes are highly interconnected to simulate the mammalian brain.

Non-Linearity Correction A signal processing technique to equalize the response from various nonuniform unit cells. It a simple matrix multiply (one correction for every pixel). This is a correction on a per pixel basis for nonlinear response to different radiance levels. This operation is a higher order implementation of the gain correction discussed above and shares the characteristics of gain correction as described above. The primary difference is the complexity of the calculation of the correction matrix, relative to the simple gain correction described above.

Nonlinear Matched Filter A signal processing technique to discriminate targets from constant background and background clutter. It is a high pass spatial frequency filter. It is a matched filter with the second derivative of the scene subtracted. The subtraction of the second derivative prevents the ringing common in LMS filters and improves performance by using local inflections of the clutter.

Nyquest Frequency A sampling frequency twice that of the minimum required resolution.

Object Space The volume outside an optical system (most of the world).

Off Axis An architecture for optical systems that positions the elements away from the axis accomplished by grinding a "tilt" into the mirror or lens. This system can have the advantages of no central obscuration, wider FOVs, and high off-axis rejection capability.

Off the Shelf A generic buzz term to show that no development is required and that the item is sold from an inventory, like a coffee cup.

On Axis An architecture for optical systems that positions the elements centered along the optical axis. This needs a central obscuration for reflecting systems.

Open Loop Pointing The pointing of the sensor's line of sight without comparing the sensors output to correct. Adequate for rough pointing of large fields of views.

Operability A term describing whether a pixel meets the stated requirements or not. When applied to a FPA is it usually given as a percentage. For example, if it is a requirement that the D* must be no lower than 5×10^{11} @ 77, with less than 10 percent uncorrected uniformity and less than 2 percent crosstalk. A 98 percent operability indicates that 98 percent of the detector pixels meet all of the requirements and the other two do not.

Optical Blur Circle The image of a point source at the focal plane including all aberrations, diffractions, and alignment effects.

Outgas The process of gas emission from surfaces. Usually used in reference to contamination or cleaning up contamination.

Parallel Processor A processor architecture having several small processors that all work on a piece of the problem simultaneously.

Parallel Scanner A technique of moving the LOS across a line of detectors to form an image. In the most literal context, there would be no interlace or scanning in the direction defined by the line of detectors.

Passband Oldspeak for bandpass.

Passivate A technique of insulating and chemically neutralizing parts of integrated circuits. This is especially important for HgCdTe FPAs.

Passive A detection process using no emission. IR sensors are passive as they collect photons and emit none to interrogate the target. This is in contrast to laser rangers and radar, which emit beams.

Peltier Effect The phenomena of absorbing or generating heat at the connection of two dissimilar metals when a current is applied. The direction of the heat flow depends upon the current, so devices based on this principle can be used to heat or cool. Thermo-electric coolers employ the Peltier effect.

Petzval Curvature The paraboloidal optical curving of the image at the focal plane caused by astigmatism.

Phase Conjunction A technique to replicate mirror images of the microscopic spatial distortions present in wavefront error. These distortions can be caused by the atmosphere or the optical system.

Photoconductor An infrared detector that uses a change in conduction to detect impinging infrared radiation.

Photo Volatic An infrared detector that produces a change in voltage to detect impinging infrared radiation.

Pixel Footprint The projected area of a pixel onto the background and/or target.

Pixel IFOV The instantaneous, non-scanned filed of view of a FPA detector pixel. This is the angular substance that an unit cell occupies. If a one degree square FOV was supported by a 128 by 128 FPA, each detector pixel would have an IFOV of 136 µrad (one degree/128) on each side.

Pixel The term literally means a "picture element." It is the smallest resolvable unit of a display. Strictly speaking, pixel should only be used in reference to a display or data, not the focal plane. However, current popular usage places it synonymous with unit cell or detector element of a FPA.

Plant The subsystem that represents the dynamics of the device that will be used to perform dynamic control. "Plant" is "servo-ese" for system.

Point Source A source of radiation that cannot be resolved by the optical system or FPA. The ideal point source would have the IR radiation contained in completely parallel rays for a system focused at infinity. A star is an example of a point source.

Pound An archaic English unit of mass equal to 453.6 grams. It is considered a deprecated unit whose should be discouraged.

Preliminary Design Review A formal examination of the design, including functional flows, requirements, flowdowns, and concepts. The design effort is usually about 1/4 to 1/3 complete at this point.

Primary Mirror The first and usually the largest mirror in a reflective optical system. It provides the light gathering and frequently sets the aperture size.

Prototype A close hardware replica of the final system. A prototype should meet the functional and performance attributes of the final concept. It should come close in characteristics such as weight, power, and reliability. The major differences (if any) should be in the way it is made, which may not be suitable for the production of the final system.

Pulse-Tube A derivative of a Stirling cooler that replaces the expander with a resonant moving gas to produce the cooling effect.

Pyroelectric An infrared detector that uses a change in magnet moment to detect impinging infrared radiation. These can provide low sensitivity detection to 50μm with little to no cooling.

Quantum Wells In the context of detectors, an infrared detector designed to trap the photons of a tuned wavelength that uses a PC effect to detect impinging infrared radiation. These can be made responsive from about 3 to 25 microns and are most appropriate for high spatial resolution and low sensitivity LWIR applications.

Radian A measure of angle, the circle divided by pi. Equal to 57.3 degrees

Radiance Radiant power per source area and solid angle in the dimensions of $W/cm^2/sr$ or $W/m^2/sr$.

Radiant Exitance The radiant flux per unit area in dimensions of W/m^2 or W/cm^2.

Radiant Intensity The angular flux from a target. It is best to use for sources that are smaller than a pixel. Expressed in units of W/sr.

Radiometry The science and engineering of energy measurement, calculation, and transfer analysis.

Rayleigh Criteria A rule for determining resolution. When the peak of a diffraction pattern (or airy disk) of a point source falls on the first dark minima of another diffraction pattern from a different point source. This minimum distance is equal to half of the diameter of the airy disk of $1.22\lambda/d$, where λ is the longest wavelength and d is the aperture diameter.

Real Time Usually refers to processing that occurs at the site, while the instrument is being used. Implies little to no delay and no post processing at a later date.

Reflection When a photon encounters a change in media or a surface, it is either absorbed, transmitted, or reflected. Reflection is a process where a photon bounces off a surface in a predictable fashion. Actually, the quantum mechanical process of reflection is complex involving both absorption and emission, but such discussions are well out of the scope of this book. The interested reader is referred to books on quantum mechanics, such as Feynman's QED.

Reflective Optics Optical elements and/or systems using the theory of reflection to control, focus, and condition light. Requires reflective optics.

Refractive Optics Optical elements and/or systems using the theory of refraction to control, focus, and condition light. Requires transmissive optics.

Relative Calibration (1) Calibration from one frame to another, or separated by a temporal unknown. Not traceable to a primary standard with a known accuracy. (2) The process of determining (or state of knowing) the sensor's output for a given input relative to another sensor or bandpass. Relative radiometric calibration accuracy between sensors, or bandpasses of a sensor, can be less than one percent.

Relativistic Speeds Velocities of particles approaching the speed of light, where relativistic phenomena begin to be observable. They are usually 20 percent (or more) of the speed of light.

Remotely Piloted Vehicle An airborne drone that has a pilot controlling the aircraft from a safe and remote location via radio links. RPV and UAV are sometimes used interchangeable, but a RPV is a subset of UAV.

Rolling Ball An image processing morphological filter that is composed of a series of erodes and dilates. This techniques has the effect of smoothing out a scene, like plotting the center of a rolling ball as it moves over bumpy terrain. It can be an effective clutter rejector.

Rolling Pin A morphological filter that is analogous to a single-axis rolling ball, except one dimension is extended beyond one pixel.

Rotary Wing Refers to aircraft that develop lift by moving air across their wings due to a spinning motion of the wing (e.g., helicopters). Rotary wing aircraft are contrasted to fixed wing aircraft, such as jets.

RS-170 A standard television video format.

Rugate A type of spectral filter consisting of materials of alternating properties. A rugate is a complicated interference filter with a continually varying refractive index profile.

Saturation A condition of overload or exceeding capacity.

Scaling Factor A number used to multiply another quantity in order to adjust the other quantity's value. Usually, used in determining a comparison. For example, if a mirror is twice the diameter of another, the scale factor is four for the area, and from Equation (3.12), 6.5 for mass. So if one had a similar mirror of the same mechanical design and material, the weight of the larger mirror could be estimated by multiplying the weight of the smaller mirror by 6.5.

Scan Converter The electronic subsystem that reformats and re-orders the data from an FPA into a data stream that is useful for a display.

Scan Efficiency The ratio of the useful portion of the scan to the total time it takes to accomplish the scan.

Scan Rate The speed that a LOS moves across a scene. It is usually expressed in angular measurements, such as 180 degrees/sec or 1 rad/sec.

Scan Registration The pixel-to-pixel placement from one scan to another. The knowledge or accuracy of the position of a pixel of data from one scan to another.

Scanner (1) The device that performs a LOS scan. (2) A sensor that employs a sweeping motion to "effectively" move the image across the focal plane to complete its field of regard.

Scene Standard Deviation A powerful statistic for setting the simple threshold level and performing other clutter rejection and target identification techniques. In fact, it can be used to determine which clutter rejection algorithms to perform. This can be performed in either the analog or digital domains, and is applicable to both starers and scanners.

Schottky Barrier A type of detector that operates via internal photoemission. They typically have low sensitivity, high uniformity and are producible in large arrays.

Secondary Mirror The second mirror in an optical system, the second surface that reflects the photon.

Sensor Level Calibration Calibration performed when the sensor is fully integrated. This is the most meaningful calibration, and encompasses the errors and uncertainties associated with the optics, baffles, filters, signal conditioners, cryosystem, and signal processors.

Sensor (1) A complete system used to create useful data from a scene. This includes the optics, gimbals, coolers, and image processors. It is used in this book in this context. (2) Sometimes used in the industry to denote the FPA only.

Serial Scanner (1) The scan mechanism that translates the LOS across a single (or few) detectors. (2) A sensor system that employs only one (or few) detectors to scan a scene.

Short Wave Infrared Generic bandpass typically from 1.1 to 3.0 microns.

SiC Chemical abbreviation for silicon carbide, a composite applicable to mirror blanks.

Signal Processing Techniques and processes that condition, format, and prepare a signal for interpretation, display, or information extraction. Usually implies operations on a single pixel, without regard to other pixels in the scene.

Signal Processor The hardware that executes the algorithms and electronic functions (like amplification) on incoming signals.

Simple Threshold Estimation An image processing process by which the threshold is established based on the total array's output. Simple threshold estimation must be considered carefully for use in scanning systems, since the "total array output" (e.g., a single scanned column) may not be the most desirable "neighborhood" to base a threshold on. This procedure can be modified to examine local averages (e.g., 3×3 or 5×5 surrounding pixels) and used to calculate adaptive threshold levels. The modifications do require more processing per pixel, since threshold levels are now established on a per pixel basis, rather than using a single threshold level for the entire image.

Single Event Upset An electronic dysfunction caused by a lone strike of nuclear radiation. The affected part starts functioning correctly immediately after the incident. Typical of such upsets are bit-flops and temporary voltage or current spikes.

Slew Rate The velocity at which a movement of the sensor or LOS can occur given gimbal limitations.

Smart Weapons Weapons that employ advanced processor operations normally attributed to humans. These are operations like ATR, aimpoint selection, decoy discrimination, damage assessment, and automatic navigation and guidance.

Space-Level A type of component and system qualification that exceeds military standards in environment, screening, and testing.

Space Qual Short for "space qualification." A USA hardware qualification procedure in which the selection of materials and parts is combined with subassembly, assembly, and system environmental testing.

Spatial Convolution A signal processing technique to discriminate moving targets from constant background and background clutter. It allows substantial clutter rejection for clutter sources that are larger than a pixel. The smaller the matrix, the fewer operations and the better the clutter rejection. The irony is that the smaller the matrix, the more of the target that gets filtered out. This is because a point source rarely has all of its energy in one pixel. Hence, most fielded systems employ a five by five convolution, not a three by three. This technique is done digitally, traditionally with starers. Even though the scanning sensor acquires data temporally, the temporal scans can be buffered (as in FLIR systems) to create a spatial image. Thus, convolution can be executed on scanning sensor systems and requires buffering of three or five columns of original sensor data.

Spatial Filter An image processing filter that works in the spatial domain, usually reducing noise and clutter by comparing one pixel (or set of neighboring pixels) to another.

Specifications Detailed, written rules, attributes, or parameters that a piece of hardware, software, or study must satisfy to meet the requirements. Specifications are derived from requirements, and usually represent one path to meet the requirements, but not necessarily the only path. Specifications are more detailed than requirements, and include precise physical envelopes, power requirements, sensitivity, and tolerances, as are flowed to each component.

SPRITE A focal plane architecture that has a bias, geometry, and readout mechanism such that integration time is increased in the scan direction. SPRITEs are popular in first generation European systems.

Split Stirling A cryocooler using the Stirling cycle with the expander connected to the compressor by a single gas line. Allowing the expander to be remotely located (while provide cooling) up to about a meter away from the compressor, see Chapter 5.

Stabilization The process of compensating for outside dynamic stimulus to the LOS. Usually this includes all jitter and vibration.

Standard Advanced Dewar Assembly The next generation FLIR FPA dewar, capable of housing second generation FPAs. A data sheet on the development is included in Chapter 5.

Starer A sensor that needs employ no scanning to complete its frame view—like your eye.

Statement of Work A contractual document issued by a customer to a contractor or a contractor to a subcontractor describing the functions they are to perform. Specification and requirements should not be included in this document. Rather, the tasks to be performed should be described in detail. Ideally, it should be released with the RFP package and modified during negotiations.

Stealth A general term applied to low observable systems. For example, the F-177 is a stealth plane; it is difficult to detect with radar.

Step-Starer A sensor that is moved in defined discrete increments (typically 0.5–0.9 of the frame FOV) to cover the field of regard.

Steradian A measurement of a solid angle defined as the area subtended by the radius of a sphere squared on its surface. The total angle subtended by sphere (everything) is 4π steradians or about 12.6 steradians. The continental USA subtends about 0.3 sr of the face of the Earth.

Stirling Cycle A refrigeration process discovered by Stirling, where the heat is removed from one end of the device and generated at another. Chapter 5 contains a detailed explanation.

Stochastic Process A process, procedure, or calculation including random variables based on and determined by probabilistic laws.

Strapdown Seeker A missile seeker with no gimbal. It is hard mounted, fixed to the body.

Stratosphere A region of the atmosphere where temperature increases with increasing altitude. Usually the lower boundary is at an altitude of 10 to 18 km and the upper boundary is about 50 km.

Stray Light Unwanted light from an off-axis source. Light that leaks into a system that is out of the field of view.

Streak Detection A signal processing technique to discriminate moving targets from constant background and background clutter. It is commonly employed only in starers, although it can also be employed in scanners with temporal memory. However, in scanners the image moves across the array so fast that it is unlikely the target will cross pixel boundaries during the integration time, thus creating a streak.

Strehl Ratio The ratio of aberrated blur spot intensity to one of no aberration. The higher the Strehl ratio, the less the effects of all aberrations.

Sub-Microradian An angular measurement less than a microradian.

Sub-Millimeter Usually, the part of the EM spectrum from 100–1,000 microns.

Sub-Pixel Less than a pixel. Usually, used in image process to extract information of a detail greater than the angular extent of a unit cell. May or may not imply super resolution.

Sunshade A “forebaffle” or specially made shield extending beyond the end of the telescope to block radiation from the sun.

Super-resolution A process of extracting information to a detail that exceeds the diffraction limit and/or the Raleigh Criteria of resolution. For example, the diffraction limit of a sensor may be 100 μrad, but through specialized signal processing and data acquisition techniques, targets are separated when they are 30 μrad apart.

Surveillants Assets or sensors doing surveillance

Temperature-Tunable Optical Filter A solid state tunable spectral filter. There are no moving parts. The principle is usually based on a shift in chromatic index of refraction or transmission with temperature.

Temporal Filter An image processing filter that works in the time domain, usually reducing noise and clutter by comparing one scene to another.

Temporal Spending Plan A schedule showing the funds irrevocably committed as a function of time. It is important that the funds spent match the inflow of funds from the customer, as well as required manloading and subcontracting. It may be impossible to complete a project for $1 billion in one funding profile, while entirely possible to accomplish the project for $200 million distributed more appropriately.

Temporal An adjective describing a function that depends upon the flow of time.

Tertiary Mirror The third mirror in an optical system.

Testbed A facility used to conduct evaluations and experiments.

Then Year Dollars Used with a price quoted with considerations for rate and inflation increases.

Thermal Parasitic A constant heat load caused by radiation, conduction, or convection.

Thermoelectric Coolers Solid state coolers that employ the Peltier effect to cool one side. Typically these devices have 1 percent or less efficiency with high reliability. See Chapter 5.

Thermography IR sensing of temperature differences. often, applied to industrial applications and the medical industry. See Chapter 8.

Three-Axis Stabilized A technique of providing compensation for external motion in three axes. The three axes are usually the azimuth, elevation, and a cross-elevation axis. The more axes providing stabilization, the merrier.

Threshold Violation A signal processing technique of comparing the output of a pixel to a set threshold. Threshold violation is a simple comparison of signal levels to identify potential targets. It can be done in either the analog or digital domain, and must be done for both starers and scanners. This in itself is not a “clutter rejection” technique, and must be done for all current FPAs.

Time Constant The time it takes a detector to rise to approximately 63 percent of its peak final value.

Time Delay and Integration It is a technique to increase effective integration and hence, sensitivity while retaining resolution by adding signal from one detector to another. Chapter 4 contains a discussion on this technique.

Touch Labor The labor directly involved in producing the end product. Usually this includes, assembly technicians, test personnel, and production engineering. Usually, this does not include management, finance, or marketing.

Trade Studies (1) Analysis conducted to compare a technique, architecture, or component to others of its class over all pertinent attributes. For example, an FPA trade for a MWIR sensor might compare HgCdTe, Pt:Si, InSb and quantum wells. The characteristics used to compare them are driven by requirements and programmatics but might include sensitivity, cost, availability, pixel size, cooling requirements, and well size. (2) Analysis conducted to accomplish a system-wide balance among subsystems and components. For example optics size might be "traded" against FPA sensitivity and cost for a given level of system NEI.

Tunable Filters Spectrally agile filters whose optical properties can be quickly changed.

Turret A rotating assembly.

Two-Axis Stabilized A technique of providing compensation for external motion in two axes. The two axes are usually the azimuth and elevation axis. The more axes, the merrier.

Uniformity Correction An electronic normalization of pixels. It is a processing technique to adjust the sensitivity so that all unit cells have a more uniform response.

Uniformity The degree of variation from one pixel or unit cell to another.

Unit Cell An individual detector unit of a FPA. A 256 by 256 FPA has 65,536 unit cells. These are also sometimes called elements, pixels, or just detectors. Unit cell sometimes refers to the readout's pixels only.

Unmanned Areal Vehicle An airborne drone that has no human pilot on board. RPV and UAV are sometimes interchangeable, but an RPV is a subset of a UAV. A UAV may or may not have a pilot in remote control.

Update Rate The speed at which a point on a scene is detected between scans or frame readouts. For staring sensors operating at 30 Hz, it is a 30 Hz update. For an IRST spinning at 60 RPM, it is once per second, or 1 Hz. For step-stare systems covering a FOR every three seconds, it is once every three seconds.

Van Allen Belts Regions of charged particles that surround the earth, first discovered by Dr. James A. Van Allen. Chapter 11 has a detailed explanation.

Very Long Wave Infrared A generic spectral region typical from about 18 to 100 microns.

Visible The portion of the electromagnet spectrum roughly detectable by the human eye. It includes from about 0.35 to 0.7 microns.

Wall Plug Efficiency Slang to denote total system electrical efficiency. This invokes the colorful notion of measuring the power consumed at the point where the device is plugged into a wall for the electrical input. As it includes the power supply, batteries, and all losses, it is the most useful measure of efficiency to use.

Wavelength The linear distance from one point on a wave to another. For example, 14μm infrared radiation has a linear distance of 14 μm from the peak of the wave to the peak on the next. Wavelength is inversely proportional to frequency, since all EM radiation travels at the same speed in a vacuum.

Well A general term to describe the holding capacity of an FPA unit cell. Usually measured in millions of electrons.

Well Saturation A condition where the pixel or unit cell has as many electrons as it can hold. Saturation (and, potentially, blooming) occurs if more free electrons are generated than the well can hold during the integration.

White-Hot A mode of black and white display where the whiter the target the more inband IR radiation it is producing. Thus, usually the brighter the object, the hotter it is. This mode supports one's intuitive scene when viewed.

Z-plane A plane located perpendicular to the optical axis. Usually meant to denote three-dimensional electronics attached to the focal plane. See Chapter 6.

Zenith A point directly overhead, opposite nadir.

Appendix C

Bibliography

CRYOCOOLERS

Walker, G. *Miniature Refrigerators for Cryogenic Sensors and Cold Electronics.* (Oxford: Claredon Press, 1989), ISBN 0-19-854815-X.

DETECTORS AND FPAs

Dereniak, E., and D. Crowe. *Optical Radiation Detectors.* (New York: John Wiley & Sons, 1984), ISBN 0-471-8979-3.

Vincent, J.D. *Fundamentals of Infrared Detector Operation and Testing.* (New York: John Wiley & Sons, 1990), ISBN 0-471-50272-3.

ELECTRONICS

Horowitz, P., and W. Hill. *The Art of Electronics.* (Cambridge: Cambridge University Press, 1983), ISBN 0-521-23151-5.

ELECTRO OPTICAL SYSTEMS

Accetta, J., and D. Shumaker, senior editors. *The Infrared and Electro-Optics Systems Handbook.* (Ann Arbor: ERIM, and Bellingham: SPIE, 1993), ISBN 0-819-41072-1.

Begunov, B. N., N. P. Zakaznov, S.I. Kiryushin, and V.I. Kuzichev. *Optical Instrumentation, Theory and Design.* (Moscow: Mir Publishers, 1988), ISBN 5-03-000008-9.

Hooton, E., and K. Munson, editors. *Jane's Battlefield Surveillance Systems.* (Surrey: Jane's, 1990), ISBN 0 7106 0913 2.

Seyrafi, K. *Electro Optical Systems Analysis.* (Los Angeles: Electro Optical Research Company, 1975), Library of Congress Catalog # 73-8681.

The Burle Electro Optics Handbook (a.k.a. the *RCA Electro Optics Handbook*). (Lancaster: Burle Industries, 1974).

The Photonics Directory (in four volumes). (Pitsfield: Laurin, annual), Library of Congress Catalog # A54-6009

Waldman, G., and J. Wootten. *Electro Optical Systems Performance Modeling* (Boston: Artech House, 1993). ISBN 0-89006-541-1.

Yakushenkov, Y. *Electro Optical Devices.* (Moscow: Mir Publishers, 1980).

INFRARED SYSTEMS

Hudson, R. *Infrared Systems Engineering.* (New York: John Wiley, 1969), Library of Congress Catalog Card # 68-8715.

Johnson, R. B., and Wolfe, W. L., editors. *Selected Papers on Infrared Design* (in two volumes). (Bellingham: SPIE, 1985), ISBN 0-89252-548-7.

Lloyd, J. M. *Thermal Imaging Systems.* (New York: Plenum Press, 1975), ISBN 0-306-30848-7.

Spiro, I., editor. *Selected Papers on Radiometry.* (Bellingham: SPIE, 1990), ISBN 0-8194-0378-4.

Spiro, I., and M. Schlessinger. *Infrared Technology Fundamentals.* (New York: Marcel Dekker, 1989), ISBN 0-8247-8134-1.

Wolfe, W., and G. Zissis, editors. *The Infrared Handbook.* (Ann Arbor: ERIM, 1985), ISBN: 0-9603590-1-X.

Wyatt, C. *Radiometric Calibration: Theory and Methods.* (Orlando: Academic Press, 1978), ISBN 0-12-766150-6.

Wyatt, C. *Radiometric System Design.* (New York: Macmillan Publishers, 1987). ISBN 0-02-948800-1.

MANAGEMENT

Augustine, N. *Augustine's Laws.* (New York: Viking, 1983), ISBN 0-670-80942-X.

Burgelman, R., and Modesto Maidique, M. *Strategic Management of Technology and Innovation.* (Homewood: Irwin,1988), ISBN 0-256-03481-8.

Fisher, A. *Small Group Decision Making.* (New York: McGraw-Hill, 1980), ISBN # 0-070211091-8.

Kepner, C., and B. Tregoe. *The New Rational Manager.* (Princeton: Princeton Research Press, 1981), Library of Congress Catalog # 80-84367.

Kerzner, H. *Project Management.* (New York: Van Nostrand Reinhold, 1984), ISBN: 0-442-24879-2.

Kezsborm, D., D. Schiling, and K. Edward. *Dynamic Project Management.* (New York: John Wiley & Sons, 1989), ISBN 0-471-85248-1.

Kimmons, R., and J. Loweree, editors. *Project Management.* (New York: Marcel Dekker Inc., 1989), ISBN 0-8247-7676-3.

Managing Projects and Programs. (Complied by the Harvard Business Review). (Boston: Harvard Business Review Books, 1989), ISBN 0-87584-213-5.

Meredith, J., and S. Mantel. *Project Management, A Managerial Approach.* (New York: John Wiley & Sons, 1985), ISBN 0-471-80964-0.

Moder, J., C. Philips, and E. Davis. *Project Management With CPM, PERT and Precedence Diagramming.* (New York: Van Nostrand Reinhold, 1983), ISBN 0-442-25415-6.

Rockwell, W.F. *The Twelve Hats of A Company President.* (Englewood Cliffs: Prentice-Hall, 1971), ISBN 0-13-934166-8.

Rosenau, M. *Project Management for Engineers.* (New York: Van Nostrand Reinhold, 1984), ISBN 0-534-03383-0.

OPTICS

Born, M., and E. Wolf. *Principles of Optics.* (Oxford: Pergamon Press, 1980).

Feynman, R., *QED.* (Princeton: Princeton University Press, 1985), ISBN 0-691-08388-6.

Hecht, E., and A. Zajac. *Optics.* (Reading: Addison-Wesley, 1974), Library of Congress Catalog 79-184159.

Malacara, D., editor. *Optical Shop Testing.* (New York: John Wiley & Sons, 1978), ISBN 0-471-01973-9.

Morelli, D. *Interference Filter Handbook.* (Santa Rosa: OCLI, 1991).

Stover, J. *Optical Scattering Measurement and Analysis.* (New York: McGraw-Hill Inc., 1990), ISBN 0-07-061814-3.

Yoder, P. *Opto-Mechanical Systems Design.* (New York: Marcel Dekker, 1986), ISBN 0-8247-7347-0.

Welford, W. *Useful Optics.* (Chicago: The University of Chicago Press, 1991), ISBN 0-226-89306-5.

Index